Biology of Cellular Transducing Signals

GWUMC Department of Biochemistry Annual Spring Symposia

Series Editors:
Allan L. Goldstein, Ajit Kumar, and J. Martyn Bailey
The George Washington University Medical Center

BIOLOGY OF CELLULAR TRANSDUCING SIGNALS
Edited by Jack Y. Vanderhoek

BIOMEDICAL ADVANCES IN AGING
Edited by Allan L. Goldstein

CARDIOVASCULAR DISEASE
Molecular and Cellular Mechanisms, Prevention, and Treatment
Edited by Linda L. Gallo

CELL CALCIUM METABOLISM
Physiology, Biochemistry, Pharmacology, and Clinical Implications
Edited by Gary Fiskum

DIETARY FIBER IN HEALTH AND DISEASE
Edited by George V. Vahouny and David Kritchevsky

EUKARYOTIC GENE EXPRESSION
Edited by Ajit Kumar

NEURAL AND ENDOCRINE PEPTIDES AND RECEPTORS
Edited by Terry W. Moody

PROSTAGLANDINS, LEUKOTRIENES, AND LIPOXINS
Biochemistry, Mechanism of Action, and Clinical Applications
Edited by J. Martyn Bailey

THYMIC HORMONES AND LYMPHOKINES
Basic Chemistry and Clinical Applications
Edited by Allan L. Goldstein

Biology of Cellular Transducing Signals

Edited by

Jack Y. Vanderhoek
The George Washington University Medical Center
Washington, D.C.

Plenum Press • New York and London

Library of Congress Cataloging in Publication Data

International Washington Spring Symposium (9th: 1989: George Washington University)

Biology of cellular transducing signals / edited by Jack Y. Vanderhoek.

p. cm.. (GWUMC Department of Biochemistry annual spring symposia)

"Proceedings of the Ninth International Washington Spring Symposium at the George Washington University, held May 8–12, 1989, in Washington, D.C."—T.p. verso.

Includes bibliographical references.

ISBN-13: 978-1-4612-7866-5 e-ISBN-13: 978-1-4613-0559-0

DOI: 10.1007/978-1-4613-0559-0

1. Cellular signal transduction—Congresses. I. Vanderhoek, Jack Y. II. Title. III. Series.

[DNLM: 1. Cell Communication—congresses. 2. Signal Transduction—congresses. QH 604.2 I61b 1989]

QP517.C45I58 1990

574.87′5—dc20

DLC 89-70924

for Library of Congress CIP

Proceedings of the Ninth International Washington Spring Symposium
at The George Washington University, held May 8–12, 1989,
in Washington, D.C.

Softcover reprint of the hardcover 1st edition 1990
A Division of Plenum Publishing Corporation
233 Spring Street, New York, N.Y. 10013

PREFACE

Intercellular communication in multicellular organisms enables cells to respond to environmental changes. Intercellular signal transduction transmitters include hormones, peptide growth factors, neurotransmitters as well as some lipid-soluble mediators. Once signalling molecules are bound to their cell surface receptors, one or more intracellular signals are generated which alter the behavior of the target cell.

The IXth International Washington Spring Symposium at the George Washington University in Washington, D.C. was organized to assess the current status of the field of signal transduction processes and regulatory mechanisms. The symposium was held on May 8-12, 1989 and was attended by more than 1000 scientists from 30 countries. Most of the papers presented at the plenary sessions have been collected in this volume.

The first section of this book details the action and regulation of receptors such as β-adrenergic receptors and receptors for EGF, insulin, leukotrienes, phosphoinositides and prostaglandins.
Section two focuses on the family of guanine nucleotide regulatory proteins (G proteins). These G proteins are known to mediate the coupling of receptor-mediated signals to several intracellular effector systems. Papers are presented describing the intracellular localization of G proteins, the utilization of G protein antibodies, the interaction of G proteins with tubulin and the involvement of G proteins in the regulation of ion channels, adenylate cyclase and in the activation of neutrophils and T cells.
In the third section, several papers describe the second messenger role of phosphoinositides as well as the regulation of phosphoinositide production.
Sections four and five are concerned with protein kinase C and phospholipases, which are pivotal enzymes in cellular regulation. Included are studies on the subcellular localization of protein kinase C and the involvement of this enzyme in mediating certain mitogenic signals. Several chapters are devoted to the mechanisms of regulation of phospholipases.
The sixth section describes mechanistic aspects of calcium mobilization in chromaffin cells, hepatocytes, mast cells and platelets when these cells are stimulated by a variety of signalling molecules such as epinephrine, hormones, IgE or vasoactive peptides.
The final two sections contain chapters on recent advances in the involvement of eicosanoids and phosphatidylinositol glycans in signal transduction processes.

The multidisciplinary nature of the papers presented in this volume will enable biochemists, cell and molecular biologists, pharmacologists and other scientists to target new areas of exploration and gain further insights into the biology of cellular transducing signals.

Jack Y. Vanderhoek

Washington, D.C.

CONTENTS

PART I — RECEPTORS

1. Real–Time Analysis of Macromolecular Assembly During Cell Activation 1
L.A. Sklar, W.N. Swann, S.P. Fay and Z.G. Oades

2. Identification of Residues Essential for the Active Conformation of the β-Adrenergic Receptor by Site–Directed Mutagenesis........ 11
Mari R. Candelore, Sandra L. Gould, Wendy S. Hill, Anne H. Cheung, Elaine Rands, Barbara A. Zemcik, Irving S. Sigal, Richard A.F. Dixon and Catherine D. Strader

3. Mechanisms for Diversification of Responses of Human Polymorphonuclear Leukocytes to Leukotrienes........ 21
Catherine H. Koo, Jeffrey W. Sherman, Laurent Baud, Winzhen Jin, Kai Mai, and Edward J. Goetzl

4. The Role of Covalent and Non–covalent Mechanisms in Insulin Receptor Action........ 29
Paul L. Rothenberg and C. Ronald Kahn

5. EGF Triggers A Similar Signalling Cascade in Different Cell Types Overexpressing the EGF Receptor........ 39
Marco Ruggiero, Timothy P. Fleming, Toshimitsu Matsui, Eddi di Marco, Christopher Molloy, Pier Paolo di Fiore, and Jacalyn H. Pierce

6. Second Messenger Receptors in the Phosphoinositide Cycle........ 49
Solomon H. Snyder, Christopher D. Ferris and Anne B. Theibert

7. Mechanisms of Receptor Regulation for the Phosphoinositide Signalling System........ 61
Frank S. Menniti, Haruo Takemura, Hiroshi Sugiya, and James W. Putney, Jr.

8. Computer Simulation of Prostaglandin–Regulated Cyclic AMP Metabolism in Platelets: Activation and Desensitization of Adenylate Cyclase Appear to be Mediated Through Separate Prostaglandin Receptors Coupled to G_s and G_i........ 73
Barrie Ashby

PART II — G PROTEINS

9. The Family of G Proteins 83
Eva J. Neer, Yung-Kang Chow, Suzanne Garen-Fazio, Thomas Michel, Carl J. Schmidt, and Seth Silbert

10. Roles of G Protein Subunits in Coupling of Receptors to Ionic Channels and Other Effector Systems 93
Lutz Birnbaumer, Atsuko Yatani, Rafael Mattera, Juan Codina and Arthur M. Brown

11. Receptor—Mediated Alterations of Calcium Channel Function in the Regulation of Neurosecretion 107
Kathleen Dunlap and George G. Holz

12. Antibodies as Probes of G—Protein Structure and Function 115
Allen M. Spiegel, William F. Simonds, Paul K. Goldsmith, and Cecilia G. Unson

13. Transducin: A Model of Biological Coupling Enzymes 125
Yee—Kin Ho

14. Co—Localization by Immunofluorescence of the α subunit(s) of G_i with Cytoplasmic Structures 133
Jean M. Lewis, Marilyn J. Woolkalis, George L. Gerton, and David R. Manning

15. "Cross-Talk" and the Regulation of Hepatocyte Adenylate Cyclase Activity: Desensitization, G_i and Cyclic AMP Phosphodiesterase Regulation 141
Miles D. Houslay

16. Activation of Human Neutrophils by Granulocyte—Macrophage Colony—Stimulating Factor: Role of Guanine—Nucleotide Binding Proteins 153
R.I. Sha'afi, J. Gomez-Cambronero, M. Yamazaki, M. Durstin, T.F.P. Molski and C.-K. Huang

17. Cytoskeletal Participation in the Signal Transduction Process: Tubulin G Protein Interactions in the Regulation of Adenylate Cyclase 163
Kun Yan and Mark M. Rasenick

18. Initiation and Modulation of Signal Output from the T cell Antigen Receptor 173
Robert T. Abraham, James A. Augustine, Janis W. Schlager, Thomas J. Barna and Paul J. Leibson

19. Selective Phosphorylation of a G Protein within Permeabilized and Intact Platelets Caused by Activators of Protein Kinase C 185
Kenneth E. Carlson, Lawrence F. Brass and David R. Manning

20. G Proteins and Toxin—Catalyzed ADP—Ribosylation 193
Martha Vaughan and Joel Moss

21. Novel Endogenous ADP—Ribosylations of Proteins in Transformed Cells: Effects of Cellular Transformation and Differentiation 201
Harjit S. Banga and Maurice B. Feinstein

PART III — INOSITOL PHOSPHATES

22. Inositol 1,4,5–Trisphosphate and Inositol 1,3,4,5–Tetrakisphosphate: Two Second Messengers in a Duet of Calcium Regulation 213
Robin F. Irvine

23. The Role of Endogenously Stimulated Protein Kinase C in Regulating Inositol Polyphosphate Accumulation in Human Platelets 227
Susan E. Rittenhouse, Warren G. King, C. Peter Downes, and Gregory L. Kucera

24. Leukotriene D_4 Increases the Metabolism of an Inositol Tetrakisphosphate-Containing Phospholipid in U937 Cells 235
Henry M. Sarau, Maritsa N. Tzimas, James J. Foley, James D. Winkler, Matthew E. Kennedy, and Stanley T. Crooke

PART IV — KINASES

25. Immunocytochemical Localization of Protein Kinase C 245
Susan Jaken, Susan C. Kiley, Theresa Klauck, Liqun Dong, and Susannah Hyatt

26. A Protein Kinase C Pseudosubstrate Peptide Blocks G–Protein and Mitogenic Lectin Induced Phosphorylation of the CD3 Antigen in Human T–Lymphocytes 253
Denis R. Alexander, Jonathan D. Graves, Susan C. Lucas, J. Mark Hexnam, Doreen A. Cantrell, and Michael J. Crumpton

27. Growth Factor–Stimulated Tyrosine Phosphorylation of Phospholipase C 263
Matthew Wahl, Nancy Olashaw, Shunzo Nishibe, Jack Pledger and Graham Carpenter

PART V — PHOSPHOLIPASES

28. Eicosanoid Formation and Regulation of Phospholipase A2 275
Eduardo G. Lapetina

29. Inhibition of Phospholipase A_2 Activation in $P388D_1$ Macrophage–like Cells 281
Keith B. Glaser, Mark D. Lister and Edward A. Dennis

30. The Role of GTP–Binding Proteins in the Regulation of Phospholipase A2 and C in Bovine Retinal Rod Outer Segments 289
Carole L. Jelsema and Alice D. Ma

31. Stimulus–Coupled Activation of Phospholipase D 301
M. Motasim Billah, John C. Anthes, Robert W. Egan, Theodore J. Mullmann and Marvin I. Siegel

PART VI — CALCIUM

32. The Orchestration of Stimulatory Signals for Secretion in Rat Basophilic Leukemia (RBL-2H3) Cells 313
Michael A. Beaven, Russell Ludowyke, Heloisa M.S. Gonzaga, Dolores Collado-Escobar, Michihiro Hide and Hydar Ali

33. Synchronized Ca^{2+} Transients induced by Glucagon in Fura2 Loaded Hepatocytes 323
Jan B. Hoek, Kathleen E. Coll, Thomas A. Rooney, and Andrew P. Thomas

34. Calcium Mobilization Mechanisms in Bovine Chromaffin Cells by Vasoactive Peptides: Angiotensin, Bradykinin and Endothelin 333
Kathy Rasmussen and Morton P. Printz

35. Influence of GDP(B)S on Agonist-Induced Calcium Mobilization and Platelet Function 343
Gundu H.R. Rao, J.S. Cox, V.G. Mahadevappa and James G. White

PART VII — EICOSANOIDS

36. Regulation of the Production of Lipoxygenase Products and the Role of Eicosanoids in Signal Transduction 353
Anthony Ford-Hutchinson

37. Eicosanoid Synthesis in Resident Macrophages: Role of Protein Kinase C 359
Volkhard Kaever, Hans-Jurgen Pfannkuche, Klaus Wessel, Henning Sommermeyer and Klaus Resch

38. Modulation of the S-K^{+} Channel of Aplysia Sensory Neurons by Lipoxygenase Metabolites of Arachidonic Acid and cAMP 371
Steven A. Siegelbaum, Andrea Volterra and N. Buttner

39. Inhibition of Leukotriene D_4 Mediated Release of Prostacyclin Using Antisense DNA 381
Michael A. Clark, Terresa M. Conway, Mike Cook, Lynne Webb, Janice Dispoto, Seymore Mong, Robert Short, Sheldon Stiener, David Littlejohn, and S.T. Crooke

PART VIII -- GLYCOSYLATED PHOSPHOINOSITIDES

40. Second Messengers of Insulin Action 391
Alan R. Saltiel

41. Control by TSH of the Production by Thyroid Cells of an Inositol Phosphateglycan, Putative Mediator of the cAMP-Independent Effects in the Gland 401
L. Martiny, B. Thuilliez, B. Lambert, F. Antonicelli, C. Jacquemin and B. Haye

Index 411

REAL-TIME ANALYSIS OF MACROMOLECULAR ASSEMBLY DURING CELL ACTIVATION

L. A. Sklar, W. N. Swann, S. P. Fay, and Z. G. Oades

Scripps Clinic and Research Foundation
10666 North Torrey Pines Road
La Jolla, CA 92037

INTRODUCTION

Cell activation reflects a sequence of transient macromolecular assemblies. For example, during signal transduction many ligand-receptor complexes interact with G proteins to form a ternary complex of ligand, receptor, and G protein. Receptors which release intracellular calcium use an activation pathway which involves the action of phospholipase C. In this situation, the sequence of activation events involves: the formation of the ligand-receptor-G protein complex; the activation of the ternary complex by the binding of GTP; an assembly of the activated G protein, the phospholipase and its substrate, PIP2; and finally, calcium elevation following the release of inositol trisphosphate from the substrate. Since calcium release is detected within two seconds, the macromolecular asembly and dissasembly within the sequence occurs on a subsecond time scale.

We have previously described a number of complimentary fluorescence and flow cytometric methods which make it possible to examine interactions among ligand, receptor and G proteins in intact as well as broken neutrophil preparations and also to characterize the processing of receptors in intact neutrophils (Sklar, 1987; Sklar, et. al. 1989A). These techniques provide time resolution in the second and subsecond time range and are beginning to provide insight into the dynamics, mechanism, and structural interactions in the transduction pathway. These biophysical approaches have the potential to be extended to many types of cell surface receptors.

RESULTS

THE SYSTEM. We have examined a family of fluorescent formyl peptides as ligands for the formyl peptide receptor on neutrophils. The neutrophils have been isolated from human blood. The formyl peptide receptors appear to recognize bacterial peptides during host defense. The fluorescein (Fl) labelled peptides are F-Met-Leu-Phe-Lys-Fl, F-Met-Leu-Phe-Phe-Lys-Fl, and F-Nle-Leu-Phe-Nle-Tyr-Lys-Fl which contain 4, 5, and 6 amino acids, resp.

The interactions among ligand (L), receptor (R), and G protein have been studied in a permeabilized cell system (Sklar et. al. 1987) as illustrated in Figure 1. In the absence of guanine nucleotide, the addition of ligand permits the formation of a tightly associated ternary complex (LRG). When guanine nucleotide is added, the receptor and G protein dissociate returning the receptor to a low affinity form (LR) from which the ligand dissociates rapidly.

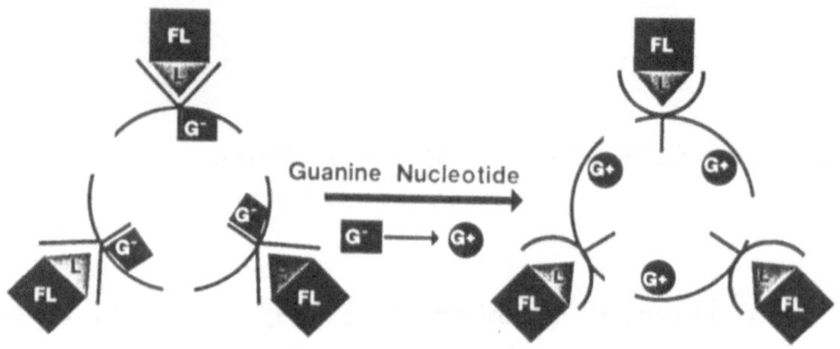

Fig. 1. Schematic Diagram of Ternary Complex Interactions in the Permeabilized Cells

The analysis of the macromolecular assembly and disassembly of the components of the ternary complex depends upon the ability to discriminate the free from the bound ligand, the ability to detect ligand in high or low affinity form, and the ability to examine the interconversion among the high and low affinity forms. Spectroscopic methods offer the possibility of detecting these interactions in a homogeneous assay.

Fluoresecence flow cytometry intrinsically discriminates free and bound ligand (Figure 2). In flow cytometry, cells are analyzed individually in a stream a rate of thousands of cells per second. The fluorescent ligand accumulates on the cells compared to its concentration in the dilute stream which surrounds the cells. It is possible to detect tens of thousands of receptors per cell in the presence of ligand

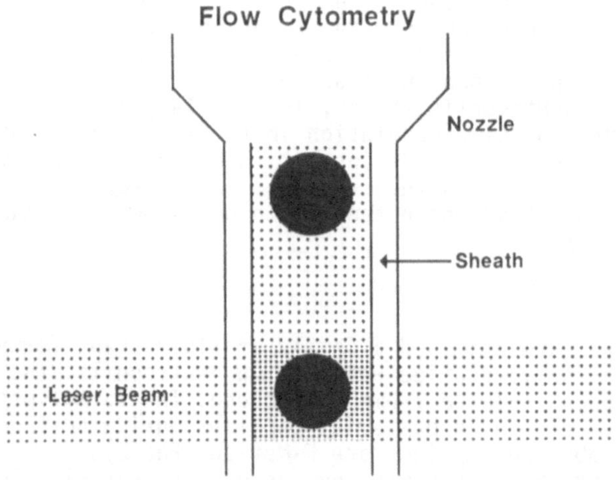

Fig. 2. Schematic Diagram of Flow Cytometric Analysis. A cell with ligand concentrated on its surface is depicted by the dark circle and the dilute ligand in the medium by the dots. Fluorescence is detected from the volume illuminated by the laser.

concentrations up to 100 nM free ligand without a wash step. The analysis of cytometric data is discussed by Šklar et. al. (1984, 1989A).

An analysis of the equilibrium binding of ligand in the absence and presence of guanine nucleotide is shown in Figure 3. Results with the hexapeptide show that the ligand binding to the ternary complex LRG is typically two orders of magnitude tighter than to receptor alone. However, the fit to the data suggest that all of the receptors can be interconverted among the high and low affinity forms and that there must be sufficient G proteins to accomodate all of the receptors. It is worth pointing out the direct analysis of binding to the low affinity complex by radioligand methods would be difficult because rapidly dissociating ligand is lost during the wash step.

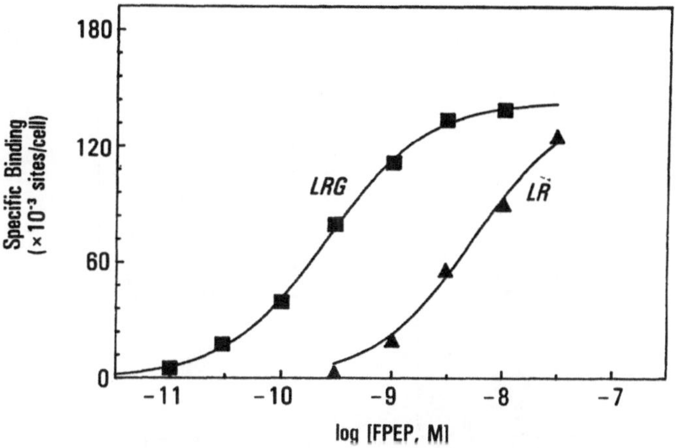

Fig. 3. Cytometric Analysis of Hexapeptide Binding to Permeabilized Cells in the Presence (LR) and the Absence (LRG) of GTP[S]. Solid lines are single site fits.

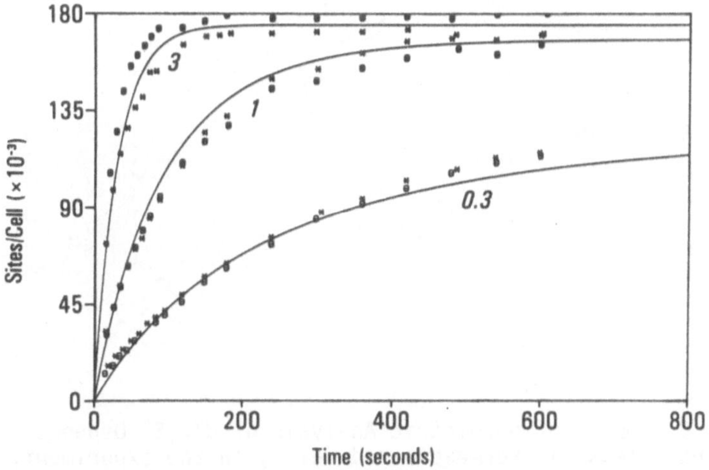

Fig. 4. Continuous Cytometric Analysis of the Binding of Hexapeptide to Permeabilized Cells in the Absence of Guanine Nucleotide. The ligand concentrations are 3, 1, and 0.3 nM. The solid lines represent single site fits.

For cell activation, there must be assembly of ternary complex. The assembly kinetics are examined in Figure 4 over the ligand concentration range where the receptors become saturated. This is the same range over which cell responses are saturated (Sklar et. al. 1985). The analysis indicates that the ternary complex appears to form in a single step at essentially a diffusion-limited rate of 2 x 1e7/Msec. Direct measurements of low affinity binding to the receptor (LR) yield the same diffusion limited rate (data not shown). The interpretation is that to form LRG, ligand first binds to receptor, then receptor finds the G protein and releases the endogenous GDP to form the stable ternary complex. Since the receptor is occupied at a diffusion limited rate and ternary complex forms without appreciable delay, the intervening steps occur rapidly in a time estimated to be no more than several hundred msecs.

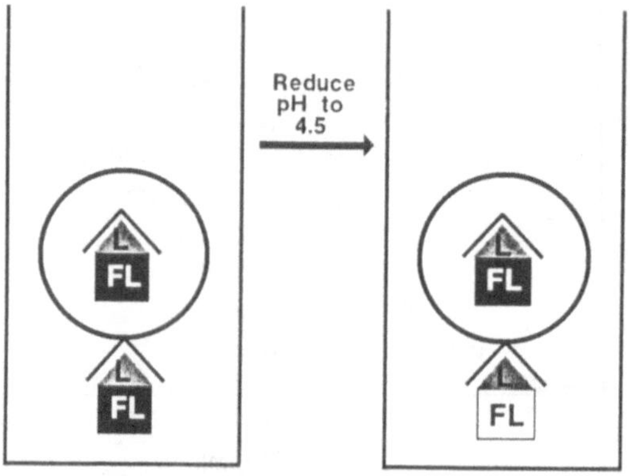

Fig. 5. Schematic Diagram of Antibody Based Detection System. The diagram represents a cell suspension in a cuvette in which free ligand is bound rapidly and quenched by antibody to fluorescein.

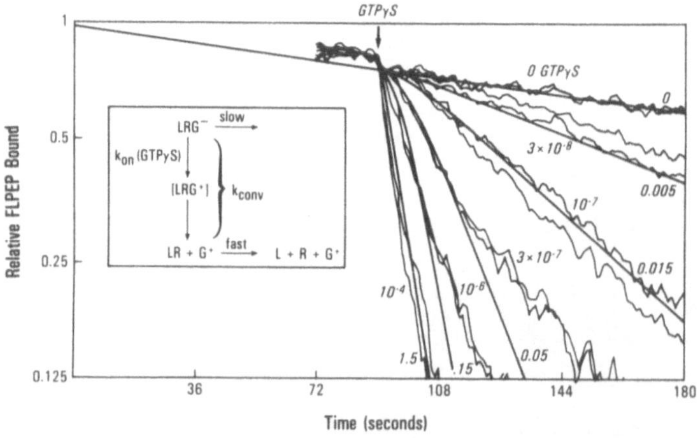

Fig. 6. Spectrofluorometric Analysis of GTP[S] Dependent Ligand Release in Permeabilized Cells. In the experiment, ligand is equilibrated with permeabilized cells and then GTP[S] is added as indicated. Solid lines are fit to the model in the inset; kconv is shown at base of each line. (Reprinted with permission from Sklar et. al. 1989A)

During cell activation, the ternary complex is activated by the binding of GTP. The rate of dissolution of the ternary complex can be modeled by examining the formation of the rapidly dissociating LR complex as a function of the concentration of guanine nucleotide. Because flow cytometry is too slow to observe this reaction, we turned to a detection system which uses antibody to fluorescein (Figure 5). Antibody binds to and quenches the free ligand within a second, but detects the receptor bound ligand at a rate three orders of magnitude more slowly, if at all.

The dissolution of LRG was measured as shown in Figure 6. In the absence of guanine nucleotide, antibody detects a slowly dissociating ternary complex. When saturating nucleotide is added there is a rapid interconversion to the unstable complex LR. The mechanism of the interconversion was probed by varying the nucleotide concentration. An analysis of the data indicates that the overall interconversion from LRG

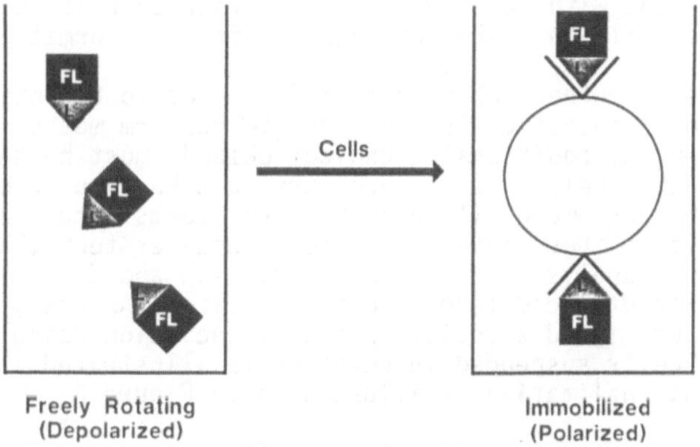

Fig. 7. Schematic Diagram of Polarization Analysis of Ligand Binding to Cells Suspended in a Fluorescence Cuvette. Cells are excited with polarized light. The emitted light remains polarized when ligand is bound to the receptor amd immobilized. The light is depolarized when ligand is free and rapidly mobile.

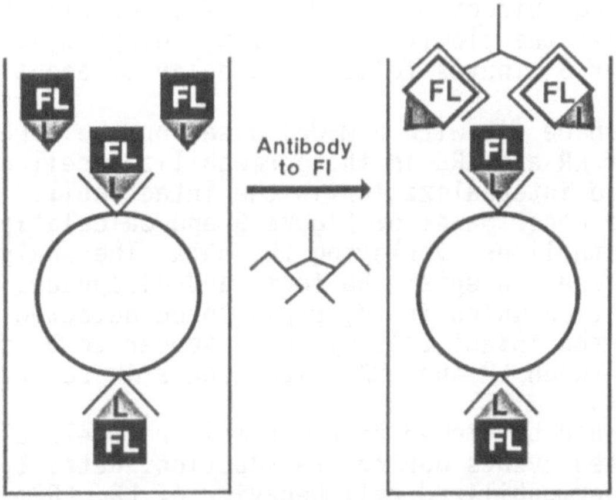

Fig. 8. Schematic Diagram of Internalization Analysis. The detection system for internalization of fluorescent ligand on cells suspended in cuvettes takes advantage of fluorescein as a pH indicator and uses a rapid pH change as a quench step to identify extracellular ligand.

to LR occurs at a rate which is proportional to the nucleotide concentration. These results imply that the binding of nucleotide is the rate-limiting step, while the release of activated G protein from LRG occurs too rapidly to be measured with this technology. Experiments with GTP indicate that the conversion occurs at a rate of ~10/sec when GTP is 500uM which is its approximate intracellular level.

THE INTACT CELL. Experiments in the intact cell indicate that three receptor forms exist (Sklar et. al. 1989B). The two forms LRG and LR have been discussed above and the third form is a slowly dissociating desensitized receptor ("LRX"). While the mechanism of LRX formation is not completely understood, some information is available. The formation of the desensitized receptor requires energy (Sklar et. al. 1989B), but direct evidence of formyl peptide receptor phosphorylation has not yet been obtained. However, experiments with ribosylation by pertussis toxin indicate that while occupancy by agonist is required for desensitization, neither G protein, nor cell activation is required to form LRX (Sklar et. al. 1989B). Since, the third receptor state forms in cells depleted of F-actin by treatment with botulinum C2 toxin (Norgauer et. al. 1989), it appears that microfilaments are not required for its formation.

The results from the broken cell studies need to be integrated with results obtained in intact cells. In order to perform measurements in intact cells, several additional technical demands must be satisfied. First, in the intact cells, it is necessary to take into account the fact that interconversions among all three receptor forms occur rapidly. At a minimum, analysis of these interconversions requires technology which permits continuous analysis of ligand association and dissociation. Second, ligand-receptor complexes are internalized. An assay system which allows analysis of ligand association and dissociation using fluorescence polarization on cells suspended in cuvettes is illustrated in Figure 7. The assay for internalization is illustrated in Figure 8.

Measurements on the intact cell using fluorescence polarization are shown in figure 9, middle panel. In contrast to the broken cell system, however, there is a rapid and spontaneous interconversion among the receptors states. The evolution among the receptor states is probed by adding antagonist at distinct times (15, 30, 60, and 120 sec) and then detecting the ligand dissociation. At early times, a rapidly dissociating form similar to LR is followed by a slowly dissociating, desensitized receptor, LRX. After a minute or so of binding only the slow dissociating receptor is detected. Direct measurements of internalization (data not shown) indicate that the slowly dissociating receptor is internalized with a half-time of 3 minutes following a delay of about 30 seconds.

These data can be fit with a model based on the direct observation of the behavior of LR and LRG in the permeabilized cell combined with behavior of LRX and internalization in the intact cell. This model is represented in the upper panel of Figure 9 and calculations based on the model are the smooth lines overlaying the data. The analysis is consistent with these concepts: LRG forms and disappears within a second during cell activation which is too rapid to be detected with the time resolution shown; the intact cell dynamics appear to be dominated by the interconversion between LR and LRX which occurs in several seconds.

The model should be viewed as a minimal or "skeletal" description of the receptor related events during transduction. Rates 1, 2, and 3 are derived from the permeabilized cell behavior of LR, LRG, and their interconversions. Rate 5 is determined from the long-time dissociation behavior of the desensitized receptor in the intact cell and rate 6 from separate internalization measurements (not shown). Rate 4 is calculated by an optimization scheme which provides the best overall fit to the data. The calculation yields a rate of interconversion from LR to LRX with a half-time on the order of 10 seconds.

The lower panel in Figure 9 shows the calculated behavior of the ternary complex, the active form of the receptor. During normal binding, LRG forms rapidly and decays slowly over a period of several minutes as binding approaches "equilibrium". When binding of the agonist is interrupted by applying a saturating concentration of an antagonist, the ternary complex decays rapidly, within a period of seconds.

The functional behavior of cells under pulse stimulation, which creates binding behavior like that illustrated in Figure 9, is illustrated in Figure 10. Normal cell stimulation is accompanied by cell responses which are initiated within a few seconds, reach maximal rates of response within tens of seconds, then decay slowly over periods of minutes. When agonist binding is interrupted, it takes a few seconds for the response to be affected, then the response decays. The cell responses decay because the transduction pathway collapses once the active states of the receptor have been lost.

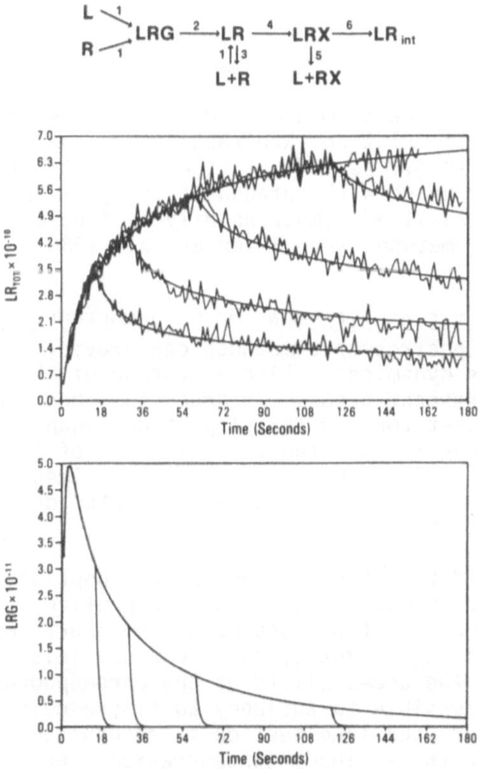

Fig. 9. Interconverting Receptor Model. In the presence of L, membrane R and G associate to form LRG, which is sequentially converted to LR and LRX prior to internalization. Middle Panel: The data are a polarization analysis of ligand binding (1 nM to 1e7 cells/ml at 37). The data reflect the sum of all occupied receptors. Dissociation is monitored after antagonist addition to the different samples at 15, 30, 60, or 120 sec. The smooth curves are a fit to the data using the model. Lower Panel: Calculated behavior of LRG in the model during binding and dissociation. (Reprinted with permission from Sklar et. al. 1989A).

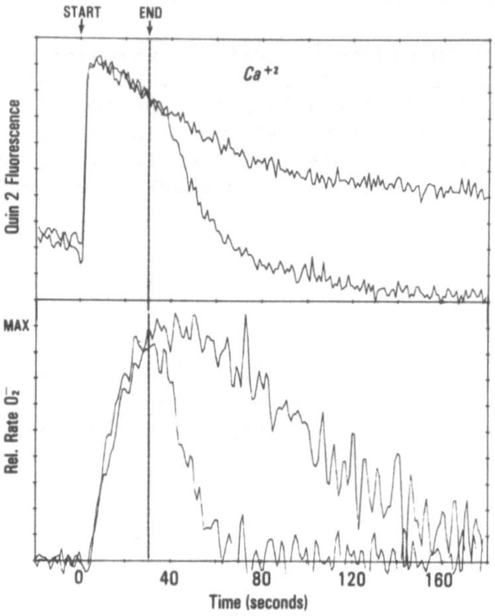

Fig. 10. Calcium and Oxidant Responses During Pulse Stimulation. Neutrophils were stimulated with formyl peptide at time 0 (Start), then interrupted with antagonist (End). Responses were examined simultaneously by fluorescent methods (see Sklar et. al. 1985)

REAL-TIME ANALYSIS OF STRUCTURE OF MACROMOLECULAR ASSEMBLY AT NATURAL ABUNDANCE. Spectroscopic methods can provide an insight into structure as well as dynamics, taking advantage of the environmental sensitivity of fluorescent probes. In order to study the size of the receptor binding pocket for the formyl peptide ligand, we have compared the fluorescent properties and the accessibility of the chromophore of peptides containing 4, 5, or 6 amino acids. Experiments are typically performed at cell densities of 1e7/ml and receptor densities of 1 nM, both similar to that found in blood.

We have observed that each of the shorter peptides is quenched upon binding to the receptor while the longer hexapeptide is not. These results suggested that the fluoresceins of the short peptides were in contact with the binding pocket of the receptor. This conclusion has been verified by probing the accesibility of the chromophores. The short peptides are not accessible to antibody to fluorescein while they are on the receptor; however, the long peptide is recognized by antibody at a rate proportional to the antibody concentration. The rate constant is approximately 1E5/Msec, typical of antibodies recognizing haptens on cell surfaces. Taken together, these observations suggest that the binding pocket is large enough to accomodate 5-6 amino acids, but that the chromophore on the hexapeptide protrudes from the pocket.

DISCUSSION

REAL-TIME TECHNOLOGY. We have described four different approaches to the analysis of ligand-receptor dynamics and molecular assembly and disassembly during signal transduction. These methods involve fluorescence flow cytometry, and three different types of fluorescence spectroscopy: polarization which discriminates free from bound ligand on the basis of rotational mobility; antibodies to fluorescein which bind free fluorescein (or fluorescent ligand) three orders of magnitude more rapidly than receptor associated ligand; and environmental sensitivity which alters the spectral properties of bound as compared to free ligands. The sensitivity of the methods is such that each analysis involves no more than 100 fmoles of receptor.

Each method has advantages and disadvantages. Flow cytometry is the most sensitive. As few as 1000 receptors per cell can be detected if the Kd for ligand binding is in the nM range. For uM affinty ligands, hundreds of thousands to millions of receptors must be present. The main weakness of the technology is the poor time resolution which is limited by the time that it takes to deliver sample into the laser beam.

Each of the spectrofluorimetric methods requires that a substantial fraction of the ligand molecules be receptor associated. In the most general case it requires that the receptor concentration is comparable to or exceeds the dissociation constant. Polarization works best for small molecules whose rates of rotation are more rapid than their fluorescence lifetime. The antibody methodology is appropriate for measuring ligand dissociation rates when the bound ligand is inaccessible to the antibody. To take advantage of environmental sensitivity as a method of analysis requires some knowledge of the ligand-receptor interaction so that probes with chromophores in the binding pocket can be designed. Taken together these methods offer opportunities to characterize ligand-receptor interactions in what are likely to be 100's of peptide receptor systems. By combining these methods with stopped-flow mixing devices, resolution in the msec time frame will also be possible.

DYNAMICS OF ASSEMBLY AND DISASSEMBLY IN THE NEUTROPHIL FORMYL PEPTIDE RECEPTOR SYSTEM. Since intracellular signals are detected within seconds of exposure to formyl peptides, the transduction events clearly must occur in a subsecond time frame. For the formyl peptide receptor, a picture emerges in which ligand binds to receptor and forms ternary complex at a diffusion limited rate. The ternary complex is activated within 100 msec by the binding of GTP and activation follows approximately single-step kinetics. In both the assembly and disassembly processes, the information has been limited by the time resolution of the technology. In subsequent work, stopped-flow mixing with polarization or environmentally sensitive probes will be used to detect subsecond events.

Several fundamental issues need to be studied: 1) the rate at which R finds G when LRG forms; 2) the steps that occur during the dissolution of ternary complex caused by the binding of guanine nucleotide. In the first case, the analysis shows that ternary complex forms as rapidly as L finds R. We cannot discriminate however whether R and G are preassembled or whether the assembly events are too fast to measure without stopped-flow methods. In the second case, since dissolution of ternary complex proceeds rapidly we are not able to detect the "transition state" which would reflect the assembly of L, R, G, and guanine nucleotide. Future studies will attempt to evaluate the lifetime of the "quaternary" complex after nucleotide binds and differences which arise during the dissolution of the complex as a consequence of the binding ofactivating(GTP) as compared to non-activating (GDP) nucleotides. The mechanism of the desensitization of the receptor needs to be resolved by biochemical rather than biophysical means. In addition, models for the intact cell will have to evaluate possible amplification steps in whichLR sequentially activates several G proteins prior to its desensitization.

ACKNOWLEDGMENTS. We would like to thank Harvey Motulsky for his advice concerning the equilibrium binding measurement shown in Figure 3, Heinz Mueller and Bruce Seligmann for their suggestions concerning the structure of the binding pocket, and Richard Freer for the synthesis of the pentapeptide. This work was supported by NIH Grant AI19032.

REFERENCES

Norgauer, J., Just, I., Aktories, K., and Sklar, L. A., 1989, The Influence of Botulinum C2 Toxin on F-Actin and N-Formyl Peptide Receptor Dynamics in Human Neutrophils. J. Cell Biol., 109, in press.

Sklar, L. A., 1987, Real-time Spectroscopic Analysis of Ligand-Receptor Dynamics. Ann. Rev. Biophys. Biophys. Chem. 16, 479-506.

Sklar, L. A., Finney, D. A., Oades, Z. G., Jesaitis, A. J., Painter, R. G, and Cochrane, C. G., 1984, The Dynamics of Ligand-Receptor Interactions: Real-Time Analyses of Association, Dissociation, and Internalization of an N-Formyl Peptide and Its Receptor on Human Neutrophils. J. Biol. Chem. 259, 5661-5669.

Sklar, L. A., Hyslop, P. A., Oades, Z. G., Omann, G. M., Jesaitis, A. J., Painter, R. G., Cochrane, C. G., 1985, Signal Transduction and Ligand-Receptor Dynamics: Transient Responses and Occupancy-Response Relations at the Formyl Peptide Receptor. J Biol. Chem. 260, 11461-11467.

Sklar, L. A., Bokoch, G. M., Button, D., and Smolen, J. E., 1987, Regulation of Ligand-Receptor Dynamics by Guanine Nucleotides: Real-Time Analysis of Interconverting States for the Neutrophil Formyl Peptide Receptor. J. Biol. Chem. 1987, 135-139.

Sklar, L. A., Mueller, H., Swann, W. N., Comstock, C., Omann, G. M. and Bokoch, G. M., 1989A, Dynamics of Interaction Among Ligand, Receptor, and G Protein. ACS Symposium Series 383, 52-69.

Sklar, L. A., Oades, Z. G., Omann, G. M., and Mueller, H., 1989B, Three States for the Formyl Peptide Receptor on Intact Cells. J. Biol. Chem., 264, in press.

IDENTIFICATION OF RESIDUES ESSENTIAL FOR THE ACTIVE CONFORMATION OF THE ß-ADRENERGIC RECEPTOR BY SITE-DIRECTED MUTAGENESIS

Mari R. Candelore*, Sandra L. Gould*, Wendy S. Hill+, Anne H. Cheung*, Elaine Rands+, Barbara A. Zemcik*, Irving S. Sigal*+, Richard A.F. Dixon+, and Catherine D. Strader*

*Department of Molecular Pharmacology & Biochemistry
+Department of Molecular Biology
Merck Sharp & Dohme Research Laboratories
Rahway, N.J. 0706 and West Point, P.A. 19486

Many hormones and neurotransmitters mediate their actions by stimulating one of a class of receptors which are functionally coupled to guanine nucleotide binding regulatory proteins (G-proteins). Activation of the G-protein by the agonist-occupied form of the receptor causes the release of GDP from the G-protein, allowing GTP to bind. The GTP-bound form of the G-protein is the active form, which dissociates from the receptor, leading to the activation or inhibition of specific effector enzymes and modulating levels of intracellular second messengers. The elucidation of the primary sequences of several G-protein coupled receptors by molecular cloning has shown them to share characteristic structural elements, presumably related to their common mechanism of action (see Dixon, et al, 1989, for review). All of these receptors whose sequences are known contain 7 stretches of conserved hydrophobic amino acids, thought to form transmembrane $\alpha-$helices, which connect 8 more divergent hydrophilic regions, postulated to form alternating extracellular and intracellular loops. In our laboratory, we have used genetic analysis of the ß-adrenergic receptor (ßAR), which is linked to G_s to stimulate adenylyl cyclase, to identify structural domains and specific amino acid residues required for receptor function. A model for the structure of the ßAR is shown in Figure 1. From these studies and from biochemical and genetic analysis of other G-protein coupled receptors, a model for receptor activity is emerging, which may serve to identify potential sites of pharmacological intervention within these signal transduction systems.

<u>Ligand binding domains</u>: Biochemical analysis of rhodopsin, which activates the G-protein transducin, and deletion mutagenesis studies of the ßAR have demonstrated the ligand binding regions of both of these receptors to involve residues within the conserved hydrophobic core domains (Findlay and Pappin, 1986; Dixon, et al, 1987a,b; Strader, et al, 1987a, 1988). In contrast, G-protein coupling involves residues within the more divergent putative 3rd intracellular loops of these receptors (Figure 1) (Dixon, et al, 1987a; 1989). More recently, site-directed mutagenesis of specific amino acid residues within the hydrophobic region of the ßAR has implicated specific amino acid side chains in receptor-ligand binding interactions. These studies suggest that the protonated amine group of the ligand interacts with the carboxylate side chain of Asp^{113} in the third hydrophobic domain of the receptor, presumably reflecting the formation of an ion pair (Figure 1) (Strader, et al, 1988). Substitution of Asp^{13} with Asn or Glu decreases the affinity of the receptor for both agonists and antagonists, indicating that this particular site of interaction is shared by both classes of compounds. Further genetic analysis suggests that catecholamine agonists also interact with the receptor <u>via</u> hydrogen bonding interactions between the catechol hydroxyl groups of the agonist and the hydroxyl side chains of Ser^{204} and Ser^{207} in ransmembrane helix 5 of the receptor (Figure 1) (Strader, et al, 1989). Substitution of

these Ser residues with Ala residues, removing the hydroxyl side chains at these positions, results in decreased affinity and efficacy of catecholamine agonists. The losses in agonist activity are quantitatively similar to the reductions observed upon removal of the catechol hydroxyl substituents from the agonist ligands, consistent with the formation of specific hydrogen bonds at these positions. Substitution of these Ser residues does not affect the affinity of the receptor for antagonist ligands. This absence of an effect on antagonist binding is consistent with the results from pharmacophore mapping of adrenergic ligands, which indicate that antagonists do not occupy the agonist-specific catechol binding site.

Further support for assignment of these specific molecular interactions to Asp^{113}, Ser^{204}, and Ser^{207} in the βAR arises from a comparison of the primary sequences of other G-

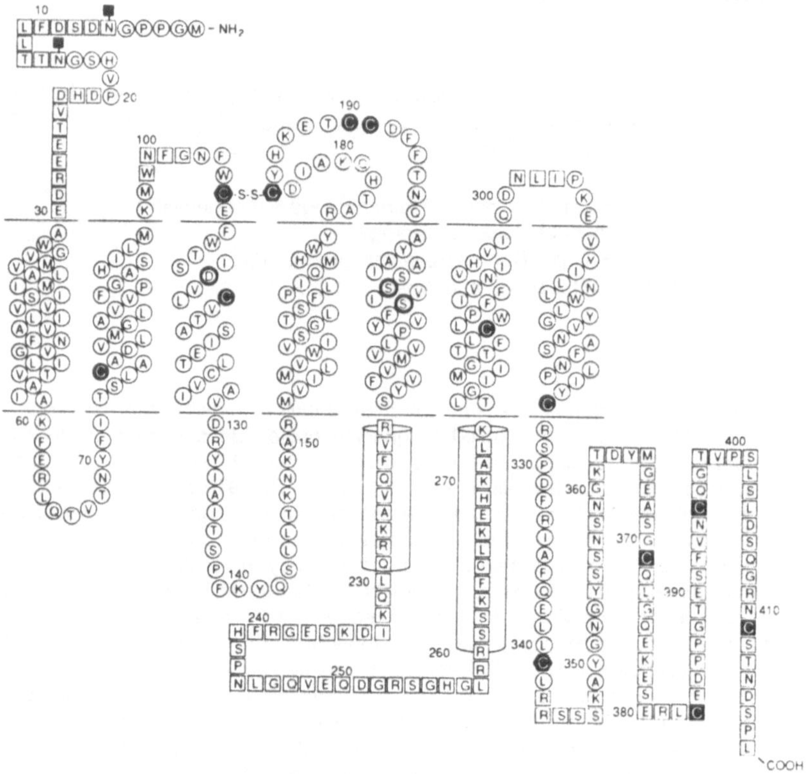

Figure 1. Structural model of the βAR. The predicted transmembrane topology of the receptor is indicated, with specific amino acid residues represented by the single letter code. Asp^{113}, Ser^{204}, and Ser^{207}, are encircled in bold. The regions in the third inner loop which are required for G-protein coupling are encased in cylinders. Solid symbols indicate Cys residues which were substituted or deleted in the work reported here. The two glycosylation sites are shown with small solid squares.

protein coupled receptors which bind similar ligands. Thus, all receptors of this class which bind protonated amine ligands, including all known subtypes of the adrenergic, dopamine, muscarinic, and serotonin receptors, have an Asp residue in the third hydrophobic domain at a position analogous to that of Asp^{113} in the ßAR, whereas receptors which do not bind amine ligands lack an acidic residue at this position. Similarly, Ser residues in transmembrane helix 5 at positions analogous to those of Ser^{204} and Ser^{207} appear to be specifically characteristic of G-protein coupled receptors which bind catecholamine agonists. The conservation of these functionally important residues among diverse G-protein coupled receptors further strengthens the hypothesis that the three-dimensional structures of these receptor proteins are similar. Knowledge of the tertiary structures of one or more G-protein coupled receptors is essential for a complete understanding of their mechanism of action. The absence of large quantities of functionally active ßAR currently precludes direct structural determination by X-ray diffraction or similar biophysical techniques. The current working model which has been developed for the structure of the ßAR and other G-protein linked receptors is based on analogies with a low-resolution electron micrographic structure which has been determined for bacteriorhodopsin (Henderson and Unwin, 1975). In order to strengthen the information available for the ßAR directly, we have utilized site-directed mutagenesis of residues which are potentially structurally significant, for example, Cys residues or potential glycosylation sites, as discussed below, to develop and test models for the tertiary structure of the ßAR.

Role of Cys residues in ßAR structure: Examination of the primary sequences of G-protein linked receptors reveals that several Cys residues are highly conserved among these receptor proteins (Figure 1). Because the predicted secondary structures of the receptors of this class are similar, these conserved Cys residues represent likely candidates for participation in structurally important intramolecular disulfide bonds. The possible role of disulfide bond formation in maintaining the structural and functional properties of these receptors has previously been explored using both biochemical and pharmacological approaches. The presence of dithiothreitol (DTT) has been shown to reduce either the number or the affinity of ß-adrenergic binding sites in various membrane preparations, indicating that the ßAR contains disulfide bonds which are essential for normal ligand binding activity (Vauquelin, et al, 1979; Clark, et al, 1983). In addition, treatment of reconstituted systems containing the ßAR and G_s with DTT increased the binding of GTP S, apparently by stimulating the coupling of the receptor to G_s (Pederson, and Ross, 1985; Moxam, et al, 1988).

The ßAR contains 15 Cys residues, 13 of which have been removed by site-directed mutagenesis studies, either by molecular substitution or by deletion. In the substitution studies, the Cys residues were variously replaced with either the isosteric residues Val or Leu (Dixon, et al, 1987b, 1989), or with the smaller residues Ser (Fraser, 1989) or Gly (O'Dowd, et al, 1989). In most of these studies, the effects of the mutations on the processing and folding of the receptor were first analyzed, either by protein immunoblotting (Dixon, et al, 1987b, 1989) or by photoaffinity labelling (O'Dowd, et al, 1989), in order to rule out nonspecific effects of the mutations on protein processing. As has been demonstrated for both rhodopsin and the ßAR (Karnik, et al, 1988; Dixon, et al, 1987b), this preliminary structural analysis is critical to the subsequent interpretation of mutagenesis results. The wild-type ßAR appears on SDS-PAGE as a broad band, with an apparent M_r centered at 67 kDa. In contrast, non-glycosylated or partially glycosylated receptors migrate as sharper bands with apparent molecular weights of 47 KDa and 55 KDa, respectively (Dixon, et al, 1987b). Glycosylation of the receptor requires the normal folding and insertion of the protein into the membrane. Therefore, the appearance of the 67 kDa fully glycosylated form of the ßAR in membranes from cells expressing a mutant ßAR may be taken as an indication of correct processing of the receptor in the cells. Analysis by protein immunoblotting of mutant receptors in which Cys residues were replaced revealed that substitution of the Cys residues at positions 77, 116, 327, and 341 by Val or Leu had only minor effects on the levels of fully processed ßAR protein present and did not affect the glycosylation of the receptor appreciably. Similar results were obtained by photoaffinity labeling of a mutant receptor in which Cys^{341} was replaced by Gly (O'Dowd, et al, 1989). In contrast, substitution of Cys^{106}, Cys^{184}, Cys^{190}, Cys^{191}, or Cys^{285} with Val resulted in a decrease in the amount of normally processed receptor protein present in the membrane, suggesting that these residues might be important for maintaining the tertiary structure of the ßAR. The Cys residues at positions 190, 191, and 285 were also replaced by Ser residues, but the effects of these mutations on the folding of the receptor were not determined (Fraser, 1989). The effects of removal of the Cys residues at positions 371, 383, 395, and 411 could not be assessed by protein immunoblotting, since the C-terminal truncation used to delete these Cys residues also removed the antigenic determinant used for protein immunoblotting.

The effects of substitutions of Cys residues on the ligand binding properties of mutant receptors in which Cys residues were substituted or deleted were determined, with the results summarized in Table 1.

Table 1. Characterization of mutant ßAR.

Mutant	Receptor Mr (kDa)	^{125}I-CYP fmol/mg	^{125}I-CYP Kd(pM)	iso Kd(M)	alp Kd(M)
βAR	67	500	50	2×10^{-7}	7×10^{-10}
[Val^{77}]βAR	67	420	135	1×10^{-7}	n.d.
[Val^{106}]βAR	55,47	180	115	3×10^{-8} 4×10^{-5}	4×10^{-10} 3×10^{-8}
[Val^{116}]βAR	67	220	125	7×10^{-8}	n.d.
[Val^{184}]βAR	67,55,47	490	65	5×10^{-9} 2×10^{-6}	3×10^{-9}
[Val^{106} Val^{184}]βAR	67,55,47	155	n.d.	4×10^{-8} 4×10^{-6}	2×10^{-9}
[Val^{190}]βAR	67,55	180	n.d.	4×10^{-8} 9×10^{-6}	6×10^{-12} 5×10^{-8}
[Val^{191}]βAR	67,55	250	n.d.	2×10^{-8} 1×10^{-5}	6×10^{-12} 5×10^{-8}
[Val^{190} Val^{191}]βAR	67,55	180	n.d.	2×10^{-5}	6×10^{-6}
[Val^{285}]βAR	55,47	280	n.d.	2×10^{-7}	n.d.
[Val^{327}]βAR	67	1700	n.d.	3×10^{-7}	n.d.
[Leu^{341}]βAR	67	340	125	2×10^{-7}	n.d.
T(354)βAR	n.d.	510	50	3×10^{-7}	n.d.

T(354)ßAR is a mutant receptor truncated at position 354, which removes the Cys residues at positions 371, 383, 395, and 411. Pharmacological analyses were performed as previously described (Dixon, et al, 1987a,b). M_r indicates the molecular weight of immunoreactive receptor.

Substitution of most of the Cys residues did not substantially affect the interaction of the ßAR with ligands. However, as previously reported (Dixon, et al, 1987b, 1989), molecular replacement of Cys^{106}, Cys^{184}, Cys^{190}, or Cys^{191} produced alterations in the ligand binding properties of the receptor. As shown in Table 1, the 6 mutant receptors in which these 4 Cys residues were substituted with Val showed decreased affinities for both the agonist isoproterenol and the antagonist alprenolol, accompanied by the appearance of multiple binding sites (Dixon, et al, 1987b, 1989; see also Fig. 5 in Fraser, 1989). Several lines of evidence suggest that these multiple binding sites could not arise from interactions of the mutant receptors with G_s. First, the multiple affinity states of the mutant receptors were measured in COS-7 cells, a system in which ßAR-G_s coupling is not observed Dixon, et al, 1987a . In addition, some of the mutant receptors also showed evidence of 2 affinity sites for the antagonist alprenolol, although antagonists do not promote receptor G-protein interactions. Finally, when these mutant receptors were expressed in L cells so that G_s coupling could be measured directly, the addition of the nonhydrolyzable GTP analog Gpp(NH)p was not able to convert all of the agonist binding to a single low affinity state (data not shown).

Effects of reducing agents on mutant ßAR: We had previously suggested that the multiple agonist affinity sites observed for [Val^{106}]ßAR, [Val^{184}]ßAR, [Val^{190}]ßAR, and

[Val^{191}]ßAR might arise from various conformations of the receptor in which different disulfide bonds are formed with a Cys side chain which is normally involved in a stable disulfide bond in the wild-type receptor (Dixon, et al, 1987b, 1989). To determine whether the multiple affinity sites of the mutant receptors could result from disulfide bond formation, the ligand binding properties of [Val^{106}] ßAR, [Val^{184}]ßAR, [Val^{106}Val^{184}] ßAR, and [Val^{190}Val^{191}] ßAR were examined after exposure to the reducing agent DTT. DTT decreased the affinity of

Table 2. Adenylyl cyclase activation by mutant ßAR.

Mutant	-DTT		+DTT	
	K_{act}(M)	V_{max}(%)	K_{act}(M)	V_{max}(%)
ßAR	3×10^{-8}	86	2×10^{-6}	93
[Val^{106}]ßAR	2×10^{-5}	37	3×10^{-5}	31
[Val^{184}]ßAR	2×10^{-6}	65	2×10^{-6}	54
[Val^{190}]ßAR	4×10^{-6}	40	n.d.	n.d.
[Val^{191}]ßAR	4×10^{-7}	10	n.d.	n.d.
[Val^{190} Val^{191}]ßAR	1×10^{-5}	55	1×10^{-5}	85
[Leu^{341}]ßAR	n.d.	100	n.d.	n.d.
T(354)ßAR	1×10^{-8}	100	n.d.	n.d.

Membranes prepared from L cells expressing the wild-type or mutant ßAR were incubated in the presence or absence of 5 mM DTT in TME buffer for 30 min at 4° C, then diluted 1:2.5 into an adenylyl cyclase assay, as previously described (Dixon, et al, 1987a). V_{max} is defined as (pmol cAMP stimulated by 10^{-4} M isoproterenol / pmol cAMP stimulated by 10 mM NaF) x 100.

the wild-type ßAR for both agonists and antagonists by 10-40 fold. A DTT-mediated decrease in affinity was also observed for the mutant receptors having substitutions for the Cys residues at positions 106, 184, 190, and 191. However, these mutant receptors showed heterogeneous agonist binding in both the presence and the absence of DTT, indicating that the formation of alternative disulfide bonds could not account for the appearance of the additional ligand binding sites in the mutant receptors. The functional significance of the various affinity states of the mutant ßARs were assessed in stable L cell lines, where their ability to stimulate adenylyl cyclase was measured. [Val^{106}]ßAR, [Val^{184}]ßAR, [Val^{190}]ßAR, [Val^{191}]ßAR, and [Val^{190}Val^{191}]ßAR all displayed increased K_{act} and decreased V_{max} for adenylyl cyclase stimulation in response to isoproterenol, in comparison with the wild-type ßAR. Interestingly, the data for the isoproterenol-stimulated adenylyl cyclase activity for each of these mutant receptors could be described by a single exponential, even though the binding isotherms indicated the presence of multiple binding sites. In the presence of DTT, the wild-type ßAR exhibited a 100-fold increase in the K_{act} for exhibited a 100-fold increase in the K_{act} for isoproterenol, with no change in the V_{max} (Figure 2). In contrast, the K_{act} values for isoproterenol-stimulated adenylyl cyclase activation by the mutant receptors were not further reduced upon exposure to DTT (Table 2), as illustrated for $[Val^{106}]$ßAR in Figure 2.

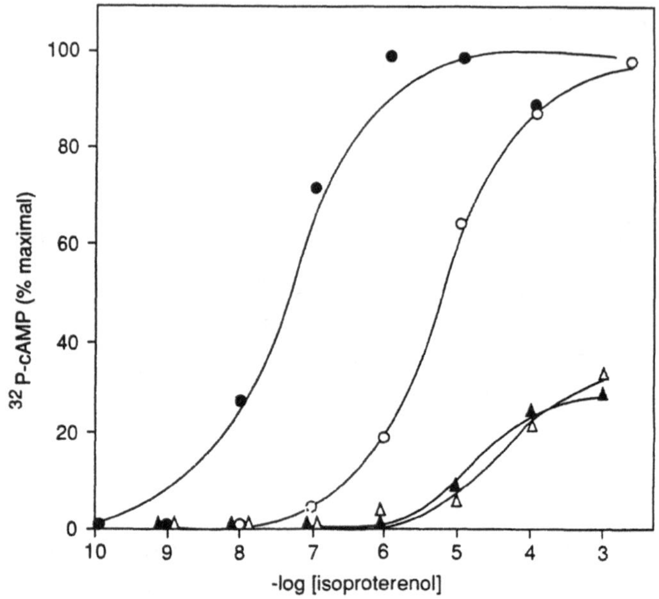

Figure 2. Effects of DTT on adenylyl cyclase stimulation. Membranes prepared from L-cells expressing wild-type ßAR (●,○) or $[Val^{106}]$ßAR (▲,△) were incu ৪ ͻd in the presence (○,△) or the absence (●,▲) of 5 mM DTT in TME buffer for 30 min at 4° C, then diluted 1:2.5 into an adenylyl cyclase assay (Dixon, et al., 1987a). The stimulation of adenylyl cyclase activity is presented as the percentage of the maximal stimulation by 10 mM NaF.

Evidence for functionally important disulfide bonds: It appears from these data that, of the multiple binding sites observed for these mutant ßARs, cnly the lowest affinity site is functionally active, correlating with the ability of the receptor to promote adenylyl cyclase activation. The molecular basis for the appearance of the higher affinity binding sites is unknown at present, although they probably involve the interactions of ligands with conformationally altered forms of the receptor. In contrast to the increase in K_{act} observed for the wild-type ßAR in the presence of reducing agents, the adenylyl cyclase stimulation by the mutant receptors is not sensitive to DTT, suggesting that the active forms of these ßAR mutants do not contain functionally important disulfide bonds. These data suggest that residues 106, 184, 190, and 191 are important in maintaining the active conformation of the wild-type ßAR. One explanation for this effect would be the involvement of these residues in intramolecular disulfide bonds. Removal of these putatiave sulfhydryl bridges, either by treatment with reducing agents or by site-directed mutagenesis, decreases the affinity of the receptor for ligands, possibly due to altered protein conformation. The similar effects of removal of receptors, argues for the existence of 2 essential disulfide bonds in the native $ß_2$AR and suggests that all 4 Cys residues are required for either of these bonds to form correctly. The 3 possibilities for disulfide bridges involving these 4 Cys residues are illustrated in

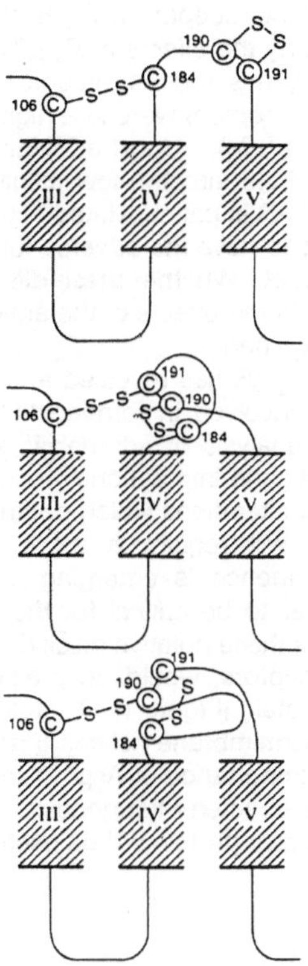

Figure 3. Models for possible intramolecular disulfide bonds in the βAR. Hydrophobic domains III, IV, and V are drawn as transmembrane helices, with the extracellular face of the protein at the top of each model and the cytoplasmic face at the bottom.

Figure 3. Each configuration would result in a covalent linkage between the putative 2nd and 3rd extracellular loops of the βAR, with an additional disulfide bond within outer loop 3. Cys^{106} and Cys^{184} are highly conserved among the G-protein linked receptors, whereas Cys^{190} and Cys^{191} are specific for the β_1- and β_2AR. In addition, the Cys residues in opsin equivalent to those at positions 106 and 184 in the βAR have recently been shown to be critical for proper folding of that protein (Karnik, et al, 1988). For these reasons, the disulfide bonding pattern depicted in the top panel of Figure 3, with 1 disulfide bond between the conserved Cys^{106} and Cys^{184} residues and another vicinal disulfide bond between Cys^{190} and Cys^{191} on the 3rd outer loop of the βAR seems the most likely of the 3 possibilities. However, the determination of the exact placement of disulfide bonds within the βAR awaits further biochemical and biophysical characterization of the wild-type and mutant receptor proteins.

Post-translational modification of the βAR: As mentioned above, the βAR is a glycoprotein, with the addition of carbohydrate resulting in characteristic alterations in the behavior of the receptor on SDS-PAGE (Lefkowitz, et al, 1983). The hamster βAR has 4 consensus sequences for N-linked glycosylation, 2 near the N-terminus and 2 near the C-terminus of the protein (Dixon, et al, 1986). Mutagenesis studies in which the Asn residues which serve as potential glycosylation sites were removed by deletion or molecular replacement with Gln residues revealed that only the 2 N-terminal glycosylation sites are modified in the native βAR (Dixon, et al, 1989; Rands, et al, 1989). Replacement of either Asn^{6} or Asn^{15} by Gln residues resulted in a partially glycosylated receptor, as assessed by SDS-PAGE. In contrast, no glycosylation was observed upon replacement of both Asn^{6} and Asn^{15}. Pharmacological analysis of these mutant receptors suggests that glycosylation of the βAR is not required for the functional activity of the receptor, but is critical for the normal membrane trafficking of the protein (Rands, et al, 1989).

It was recently demonstrated that the human β_2AR is palmitoylated and that substitution of Cys^{341} with Gly prevents this post-translational modification of the receptor (O'Dowd, et al, 1989). Furthermore, [Gly^{341}]βAR showed a reduced ability to interact with G-proteins and a greatly reduced ability to activate adenylyl cyclase, characterized by a 70% reduction in V_{max}, suggesting that palmitoylation of the βAR is critical for receptor activity (O'Dowd, et al, 1989). In contrast, substitution of Cys^{341} in the hamster β_2AR with Leu did not affect the ability of the receptor to stimulate adenylyl cyclase (Table 2). The coupling of the receptor to G_s also did not appear to be altered by this molecular substitution, as assessed by the effects of Gpp(NH)p on binding of the agonist isoproterenol. In the absence of GTP, this mutant receptor bound isoproterenol with 2 distinct classes of binding sites: 57% of the receptors were in a high affinity state ($K_d = 1x10^{-9}$ M), and 43% in a low affinity state ($K_d = 1x10^{-7}$ M). Upon addition of the nonhydrolyzable GTP analog Gpp(NH)p, all of the receptors shifted into the low affinity state ($K_d = 1x10^{-7}$ M). We have not yet determined whether [Leu^{341}]ßAR is palmitoylated. However, substitution of Cys^{341} with Leu in the hamster ßAR appears not to have the adverse functional consequences of substitution of Cys^{341} with Gly in the human ßAR. Whether these differences reflect species-specific properties of the receptors or differences in the effects of the amino acid residues used for the molecular replacements remains to be determined.

Conclusions: Genetic and biochemical analysis of the $ß_2$AR has revealed amino acid residues within the hydrophobic core of the protein which are involved in forming the ligand binding site of the receptor. Combined with structure-activity analysis of adrenergic ligands, these data provide insights into the molecular basis for receptor-ligand interactions. However, a full understanding of the specific interactions and conformational alterations which contribute to the activities of agonists and antagonists on the receptor will require a more precise determination of receptor structure. As described above, evidence is emerging from the existence of 2 disulfide bonds in the ßAR, both of which appear to be critical for the correct tertiary structure and full functional activity of the receptor. One of these putative disulfide bonds, which is probably conserved among all G-protein coupled receptors, would be predicted to covalently link the second and third outer loops of the receptor protein (Figure 1). Therefore, this disulfide bond might also stabilize the interactions between transmembrane helices 3 and 4 by constraining the conformation of this region of the protein. The importance of Asp^{113} in helix 3 to the ligand binding site of the ßAR has been established by site-directed mutagenesis. Further determination of the structural properties of this region of the receptor protein will be useful in the refinement of the working model for receptor-ligand interactions.

REFERENCES

Clark, R.B., Green, D.A., Rashidbaigi, A, and Ruoho, A., 1983, Effect of dithiothreitol on the ß-adrenergic receptor of S49 wild-type and cyc- lymphoma cells: decreased affinity of the ligand-receptor interaction. J. Cyclic Nucleotide Res. 9, 203-220.

Dixon, R.A.F., Kobilka, B.K., Strader, D.J., Benovic, J.L., Dohlman, H.G., Frielle, T., Bolanowski, M.A., Bennett, C.D., Rands, E., Diehl, R.E., Mumford, R.A., Slater, E.E., Sigal, I.S., Caron, M.G., Lefkowitz, R.J., and Strader, C.D., 1986, Cloning of the gene and cDNA for mammalian ß-adrenergic receptor and homology with rhodopsin. Nature 321: 75-79.

Dixon, R.A.F., Sigal, I.S., Rands, E., Register, R.B., Candelore, M.R., Blake, A.D., and Strader, C.D., 1987a, Ligand binding to the ß-adrenergic receptor involves its rhodopsin-like core. Nature 326: 73-77.

Dixon, R.A.F., Sigal, I.S., Candelore, M.R., Register, R.B., Scattergood, W., Rands, E., and Strader, C.D., 1987b, Structural features required for ligand binding to the ß-adrenergic receptor. EMBO J. 6: 3269-3275.

Dixon, R.A.F., Sigal, I.S., and Strader, C.D., 1989, Structure-function analysis of the ß-adrenergic receptor. Cold Spring Harbor Symp. Quant. Biol. 53, 487-489.

Findlay, J.B.C. and Pappin, D.J.C., 1986, The opsin family of proteins. Biochem. J. 238: 625-642.

Fraser, C.M., 1989, Site-directed mutagenesis of ß-adrenergic receptors. J. Biol. Chem. 264: 9266-9270.

Henderson, R. and Unwin, P.N., 1975, Three-dimensional model of purple membrane obtained by electron microscopy. Nature 257: 28-32.

Karnik, S.S., Sakmar, T.P. Chen, H-B, and Khorana, H.G., 1988, Cysteine residues 110 and 187 are essential for the formation of correct structure in bovine rhodopsin. Proc. Natl. Acad. Sci. (USA) 85: 8459-8463.

Lefkowitz, R.J., Stadel, J.M., and Caron, M.G., 1983, Adenylate cyclase coupled ß-adrenergic receptors. Ann. Rev. Biochem. 52: 159-186.

Moxham, C.P., Ross, E.M., George, S.T., and Malbon, C.C., 1988, ß-Adrenergic receptors display intramolecular disulfide bridges in situ: analysis by immunoblotting and functional reconstitution. Mol. Pharm. 33, 486-492.

O'Dowd, B.F., Hnatowich, M., Caron, M.G., Lefkowitz, R.J., and Bouvier, M., 1989,

Palmitoylation of the human β_2-adrenergic receptor. J. Biol. Chem. 264: 7564-7569.

Pederson, S.E. and Ross, E.M., 1985, Functional activation of β-adrenergic receptors by thiols in the presence or absence of agonists. J. Biol. Chem. 260, 14150-14157.

Rands, E., Candelore, M.R., Cheung, A.H., Hill, W.S., Strader, C.D., and Dixon, R.A.F., submitted for publication.

Strader, C.D., Sigal, I.S., Candelore, M.R., Rands, E., Hill, W.S., and Dixon, R.A.F., 1988, Conserved aspartic acid residues 79 and 113 of the β-adrenergic receptor have different roles in receptor function. J. Biol Chem. 263: 10267-10271.

Strader, C.D., Candelore, M.R., Hill, W.S., Sigal, I.S., and Dixon, R.A.F., 1989, Identification of two serine residues involved in agonist activation of the β-adrenergic receptor. J. Biol. Chem., in press.

Vauquelin, G., Bottari, S., Kanarelik, L. and Strosberg, A.D., 1979, Evidence for essential disulfide bonds in β_1-adrenergic receptors of turkey erythrocyte membranes. J. Biol. Chem. 254, 4462-4469.

MECHANISMS FOR DIVERSIFICATION OF RESPONSES OF HUMAN POLYMORPHONUCLEAR LEUKOCYTES TO LEUKOTRIENES

Catherine H. Koo, Jeffrey W. Sherman, Laurent Baud, Winzhen Jin, Kai Mai, and Edward J. Goetzl

Howard Hughes Medical Institute
Division of Allergy and Immunology
University of California Medical Center
Room U-426
San Francisco, California 94143-0724

INTRODUCTION

The oxidative metabolism of arachidonic acid proceeds by complex pathways resulting in the generation of an array of structurally distinct and biologically potent mediators of physiological, as well as pathological events (Samuelsson, 1983; Goetzl & Scott, 1984). The leukotrienes (LTs) are products of the 5-lipoxygenase pathway found in mast cells, polymorphonuclear (PMN) leukocytes and macrophages (Samuelsson, 1983), and exhibit two classes of activities (Goetzl, 1980). The sulfidopeptide LTs (LTC_4, LTD_4, LTE_4) are smooth muscle contractile and vasoactive factors, that account for the activities of slow-reacting substance of anaphylaxis or SRS-A (Goetzl et al., 1983; Hedqvist et al., 1985; Piper & Samboun, 1982; Krell et al., 1985). In addition, the sulfidopeptide LTs stimulate glandular and epithelial cell secretion and alter some proliferative and transport processes (Samuelsson, 1983; Goetzl, 1980; Lewis & Austen, 1984; Dahlen et al., 1980; Pologe et al., 1984; Johnson & Torres, 1984; Pek & Walsh, 1984). The specific di-hydroxy-LT or LTB_4, on the other hand, elicits PMN leukocyte and eosinophil chemotaxis, initiates PMN leukocyte lysosomal enzyme secretion, and stimulates the respiratory burst (Goetzl & Pickett, 1981; Nagy et al., 1982; Bray et al., 1981). In addition, LTB_4 also inhibits proliferative responses of mixed T-lymphocytes to mitogen and antigens, by separate and opposing effects on the helper and suppressor subsets of T-cells (Payan & Goetzl, 1983; Payan et al., 1984; Gaulde et al., 1985).

Leukocytes possess enzymatic pathways for both the generation and metabolic degradation of LTs (Baud et al., 1987). PMN leukocytes synthesize and release LTB_4 and enzymatically convert LTC_4 to LTD_4 and LTD_4 to LTE_4 (Baud et al., 1987). PMN leukocytes also express specific populations of receptors, that recognize and transduce responses to both LTB_4 and the sulfidopeptide LTs.

Leukocytic expression and characteristics of leukotriene receptors

The LTB_4 receptors on human PMN leukocytes are present in two affinity states (Goldman & Goetzl, 1984a; Kreisle & Parker, 1983). Of the total population of LTB_4 receptors, approximately 4400 high affinity receptors bind [^{3}H]LTB_4 with a Kd of 0.4 nM and the remaining 270,000 receptors bind LTB_4 with a Kd of 61 nM (Goldman & Goetzl, 1984a). The high-affinity subset transduces the chemotactic, chemokinetic and increased adherence responses that occur at 0.2-10 nM LTB_4 (Goldman & Goetzl, 1984a; Palmbad et al., 1984). The low-affinity subset appears to be coupled to the responses, such as lysosomal enzyme secretion, which are observed at higher concentrations of (20-200 nM) LTB_4 (Goldman & Goetzl, 1984a; Palmbad et al., 1984).

LTB_4 receptors are located primarily on the plasma membrane. Isolated human PMN leukocyte plasma membranes display both high- and low-affinity receptors for LTB_4, at respective densities of 3.4 and 27.9 pmoles/mg of protein with a Kd of 0.07 and 28 nM (Sherman et al., 1988a). The stereospecificities of these binding sites mirror those observed for the corresponding LTB_4 receptors on intact cells. The HL-60 cultured line of promyelocytic leukemia cells and the U937 line of human monocytic leukemia cells have proven to be useful models for the study of LT receptors and transductional pathways. Both lines fail to display LTB_4 receptors until the induction of differentiation into mature leukocytes (Goldman et al., 1986; Sarau et al., 1989). The 1α,25-dihydroxy-vitamin D_3 differentiated HL-60 cells express LTB_4 receptors of the same specificity for LTB_4 and its analogues as the native receptors on human PMN leukocytes. Following a 5-7 day incubation with 1α, 25-dihydroxy-vitamin D_3, HL-60 cells display both a high- and a low-affinity subset of receptors for LTB_4 with Kds of 2.3 and 600 nM, respectively, that are capable of transducing the two different sets of responses to LTB_4 (Goldman et al., 1986). In contrast, U937 cells differentiated with DMSO for 3 days display a single homogeneous population of high-affinity receptors with a Kd of 0.15 nM (Sarau et al., 1989) (Table I).

Although the biological response of PMN leukocytes to the sulfidopeptide LTs is restricted to enhanced adherence to surfaces both in vivo and in vitro, specific receptors for LTC_4 on PMN leukocytes have been demonstrated by direct binding studies (Baud, Koo & Goetzl, 1987). Approximately 10,000 receptors for [^{3}H]LTC_4 have been detected on intact PMN leukocytes, with a mean Kd of 30 nM, which is similar to that of the low-affinity subset of LTB_4 receptors, but is approximately 10-fold higher than that of LTC_4 receptors detected in endothelial and smooth muscle cells (Table I). The binding sites for [^{3}H]LTC_4 are selective for LTC_4, since LTD_4 and LTE_4 were 35- and 120-fold weaker competitors for [^{3}H]LTC_4 binding to human leukocytes. In contrast to LTB_4 receptors, approximately 3/4 of the cellular receptors for [^{3}H]LTC_4 in PMN leukocytes reside on intracellular granules, while the remaining 1/4 are located on the cell surface. The stereoselectivity of the two populations of receptors are similar, but the Kds of 13.4 $\pm$1.3 nM for lysosomal granules receptor and 38.3 $\pm$ 1.1 nM for plasma membrane receptors are significantly different. [^{3}H]LTD_4 binds to intact human PMN leukocytes with an apparent Kd of 1-5 nM, which is comparable to that of [^{3}H]LTD_4 receptors on other cell types (Sarau et al., 1989; Koo et al., 1988a). The HL-60 and U937 cultured cell lines again served as models for the studies of LTC_4 and LTD_4 receptors. HL-60 cells that were induced to differentiate into PMN leukocytic lineage by incubation with DMSO, displayed approximately 14,400 [^{3}H]LTD_4 binding sites/cell with a mean Kd of 33 nM, which was similar to that of kD human PMN leukocytes, with identical stereoselectivity (Fig. 1). Recent studies by Sarau et al. (1989) demonstrated that preparations of membranes from U937 cells exhibit [^{3}H]LTD_4 receptors, which increased in association with the induction of differentiation (Table I).

Structural Properties of Leukotriene Receptors

Human PMN leukocyte receptors have been solubilized by the zwitterionic detergent CHAPS (Koo, Healy & Goetzl, 1988). The solubilized receptors retain the same stereoselectivity as that observed in whole cells and membranes. Affinity purification on a LTB_4-Sepharose column isolated a protein of approximately 80 kD that was specifically eluted from the LTB_4 affinity column by excess LTB_4. Preincubation of the solubilized receptor with LTB_4 prevented the retention of this 80 kD protein by the LTB_4 affinity column. A mouse monoclonal antibody raised against partially purified LTB_4 receptors specifically precipitated an 80 kD protein from PMN leukocyte and differentiated HL-60 cell membranes, thus lending support to the identity of the 80 kD protein as the LTB_4 receptor on PMN leukocyte membranes.

Biochemical Pathways Involved in Leukotriene Receptor Signal Transduction

Unique populations of receptors for each class of LTs exist on PMN leukocytes and occupation of these receptors by their respective agonist elicits a series of related biochemical responses. Incubation of leukocytes with LTB_4, LTD_4, and the LTD_4 generated by the leukocyte from LTC_4 evokes transient increases in the cytosolic calcium concentration, $[Ca^{2+}]_i$ (Goldman et al., 1985a; Baud, Goetzl & Koo, 1987). The response of $[Ca^{2+}]_i$ to LTB_4 is rapid, with peak concentrations achieved within 1 min, and the levels return to basal concentrations by 4 min. The increases in $[Ca^{2+}]_i$ in response to LTB_4 are concentration-dependent and are achieved by both an increased influx of extracellular calcium and calcium-release from intracellular stores (Goldman et al., 1985a). Both LTC_4 and LTD_4 evoke transient increases in $[Ca^{2+}]_i$, but the response to LTC_4 is much slower in onset and more sustained with peak concentrations occurring at 1-2 min and the return to basal concentrations only after 10 min. This slower response is due to the required conversion of LTC_4 to LTD_4 (Baud, Goetzl & Koo, 1987). Although specific LTC_4 binding sites are present on leukocytes, the active mediator appears to be LTD_4. The response to LTD_4 is more rapid than LTC_4, but the return to basal level is still slower than that observed with LTB_4. The calcium response to LTD_4 is dose-dependent and is due solely to an influx of extracellular calcium (Baud, Goetzl & Koo, 1987).

Leukocyte receptors for LTB_4 and LTD_4 utilize guanine triphosphate (GTP)-binding proteins in their transductional pathway. LTB_4 receptors are coupled to a pertussis toxin-sensitive GTP-binding protein. Pretreatment of intact leukocytes with pertussis toxin selectively decreases the number of high-affinity receptors and also inhibits LTB_4-induced increases in $[Ca^{2+}]_i$, and subsequent biological responses (Goldman et al., 1985b). An association of LTB_4 receptors with pertussis-toxin sensitive GTP-binding proteins is suggested by results of studies of the modulation of $[^3H]LTB_4$ binding by guanine nucleotides and the modulation of $[^3H]$ GMP-PNP binding by LTB_4 (Sherman et al, 1988a). The addition of GMP, GDP and GTP analogues to leukocyte membranes converts 50% of the high-affinity receptors to a low-affinity state and introduction of LTB_4 to the membranes specifically increases $[^3H]$ GMP-PNP binding activity (Sherman et al., 1988a). This mutual modulation of LTB_4 and GTP binding suggests the involvement of an agonist-induced complex of the LTB_4 receptor protein and a GTP-binding protein, which is supported by the studies of separation of solubilized receptors by gel exclusion chromatography. When solubilized PMN leukocyte membrane proteins are separated by Sephacryl S300 column chromatography, $[^3H]LTB_4$ binding activity eluted primarily at the 65_3kD position. However, if intact membranes are first preincubated with [^{3}H]LTB_4 prior to solubilization, the [^{3}H]LTB_4 binding activity eluted principally at the 140 kD position. This LTB_4-induced increase in apparent receptor size is inhibited by pertussis toxin, but not cholera toxin treatment (Sherman et al., 1988b).

LTD_4 receptors appear to be coupled to more than one class of GTP-binding proteins. Pertussis toxin pretreatment of leukocytes inhibited LTD_4-induced $[Ca^{2+}]_i$ increases by 85%[31], as compared to 98% for the LTB_4-induced responses (Baud et al., 1987b). Sarau et al. (1989) recently showed that pertussis toxin treatment of differentiated U937 cells completely inhibited the LTB_4-induced $[Ca^{2+}]_i$ increase, but only blocked the LTD_4-induced calcium response by 50%. Thus, it appears that LTD_4 receptor transduction involves both pertussis toxin- sensitive and -insensitive GTP-binding proteins.

Distinctive Mechanisms of Regulation of Leukotriene Receptors

Leukotriene receptors are subject to homologous desensitization. Prior exposure of leukocytes to nM concentrations of LTB_4 reduced the EC_{50} for a calcium response from 2×10^{-10} M to 2×10^{-9} M, but did not affect the formyl-peptide chemoattractant elicited calcium response. Functional deactivation is accompanied by a decrease in receptor-binding activity. The time-course of deactivation is rapid, with the maximum effect achieved by 2 min at 37^0C (unpublished observations, K. Mai et al.). Preincubation of PMN leukocytes with a concentration of LTB_4 that completely eliminated the chemotactic response, but did not affect the lysosomal degranulation response, eliminated all detectable high-affinity receptors for LTB_4 (Goldman & Goetzl, 1984b) and did not alter the expression of low-affinity receptors.

SUMMARY

PMN leukocytes are unique in that they express enzymatic pathways for the synthesis and degradation of LTs and at the same time display high affinity receptors for the LTs. The diversity of biological responses to LTs appears to involve molecular heterogeneity of receptors as well as transductional pathways associated with each class of receptors.

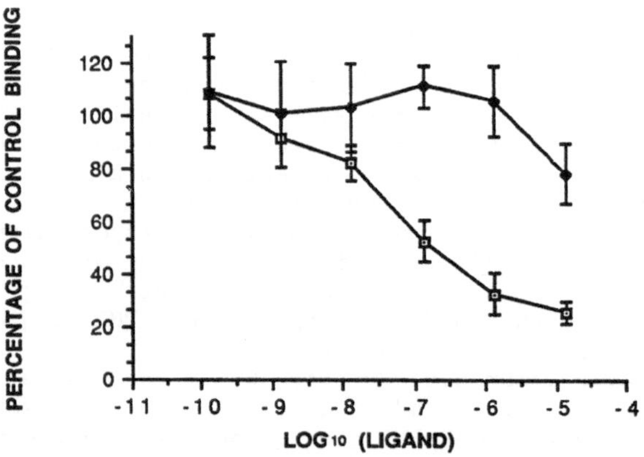

Fig 1. Competition by LTC_4 (◇) and LTD_4 (◆) for [^{3}H]LTC_4 binding to HL-60 cells. HL-60 cells were incubated at 4^0C in the presence of serine-borate with [^{3}H]LTC_4 and a range of concentrations of unlabeled competitors. Results are expressed as % of [^{3}H]LTC_4 binding in the absence of any competitors.

TABLE I

CHARACTERISTICS OF LEUKOTRIENE RECEPTORS ON LEUKOCYTES

LIGAND	LTB_4			LTC_4		LTD_4	
Target Cell	Human PMNL[a]	HL60	U-937	Human PMNL	HL-60	Human PMNL	U-937
kD (nM)	0.4/61	23/680	0.15	26	33	~1	0.35
Number of receptors/ cell (x10^3)	4.5/270	6.4/2200	0.300[b]	45	14	~40	0.44

[a]: PMN leukocytes
[b]: pmoles receptor long protein

REFERENCES

Baud, L., Goldman, D.W., Koo, C.H., Marotti, T., Harvey, J.P. and Goetzl, E.J. Molecular and cellular diversity of the polymorphonuclear leukocyte receptors for leukotrienes. IN Advances in Prostaglandin, Thromboxane and Leukotriene Research, Vol 17. Samuelsson, B., Paoletti, R. and Ramwell, P.W. Raven Press, N.Y. p. 163, 1987.

Baud, L., Goetzl, E.J., and Koo, C.H. Stimulation by leukotriene D_4 of increases in the cytosolic concentration of calcium in dimethylsulfoxide-differentiated HL-60 cells. J. Clin. Invest. 80:983:, 1987.

Baud, L., Koo, C.H. and Goetzl, E.J. Specificity and cellular distribution of human polymorphonuclear leukocyte receptors for leukotriene C_4. Immunol. 62:53, 1987.

Bray, M.A., Cunningham, F.M., Ford-Hutchinson, A.W. and Smith, M.J.H. Leukotriene B_4: A mediator of vascular permeability. Br. J. Pharmacol. 72:483, 1981.

Dahlen, S.E., Hedqvist, P., Hammarstrom, S. and Samuelsson, B. Leukotrienes are potent constrictor of human ronchi. Nature 288:484, 1980.

Gaulde, N., Atluru, D., and Goodwin, J.S. Effects of lipoxygenase metabolites of arachidonic acid on proliferation of human T cells and T cell subsets. J. Immunol. 134:1125, 1985.

Goetzl, E.J. Mediators of immediate hypersensitivity derived from arachidonic acid. N. Engl. J. Med. 303:822, 1980.

Goetzl, E.J., Brindley, L.L. and Goldman, D.W. Enhancement of human neutrophil adherence by synthetic leukotriene constituents of the slow reacting substance of anaphylaxis. Immunol. 50:35, 1983.

Goetzl, E.J. and Pickett, W.C. Novel structural determinants of the human neutrophil chemotactic activity of leukotriene B. J. Exp. Med. 153:482, 1981.

Goetzl, E.J. and Scott, W.A. III (eds). Proceedings of a conference on regulation of cellular activities by leukotrienes and other lipoxygenas products of arachidonic acid. J. Allergy Clin. Immunol. 74 (Suppl.):309, 1984.

Goldman, .W., Chang, F.H., Gifford, L.A., Goetz , E.J., and Bourne, H.R. Pertussis toxin inhibition of chemotactic factor-induced calcium mobilization and function in human polymorphonuclear leukocytes. J. Exp. Med. 162:145, 1985b.

Goldman, D.W., Gifford, L.A., Olson, D.M. and Goetzl, E.J. Transduction by leukotriene B_4 receptors of increases in cytosolic calcium in human polymorphonuclear leukocytes. J. Immunol. 135:525, 985a.

Goldman, D.W., and Goetzl, E.J. Heterogeneity of human polymorphonuclear leukocyte receptors for leukotriene B_4. Identification of a subset of high affinity receptors that transduce the chemotactic response. J. Exp. Med. 159:1027, 1984a.
Goldman, D.W. and Goetzl, E.J. Selective transduction of human polymorphonuclear leukocyte functions by subsets of receptors for leukotriene B_4. J. Allergy Clin. Immunol. 74:373, 1984b.
Goldman, D.W., Olson, D.M., Payan, D.G., Gifford, L.A. and Goetzl, E.J. Development of receptors for leukotriene B_4 in HL-60 cells induced to differentiate by 1 alpha, 25-dihydroxyvitamin D3. J. Immunol. 136:4631, 1986.
Hedqvist, P., Dahlen, S.E. and Palmertz, U. Leukotrienes as mediators of airway anaphylaxis. IN Advances in Prostaglandin, Thromboxane and Leukotriene Research. Hayaishi, O. and Yamamoto, S. (eds). Raven Press, N.Y., 1985, p. 345.
Johnson, H.M. and Torres, B.A. Leukotrienes: Positive signals for regulation of α-interferon production. J. Immunol. 132:413, 1984.
Koo, C.H., Baud, L., Sherman, J.W., Harvey, J.P., Goldman, D.W. and Goetzl, E.J. Molecular properties of leukocyte receptors of leukotrienes. IN Cellular and Molecular Aspects of Inflammation. Post, G. and Crooke, S. (eds). Plenum Press, N.Y., 1988.
Koo, C.H., Healy, A. and Goetzl, E.J. Ligand-affinity purification of solubilized human leukocyte receptors for leukotriene B_4. FASEB J. 2: A668, 1988.
Kreisle, R.A. and Parker, C.W. Specific binding of leukotriene B_4 to a receptor on human polymorphonuclear leukocytes. J. Exp. Med. 157:628, 1983.
Krell, R.D., Tsai, R.S. and Giles, R.G. Pharmacologic aspects of leukotriene receptors. IN Leukotrienes and Other Lipoxygenase Products. Piper, P.J. (ed.) Wiley, Chichester, N.Y. 1985, p. 222..
Lewis, R.A. and Austen, K.F. The biologically active leukotrienes. Biosynthesis, metabolism, receptors, functions and pharmacology. J. Clin. Invest. 73:889, 1984.
Nagy, L, Lee, T.H., Goetzl, E.J., Pickett, W.C. and Kay, A.B. Complement receptor enhancement and chemotaxis of human neutrophils and eosinophils by leukotrienes and other lipoxygenase products. Clin. Exp. Immunol. 47:541, 1982.
Palmbad, J., Gyllenhammer, H., Lindgren, J.A. and Malmsten, C.L. Effects of leukotrienes and F-met-leu-phe on oxidative metabolism of neutrophils and eosinophils. J. Immunol. 132:3041, 1984.
Payan, D.G. and Goetzl, E.J. Specific suppression of human T-lymphocyte function by leukotriene B_4. J. Immunol. 131:551, 1983.
Payan, D.G., Missirian-Bastian, A. and Goetzl, E.J. Human T-lymphocyte subset specificity of the regulatory effects of leukotriene B_4. Proc. Natl Acad. Sci USA. 81:350, 1984.
Pek, S.B. and Walsh, M.F. Leukotrienes stimulate insulin release from the rat pancreas. Proc. Natl. Acad. Sci USA. 81:2199, 1984.
Piper, P.J. and Samboun, M.N. Stimulation of arachidonic acid metabolism and generation of thromboxane A_2 by leukotrienes B_4, C_4 and D_4 in guinea-pig lung *in vitro*. Br. J. Pharmacol. 77:267, 1982.
Pologe, L.G., Cramer, E.B., Pawlowski, N.A., Abraham, E., Cohn, Z.A. and Scott, W.A. Stimulation of human endothelial cell prostacyclin synthesis by select leukotrienes. J. Exp. Med. 160:1043, 1984.
Samuelsson, B. Leukotrienes: Mediators of immediate hypersensitivity reactions and inflammation. Science. 220:568, 1983.
Sarau, H.M., Foley J.J., Wu, H.L. Crooke, S.T. and Mong, S. Co-expression of leukotriene B_4 receptors on human monocytic leukemia U-937 cells: Characterization of the receptors and signal transduction processes. Mol. Pharmacol. (in press) 1989.

Sherman, J.W., Goetzl, E.J. and Koo, C.H. Selective modulation by guanine nucleotides of the high affinity subset of plasma membrane receptors for leukotriene B_4 on human polymorphonuclear leukocytes. J. Immunol. 140:3900, 1988a.
Sherman, J.W., Goetzl, E.J. and Koo, C.H. Ligand-induced formation of a leukotriene B_4 receptor-guanine nucleotide binding protein complex. Clin. Res. 36:470A, 1988b.

THE ROLE OF COVALENT AND NON-COVALENT MECHANISMS IN INSULIN RECEPTOR ACTION

Paul L. Rothenberg and C. Ronald Kahn

Departments of Pathology and Medicine, Brigham and Women's Hospital, Harvard Medical School and Joslin Diabetes Center
Boston, MA 02215

Diminished cellular responsiveness to the metabolic and growth promoting effects of insulin occurs in several pathophysiological conditions -- most commonly in insulin and non-insulin dependent diabetes mellitus, but also in obesity, uremia, glucocorticoid and growth hormone excess, and in certain rare syndromes such as lipoatrophic diabetes and leprechaunism (Kahn, 1985; Reaven, 1988). Although insulin action has been intensively investigated at the physiological and biochemical level for close to half a century, we are only beginning to obtain an in-depth understanding of the molecular mechanism of insulin signal transduction. A complete molecular analysis of this process will be necessary for a rational therapeutic approach to insulin-resistant disease states.

Insulin responses begin upon insulin binding to the cell surface insulin receptor (Freychet, 1971). The insulin receptor is an integral membrane protein of approximately 350 kDa. It is a symmetrical oligomer composed of two α-subunits (each 135 kDa), and two β-subunits (each 95 kDa) which are linked by disulfide bonds (Massague, 1981; Kasuga, 1982). The current working model of the insulin receptor is shown in Figure 1.

The α- and β-subunits of the insulin receptor are initially synthesized together as a single chain proreceptor with a molecular weight of about 160 kDa (Hedo, 1983). In vivo, the proreceptor is processed by specific proteolytic cleavage and glycosylation to give species of 120-180 kDa, which are then disulfide linked, further glycosylated and inserted into the plasma membrane as the mature receptor (Olsen and Lane, 1988). The half-life of the receptor is between 8 and 12 hours (Deutsch, 1983). The insulin receptor cDNA has been cloned from human (Ullrich, 1985; Ebina, 1985) and Drosophila (Petruzelli, 1986) tissues. Two closely related human cDNAs have been obtained which predict a highly conserved receptor containing 1346 or 1358 amino acids. The 5'-end of the human cDNA bears a signal sequence, followed by the α-subunit coding sequence, and then the β-subunit coding sequence. The difference in the two human sequences which have thus far been identified is a 36 base coding region of the α-subunit which allows for addition of 12 amino acids near the C-terminus. Preliminary data suggests this may relate to the existence of two receptor isoforms of differing insulin binding affinity.

The extracellular α-subunits contain the insulin binding domain which can be affinity labelled with insulin and bifunctional cross-linking reagents (Yip, 1980; Pilch, 1980). This binding domain discriminates insulin specifically from among other extracellular polypeptide hormones. Binding of insulin is complex and gives rise to curvilinear Scatchard plots. This may be related to high (K_D 0.1nM) and low (K_D 8nM) affinity forms of the receptor or negative cooperative interactions among binding sites. Hormone binding affinity is directly proportional to biological effect. Both naturally occurring and synthetically produced insulin analogues of higher affinity are proportionally more potent than low affinity derivatives. Although there are two α-subunits in each receptor, it remains uncertain whether the binding of one or two insulin molecules are required to activate an individual receptor (Sweet, 1987; Fujita-Yamagúchi, 1983).

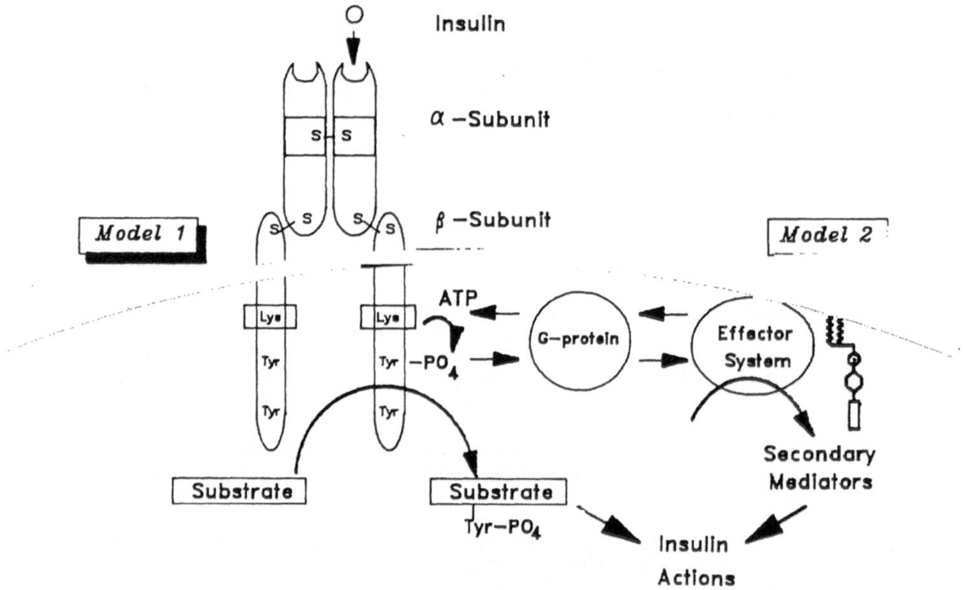

Figure 1. Models of post-receptor mechanisms of insulin singal transduction. Model 1-substrate phosphorylation. Model 2-non-covalent pathways. Details are discussed in the text.

The β-subunit has a small extracellular segment, which is also glycosylated, and a single hydrophobic segment of 23 amino acids which is the inferred, membrane-spanning region The intracellular portion of the β-subunit contains regions highly homologous to several protein tyrosine kinases and includes a consensus ATP binding site.

A major advance in our understanding of the mechanism of insulin action came in 1982 when it was first shown that the insulin receptor is an insulin-activated tyrosine kinase (Kasuga, 1982). Insulin binding to the extracellular α-subunit stimulates the protein kinase activity intrinsic to the β-subunit, such that the β-subunit undergoes an intramolecular, self-catalyzed ATP-dependent autophosphorylation on at least five tyrosine residues (for review, see White, 1986). This kinase activity occurs in wheat germ agglutinin purified receptor preparations, in immunoprecipitated receptors, in receptors purified to homogeneity as well as in intact cells (Petruzzelli, 1982; Avruch, 1982). Three of these tyrosines occur in a domain which includes residues 1146, 1150 and 1151 of the β-subunit. Autophosphorylation occurs in a specific sequence among these tyrosines, and importantly, is associated with the pronounced activation of the tyrosine kinase catalytic activity towards other polypeptide substrates (White, 1988). This enhanced kinase activity persists even if insulin is dissociated from the receptor, but can be terminated by dephosphorylation of the receptor by exogenous phosphotyrosine phosphatase activity (Herrera, 1988). A phosphorylation induced conformational change in the β-subunit most probably underlies the activation of the kinase, and such a conformational change can be detected by specific antibodies which have been generated against defined peptide segments of the receptor (Perlman, 1989).

In vitro, purified insulin receptors undergo insulin-stimulated phosphorylation only on tyrosine residues. But in intact cells insulin causes the appearance within the β-subunit of phosphoserine and phosphothreonine, in addition to phosphotyrosine (Kasuga, 1982b). Other, as yet unidentified serine/threonine kinases must be activated in the intact cell to secondarily phosphorylate the insulin receptor. The biochemical consequences for insulin receptor action of such reciprocal serine/threonine phosphorylations remains largely unknown.

Insulin receptor kinase activity appears essential in the mechanism of insulin signal transduction. If the lysine residue contained within the ATP binding site of the β-subunit at position 1018 is specifically replaced by any one of several amino acids and the mutant receptors expressed in cells, such receptors completely lack tyrosine kinase activity and are also totally ineffective in mediating insulin stimulation of cellular metabolism and growth (Ebina, 1987; Chou, 1989). The pivotal role of receptor kinase activity is also supported by the blockade of insulin receptor kinase activation and insulin bioeffects upon microinjection into cells of antibodies raised against intracellular portions of the β-subunit (Morgan, 1987), or against phosphotyrosine (Takayama,1988). Consistent with these observations, insulin receptor kinase activity has been found to be diminished in some insulin resistant states. In both Type I and Type II diabetes there is a decreased insulin-stimulated receptor autophosphorylation and kinase activity, at least when assayed in vitro (Caro, 1986; Comi, 1987; Freidenberg, 1987, for review, see Reddy, 1989). In both types of diabetes this receptor dysfunction may be partly alleviated by therapeutic treatment of the disease. This implies an important regulatory feature for the kinase activity. The biochemical basis for this reversible dysfunction is unknown, but a candidate mechanism is suggested by the observation that protein kinase C activation in vivo, e.g., by application of phorbol esters, impairs both insulin receptor kinase activity and insulin bioeffects (Takayama, 1984). In vivo, purified protein kinase C can phosphorylate the insulin receptor, reducing its activity (Bollag, 1987). Whether protein kinase C is a physiologically relevant modulator of insulin receptor kinase activity is currently under investigation. However, in many insulin-resistant states, no defect in insulin receptor kinase activity can be identified. This has been termed "post-receptor" resistance.

The biochemical events which follow the activation of the insulin receptor kinase are largely unknown. Insulin receptor kinase activity towards diverse polypeptides and proteins can be demonstrated in vitro, e.g., histones, or synthetic copolymers of Glu-Tyr. But the crucial identification of any known cellular protein which is both a substrate for the insulin receptor kinase and whose activity in vivo is modulated by insulin is lacking. Nevertheless, a widely held hypothesis posits a series or cascade of regulatory phosphorylations of critical cellular substrates. Because insulin induces serine phosphorylation and dephosphorylation of several cellular enzymes and thereby modulates their activity, it is plausible that direct substrates of the insulin receptor kinase might themselves be serine kinases or phosphoprotein phosphatases. Thus far, direct tests of this hypothesis have been inconclusive.

An alternative approach to elucidating the post-receptor mechanism of insulin action is the isolation of endogenous substrates of the receptor kinase. Several such proteins have been identified. The first, and most well studied, is a protein of approximately 185 kDa termed pp185 (White, 1985). This soluble, cytosolic protein is extremely rapidly phosphorylated on tyrosine in response to insulin or IGF-I, but not by other receptor tyrosine kinases (White, 1987; Kadowaki, 1987 Shemer, 1987). This very rapid phosphorylation is consistent with a role in mediating many of insulin's early metabolic effects, like alteration of transmembrane ion flux or activation of intracellular enzymes. pp185 tyrosine phosphorylation is transient, occurring maximally at about 30 seconds after insulin application, and then almost complete dephosphorylation of tyrosines ensues over the next few minutes, even with continued insulin stimulation (Figure 2; Rothenberg, in preparation). pp185 phosphorylation has been observed in all insulin-sensitive cells and

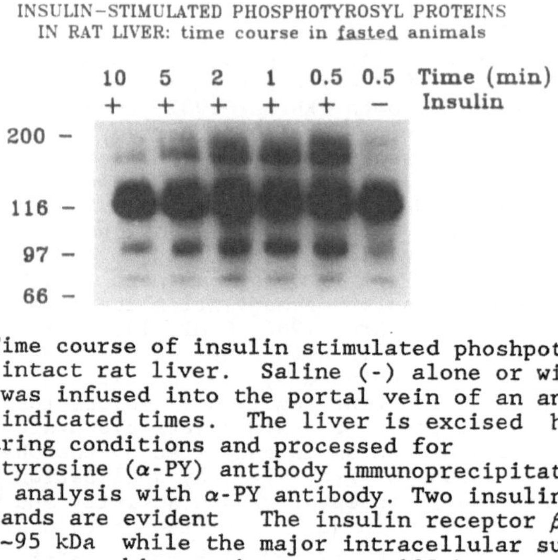

Figure 2. Time course of insulin stimulated phoshpotyrosyl proteins in intact rat liver. Saline (-) alone or with 10^{-6}M insulin (+) was infused into the portal vein of an anesthetized rat for the indicated times. The liver is excised homogenized under denaturing conditions and processed for anti-phosphotyrosine (α-PY) antibody immunoprecipitation and Western blot analysis with α-PY antibody. Two insulin stimulated bands are evident The insulin receptor β-subunit migrates at ~95 kDa while the major intracellular substrate of the insulin receptor kinase migrates at ~185 kDa.

tissues which have been examined. The concentrations of insulin required for its phosphorylation are essentially identical to those activating the insulin receptor kinase. Recently pp185 has been demonstrated to be a bona fide substrate of the insulin receptor which can be directly phosphorylated by the insulin receptor in reconstituted _in vitro_ assays (Tashiro-Hashimoto, 1989). The level of tyrosine-phosphorylated pp. 185 is clearly modulated in the intact, normal rodent liver by fasting and feeding ie., states associated with physiologically relevant alterations in tissue insulin sensitivity. Whether this is due to a change in the amount of pp185 present in the cell or altered susceptibility to tyrosine phosphorylation remains to be clarified.

More evidence supporting a role of pp185 phosphorylation in insulin action is derived from recent transfection studies employing a mutant insulin receptor which is altered at tyrosine 960 in the β-subunit (White, 1988). Tyrosine 960 is not autophosphorylated during receptor kinase activation, but may be important in substrate recognition. While these mutant receptors are normal, insulin-sensitive tyrosine kinases in vitro, upon transfection into cells, these altered receptors do not phosphorylate pp185 and also do not mediate insulin's bioeffects. pp185 is perhaps the most abundant endogenous substrate of the insulin receptor kinase, but additional, lower Mr proteins have been identified (Bernier, 1987; Haring, 1987) and purification attempts are proceeding in several laboratories.

Insulin receptor kinase activity is required for insulin action on cells, but perhaps not all of insulin signal transduction proceeds via phosphorylation cascades involving endogenous substrate proteins. For example, insulin stimulation of glucose transport has not been associated with any phosphorylation changes of the glucose transporter or in any of the associated proteins in the glucose transport pathway (Gibbs, 1986). Moreover, certain monoclonal antibodies raised against the insulin receptor have been shown to be insulinomimetic when applied to whole cells, but are not demonstrably able to activate the insulin receptor kinase (Hawley, 1989). These observations have fostered an alternative hypothesis of insulin signal transduction wherein insulin receptor autophosphorylation leads to a rearrangement of β-subunit conformation. This conformational change has been demonstrated by domain-specific antibodies, as discussed previously. Such conformational change could enable the insulin receptor to then productively interact non-covalently with some other signal transducing components of the plasma membrane.

Secondary effectors of the insulin receptor might include some component of the guanine nucleotide binding protein (G-protein) system, or perhaps enzymes involved in lipid-related signalling pathways such as phosphatidylinositol kinases (Sale, 1986), or phospholipases (Saltiel, et.al). A role of G-proteins in insulin action is suggested by observations that pertussis toxin, a bacterial enzyme which enters cells and inactivates certain G-proteins by catalyzing their ADP-ribosylation, blocks certain of insulin's biological activities in several cell types (Goren, 1985; Heyworth, 1986; Luttrell, 1988). Moreover, we have recently observed that insulin inhibits pertussis toxin catalyzed ADP-ribosylation in isolated liver and fat cell plasma membranes (Figure 3; Rothenberg, 1988). Also, antibodies to ras-related proteins also block insulin actions when introduced into cells (Korn, 1987). Ras proteins are partly homologous to the γ-subunits of the G-protein heterotrimer (Gautam, 1989). This suggests a direct, non-covalent interaction between G-proteins and the insulin receptor. In vitro, certain G-proteins such as G_i and transducin can be phosphorylated by the insulin receptor, but this occurs with low stoichiometry and does not appear to occur in intact cells (Zick, 1986; O'Brian, 1987). Insulin inhibition of pertussis toxin catalyzed ADP-ribosylation of G_i appears to occur without requiring tyrosine kinase action.

The insulin action pathway does not involve any of the well characterized G-protein coupled secondary effector systems, such as cAMP production through adenylate cyclase or inositol phosphate production via PI-specific phospholipase C. But insulin action on cells may involve the activation of a special phospholipase C which catalyzes release of a novel inositol-glycan compound (for review, Saltiel, 1988). This "mediator" is capable of mimicking some insulin actions in whole cells (Gaulton,1988). Insulin also causes the rapid release of certain ecto-enzymes from plasma membranes. Such enzymes are anchored to the lipid bilayer by a glucosyl-phosphatidylinositol moiety, and these enzymes can be released by the application of exogenous PI-specific phospholipase C (Low, 1988). Although the epidermal growth factor receptor has been recently demonstrated to directly phosphorylate PI-specific phospholipase C in vivo (Wahl, 1989) the insulin receptor kinase apparently does not (White, M.F., unpublished observations).

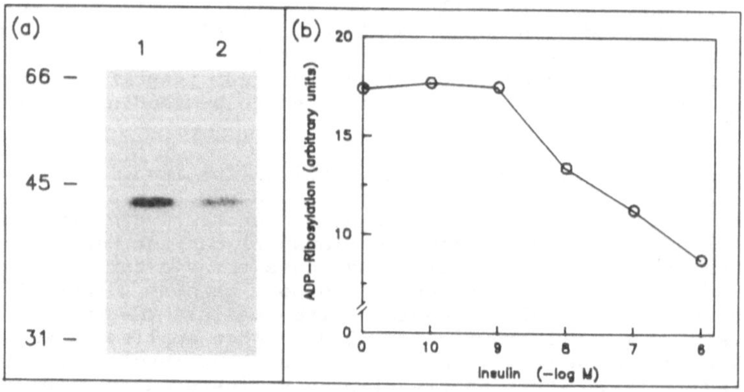

Figure 3. Insulin inhibition of pertussis toxin catalyzed ADP-ribosylation of G-proteins in isolated adipocyte plasma membranes. Panel (a) autoradiogram of SDS-PAGE resolved proteins. Membranes (7.5 ug protein) were incubated with pre-activated pertussis toxin (1 ug) and [α-^{32}P]NAD (5 uM NAD, 0.5 x 10^5 cpm) for 30 min at 30°C in the absence (lane 1) and presence (lane 2) of 10^{-6} insulin. Numbers at the left indicate M_r calibration standards x 10^{-3}. (b) Insulin concentration dependence. The bands were quantitated by sensitometric scanning.

In summary, although the preliminary evidence supporting a non-covalent post-receptor signalling pathway in the mechanism of insulin signal transduction is circumstantial, there is sufficient reason to consider the hypothesis that not all of insulin's action may be mediated by an intracellular phosphorylation cascade. Perhaps both types of mechanisms will be found necessary to ultimately explain the myriad biological effects of this important hormone.

REFERENCES

Avruch, J., Nemenoff, R.A., Blackshear, P.J., Pierce, M.W., Osathanondh, R., 1982, Insulin-stimulated tyrosine phosphorylation of the insulin receptor in detergent extracts of placental membranes, J Biol Chem, 257:15162.

Bernier, M., Laird, D.M., Lane, M.D., 1987, Insulin-activated tyrosine phosphorylation of a 15-kilodalton protein in intact 3T3-L1 adipocytes, Proc Natl Acad Sci USA, 84:1844.

Bollag, G.E., Roth, R.A., Beaudoir, J., Mochly-Rosen, D., Koshland, D.E., Jr., 1987, Protein kinase C directly phosphorylates the insulin receptor in vitro and reduces its protein kinase activity, Proc Natl Acad Sci USA, 83:5822.

Caro, J.F., Ittoop, O., Pories, W.J., Meelhein, D., Flickinger, E.G., Thomas, F., Jenquin, M., Silverman, J.F., Khazanie, P.G. and Sinha, M.K., 1986, Studies on the mechanism of insulin resistance in the liver from humans with noninsulin-dependent diabetes. Insulin action and binding in isolated hepatocytes, insulin receptor structure and kinase activity, J Clin Invest, 78:249.

Chou, C.K., Dull, T.J., Russell, D.S., Gherzi, R., Lebwohl, D., Ullrich, A., Rosen, O.M., 1987, Human insulin receptors mutated at the ATP-binding site lack protein tyrosine kinase activity and fail to mediate postreceptor effects of insulin, J Biol Chem, 262:1842.

Comi, R.J., Grunberger, G., Gordon, P., 1987, Relationship of insulin binding and insulin-stimulated tyrosine kinase activity is altered in type II diabetes, J Clin Invest, 79:453.

Deutch, P.J., Wan, C.F., Rosen, O.M., Rubin, C.S., 1983, Latent insulin receptors and possible receptor precursors in 3T3-LI adipocytes, Proc Natl Acad Sci USA, 80:133.

Ebina, Y., Ellis, L. Jarnagin, K., Edery, M., Graf, L., Clauser, E., Ou, J-H., Masiar, F., Kan, Y., Goldfine, I., Roth, R., Rutter, W., 1985, The human insulin receptor cDNA: The structural basis for hormone activated trans-membrane signalling, Cell 40:747.

Ebina, Y., Araki, E., Taira, M., Shimada, F., Mori, M., Craik, C.S., Siddle, K., Pierce, S.B., Roth, R.A., Rutter, W.J., 1987, Replacement of lysine residue 1030 in the putative ATP-binding region of the insulin receptor abolishes insulin- and antibody-stimulated glucose uptake and receptor kinase activity, Proc Natl Acad Sci USA, 84:704.

Freidenberg, G.R., Henry, R.R., Klein, H.H., Reichart, D.R., Olefsky, J.M., 1987, Decreased kinase activity of insulin receptors from adipocytes of non-insulin-dependent diabetic subjects, J Clin Invest, 79:240.

Freychet, P.J., Roth, J., Neville, D.M., Jr., 1971, Insulin receptors in the liver: Specific binding of [^{125}I] insulin to the plasma membrane and its relation to insulin bioactivity, Proc Natl Acad Sci USA, 68:1833.

Fujita-Yamaguchi, Y., Cho, S., Sakamoto, Y., Itakura, K., 1983, Purification of insulin receptors with full binding activity, J Biol Chem, 258:5045.

Gaulton, G., Kelly, K., Pawlowski, J., Mato, J.M., Jarett, L., 1988, Regulation and Functions of an Insulin-Sensitive Glycosyl-Phosphatidylinositol during T-lymphocyte activation, Cell 53:963.

Gautam, N., Baetscher, M., Aebersold, R., Simon, M.I., 1989, A G-protein gamma subunit shares homology with <u>ras</u> proteins, Science, 244:971.

Gibbs, E.M., Allard, W.J., Lienhard, G.E., 1986, The glucose transporter in 3T3-L1 adipocytes is phosphorylated in response to phorbol ester but not in respose to insulin, J Biol Chem, 261:16597.

Goren, H.J., Northup, J.K., Hollenberg, M.D., 1985, Action of insulin modulated by pertussis toxin in rat adipocytes, Can J Physiol Pharmacol, 63:1017.

Haring, H.U., White, M.F., Machicao, F. Ermel, B., Schleicher, E., Obermaier, B., 1987, Insulin rapidly stimulates phosphorylation of a 46-kDa membrane protein on tyrosine residue as well as phosphorylation of several soluble proteins in intact fat cells, Proc Natl Acad Sci USA, 84:113.

Hawley, D.M. Maddox, B.A. Patel, R., Wong, K.Y., Mamula, P.W., Firestone, G.L., Brunetti, A., Verspohl, E., Goldfine, I., 1989, Insulin receptor monoclonal antibodies that mimic insulin action without activating tyrosine kinase, J Biol Chem, 264:2438.

Hedo, J., Kahn, C.R., Hayoshi, M., Yamada, K.M., Kasuga, M., 1983, Biosynsthesis and glycosylation of the insulin receptor: Evidence for a single polypeptide precursor of the two major subunits, J Biol Chem, 258:10020.

Herrera, R., Rosen, O.M., 1982, Autophosphorylation of the insulin receptor _in vitro_. Designation of phosphorylation sites and correlation with receptor kinase activation, J Biol Chem, 261:11989.

Heyworth, C.M., Grey, A.M., Wilson, S.R., Hanski, E., Houslay, M.D., 1989, The action of isle 1 activating protein on insulin's ability to inhibit adenylate cyclase and activate cyclic AMP phosphodiesterases in hepatocytes. Biochem J, 235:145.

Kadowaki, T., Koyasu, S., Nishida, E., Tobe, K., Izumi, T., Takaku, F., Sakai, H., Yahara, I., Kasuga, M., 1987, Tyrosine phosphorylation of common and specific sets of cellular proteins induced by insulin, IGF-I and EGF in the intact cell, J Biol Chem, 262:7342.

Kahn, C.R., 1975, Membrane Receptors for Polypeptide Hormones, _in_: Methods in Membrane Biology, Vol 3, E.D. Korn, ed., Plenum Press, New York.

Kahn, C.R., 1985, The molecular mechanism of insulin action, Annual Rev Med, 36:429.

Kasuga, M., Hedo, J.A., Yamada, K.M., Kahn, C.R., 1982, Structure of the insulin receptors and its subunits, J Biol Chem 257:10392.

Kasuga, M., Karlsson, F.A., Kahn, C.R., 1982a, Insulin stimulates the phosphorylation of the 95,000-dalton subunit of its own receptor, Science, 215:185.

Kasuga, M., Zick, Y., Blithe, D.L., Karlsson, F.A., Haring, H.U., Kahn, C.R., 1982b, Insulin stimulation of phosphorylation of the β-subunit of the insulin receptor: Formation of both phosphoserine and phosphotyrosine, J Biol Chem, 257:9891.

Korn, L.J., Siebel, C.W., McCormick, F., Roth, R.A., 1987, Ras p21 as a potential mediator of insulin action in Xenopus oocytes, Science, 236:840.

Luttrell, L.M., Hewlett, E., Romero, G., Rogol, A.D., 1988, Pertussis toxin treatment attenuates some effects of insulin in BC3H1 murine myocytes, J Biol Chem, 263:6134.

Low, M.G., Saltiel, A.R., 1988, Structural and functional roles of glycosyl-phosphatidylinositol in membranes, Science 239:268.

Massague, J., Pilch, P.F., Czech, M.P., 1981, Electrophoretic resolution of three major insulin receptors with unique subunit stoichiometries, Proc Natl Acad Sci USA, 77:7137.

Morgan, D.O., Roth, R.A., 1987, Acute insulin action requires insulin receptor kinase activity: Introduction of an inhibitory monoclonal antibody into mammalian cells blocks the rapid effects of insulin, Proc Natl Acad Sci USA 84:41.

O'Brian, R.M., Houslay, M.D., Milligan, G., Siddle, K., 1987, The insulin receptor tyrosine kinase phosphorylates holomeric forms of the guanine nucleotide regulatory proteins G_i and G_o, FEBS Letters, 212:281.

Petruzzelli, L., Herrera, R., Arrenas-Garcia, R., Fernandez, R., Birnbaum, M.J., Rosen, O.M., 1986, Isolation of a Drosphilia genomic sequence homologous to the kinase domain of the human insulin receptor and detection of the phosphorylated Drosphilia receptor with an anti-peptide antibody, Proc Natl Acad Sci USA 83:4710.

Petruzzelli, L.M., Gangaly, S., Smith, C.R., Cobb, M.H., Rubin, C.S., Rosen, O.M., 1982, Insulin activates a tyrosine-specific protein kinase in extracts of 3T3-L1 adipoytes and human placenta, Proc Natl Acad Sci USA 79:6792.

Perlman, R., Bottaro, D., White, M.F., Kahn, C.R., 1989, Conformational changes in the α and β-subunits of the insulin receptor identified by anti-peptide antibodies, J Biol Chem, 264:8946.

Pilch, P.F., Czech, M.P., 1980, Purification of the adipocyte insulin receptor by immunoaffinty chromatography, J Biol Chem, 255:1722.

Reaven, G.M., 1988, The role of insulin resistance in human disease, Diabetes 37:1595.

Reddy, S.K., Kahn, C.R., 1989, Insulin resistance: A look at the role of insulin receptor kinase, Diab Med, 5:621.

Ronnett, G.V., Knutson, V.P., Kohanski, R.A., Simpson, T.L., Lane, M.D., 1984, Role of glycosylation in processing of the newly translated insulin proreceptors in 3T3-L1 adipocytes, J Biol Chem, 259:4566.

Rosen, O.M., 1987, After insulin binds, Science, 237:1452.

Takayama, S., White, M.F., Lauris, V., Kahn, C.R., 1984, Phorbol esters modulate insulin receptor phosphorylation and insulin action in cultured hepatoma cells, Proc Natl Acad Sci USA, 81:7797.

Tashiro-Hashimoto, Y., Tobo, K., Koshio, O., Izumi, T., Takaku,F., Akanuma, Y., Kasuga, M., 1989, Tyrosine phosphorylation of pp185 by insulin receptor kinase in a cell-free system, J Biol Chem, 264:6879.

Ullrich, A., Bell,J., Chen, E., Herrera, R., Petruzzelli, L., Dull, T., Gray, A., Coussens, L., Liao, Y-C, Tsubokawa, M., Mason, A., Seeburg, P., Grunfeld, C., Rosen, O., Ramachandran, J., 1985, Human insulin receptor and its relationship to the tyrosine kinase family of oncogenes, Nature 313:756.

White, M.F., Kahn, C.R., 1986, The insulin receptor and tyrosine phosphorylation, in: The Enzymes, Vol XVII, P.D. Boyer, E.G. Krebs, eds., Academic Press, Inc., Orlando.

White, M.F., Haring, H.U., Kasuga, M., Kahn, C.R., 1984, Kinetic properties and sites of autophosphorylation of the partially purified insulin receptor from hepatoma cells, J Biol Chem 259:255.

White, M.F., Maron, R., Kahn, C.R., 1985, Insulin rapidly stimulates tyrosine phosphorylation of a Mr 185,000 protein in intact cells, Nature, 318:183.

White. M.F., Shoelson, S.E., Keutmann, H., Kahn, C.R., 1988, Cascade of tyrosine autophosphorylation in the β-subunit activates the insulin receptor, J Biol Chem, 263:2969.

White, M.F., Stegmann, E.W., Dull, T.J., Ullrich, A., Kahn, C.R., 1981, Characterization of an endogenous substrate of the insulin receptor in cultured cells, J Biol Chem, 262:9769.

Yip, C.C., Yeung, C.W.T., Moule, M.L., 1980, Photoaffinty labelling of the insulin receptor proteins of liver plasma membrane, Biochemistry, 19:70.

Zick, Y., Sagit-Eisenberg, R., Pines, M., Gierschik, P., Spiegel, A., 1986, Muti-site phosphorylation of the α-subunit of transducin by insulin receptors and protein kinase C, Proc Natl Acad Sci USA, 83:9194.

EGF TRIGGERS A SIMILAR SIGNALLING CASCADE IN DIFFERENT CELL TYPES OVEREXPRESSING THE EGF RECEPTOR

Marco Ruggiero, Timothy P. Fleming, Toshimitsu Matsui, Eddi di Marco, Christopher Molloy Pier Paolo di Fiore and Jacalyn H. Pierce

LCMB, National Cancer Institute, National Institutes of Health, Bethesda, Maryland 20892, USA

INTRODUCTION

The mechanism by which growth factors stimulate mitogenesis upon binding to their receptors involves several biochemical events and the generation of intracellular second messengers. EGF[1] is regarded as a classical example of a growth factor which acts on a receptor harboring tyrosine kinase activity (Berridge, 1987 a,b; Carpenter, 1987). Recently, several reports have indicated that in certain cell types, EGF stimulates inositol lipid turnover with the formation of Ca^{2+}-mobilizing inositol polyphosphates and 1,2-diacylglycerol activating protein kinase C (Hepler et al, 1987; Pandiella et al., 1987; Johnson and Garrison, 1987; Earp et al., 1988; Takasu et al., 1988; Moscat et al., 1988). However, other studies demonstrated that EGF was unable to trigger phosphoinositide metabolism in 3T3 fibroblasts and in a clone of A431 carcinoma cells (Besterman et al, 1986; Macara, 1986; Wakelam et al., 1986; Pandiella et al, 1987). Therefore, EGF-induced inositol lipid turnover might be considered lineage-specific, or at least cell type-specific (Earp et al., 1988), and the role of phosphoinositide metabolism in EGF-induced mitogenesis remains to be established.

In order to address this question we studied the effect of EGF on inositol lipid and arachidonic acid metabolism in cell lines of different origin, into which a LTR-driven expression vector containing the normal human EGF receptor cDNA was transfected. In the cell lines tested, EGF was mitogenic and gave rise to the full spectrum of responses related to phosphoinositide metabolism stimulation. Thus, our findings demonstrate that the normal human EGF receptor is able to couple to the phosphoinositide signalling machinery in cells of different origin and the magnitude of inositol lipid metabolism correlates well with the mitogenic response.

---
[1]See page 47

MATERIALS AND METHODS

The continuous mouse NIH/3T3 (Jainchill et al., 1969), NR6 (Pruss and Herschman, 1977) fibroblast, and the immature myeloid 32D (Greenberger et al., 1983) cell lines have been previously described. EGF receptor overexpressing NIH/3T3, NR6 fibroblasts (Di Fiore et al., 1987), and 32D (Pierce et al., 1988) lines have been recently described. NIH/3T3 and NR6 fibroblasts were routinely grown in Dulbecco's Modified Eagle Medium (Gibco), supplemented with 10% calf serum. 32D cells and their EGF receptor overexpressing counterpart, were grown in RPMI medium (Gibco) supplemented with 15% calf serum and interleukin 3 (Genzyme).

32D cells, NIH/3T3, NR6 fibroblasts and their EGF receptor overexpressing counterpart, were prelabelled for 48 h with 10 mCi/ml of [^{3}H]myo-inositol (NEN Dupont) following the procedure described in Lacal et al. (1987) and Chiarugi et al. (1987). In some experiments, [^{3}H]thymidine incorporation and [^{3}H]inositol phosphate formation were assayed in parallel, in control and EGFR-NIH/3T3 fibroblasts (Table 1). Thus, in order to follow exactly the same procedure for both assays, the cells were labelled for 24 h in the presence of serum, and for the following 24 h in serum-free medium. The final incubation with EGF (Collaborative Research), or PDGF (R&D) was carried out in 35-mm dishes in the presence of LiCl (10 mM). The reaction was terminated by the addition of acid ice-cold methanol, and inositol phosphates were extracted and separated according to Lapetina et al. (1985) and Palmer et al. (1986). Total inositol phosphates were eluted from Dowex AG 1X8 columns (Bio-Rad) with 10 ml of 1.2 M ammonium formate/0.1 M formic acid. In agreement with previous studies (Besterman et al., 1986), preliminary experiments demonstrated that the accumulation of total inositol phosphates increased as a function of time for up to 6 hours after stimulation (not shown). Following Lacal et al. (1987) and Moscat et al. (1988), we chose to study total inositol phosphate formation at 60 (Table 1) and 10 min (Tables 2 and 3) after growth factor addition. However, inositol trisphosphate isomers were studied at earlier time-points (15 s - 10 min). These compounds were separated by h.p.l.c. following the method described by Irvine et al. (1985).

In experiments in which cell were prelabelled with [^{14}C]arachidonic acid, the lipids were extracted with neutral chloroform/methanol/water (final concentration, 200/200/100, v/v). Arachidonic acid was then separated on t.l.c. plates (Whatman, pre-coated LK6D) using as solvent the upper phase of ethylacetate/2,2,4-trimethyl-pentane/acetic acid/water (90/50/20/100, v/v) (Ruggiero et al., 1985). Diacylglycerol and arachidonic acid were also separated by t.l.c. using as solvent ethyl ether/hexanes/acetic acid (40/60/1, v/v). The lipids were visualized by autoradiography and the corresponding areas on the silica gel plates were then scraped, transferred into scintillation vials, and [^{14}C]-radioactivity measured by liquid scintillation counting.

The fluorescent indicator fura-2 was used to determine the intracellular Ca^{2+} concentration ([Ca^{2+}]i) in EGFR-32D cells in suspension. Measurements were carried out on 2 ml samples of cells in suspension (2 x 10^6 cells/ml) as previously described (Grynkiewicz et al., 1985; Malgaroli et al., 1987). [Ca^{2+}]i increase in response to EGF was measured in cells in suspension in the complete incubation medium (containing 1 mM Ca^{2+}) and in the presence of 5 mM EGTA.

Determination of DNA synthesis in response to EGF and PDGF was monitored as [^{3}H]thymidine incorporation. We followed the method reported in Finzi et al. (1987) and Di Fiore et al. (1987) for NIH/3T3 and NR6 fibroblasts and their EGF receptor overexpressing counterpart.

RESULTS

We initially studied the effects of EGF on inositol lipid metabolism and DNA synthesis in EGF receptor overexpressing NIH/3T3 fibroblasts. In order to compare the effect of EGF with that of a mitogen known to stimulate inositol lipid turnover, we also measured [^{3}H]inositol phosphate accumulation and [^{3}H]thymidine incorporation in response to PDGF in the same cell line.

TABLE 1

DNA SYNTHESIS AND INOSITOL PHOSPHATE FORMATION INDUCED BY EGF AND PDGF

	DNA synthesis	
	EGF	PDGF
NIH/3T3	1.85 + 0.09	7.25 + 0.32
EGFR-NIH/3T3	8.54 + 0.44	7.19 + 0.37
	Inositol phosphates	
	EGF	PDGF
NIH/3T3	1.05 + 0.05	25.26 + 1.03
EGFR-NIH/3T3	22.30 + 1.01	24.59 + 1.08

DNA synthesis and inositol phosphate formation are expressed as fold stimulation over control ± S.E. (n=3). The cells were incubated with EGF (100 ng/ml) or PDGF (100 ng/nl).

Table 1 shows that both EGF and PDGF induced stimulation of inositol lipid turnover and DNA synthesis in EGFR-NIH/3T3 fibroblasts. The relative potencies of the two growth factors appeared to be similar. Table 1 also shows that EGF induced relatively little stimulation of DNA synthesis and no detectable inositol phosphate formation in control NIH/3T3 cells. These results indicate that the ability of EGF and PDGF to stimulate phosphoinositide turnover correlated well with their mitogenic effect, strongly suggesting a relationship between inositol lipid metabolism and cell growth.

To determine whether the effect of EGF on inositol lipid metabolism was a phenomenon generalizable to other cell types, we measured the rapid (10 min) accumulation of radioactive inositol phosphates in control and EGF receptor overexpressing cell lines. Table 2 shows that EGF induced a significant increase of inositol phosphate accumulation in the lines overexpressing the receptor, whereas no effect was detected in the parental lines.

TABLE 2

FORMATION OF INOSITOL PHOSPHATES IN RESPONSE TO EGF

	No addition	EGF
32D	3677+151	3589+211
NIH/3T3	5717+269	5896+312
NR6	7351+344	7487+412
EGFR-32D	3549+98	12697+873
EGFR-NIH/3T3	5284+251	12702+130
EGFR-NR6	7058+229	17157+439

Cells were labelled with 10 mCi/ml of [^{3}H]myo-inositol for 48 h as described under Materials and Methods. Monolayers of serum-starved cells in 35-mm dishes, were incubated with EGF (100 ng/ml) for 10 min at 37°C in the presence of LiCl (10 mM). Incubation was terminated by the addition of ice-cold methanol, and inositol phosphates were extracted and analysed as described. Data, expressed as c.p.m., are means ± S.E. (n = 3).

In order to further characterize the inositol trisphosphates formed in response to EGF, inositol (1,4,5)trisphosphate and inositol (1,3,4)trisphosphate were separated by h.p.l.c. and identified by co-elution with reference standards. In all the cell line tested inositol (1,4,5)trisphosphate was formed rapidly (15-30 s), and transiently, returning to basal level in about 5 min. Inositol (1,3,4)trisphosphate, the product of successive phosphorylation and dephosphorylation of the 1,4,5-isomer (Irvine et al., 1986; Hawkins et al., 1986), showed a slower increase, although remained elevated for up to 5 min. Inositol monophosphates continued to increase for the entire length of the experiment, and, 10 min after EGF addition, they represented more than 90% of the total inositol phosphates formed.

It is generally accepted that hydrolysis of phosphatidylinositol 4,5-bisphosphate by phosphoinositidase C is a Ca^{2+}-independent phenomenon (Berridge, 1987a), which occurs before activation of phospholipase A_2 (Lapetina, 1986). However, increasing intracellular Ca^{2+} concentration, and formation of arachidonic acid and its metabolites are known to interfere with the agonist-induced inositol lipid turnover (Majerus et al., 1984; Nishizuka, 1984). EGF is known to induce the influx of extracellular Ca^{2+} and the activation of arachidonate metabolism (Hepler et al., 1987; Casey and Mitchell, 1987). Therefore, we sought to determine whether chelation of extracellular Ca^{2+} or inhibition of cyclooxygenase activity could impair the ability of the growth factor to stimulate phosphoinositide turnover.

Table 3, shows the formation of inositol phosphates in EGFR-32D and EGFR-NIH/3T3 cell lines pretreated with EGTA (2 mM), or indomethacin (30 mM), and stimulated with EGF for 10 min at 37°C. Neither chelation of extracellular Ca^{2+} by EGTA, nor inhibition of prostaglandin synthesis by indomethacin, prevented EGF from stimulating inositol lipid turnover as monitored by inositol phosphate formation. However, under these conditions, inositol phospholipid turnover was significantly reduced. These results suggest that influx of extracellular Ca^{2+} and prostaglandin formation may enhance EGF-stimulated inositol lipid metabolism by acting as auto-amplificative second messengers. A positive feed-back mechanism of this type has been characterized in detail in human platelets (Lapetina, 1986). Qualitatively identical results were obtained in EGFR-NR6 cells.

TABLE 3

EFFECT OF EGTA OR INDOMETHACIN ON EGF-INDUCED INOSITOL PHOSPHATE FORMATION

	EGFR-32D	EGFR-NIH/3T3
No addition	3744+180	4977+202
EGF	11863+959	12097+387
EGTA+EGF	8192+343	7988+145
INDO+EGF	8073+421	8244+213

EGFR-32D cells, prelabelled with [^{3}H]myo-inositol, were incubated with EGF (100 ng/ml) for 10 min at 37°C, in the presence of LiCl (10 mM). EGTA (2 mM), or indomethacin (INDO; 30 mM) were added 5 min before EGF. [^{3}H]-inositol phosphates were analysed by ion-exchange chromatography. Data are expressed as c.p.m., and are means $\pm$ S.E. of three replicate samples in a single experiment, one out of three that gave almost identical results.

Stimulation of inositol lipid turnover is often followed by the release and metabolism of arachidonic acid. Sequential activation of phosphoinositidase C, phospholipase A_2 and cyclooxygenase has also been described (Lapetina, 1986). Accumulation of radioactive arachidonic acid was investigated in control and EGF receptor-overexpressing cells, after prelabelling the cultures with [^{14}C]arachidonic acid. Table 4 shows that EGF induced the formation of diacylglycerol and the liberation of arachidonic acid from [^{14}C]arachidonic acid-labelled phospholipids in EGF receptor-expressing 32D cells and NR6 fibroblasts. Chelation of extracellular Ca^{2+} by EGTA significantly decreased the extent of diacylglycerol and arachidonic acid formation in response to EGF in EGFR-32D cells. This seems to indicate that intracellular Ca^{2+} mobilization by inositol (1,4,5)trisphosphate was sufficient to activate phosphoinositidase C and phospolipase A_2. Analogous results were obtained in EGFR-NIH/3T3 cells. In contrast, none of the parental cell lines showed a detectable response to EGF in terms of diacylglycerol or arachidonic acid formation.

TABLE 4

FORMATION OF ARACHIDONIC ACID AND DIACYLGLYCEROL INDUCED BY EGF

EGFR-32D

	Arachidonic acid	DAG
No addition	2945+312	1480+10
EGF	6478+397	2245+48
EGTA + EGF	4843+418	1969+22

EGFR-NR6

	Arachidonic acid	DAG
No addition	6254+392	3459+187
EGF	9588+251	4897+211

Cells were labelled with [^{14}C]arachidonic acid for 48 h as described. Monolayers of serum-starved cells in 35-mm dishes were incubated with EGF (100 ng/ml) for 10 min at 37°C. When present, EGTA (2mM) was added 5 min before EGF. Incubation was terminated by the addition of ice-cold methanol, and lipids were extracted and separated on thin-layer chromatography as described. Data, expressed as c.p.m., are means $\pm$ S.E. (n = 3). DAG: diacylglycerol.

Diacylglycerol and arachidonic acid have been shown to activate protein kinase C in different cell types (Nishizuka, 1988). We measured protein kinase C activation in response to EGF, in EGF receptor-overexpressing fibroblasts (NIH/3T3 and NR6), by monitoring the phosphorylation of a specific endogenous substrate, an acidic 80 kd protein (p80) (Blackshear et al., 1985; Rodriguez-Pena and Rosengurt, 1986). In cells prelabelled with ^{32}Pi and assayed for p80 phos-phorylation as described in Moscat et al. (1988), EGF (100 ng/ml for 10 min at 37°C) caused significant increase of p80 phosphorylation (not shown). These results indicate that EGF stimulation led to activation of protein kinase C in cells overexpressing the receptor. In contrast, EGF did not detectable increase p80 phosphorylation in control fibroblasts.

Finally, we measured intracellular Ca^{2+} mobilization in EGFR-32D cells. As with inositol lipid and arachidonic acid metabolism, EGF was found to have a marked effect on [Ca^{2+}]i in EGFR-32D, but not in the parental 32D line. In EGFR-32D cells in suspension loaded with the fluorescent [Ca^{2+}]i indicator fura 2, [Ca^{2+}]i increased from 89 to 339 nM within 1-2 s after the addition of EGF (100 ng/ml). This time-course reflected inositol (1,4,5)trisphosphate formation induced by EGF. When these studies were performed in the presence of 5 mM EGTA, the magnitude of the EGF-induced [Ca^{2+}]i peak was reduced. These results indicate that EGF increased [Ca^{2+}]i in EGFR-32D cells by a dual mechanism, inducing both mobilization from intracellular store(s) and influx from the extracellular medium.

In conclusion, our study demonstrates that the normal human EGF receptor expressed at high level, is able to trigger a similar signalling cascade in cell lines that are mitogenically stimulated by EGF. Therefore, the ability to induce phosphoinositide turnover, arachidonate metabolism, protein kinase C activation and Ca^{2+} mobilization is mediated by the EGF receptor in different cell types into which it was transfected.

REFERENCES

Berridge, M. J. (1987) Inositol lipids and cell proliferation. Biochim. Biophys. Acta, 907:33.

Berridge, M. J. (1987) Inositol lipids and DNA replication. Phil. Trans. R. Soc. Lond., 317:525.

Besterman. J. M., Watson, S. P., and Cuatrecasas, P. (1986) Lack of association of epidermal growth factor-, insulin-, and serum-induced mitogenesis with stimulation of phosphoinositide degradation in BALB/c 3T3 fibroblasts. J. Biol. Chem., 261:723.

Blackshear, P. J., Witters, L. A., Girard, P. R., Kuo, J. F., and Quamo, S. N. (1985) Growth factor-stimulated protein phosphorylation in 3T3 cells: evidence for protein kinase C-dependent and -independent pathways. J. Biol. Chem., 260:13304.

Carpenter, G. (1987) Receptors for epidermal growth factor and other polypeptide mitogens. Annu. Rev. Biochem., 56:881.

Casey, M. L., and Mitchell, M. D. (1987) Epidermal growth factor-stimulated prostaglandin E_2 production in human amnionic cells:specificity and nonesterified arachidonic acid dependency. Mol. Cell Endocrinol., 53:169.

Chiarugi, V., Porciatti, F., Pasquali, F., Magnelli, L., Giannelli, S., and Ruggiero, M. (1987) Polyphosphoinositide metabolism is rapidly stimulated by activation of a temperature-sensitive mutant of Rous sarcoma virus in rat fibroblasts. Oncogene, 2:37.

Di Fiore, P. P., Pierce, J. H., Fleming, T. P., Hazan, R., Ullrich, A., King, C. R., Schlessinger, J., and Aaronson, S. A. (1987) Overexpression of the human EGF receptor confers an EGF-dependent transformed phenotype to NIH/3T3 cells. Cell, 51:1063.

Earp, H. S., Hepler, J.R., Petch, L. A., Miller, A., Berry, A., Harris, J., Raymond, V.W., McCune, B.K., Lee, L. W., Grisham, J.W., and Harden, T. K. (1988) Epidermal growth factor (EGF) and hormone stimulate phosphoinositide hydrolysis and increase EGF receptor protein synthesis and mRNA levels in rat liver epithelial cells. J. Biol. Chem., 263:13868.

Finzi, E., Fleming, T. P., Segatto, O., Pennington, C. Y. Bringman, T. S., Derynck, R., and Aaronson, S. A. (1987) The human transforming growth factor type alpha coding sequence is not a direct-acting oncogene when overexpressed in NIH/3T3 cells. Proc. Natl. Acad. Sci. USA, 84:3733.

Greenberger, J. S., Sakakeeny, M. A., Humphries, R. K., Eaves, C. J., and Eckner, R. J. (1983) Demonstration of permanent factor-dependent multipotential (erythroid/neutrophil/basophil) hematopoietic progenitor cell lines. Proc. Natl. Acad. Sci. USA, 80:2931.

Grynkiewicz, G., Poenie, M., and Tsien, R. Y. (1985) A new generation of Ca^{2+} indicators with greatly improved fluorescence properties. J. Biol. Chem., 260:3440.
Hawkins, P. T., Stephens, L., and Downes, C. P. (1986) Rapid formation of inositol 1,3,4,5-tetrakisphosphate and inositol 1,3,4-trisphosphate in rat parotid glands may both result indirectly from receptor-stimulated release of inositol 1,4,5-trisphosphate from phosphatidylinositol 4,5-bisphosphate. Biochem. J., 238, 507.
Hepler, J. R., Nakahata, N., Lovenberg, T. W., DiGuiseppi, J., Herman, B., Earp, H. S., and Harden T. K. (1987) Epidermal growth factor stimulates the rapid accumulation of inositol (1,4,5)-trisphosphate and a rise in cytosolic calcium mobilized from intracellular stores in A431 cells. J. Biol. Chem., 262:2951.
Irvine, R. F., Anggard, E. E., Letcher, A. J., and Downes, C. P. (1985) Metabolism of inositol 1,4,5-trisphosphate and inositol 1,3,4-trisphosphate in rat parotid glands. Biochem. J., 229:505.
Irvine, R. F., Letcher, A. J., Heslop, J. P., and Berridge, M. J. (1986) The inositol tris/tetrakisphosphate pathway-demonstration of Ins(1,4,5)P3-kinase activity in animal tissues. Nature, 320:631.
Jainchill, J. L., Aaronson, S. A., and Todaro, G. J. (1969) Murine sarcoma and leukemia viruses: assay using clonal lines of contact-inhibited mouse cells. J. Virol., 4:549..
Johnson, R. M. and Garrison, J. C. (1987) Epidermal growth factor and angiotensin II stimulate formation of inositol 1,4,5- and inositol 1,3,4-trisphosphate in hepatocytes. J. Biol. Chem., 262:17285.
Lacal, J. C., Moscat, J., and Aaronson, S. A. (1987) Novel source of 1,2-diacylglycerol elevated in cells transformed by Ha-ras oncogene. Nature, 330:269.
Lapetina, E.G.: Inositide-dependent and independent mechanisms in platelet activation. In "Phosphoinositides and Receptor Mechanisms" pp. 271-286, 1986, Alan R. Liss Inc. (New York, NY USA).
Lapetina, E. G., Silio, J., and Ruggiero, M. (1985) Thrombin induces serotonin secretion and aggregation independently of inositol phospholipids hydrolysis and protein phosphorylation in human platelets permeabilized with saponin. J. Biol. Chem., 260:7078.
Macara I. (1986) Activation of $^{45}Ca^{2+}$ influx and $^{22}Na^{+}/H^{+}$ exchange by epidermal growth factor and vanadate in A431 cells is independent of phosphatidylinositol turnover and is inhibited by phorbol esters and diacylglycerol. J. Biol. Chem., 261:9321.
Majerus, P. W., Neufeld, E. J., and Wilson, D. B. (1984) Production of phosphoinositide-derived messengers. Cell, 37:701.
Malgaroli, A., Milani, D., Meldolesi, J., and Pozzan, T. (1987) Fura-2 measurement of cytosolic free Ca^{2+} in monolayers and suspensions of various types of animal cells. J. Cell Biol., 105:2145.
Moscat, J., Molloy, C. J., Fleming, T. P., and Aaronson, S. A. (1988) Epidermal growth factor activates phosphoinositide turnover and protein kinase C in BALB/MK keratinocytes. Molec. Endocrinol., 2:799.
Nishizuka, Y. (1984) The role of protein kinase C in cell surface signal transduction and tumor promotion. Nature, 308:693.

Nishizuka, Y. (1988) The molecular heterogeneity of protein kinase C and its implications for cellular regulation. Nature, 334:661.

Palmer, S., Hawkins, P. T., Michell, R. H., and Kirk, C. J. (1986) The labelling of polyphosphoinositides with [^{32}P]Pi and the accumulation of inositol phosphates in vasopressin-stimulated hepatocytes. Biochem. J., 238:491.

Pandiella, A., Malgaroli, A., Meldolesi, J., and Vicentini, L.M. (1987) EGF raises cytosolic Ca^{2+} in A431 and Swiss 3T3 cells by a dual mechanism. Exp. Cell Res., 170:175.

Pierce, J. H., Ruggiero, M., Fleming, T. P., Di Fiore, P. P., Greenberger, J. S., Varticovski, L., Schlessinger, J., Rovera, G., and Aaronson, S. A. (1988) Signal transduction through the EGF receptor transfected in IL3-dependent hematopoietic cells. Science, 239:628.

Pruss, R. M., and Herschman, H. R. (1977) Variants of 3T3 cells lacking mitogenic response to epidermal growth factor. Proc. Natl. Acad. Sci. USA, 74:3918.

Rodriguez-Pena, A., and Rozengurt E. (1986) Phosphorylation of an acidic mol. wt. 80000 cellular protein in a cell-free system and intact Swiss 3T3 cells: a specific marker of protein kinase C activity. EMBO J., 5:77.

Ruggiero, M., Zimmermann, T. P., and Lapetina, E. G. (1985) Saponin suppression of protein phosphorylation in human platelets is related to ATP depletion. Biochem. Biophys. Res. Comm., 131:620.

Takasu, N., Takasu, M., Yamada, T. and Shimizu, T. (1988) Epidermal growth factor (EGF) produces inositol phosphates and increases cytoplasmic free calcium in cultured porcine thyroid cells. Biochem. Biophys. Res. Comm., 151:530.

Wakelam, M. J. O., Davies, S. A., Houslay, M. D., McKay, I., Marshall, C. J., and Hall, A. (1986) Normal p21$^{N\text{-}ras}$ couples bombesin and other growth factor receptors to inositol phosphate production. Nature, 323:173.

ACKNOWLEDGEMENTS

The authors wish to thank Dr. Stuart A. Aaronson for helpful discussion and continuing support
Marco Ruggiero was supported in part by a fellowship from Associazione Italiana per la Ricerca sul Cancro (AIRC).

1 ABBREVIATIONS USED IN THE TEXT

EGF: Epidermal Growth Factor
PDGF: Platelet-derived Growth Factor
EGFR-lines: cell lines overexpressing the Epidermal Growth Factor Receptor
h.p.l.c.: high-pressure liquid chromatography
t.l.c.: thin-layer chromatography
EGTA: Ethyleneglycol-bis-(b-aminoethyl ether)-N,N,N',N'-Tetraacetic Acid
TCA: Trichloroacetic acid

SECOND MESSENGER RECEPTORS IN THE PHOSPHOINOSITIDE CYCLE

Solomon H. Snyder, Christopher D. Ferris and Anne B. Theibert

Departments of Neuroscience, Pharmacology and Molecular Sciences, and Psychiatry and Behavioral Sciences, The Johns Hopkins University School of Medicine, 725 North Wolfe Street, Baltimore, Maryland 21205 - (301) 955-3024

Abundant evidence suggests that the phosphoinositide (PI) system is a prominent and prevalent second messenger system. In addition to mediating short term changes in humoral transmission, the PI cycle may regulate long term alterations evoked by growth factors, hormones and in learning and memory, as is exemplified by long term potentiation. The PI cycle influences cellular function through the generation of diacylglycerol (DAG) and inositol trisphosphate (IP_3) as well as other inositol phosphates. DAG stimulates protein kinase C (PKC), whereas the primary function of IP_3 is to release calcium from intracellular stores. It is easy to conceptualize how IP_3-induced release of calcium would activate short term alterations such as neuronal firing or muscle contraction. One of the challenges of PI research is to determine mechanisms whereby growth, differentiation and synaptic plasticity might be modulated by this second messenger system. Another challenge is to explain how discrete processing of information can be handled by a single second messenger system which is responsive to so many different primary messengers. Neurotransmitters achieve their selectivity, in part, by acting at subtypes of receptors. Accordingly, one might expect heterogeneity among the various enzymes and binding proteins involved in the PI cycle (for a comprehensive review of the PI system, see Berridge 1987).

We have attempted to clarify some of these questions by two approaches. One involves a search for differential localizations throughout the brain of PI components, such as enzymes, for which multiple forms have been demonstrated. Another approach involves a molecular characterization of key second messenger receptor molecules, such as the binding sites for IP_3 and inositol-tetrakisphosphate (IP_4). Purification of the IP_3 receptor has permitted high resolution localization of the receptor at the electron microscope level and enabled us to reconstitute function of the isolated receptor proteins.

LOCALIZATION STUDIES

A system as widely distributed as the PI cycle might be present in almost all neuronal and glial cells in the brain, with homogeneous distributions for PI elements such as phospholipase C (PLC), protein kinase C (PKC) or the IP_3 receptor. Instead, our studies reveal marked heterogeneity. Our first investigations compared the localization of PKC labeled by [^{3}H]phorbol esters with the localization of [^{3}H]IP_3 binding sites and adenylate cyclase labeled with [^{3}H]forskolin (Worley et al., 1987a, 1989). Generally, IP_3 receptors and PKC have similar localizations with highest levels in the cerebellum and hippocampus. However, there are some important differences. For instance, IP_3 receptor density is higher in the cerebellum, by a factor of up to a hundred, than in most other brain regions, while PKC is somewhat more uniformly distributed. Also, certain areas of the CNS possess abundant PKC with no detectable IP_3 receptor binding, as exemplified by the external plexiform layer of the olfactory bulb and the substantia gelatinosa of the spinal cord. This suggests that in such areas DAG may be produced from sources other than phosphatidylinositol-bisphosphate (PIP_2). Like PI cycle elements, forskolin binding is also concentrated in the cerebellum and caudate. However, in the cerebellum PKC and IP_3 receptors are highly localized to Purkinje cells, while adenylate cyclase is concentrated in parallel fibers of granule cells which synapse upon the Purkinje cells. Though both IP_3 receptors and PKC, as well as adenylate cyclase, occur in descending striatonigral pathways, adenylate cyclase and the PI elements occur in distinct pathways.

The reciprocal localizations of adenylate cyclase and PI cycle elements imply reciprocal interactions between neurons utilizing the same systems. Within individual neurons there also may be "cross-talk" between the two systems. For instance, PKC phosphorylates and thereby inactivates the G_i protein, responsible for synaptic inhibition of adenylate cyclase (Katada, 1985). In hippocampal slices, field recordings show cholinergic-PI influences on adenosinergic-G_i linked transmission. Muscarinic cholinergic neuronal stimulation, by augmenting PKC activity, blocks adenosine induced synaptic inhibition, which is known to be mediated by adenylate cyclase inhibition via a G_i protein (Worley et al., 1987b). A dampening of the PI cycle by therapeutic concentrations of lithium reverses this effect of muscarinic stimulation; this might account for the therapeutic actions of lithium (Worley et al., 1988).

Discrimination in synaptic transmission by elements of the PI cycle may stem also from the discrete localization of isozymes. For instance, multiple subtypes of PKC have been identified (Nishizuka, 1988). Type I is selectively concentrated in Purkinje cells of the cerebellum, resembling the distribution of PKC monitored by autoradiography with [^{3}H]phorbol esters. On the other hand, Type II PKC in the cerebellum is concentrated in granule cells. Furthermore, four subtypes of PLC have been discriminated (Suh et al., 1988; Bennett et al., 1988). We have differentiated the localization of mRNA for these subtypes by in situ hybridization (Ross et al., 1989a). An isozyme that we designate PLC-A is highly concentrated in Purkinje cells, much like IP_3 receptors and PKC. PLC-A is also selectively concentrated in the choroid plexus and raphe nuclei resembling the localization of serotonin-1C receptors, which are known to act through the PI cycle. PLC I mRNA resembles the localization of muscarinic receptor subtypes 1 and 3 and therefore may mediate their actions. PLC II has sequence homologies to non-catalytic areas of oncogene-related tyrosine kinases such as c-src. Interestingly, the localization of mRNA for PLC II resembles localizations we observed by in situ hybridization for mRNA of c-src (Ross et al., 1988). G proteins which mediate the actions of PLC also display a heterogeneous localization of their mRNA (Largent et al., 1988). The localization of mRNA for G_o and localizations of $G_{\alpha o}$ protein by immunohistochemistry (Worley et al., 1986) resemble elements of the PI cycle, especially PLC A and PLC I. These localizations suggest that G_o may be the G protein uniquely associated with the PI cycle.

To clarify how the PI cycle might influence a multiplicity of cellular functions, we localized the IP_3 receptor protein within cerebellar Purkinje cells by immunohistochemistry at the electron microscope level (Ross et al., 1989b). IP_3 receptors are selectively localized to the endoplasmic reticulum (ER), as had been suggested by numerous biochemical studies monitoring IP_3 release of Ca^{2+}. The majority of receptor is associated with rough ER. Rough ER with IP_3 receptors often lies close to the nuclear membrane, which is also heavily labeled. Association of IP_3 receptors with nuclei and the rough ER, which is involved in protein synthesis, suggests ways whereby the PI system can influence nuclear function. Receptors are also contained on smooth ER. A number of the receptors occur in cisternae close to the plasma membrane near presynaptic terminals, a localization appropriate for mediating the rapid calcium mobilization induced by IP_3 in postsynaptic cells.

What accounts for the extraordinary enrichment of the PI cycle in Purkinje cells? Purkinje cells receive input from the parallel fibers of granule cells, whose number is so great that granule cells may be the most plentiful class of neuron in the brain. Since each parallel fiber makes multiple contacts with Purkinje cells, the parallel fiber-Purkinje cell contact is likely the most abundant CNS synapse. Accordingly, the enrichment of the PI system in Purkinje cells may reflect the synaptic influences of parallel fibers for which the neurotransmitter is glutamate. The three major classes of glutamate receptors, quisqualate, kainate and NMDA, have been thought to be exclusively ionotropic with no direct biochemical second messenger. Recently a fourth subtype of glutamate receptor has been described which is most potently stimulated by quisqualate and acts through the PI system (Sugiyama et al., 1987; Nicoletti et al., 1986; Sladeczek et al., 1985). By examining the influence of numerous transmitters upon cerebellar PI turnover, we established that this quisqualate-type, PI-linked glutamate receptor represents a major synaptic system for parallel fiber influences on Purkinje cells (Blackstone et al., 1989).

MOLECULAR CHARACTERIZATION OF INOSITOL PHOSPHATE RECEPTORS

The great enrichment of IP_3 receptors in the cerebellum facilitated our identifying IP_3 binding sites in cerebellar membranes (Worley et al., 1987a,c). Binding in membranes has selective high affinity for IP_3. K_D values initially appeared to be about 80 nM, but with improved ligands and binding conditions, we now feel a more accurate value is about 5 nM. Calcium inhibits binding with an IC_{50} of about 300 nM corresponding to physiological intracellular concentrations of calcium. This suggests that calcium normally exerts a feedback inhibition on IP_3 receptors such that released calcium prevents further release of calcium by IP_3. Such a mechanism may account in part for the dramatic oscillations in intracellular calcium concentrations observed following stimuli that release calcium.

IP_3 and IP_4 binding activities are readily solubilized with detergents. The IP_3 binding protein has been purified to homogeneity in several steps (Supattapone et al., 1988a). About a thousand fold purification results in a homogeneous receptor protein with a subunit molecular weight of about 270 kD. Gel fractionation experiments indicate that the native protein has a molecular weight of about one million, suggesting that the native IP_3 receptor is composed of four identical subunits. Characterization of the binding activity shows that the purified receptor protein retains the same affinity and selectivity for IP_3 over other inositol phosphates as in the membrane state. Recently we have been able to solubilize and extensively purify an IP_4 binding protein which is likely the physiological IP_4 receptor (Theibert et al., 1987, 1989; Sklar et al., 1988). In crude membrane preparations IP_4 binding is heterogeneous, but in solubilized preparations IP_4 binds to a homogeneous receptor protein. The inositol phosphate specificity of IP_4 and IP_3 receptors differ markedly (Table 1). For instance, IP_3 is less than 3% as potent as IP_4 in competing for $[^3H]IP_4$ binding while IP_4 is several orders of magnitude weaker than IP_3 in competing for $[^3H]IP_3$ binding.

The IP_3 signal generated by PIP_2 hydrolysis is terminated by the action of an IP_3 phosphatase and IP_3 kinase. The phosphatase products are presumed to be inactive as second messengers since they do not release Ca^{2+}, while the kinase generates another second messenger, IP_4. The regional distribution of the phosphatase is similar to IP_3 receptors in that it is greatly enriched in the cerebellum (Theibert et al., 1989). IP_4 binding and the kinase likewise have a similar distribution in brain with hippocampus, striatum and cerebellum being approximately three times more enriched than other regions. In addition to solubilizing the IP_3 and IP_4 binding proteins, we have recently solubilized the membrane-associated IP_3 phosphatase and IP_3 kinase. These enzymes and binding sites must be discrete molecular species since they can be separated by a series of chromatography steps (Table 2). The IP_3 and IP_4 binding proteins and IP_3 kinase selectively associate with heparin agarose, while the IP_3 phosphatase and most other solubilized proteins do not. The IP_3 receptor is separated from the IP_4 receptor and kinase by its specific affinity for concanavalin-A. Kinase activity selectively interacts with calmodulin-agarose and can be separated from the IP_4 binding site.

Table 1. Inositol Phosphate Specificity of Purified $InsP_3$ and Partially Purified $InsP_4$ Receptors

	Affinity at Receptor (IC_{50}, nM)	
	$InsP_3$	$InsP_4$
$Ins(1,4,5)P_3$	10	>300
$Ins(1,3,4)P_3$	>1000	N.D.
$Ins(2,4,5)P_3$	500	>300
$Ins(1,3,4,5)P_4$	>1000	10
$Ins(1,4,5,6)P_4$	N.D.	300
$InsP_6$	>1000	50

Binding of $[^3H]InsP_3$ and $[^3H]InsP_4$ was determined as described in Supattapone et al., 1988a and Theibert et al., 1989.

REGULATION OF THE IP_3 RECEPTOR

The purified IP_3 receptor differs in one striking way from the particulate receptor, namely in the effects of calcium. In crude detergent solubilized membranes calcium inhibits IP_3 binding with the same 300 nM IC_{50} as in the particulate state. Heparin-agarose fractionation results in a loss of the inhibition by calcium with the purified receptor being resistant to calcium concentrations as high as 1.5 mM. Sensitivity to calcium can be restored to the purified receptor by adding back a preparation of detergent solubilized membranes. The membrane associated protein, which conveys the inhibitory effects of calcium, has been designated "calmedin" (Danoff et al., 1988, 1989). Whereas the IP_3 receptor shows marked differences in concentration throughout the brain, calmedin activity is more uniformly distributed and so may have functions other than the regulation of IP_3 receptors. We obtained evidence that the calcium which inhibits IP_3 binding is first bound to calmedin, which thus is a novel membrane associated calcium binding protein.

Following the isolation of the IP_3 receptor as a protein with a subunit weight of 270 kD and great enrichment in the cerebellum, we noted a similarity to other proteins enriched in the cerebellum and of a similar size. One of these, PCPP-260, was detected as a phosphorylated protein, a substrate for cyclic AMP protein kinase (PKA) but not cyclic GMP dependent protein kinase (Walaas et al., 1986). The IP_3 receptor is stoichiometrically phosphorylated by PKA but not PKC or

Table 2. Chromatographic Separation of Solubilized Phosphatase, Kinase and Binding Activities

Fraction	Membrane-associated IP_3 Phosphatase	Membrane-associated IP_3 Kinase	$[^3H]IP_4$ Binding	$[^3H]IP_3$ Binding
		°/∘ RECOVERY		
1°/∘ CHAPS Solubilized	100	100	100	100
Heparin-agarose				
Flow Through	88.4	1.4	3.3	1.0
0.5 M NaCl Eluate	3.6	30.1	63.7	47.0
Concanavalin-A Sepharose				
Flow Through	3.4	23.9	29.3	0.4
Alpha Me-Man Eluate	<0.1	<1.0	<1.0	24.9
Calmodulin-Agarose				
Flow Through	nd	<0.5	14.2	nd
EGTA eluate	nd	12.1	<0.5	nd

IP_3 phosphatase and IP_3 kinase activities, IP_4 and IP_3 binding were assayed as described in Theibert et al., 1989. nd = not determined. Alpha Me-Man refers to alpha-methylmannoside.

calcium-calmodulin kinase. The selective phosphorylation by PKA suggests that the IP_3 receptor may be identical to PCPP-260.

Does phosphorylation of the receptor alter its binding properties? No change in K_D is apparent but there is a 15°/∘ decrease in the B_{max} of IP_3 binding upon phosphorylation. To determine whether phosphorylation alters the calcium releasing functions of IP_3, we measured $^{45}Ca^{2+}$ released from brain microsomal preparations (Supattapone et al., 1988b). Phosphorylation of cerebellar microsomes by PKA shifts the dose-response curve to the right by an order of magnitude so that the IP_3 concentration required to produce half-maximal release of calcium is increased from 100 nM to 1 µM upon phosphorylation. The decrease in IP_3 release of calcium upon phosphorylation of the receptor suggests further evidence for cross-talk between the cyclic AMP and PI second messenger systems. Cyclic AMP generated would decrease the potency of IP_3 in releasing calcium. Interestingly, cyclic AMP also stimulates the calcium regulated ATPase pump of ER. Indeed, in our experiments with cerebellar microsomes, the total accumulation of calcium is tripled following phosphorylation of the receptors by PKA. As a result, the maximal calcium released by IP_3 is at least doubled upon phosphorylation. Therefore, formation of cyclic AMP may increase the pool of calcium within the ER accessible for release by IP_3. Perhaps the decreased calcium releasing potency of IP_3 following phosphorylation is a compensatory mechanism preventing excessive release of calcium from the enlarged pool.

RECONSTITUTION OF PURIFIED IP_3 RECEPTOR

To determine whether the IP_3 binding protein is the functional receptor responsible for IP_3 release of calcium, we have carried out reconstitution studies (Ferris et al., 1989). Purified IP_3 binding protein has been incorporated into liposomes. IP_3 enhances calcium flux in these liposomes (Table 3). About a 50°/∘ maximal response occurs with 100 nM IP_3. By contrast, IP_4 and IP_6 are inactive at the highest concentration tested and 2,4,5-IP_3 is substantially weaker. These experiments confirm that the IP_3 binding protein represents the physiological receptor associated with calcium release. Of particular importance is the conclusion that a single protein molecule contains both the IP_3 recognition site and the apparatus to release calcium. In this sense, the IP_3 receptor, a unique "second messenger receptor," resembles ionotropic neurotransmitter receptors, such as the nicotinic cholinergic, GABA and glycine receptors, in which

Table 3. IP_3 Receptor Protein Reconstituted in Liposomes Mediates $^{45}Ca^{2+}$ Flux

Ligand		$^{45}Ca^{2+}$ Flux (°/° Maximal)
1,4,5-IP_3	1 nM	5
	10 nM	12
	100 nM	54
	1 μM	74
	10 μM	100
	20 μM	98
	200 μM	95
2,4,5-IP_3	300 nM	0
	2 μM	56
	20 μM	62
IP_4	20 μM	0
IP_6	200 μM	0
100 μg/ml Heparin + 10 μM IP_3		5

$^{45}Ca^{2+}$ flux was determined as described in Ferris et al., 1989.

the recognition site for the neurotransmitter and the associated ion channel are part of the same protein complex. Whether the IP_3 receptor protein sequence resembles that of these receptors or of some other membrane protein awaits completion of molecular cloning studies.

Acknowledgements: Supported by USPHS grants MH-18501, DA-00266, and Research Scientist Award DA-00074 to S.H.S.

References

Bennett, C.F., Balcarek, J.M., Varrichio, A., and Crooke, S.T., 1988, Molecular cloning and complete amino-acid sequence of form-I phosphoinositide-specific phospholipase C, Nature, 334:268.

Berridge, M.J., 1987, Inositol trisphosphate and diacylglycerol: two interacting second messengers, Ann. Rev. Biochem., 56:159.

Blackstone, C.D., Supattapone, S., and Snyder, S.H., 1989, Phosphoinositide-linked glutamate receptors mediate cerebellar parallel fiber-purkinje cell synaptic transmission, Proc. Natl. Acad. Sci. USA, in press.

Danoff, S.K., Supattapone, S., and Snyder, S.H., 1988, Characterization of a membrane protein from brain mediating the inhibition of inositol 1,4,5-trisphosphate receptor binding by calcium, Biochem. J., 254:701.

Danoff, S.K., Theibert, A., Evans, R., and Snyder, S.H., 1989, Ca^{2+}-dependent regulation of Ins-1,4,5-P_3 binding to rat cerebellar receptor, Neurosci. Abstracts, in press.

Ferris, C., Huganir, R., Supattapone, S., S., and Snyder, S.H., 1989, Functional reconstitution of purified inositol 1,4,5-trisphosphate receptor from rat cerebellum, Neurosci. Abstracts, in press.

Katada, T., Gilman, A., Watanabe, Y., Bauer, S. and Jakobs, K., 1985, Protein kinase C phosphorylates the inhibitory guanine-nucleotide-binding regulatory component and apparently suppresses its function in hormonal inhibition of adenylate cyclase, Eur. J. Biochem., 151:431.

Largent, B.L., Jones D.T., Reed, R.R., Pearson, R.C.A., and Snyder, S.H., 1988, G protein mRNA mapped in rat brain by in situ hybridization, Proc. Natl. Acad. Sci. USA, 85:2864.

Nicoletti, F., Wrobleski, J.T., Novelli, A., Alho, H., Guidotti, A., and Costa, E., 1986, The activation of inositol phospholipid metabolism as a signal-transducing system for excitatory amino acid in primary cultures of cerebellar granule cells, J. Neuroscience, 6:1903.

Nishizuka, Y., 1988, The molecular heterogeneity of protein kinase C and its implications for cellular regulation, Nature, 334:661.

Ross, C.A., Wright, G.E., Resh, M.D., Pearson, R.C.A., and Snyder, S.H., 1988, Brain-specific src oncogene mRNA mapped in rat brain by in situ hybridization, Proc. Natl. Acad. Sci. USA, 85:9831.

Ross, C.A., MacCumber, M.W., Glatt, C.E., and Snyder, S.H. 1989a, Brain phospholipase C isozymes: differential mRNA localizations by in situ hybridization, Proc. Natl. Acad. Sci. USA, 86:2923.

Ross, C.A., Meldolesi, J., Milner, T.A., Satoh, T., Supattapone, S., and Snyder, S.H., 1989b, Inostiol (1,4,5) trisphosphate receptor localized to endoplasmic reticulum in cerebellar Purkinje neurons, Nature, in press.

Sklar, P., Theibert, A., Supattapone, S., Mourey, R., and Snyder, S.H., 1988, Characterization of the solubilized $InsP_4$ binding site in rat cerebellum, Neurosci. Abstracts, 14:1204.

Sladeczek, F., Pin, J.-P., Recansens, M., Bockaert, J., and Weiss, S., 1985, Glutamate stimulates inositol phosphate formation in striatal neurones, Nature, 317:717.

Sugiyama, H., Ito, I., and Hirono, C., 1985, A new type of glutamate receptor linked to inositol phospholipid metabolism, Nature, 325:531.

Suh, P., Ryu, S.H., Moon, K.H., Suh, H.W., and Rhee, S.G., 1988, Inositol phospholipid-specific phospholipase C: complete cDNA and protein sequences and sequence homology to tyrosine kinase-related oncogene products, Proc. Natl. Acad. Sci. USA, 85:5419.

Supattapone, S., Worley P.F., Baraban, J.M., and Snyder, S.H., 1988a, Solubilization, purification, and characterization of an inositol trisphosphate receptor, J. Biol. Chem., 263:1530.

Supattapone, S., Danoff, S.K., Theibert, A, Joseph, S.K., Steiner, J., and Snyder, S.H., 1988b, Cyclic AMP-dependent phosphorylation of a brain inositol trisphosphate receptor decreases its release of calcium, Proc. Natl. Acad. Sci. USA, 85:8747.

Theibert, A.B., Supattapone, S., Worley, P.F., Baraban, J.M., Meek, J.L., and Snyder, S.H., 1987, Demonstration of inositol 1,3,4,5-tetrakisphosphate receptor binding, Biochem. Biophys. Res. Commun., 148:1283.

Theibert, A., Ferris, C., Danoff, S., and Snyder, S., 1989, Solubilization and separation of inositol polyphosphate binding proteins and metabolizing enzymes in rat brain, Biochem. J., submitted.

Walaas, S.E., Nairn, A.D., and Greengard, P., 1986, PCPP-260, a Purkinje cell-specific cyclic AMP-regulated membrane phosphoprotein of M_r 260,000, J. Neuroscience, 6:954.

Worley, P.F., Baraban, J.M., Van Dop, C., Neer, E.J., and Snyder, S.H., 1986, G_o, a guanine nucleotide-binding protein: immunohistochemical localization in rat brain resembles distribution of second messenger system, Proc. Natl. Acad. Sci. USA, 83:4561.

Worley, P.F., Baraban, J.M., Colvin, J.S., and Snyder, Solomon H., 1987a, Inositol trisphosphate receptor localization in brain: variable stoichiometry with protein kinase C, Nature, 325:159.

Worley, P.F., Baraban, J.M., McCarren, M., Snyder, S.H., and Alger, B.E., 1987b, Cholinergic phosphatidylinositol modulation of inhibitory, G protein-linked, neurotransmitter actions: electrophysiological studies in rat hippocampus, Proc. Natl. Acad. Sci. USA, 84:3467.

Worley, P.F., Baraban, J.M., Supattapone, S., Wilson, V.S., and Snyder, S.H., 1987c, Characterization of inositol trisphosphate receptor binding in brain: regulation by pH and calcium, J. Biol. Chem., 262:12132.

Worley, P.F., Heller, W.A., Snyder, S.H., and Baraban, J.M., 1988, Lithium blocks a phosphoinositide-mediated cholinergic response in hippocampal slices, Science, 239:1428.

Worley, P.F., Baraban, J.M., and Snyder, S.H., 1989, Inositol 1,4,5-trisphosphate receptor binding: autoradiographic localization in rat brain, J. Neuroscience, 9:339.

MECHANISMS OF RECEPTOR REGULATION FOR THE PHOSPHOINOSITIDE SIGNALLING SYSTEM

Frank S. Menniti, Haruo Takemura, Hiroshi Sugiya, and James W. Putney, Jr.

Laboratory of Cellular and Molecular Pharmacology
National Institute of Environmental Health Sciences
Research Triangle Park, North Carolina 27709

It is now well established that the stimulation of certain cell surface receptors is transduced to the cell interior via the activation of phospholipase C (Putney, 1988). Most, if not all, receptors activate phospholipase C through an intermediary guanine nucleotide-dependent regulatory protein (G-protein) (Harden, 1989). Activated phospholipase C then catalyzes the breakdown of the minor membrane phospholipid, phosphatidylinositol 4,5-bisphosphate (PIP_2), resulting in the generation of two intracellular messenger cascades: the inositol phosphate/Ca^{2+} cascade and the diacylglycerol/protein kinase C cascade. Inositol 1,4,5-trisphosphate [(1,4,5)IP_3], the soluble product of PIP_2 breakdown, causes the release of Ca^{2+} from specialized intracellular stores and initiates the process of Ca^{2+} mobilization. Furthermore, the intracellular metabolism of (1,4,5)IP_3 results in the formation of numerous soluble inositol phosphates. These may also be involved in the regulation of intracellular calcium or may have other intracellular functions. Diacylglycerol activates protein kinase C (Nishizuka, 1984), resulting in the cell specific phosphorylation of intracellular proteins.

The focus of this chapter is the role that receptor desensitization, particularly homologus receptor desensitization, plays in the overall regulation of the phospholipase C signalling system. A pivotal role for regulation of this system at the level of the cell surface receptor is suggested by two general observations: 1) many of the tissues and cell lines examined to date have more than one phospholipase C-linked receptor, and 2) activation of different receptors on a given cell often results in different patterns of inositol phosphate formation (see for examples: Pachter et al., 1988; Hasegawa-Sasaki et al., 1988; McMillian et al., 1987; Morrison & Shukla, 1988; Hepler et al., 1988; Nakahata & Harden, 1987). With regard to receptor regulation, the most widely observed mechanism is that of receptor desensitization. Desensitization is operationally defined as a decrease in a biological response to successive receptor stimulations of equal magnitude. It is further categorized as one of two types. Homologous desensitization results in the blunting of the response only to the stimulated receptor. In heterologous desensitization, responsiveness is reduced to one or more additional stimuli. In the following discussion, we first summarize evidence indicating that desensitization of phospholipase C-linked receptors is a key element in the regulation of phospholipase C

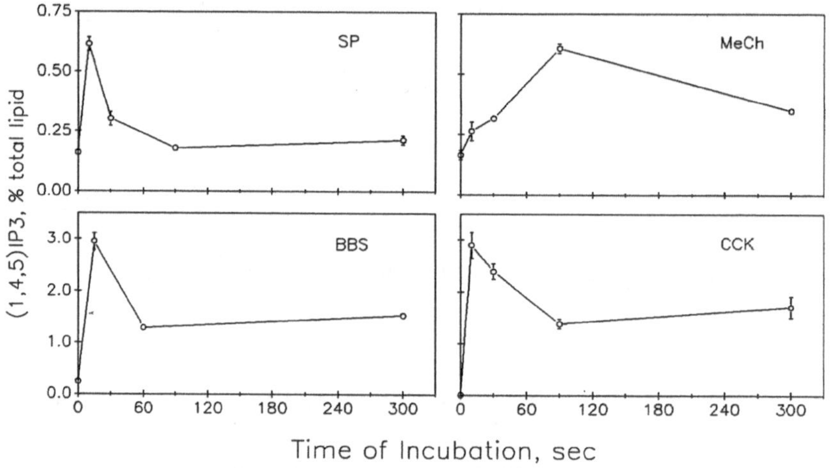

Figure 1. (1,4,5)IP_3 levels as a function of time after stimulation with SP (1 μM), MeCh (100 μM), BBS (200 nM), or CCK (1 μM). AR4-2J cells were labeled for three days with [^{3}H]*myo*-inositol prior to agonist exposure. [^{3}H](1,4,5)IP_3 was determined by HPLC and flow-type liquid scintillation counting. Values are expressed as percentage of the total [^{3}H]phosphoinositides.

activation. We then discuss possible mechanisms underlying the desensitization process. Finally, we turn to the physiological consequences of phospholipase C-linked receptor desensitization downstream from PIP_2 hydrolysis.

We have studied the desensitization of phospholipase C-linked receptors in the rat pancreatic tumor cell line, AR4-2J (Horstman et al. 1988; Menniti, Takemura, and Putney, in preparation) and in isolated rat parotid acinar cells (Sugiya et al., 1987; Sugiya & Putney, 1988a,b,c). The AR4-2J cells have at least 4 types of phospholipase C-linked receptors which when stimulated give rise to distinct patterns of phospholipase C activity. This is indicated by the changing levels of (1,4,5)IP_3 with time of stimulation (Fig 1). With maximal stimulation for each of the peptide agonists, bombesin (BBS), chole-cystokinin-8 SO_4 (CCK), or substance P (SP), the peak level of (1,4,5)IP_3 occurs within 10-20 sec. This peak level is similar after BBS or CCK but is typically 50 to 80 % lower after SP. Within 90 sec in the continued presence of BBS or CCK, (1,4,5)IP_3 declines from the peak to a steady state level several times greater than that before stimulation. However, in the continued presence of SP, (1,4,5)IP_3 returns to the prestimulation level within 90 sec. In contrast to the peptide agonists, muscarinic receptor stimulation by methacholine (MeCh) induces a slower rise in (1,4,5)IP_3 levels to a peak in 60-90 sec (Fig 1). Thereafter, (1,4,5)IP_3 declines slowly to basal level within 40-50 min in the continued presence of MeCh (not shown). The peak (1,4,5)IP_3 level achieved after maximal muscarinic receptor stimulation is similar to that after SP.

The different patterns of (1,4,5)IP_3 formation after stimulation of the different phospholipase C-linked receptors in the AR4-2J cells are at least partially accounted for by the homologous desensitization of these receptors. This is most evident for the SP and muscarinic receptor mediated responses, which completely desensitize. Desensitization of the response to SP develops very rapidly. Exposure of the AR4-2J cells to a maximal (1 μM) concentration of SP for as little as 60 sec completely

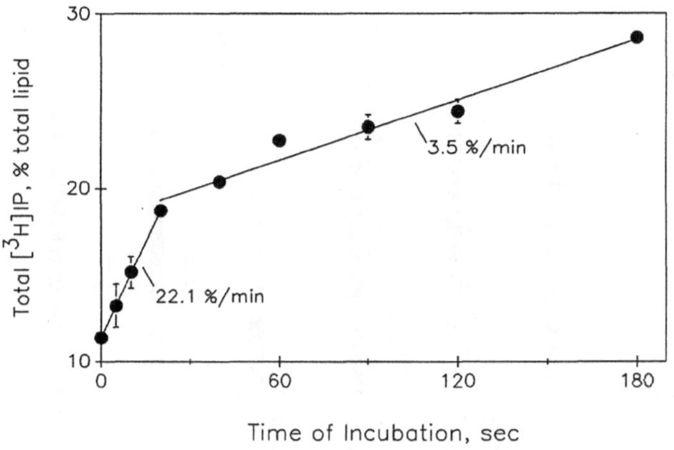

Figure 2. The accumulation of total [^{3}H]inositol phosphates ([^{3}H]IP)in AR4-2J cells stimulated with BBS in the presence of LiCl. AR4-2J cells were incubated for three days with [^{3}H]*myo*-inositol. The cells were then preincubated for 10 min with 10 mM LiCl and exposed to 200 nM BBS. Total [^{3}H]IP was determined at the indicated times by Dowex chromatography. Total [^{3}H]IP is expressed as a percent of total [^{3}H]phosphatidylinositol. Since LiCl blocks the dephosphorylation of the inositol phosphates, the accumulation of total [^{3}H]IP after stimulation in LiCl-treated cells indexes the rate of phospholipase C activity. The rates of phospholipase C activity for 0-20 and 20-180 sec after BBS addition are depicted. There is no [^{3}H]IP accumulation over basal in cells incubated with LiCl in the absence of BBS.

inhibits the accumulation of (1,4,5)IP_3 in response to a second SP exposure. In contrast, desensitization of the muscarinic response develops slowly. With continued exposure to a muscarinic agonist, the cells become completely refractory to muscarinic receptor stimulation only after 20-40 min. However, partial desensitization of the muscarinic response does occur in cells exposed to MeCh for as little as 30-60 sec. The desensitization of the inositol phosphate response to either of these two agonists is homologous, as indicated by the fact that cells made completely refractory to either SP or MeCh still respond fully to the other agonists. Thus, the degree and time course of SP and muscarinic receptor desensitization fully accounts for the changes in (1,4,5)IP_3 levels after stimulation of these receptors. The fact that desensitization is complete accounts for the return of (1,4,5)IP_3 to basal levels in the continued presence of either SP or MeCh. The respective time courses for the development of desensitization and the decay in (1,4,5)IP_3 levels are also well correlated. Furthermore, the fact that desensitization of the muscarinic receptor begins before the peak (1,4,5)IP_3 level is achieved suggests that the time course with which desensitization develops may also be a critical determinate of the peak level of (1,4,5)IP_3 achieved upon receptor activation.

In contrast to the SP and muscarinic receptors, the receptors for BBS and CCK only partially desensitize, giving rise to a biphasic pattern of phospholipase C activity. Exposure of the cells to maximal concentrations of BBS or CCK induces an initial, rapid phase of phospholipase C activity which abruptly terminates within 20 sec (as illustrated in Fig 2 for BBS). This is followed by a second, much slower phase of activity which is maintained for several hours in the continued

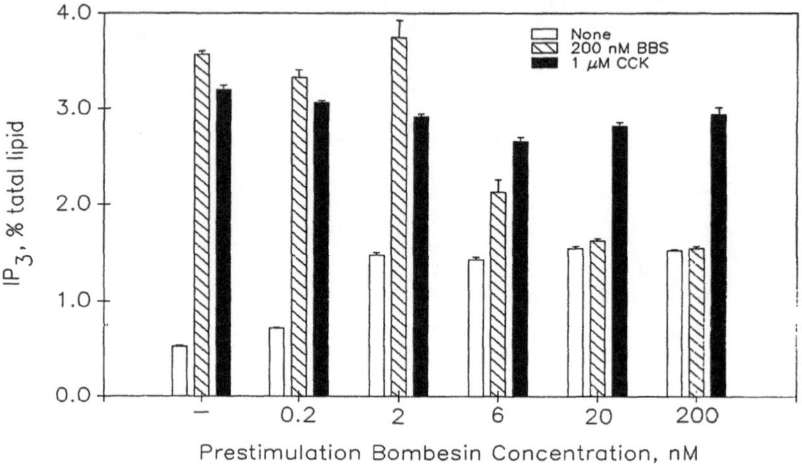

Figure 3. The effect of preincubation with different concentrations of BBS on $[^3H]IP_3$ formation in response to maximal concentrations of BBS or CCK. AR4-2J cells were labeled for three days with $[^3H]$*myo*-inositol. They were then exposed for 10 min to 0-200 nM BBS as indicated along the abscissa. After 10 min, the cells were challenged with BBS (200 nM, 10 sec), CCK (1 uM, 10 sec), or no agonist. $[^3H]IP_3$ was then determined by Dowex chromatography and is expressed as percent of total $[^3H]$phosphatidylinositol The durations chosen for the second stimulations correspond to the time to peak accumulation of $[^3H]IP_3$ for BBS and CCK in naive cells.

presence of either of these agonists. This pattern of phospholipase C activity accounts for the initial rapid accumulation of $(1,4,5)IP_3$ after BBS or CCK which then decays to the lower, sustained steady state level (Fig 1). It appears to reflect a partial homologous desensitization of the stimulated receptor, as illustrated in Fig 3. In this experiment, cells were incubated with different concentrations of BBS for 10 min. This resulted in a concentration dependent increase in the steady state level of IP_3. A further brief stimulation with a maximal concentration of the heterologous ligand CCK induces a second burst of IP_3 formation; the peak level attained is identical to that stimulated by CCK in non-BBS pretreated cells. However, the ability of a maximal concentration of BBS to stimulate a further increase in IP_3 levels is progressively diminished with increasing concentration of BBS pretreatment. Thus, the initial burst of IP_3 formation induced by BBS, but not by CCK, becomes desensitized by BBS pretreatment.

We also note the fact that in the AR4-2J cells, the activation of phospholipase C by different agonists is not additive. As mentioned above, the increase in IP_3 induced by CCK in BBS-pretreated cells reaches only the level achieved by CCK in naive cells. In the cases of MeCh and SP, these agonist have no effect on IP_3 levels in cells pretreated with BBS concentration greater than 2 nM since the sustained IP_3 level induced by higher BBS concentrations is greater than the maximum increase after MeCh or SP alone (data not shown). These findings suggest that this signalling pathway converges at the level of the G-protein/phospholipase C in these cells.

To summarize the findings in the AR4-2J cells, there is a common pool of phospholipase C which is activated by at least four different cell surface receptor types; each receptor stimulates a distinct pattern of phospholipase C activity because each receptor is differentially regulated. A key element in this regulatory process is homologous receptor desensitization. Thus, to fully understand the regulation of this signalling system it is critical to determine the mechanisms underlying phospholipase C-linked receptor desensitization. These mechanisms must account for the specificity of the homologous desensitization process, the differences in time course and degree of desensitization, and the fact that quantitatively distinct desensitization processes can occur in the same cell type.

The process of phospholipase C-linked receptor desensitization appears to involve multiple steps. Evidence summarized below suggests that the receptor is first 'uncoupled' from phospholipase C after which the receptor may be internalized. We have studied the process for the desensitization of the phospholipase C-linked SP receptor in rat parotid acinar cells in some detail (Sugiya et al., 1987; Sugiya & Putney, 1988b). Exposure of rat parotid cells to a maximal concentration of SP causes an increase in $(1,4,5)IP_3$ accumulation for approximately 60 sec; after this time there is no further increase in $(1,4,5)IP_3$ formation, indicating that the response partially desensitizes. The desensitization of the SP response is closely paralleled by a decrease in the number of SP-binding sites (determined with the radioligand, $[^3H]$SP) with no change in apparent affinity for SP. Thus, in the rat parotid cell, SP receptor desensitization corresponds to a rapid loss of the cell surface receptor for this peptide. However, in other cells and for other receptors, an 'uncoupling' step is clearly discernable before the initiation of receptor internalization. Xu and Chuang (1987) reported that the activation of phospholipase C by muscarinic receptor stimulation in cultured cerebellar granule cells decreases approximately 60 % after a 2 h preexposure of the cells to 100 μM carbachol with little change in the number of cell surface muscarinic receptors. More prolonged exposure to carbachol leads to a slower decrement in phospholipase C activation which correlates with a slow loss of receptors. Similarly, in WRK1 cells, phospholipase C activation by vasopressin receptor stimulation desensitizes with a faster time course than the internalization of the vasopressin receptor (Cantau et al., 1988). In this study, internalization of the vasopressin receptors was blocked by reducing the formation of functional coated pits (via depleting the cells of potassium); however, this manipulation did not disrupt the desensitization of vasopressin-stimulated phospholipase C activation. Finally, the LTB_4 receptor of U-937 cells undergoes homologous desensitization by a mechanism that involves a decrease in receptor affinity with little change in the number of cell surface receptors (Winkler et al., 1988). Thus, it would appear that the first, critical step in desensitization of phospholipase C-linked receptors disrupts the ability of the ligand-bound receptor to activate phospholipase C. Internalization of the receptor may then occur secondarily and with a variable time course.

What is the intracellular mechanism(s) which mediates the uncoupling of the receptor from phospholipase C and initiates the desensitization process? The most thoroughly studied mechanism for receptor desensitization is that for the ß-adrenergic receptor linked to adenylate cyclase. Lefkowitz and coworkers (Sibley et al., 1988; Lefkowitz & Caron, 1988) have demonstrated that phosphorylation of the $ß_2$-adrenergic receptor is the initiating signal for its homologus desensitization. Phosphorylation is catalyzed by a specific ß-adrenergic receptor kinase; this reaction is completely dependent on agonist occupancy of the receptor. In addition, the ß-adrenergic receptor is substrate for cAMP-dependent protein kinase and protein kinase C. However, the role of these kinases in the homologous and/or heterologous desensitization of this receptor is not clear at present. By analogy with the ß-adrenergic receptor, research has focused on the role of receptor phosphorylation as the signal for desensitization of phospholipase C-linked receptors. Of particular interest has been the role of protein kinase C since this kinase is known to be activated by diacylglycerol formed during the phospholipase C-catalyzed breakdown of PIP_2. Experimental support for an involvement of protein kinase C has come from studies of the effects of phorbol esters on the phospholipase C signalling system. The phorbol esters, which substitute for diacylglycerol and activate protein kinase C directly, have been shown in a variety of systems to attenuate receptor-stimulated phospholipase C activity (Pearce et al., 1988; Tohmatsu et al., 1986) and Ca^{2+} mobilization (Cooper et al., 1987; Vegesna et al., 1988; Tohmatsu et al., 1986). The phorbol esters also have been shown to induce the internalization of a phospholipase C-linked muscarinic receptor (Liles et al., 1986). As discussed in detail below, it appears that protein kinase C is involved in the desensitization process at several levels and that its exact role differs in different cell types. However, there are also clear instances in which protein kinase C is not involved in phospholipase C-linked receptor desensitization, indicating that other mediators are involved in this process as well.

A physiological role for protein kinase C in homologous desensitization has been demonstrated in rat renal mesangial cells. These cells have a phospholipase C-linked angiotensin II receptor which becomes completely desensitized within 4 min of exposure to this peptide (Pfeilschifter, 1988). Activation of protein kinase C with TPA and OAG inhibits angiotensin II-induced inositol phosphate formation in intact mesangial cells (Pfeilschifter, 1986) and in a plasma membrane preparation made from these cells (Pfeilschifter & Bauer, 1987). However, these agents do not affect PIP_2 hydrolysis induced by GTPγS in the membranes, indicating that the effect of protein kinase C activation is not the result of direct inhibition of phospholipase C or disruption of the ability of activated G-protein to stimulate the lipase. The protein kinase C inhibitors H-7, sphingosine, and cytotoxin I blocked the effect of TPA on the angiotensin II-stimulated phospholipase C response in intact cells (Pfeilschifter, 1988). Importantly, in this same study, these protein kinase C inhibitors were found to block the desensitization of the angiotensin II receptor induced by the peptide. Thus, in this cell type, the homologous desensitization of the angiotensin II response would appear to be mediated by an effect of protein kinase C at the level of the angiotensin II receptor. A role for protein kinase C in the desensitization of angiotensin II receptors, in this case in WB cells, is also indicated by the findings of Hepler et al. (1988). In the WB cells, down regulation of protein kinase C by prolonged exposure to phorbol esters selectively sensitizes the inositol phosphate response to angiotensin II, suggesting that the response is normally regulated by feedback inhibition via activation of protein kinase C.

The direct phosphorylation of a phospholipase C-linked receptor by protein kinase C concomitant with desensitization has also been demonstrated. In DDT_1 MF-2 smooth muscle cells, either agonist activation of the α_1-adrenergic receptor or activation of protein kinase C with phorbol esters leads to the phosphorylation of this receptor (Leeb-Lundberg et al., 1987). Receptor phosphorylation correlates with desensitization of the α_1-induced phospholipase C response. Furthermore, agonist occupancy of this α_1-adrenergic receptor specifically increased the rate, though not the extent, of receptor phosphorylation by protein kinase C (Bouvier et al., 1987). This latter finding suggests a mechanism whereby protein kinase C could mediate a specific (homologous) desensitization in a system containing several receptor types.

An effect of protein kinase C on phospholipase C activation is not restricted to phosphorylation of the phospholipase C-linked receptor. Orellana et al. (1987) found that activation of protein kinase C in intact 1321N1 astrocytoma cells, or addition of purified protein kinase C to 1321N1 cell membranes, inhibited the formation of inositol phosphates induced by incubating the membranes with GTPγS or GTPγS plus carbachol. Protein kinase C activation did not, however, affect the muscarinic receptor affinity for carbachol in the presence or absence of GTP, nor did it affect the activation of phospholipase C by calcium in the membrane preparation. Taken together these results suggest that protein kinase C mediated a decrease in the ability of an activated G-protein to stimulate phospholipase C activity.

The above findings clearly suggest a role for protein kinase C in the desensitization of phospholipase C-linked receptors in the studied cell types. However, in other systems the involvement of protein kinase C is less clear. We have examined the role of protein kinase C in the homologous desensitization of the rat parotid SP receptor (Sugiya & Putney, 1988; Sugiya et al., 1988a). Activation of protein kinase C with phorbol ester inhibits the activation of phospholipase C in response to SP in these cells; however, several lines of evidence indicate that this is not the mechanism for the homologous desensitization of the SP response. The effect of phorbol ester on SP-induced IP_3 formation in these cells is blocked by the protein kinase C inhibitors, H-7 and K-252a. However, these inhibitors do not modify the time course for IP_3 formation in response to SP nor do they prevent the desensitization of the SP response induced by exposure of the cells to SP. In fact, we have found that the desensitization of the SP receptor, as indexed by the decrease in SP binding sites, is not modified by complete inhibition of SP-induced phospholipase C activation. That is, in parotid cells depleted of cellular ATP by preincubation with antimycin, SP-induced IP_3 (and, presumably, diacylglycerol) formation is completely inhibited. However, this manipulation does not affect the SP-induced downregulation of SP-binding sites. Thus, these studies indicate that in the rat parotid, a SP-induced activation of protein kinase C does not mediate the homologous desensitization of the inositol phosphate response to SP nor does it affect the phospholipase C activity in SP-stimulated cells to any significant extent. The phorbol esters may affect the SP response by activating protein kinase C to an extent which does not normally occur with SP receptor stimulation. Alternatively, they may act by a non-protein kinase C-mediated mechanism. In other systems as well, activation of protein kinase C has no role in homologous receptor desensitization. For example, the phospholipase C activation induced by TRH in GH_4C_1 cells is not affected by TPA nor is the desensitization of the TRH receptor inhibited by the protein kinase C inhibitor, polymyxin B (Torjesen et al., 1988). In the chinese hamster lung fibroblast cell line CCL39, desensitization of the inositol phosphate response to thrombin is not modified by downregulation of protein kinase C (Paris et al., 1988). As with the rat parotid, desensitization of the thrombin receptor in these cells also occurs under conditions in which the inositol phosphate response is completely inhibited; i.e., when cells are incubated with thrombin at 4°C.

In summary, a physiological role for protein kinase C in the desensitization of phospholipase C-linked receptors is clearly implicated in several systems. In these systems, it appears that protein kinase C may exert its effect at one of several levels. The fact that the phorbol esters affect the inositol phosphate response to various stimuli in a large number of tissues suggests that this role may be widespread. However, we have found that an effect of phorbol ester *per se* is not sufficient evidence to assign a physiological role to this kinase in the feedback regulation of phospholipase C. In fact, it is clear that activation of protein kinase C is *not* universally involved in the desensitization process. Thus, there must be other mediators of desensitization in addition to this kinase.

At this time, other candidates for mediating the desensitization of phospholipase C-linked receptors are not well defined. Again, by analogy with the desensitization of the ß-adrenergic receptor, there may be specific receptor kinases involved in the desensitization of phospholipase C-linked receptors. It now appears that the ß-adrenergic receptor kinase catalyzes the agonist-dependent phosphorylation of adenylate cyclase-linked receptors other than the $ß_2$-adrenergic receptor (Mayor et al., 1987; Strasser et al., 1986). However, this kinase does not act on the agonist-occupied α_1-adrenergic receptor which is coupled to phospholipase C (Benovic et al., 1987), suggesting it may not be generally involved in phosphorylation of phospholipase C-linked receptors. On the other hand, Kwatra et al. (Kwatra & Hosey, 1986; Kwatra et al., 1987) have shown that the muscarinic receptor of chick heart is phosphorylated in an agonist-dependent fashion by a kinase distinct from the cyclic nucleotide dependent or Ca^{2+}/calmodulin dependent protein kinases or protein kinase C. Furthermore, the concentration of agonist required to induce muscarinic receptor phosphorylation correlates closely with the agonist concentration required to induce desensitization of muscarinic receptor-stimulated inositol phosphate formation in this preparation. Thus, it is not unreasonable to speculate that a phospholipase C-linked receptor-specific kinase(s) participates in the desensitization of these receptors; it is hoped that the identification of such a kinase will not prove to be as difficult as has been the identification of the phospholipase C-specific G-protein. Finally, desensitization of phospholipase C-linked receptors may be mediated by mechanisms which do not depend on phosphorylation. As mentioned above, the desensitization of the rat parotid SP receptor (Sugiya & Putney, 1988b) and the CCL39 cell thrombin receptor (Paris et al., 1988) occur under conditions which would not likely support receptor phosphorylation. Just as agonist occupancy facilitates receptor phosphorylation, it may also induce conformational changes which render the receptor accessible to other processes which diminish the ability of the receptor to activate phospholipase C. Such other mechanisms also await identification.

The final section of this chapter illustrates some of the physiological sequelae of phospholipase C-linked receptor desensitization, using the AR4-2J cells as a model system. In addition to the effects on phospholipase C activation noted above, the differential desensitization of the phospholipase C-linked receptors in these cells affects the pattern of $(1,4,5)IP_3$ metabolism and the pattern of change in intracellular Ca^{2+} concentration ($[Ca^{2+}]_i$) when these cells are stimulated with one or more agonist. In our initial characterization of inositol phosphate formation in the AR4-2J cells (Horstman et al., 1988), we noted that there was essentially no $(1,3,4,5)IP_4$ accumulation in cells stimulated with SP. However, further study has revealed that stimulation with BBS, CCK, or MeCh induces a slow accumulation of $(1,3,4,5)IP_4$ (Menniti, Oliver, and Putney, in preparation). The differential effect of SP vs the other agonists on $(1,4,5)IP_3$ phosphorylation derives from the fact that the flux through the $(1,4,5)IP_3$ 3-kinase pathway is only approximately 10 % of that through the $(1,4,5)IP_3/(1,3,4,5)IP_4$ 5-phosphatase pathway. Thus, the rapid desensitization of the SP receptor and the slow kinetics of the 3-kinase reaction interact such that the accumulation of $(1,3,4,5)IP_4$ is negligible after stimulation with SP. The more prolonged stimulation of

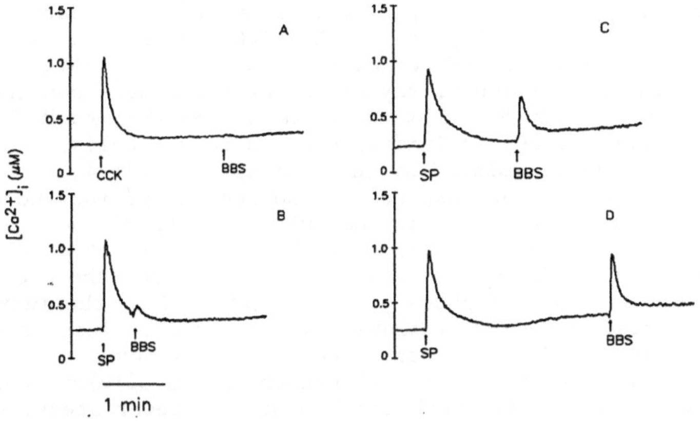

Figure 4. The effect of Ca^{2+} mobilizing agonists on $[Ca^{2+}]_i$ in AR4-2J cells. $[Ca^{2+}]_i$ was measured using the fluorescent dye, fura-2, as previously described (Horstman et al., 1988). BBS (200 nM), CCK (1 μM), and SP (1 μM) were added to cell suspensions as indicated.

$(1,4,5)IP_3$ formation induced by the other agonists allows for the slow accumulation of the 3-kinase product. A similar phenomenon has been observed in rabbit platelets (Morrison & Shukla, 1988). In these cells, IP_3 formation after stimulation with platelet activating factor rapidly desensitizes and results in no accumulation of IP_4. However, IP_4 does accumulate with the more prolonged stimulation induced by thrombin. That such differences in $(1,4,5)IP_3$ metabolism may have physiological significance is based on the widely held belief that phosphorylated derivatives of $(1,4,5)IP_3$ play a role in the regulation of intracellular calcium and/or subserve other cellular functions. The precise function(s) of these inositol phosphates are still under study.

The desensitization of phospholipase C-linked receptors also critically affects the regulation of $[Ca^{2+}]_i$ in response to the Ca^{2+} mobilizing agonists of AR4-2J cells (Takemura, Menniti, Thastrup, and Putney, in preparation; Menniti, Takemura, and Putney, in preparation). In the presence of 1 mM extracellular Ca^{2+}, stimulation of these cells with any of the four receptor agonists induces a rapid rise in $[Ca^{2+}]_i$ from a resting level of approximately 250 nM to approximately 1000-1500 nM (see Fig 4). The initial rise in $[Ca^{2+}]_i$ is attenuated but not abolished in the absence of extracellular Ca^{2+}, indicating that it is due to a combination of Ca^{2+} release from intracellular stores as well as Ca^{2+} influx. Within 90 sec after stimulation with BBS, CCK, or MeCh, $[Ca^{2+}]_i$ falls to a sustained level approximately 100-200 nM above basal and remains at this level for at least 5 min, the longest time point examined. Addition of 3 mM EGTA during this sustained phase of the Ca^{2+} response completely reverses the elevation of $[Ca^{2+}]_i$, indicating that this phase is due principally to Ca^{2+} influx. On the other hand, in the continued presence of SP, $[Ca^{2+}]_i$ declines to basal levels within 90 sec regardless of the presence of extracellular Ca^{2+}. Thus, whereas each of these agonists induce an initial Ca^{2+} response comprising release of Ca^{2+} from intracellular stores and influx of extracellular Ca^{2+}, the more prolonged influx phase is sustained only in the continued presence of BBS, CCK, and MeCh and not in the continued presence of SP. This closely correlates with the sustained inositol phosphate response induced by BBS, CCK, and MeCh as opposed to the rapid desensitization of this response to SP. The consequences of the differential desensitization of the Ca^{2+}

mobilizing receptors are also evident when the cells are exposed to different combinations of these agonists. Representative data from these experiments is presented in Figure 4. When CCK addition is followed by BBS, there is essentially no effect of the BBS on $[Ca^{2+}]_i$ (Fig 4, panel A). In general, a second addition of any of the agonists does not increase $[Ca^{2+}]_i$ if preceded within 5 min by addition of BBS, CCK, or MeCh. When SP (1 μM, final concentration) is initially added to the cells, a second increase in $[Ca^{2+}]_i$ is stimulated by addition of BBS (Fig 4, panels B-D), CCK or MeCh (not shown); the response to the second agonist increased as a function of time after SP and is maximal within 3 min of the initial SP addition. However, a second addition of SP to SP-stimulated cells has no further effect on the $[Ca^{2+}]_i$ (not shown). We interpret these data as follows: during the initial phase of Ca^{2+} mobilization, the intracellular Ca^{2+} pool is emptied and a Ca^{2+} influx phase commences. These two conditions are sustained by the continued $(1,4,5)IP_3$ formation in the presence of BBS, CCK, or MeCh, even though the intracellular level of $(1,4,5)IP_3$ decreases from its peak level to a much lower steady state level (Fig 1). Thus, a second burst of $(1,4,5)IP_3$ accumulation, as occurs in CCK stimulated cells exposed to BBS, has no further effect on $[Ca^{2+}]_i$ (Fig 4, panel A). In SP stimulated cells, $(1,4,5)IP_3$ levels rapidly return to basal levels with the dual consequences that Ca^{2+} influx rapidly wanes and the intracellular Ca^{2+} pool begins to refill. In this case, a second burst of $(1,4,5)IP_3$ accumulation induced by a heterologous agonist releases the intracellular Ca^{2+} pool even if it is only partially refilled (Fig 4, panels B-D). However, a second exposure to SP has no effect because the response is desensitized at the level of the SP receptor.

In summary, the data on AR4-2J cells presented above, as well as the findings from numerous studies by ourselves and others, establish that the desensitization of phospholipase C-linked receptors plays a critical role in the regulation of the phospholipase C-signalling cascade. Our full understanding of the regulation of this signalling system, however, awaits the elucidation of the mechanisms which underlie the desensitization process. As indicated above, desensitization of the phospholipase C-linked receptors appears to involve, as a general step, the uncoupling of the receptor and the lipase. However, the variable patterns of desensitization and the variable involvement of protein kinase C and receptor phosphorylation in this process suggest that the mechanisms which mediate this uncoupling may be diverse and, in fact, may be cell and/or receptor specific. As always, only further research will promote our understanding of this important phenomenon.

REFERENCES

Benovic, J.L., Regan, J.W., Matsui, H., Mayor, F., Jr., Cotecchia, S., Leeb-Lundberg, L.M.F., Caron, M.G., and Lefkowitz, R.J. (1987) Agonist-dependent phosphorylation of the α_2-adrenergic receptor by the ß-adrenergic receptor kinase. J. Biol. Chem., *262*:17251-17253.

Bouvier, M., Leeb-Lundberg, L.M.F., Benovic, J.L., Caron, M.G., and Lefkowitz, R.J. (1987) Regulation of adrenergic receptor function by phosphorylation. II. Effects of agonist occupancy on phosphorylation of α1- and β2-adrenergic receptors by protein kinase C and the cyclic AMP-dependent protein kinase. J. Biol. Chem., *262*:3106-3113.

Cantau, B., Guillon, G., Alaoui, M.F., Chicot, D., Balestre, M.N., and Devilliers, G. (1988) Evidence of two steps in the homologous desensitization of vasopressin-sensitive phospholipase C in WRK1 cells. Uncoupling and loss of vasopressin receptors. J. Biol. Chem., *263*:10443-10450.

Cooper, R.H., Coll, K.E., and Williamson, J.R. (1987) Differential effects of phorbol ester on phenylephrine and vasopressin-induced Ca^{2+} mobilization in isolated hepatocytes. J. Biol. Chem., *260*:3281-3288.

Harden, T.K. (1989) The role of guanine nucleotide regulatory proteins in receptor-selective direction of inositol lipid signalling. In: Inositol lipids in cell signalling. R.H. Michell, A.H. Drummond, and C.P. Downes, eds. Academic Press Limited, London, pp. 113-134.

Hasegawa-Sasaki, H., Lutz, F., and Sasaki, T. (1988) Pathway of phospholipase C activation initiated with platelet-derived growth factor is different from that initiated with vasopressin and bombesin. J. Biol. Chem., *263*:12970-12976.

Hepler, J.R., Earp, H.S., and Harden, T.K. (1988) Long-term phorbol ester treatment down-regulates protein kinase C and sensitizes the phosphoinositde signaling pathway to hormone and growth factor stimulation. J. Biol. Chem., *263*:7610-7619.

Horstman, D.A., Takemura, H., and Putney, J.W., Jr. (1988) Formation and metabolism of ^{3}H-inositol phosphates in AR42J pancreatoma cells. J. Biol. Chem., *263*:15297-15303.

Kwatra, M.M., and Hosey, M.M. (1986) Phosphorylation of the cardiac muscarinic receptor in intact chick heart and its regulation by a muscarinic agonist. J. Biol. Chem., *261*:12429-12432.

Kwatra, M.M., Leung, E., Maan, A.C., McMahon, K.K., Ptasienski, J., Green, R.D., and Hosey, M.M. (1987) Correlation of agonist-induced phosphorylation of chick heart muscarinic receptors with receptor desensitization. J. Biol. Chem., *262*:16314-16321.

Leeb-Lundberg, L.M.F., Cotecchia, S., DeBlasi, A., Caron, M.G., and Lefkowitz, R.J. (1987) Regulation of adrenergic receptor function by phosphorylation. I. Agonist-promoted desensitization and phosphorylation of α1-adrenergic receptors coupled to inositol phospholipid metabolism in DDT1 MF-2 smooth muscle cells. J. Biol. Chem., *262*:3098-3105.

Lefkowitz, R.J., and Caron, M.G. (1988) Adrenergic receptors. Models for the study of receptors coupled to guanine nucleotide regulatory proteins. J. Biol. Chem., *263*:4993-4996.

Liles, W.C., Hunter, D.D., Meier, K.E., and Nathanson, N.M. (1986) Activation of protein kinase C induces rapid internalization and subsequent degradation of muscarinic acetylcholine receptors in neuroblastoma cells. J. Biol. Chem., *261*:5307-5313.

Mayor, F., Jr., Benovic, J.L., Caron, M.G., and Lefkowitz, R.J. (1987) Somatostatin induces translocation of the β-adrenergic receptor kinase and desensitizes somatostatin receptors in S49 lymphoma cells. J. Biol. Chem., *262*:6468-6471.

McMillian, M.K., Soltoff, S.P., and Talamo, B.R. (1987) Rapid desensitization of substance P- but not carbachol-induced increases in inositol trisphosphate and intracellular Ca++ in rat parotid acinar cells. Biochem. Biophys. Res. Commun., *148*:1017-1024.

Morrison, W.J., and Shukla, S.D. (1988) Desensitization of receptor-coupled activation of phosphoinositide- specific phospholipase C in platelets: evidence for distinct mechanisms for platelet-activating factor and thrombin. Mol. Pharmacol., *33*:58-63.

Nakahata, N., and Harden, T.K. (1987) Regulation of inositol trisphosphate accumulation by muscarinic cholinergic and H_1-histamine receptors on human astrocytoma cells. Differential induction of desensitization by agonists. Biochem. J., *241*:337-344.

Nishizuka, Y. (1984) The role of protein kinase C in cell surface signal transduction and tumour promotion. Nature, *308*:693-698.

Orellana, S., Solski, P.A., and Brown, J.H. (1987) Guanosine 5'-O-(thiophosphate)-dependent inositol trisphosphate formation in membranes is inhibited by phorbol ester and protein kinase C. J. Biol. Chem., *262*:1638-1643.

Pachter, J.A., Law, G.J., and Dannies, P.S. (1988) Bombesin stimulates inositol polyphosphate production in GH_4C_1 pituitary tumor cells: Comparison with TRH. Biochem. Biophys. Res. Comm., *154*:654-659.

Paris, S., Magnaldo, I., and Pouyssegur, J. (1988) Homologous desensitization of thrombin-induced phosphoinositide breakdown in

hamster lung fibroblasts. J. Biol. Chem., *263*:11250-11256.
Pearce, B., Morrow, C., and Murphy, S. (1988) Characteristics of phorbol ester- and agonist-induced down-regulation of astrocyte receptors coupled to inositol phospholipid metabolism. J. Neurochem., *50*:936-944.
Pfeilschifter, J. (1986) Tumor promotor 12-O-tetradecanoylphorbol 13-acetate inhibits angiotensin II-induced inositol phosphate production and cytocolic Ca2+ rise in rat renal mesangial cells. FEBS Lett., *203*:262-266.
Pfeilschifter, J. (1988) Protein kinase C from rat renal mesangial cells: its role in homologous desensitization of angiotensin II-induced polyphosphoinositide hydrolysis. Biochim. Biophys. Acta, *969*:263-270.
Pfeilschifter, J., and Bauer, C. (1987) Different effects of phorbol ester on angiotensin II- and stable GTP analogue-induced activation of polyphosphoinositide phosphodiesterase in membranes isolated from rat renal mesangial cells. Biochem. J., *248*:209-215.
Putney, J.W., Jr. (1988) The role of phosphinositide metabolism in signal transduction in secretory cells. J. Exp. Biol., *139*:135-150.
Sibley, D.R., Benovic, J.L., Caron, M.G., and Lefkowitz, R.J. (1988) Phosphorylation of cell surface receptors: a mechanism for regulating signal transduction pathways. Endocr. Rev., *9*:38-56.
Strasser, R.H., Benovic, J.L., Caron, M.G., and Lefkowitz, R.J. (1986) β-agonist- and prostaglandin E_1-induced translocation of the β-adrenergic receptor kinase: Evidence that the kinase may act on multiple adenylate cyclase-coupled receptors. Proc. Natl. Acad. Sci. USA, *83*:6362-6366.
Sugiya, H., Obie, J.F., and Putney, J.W., Jr. (1988) Two modes of regulation of the phospholipase C-linked substance P receptor in rat parotid acinar cells. Biochem. J., *253*:459-446.
Sugiya, H., and Putney, J.W., Jr. (1988) Substance P receptor desensitization requires activation of receptor, but not phospholipase C. Am. J. Physiol., *255*:C149-C154.
Sugiya, H., and Putney, J.W., Jr. (1988) Protein kinase C-dependent and -independent mechanisms regulating the parotid substance P receptor as revealed by differential effects of protein kinase C inhibitors. Biochem. J., *256*:677-680.
Sugiya, H., Tennes, K.A., and Putney, J.W., Jr. (1987) Homologous desensitization of substance-P-induced inositol polyphosphate formation in rat parotid acinar cells. Biochem. J., *244*:647-653.
Tohmatsu, T., Hattori, H., Nagao, S., Ohki, K., and Nozawa, Y. (1986) Reversal by protein kinase C inhibitor of suppressive actions of phorbol-12-myrstate-13-acetate on polyphosphoinositide metabolism and cytosolic Ca^{2+} mobilization in thrombin-stimulated human platelets. Biochem. Biophys. Res. Comm., *134*:868-875.
Torjesen, P.A., Bjoro, T., Ostberg, B.C., and Haug, E. (1988) Thyrotropin-releasing hormone-stimulated inositol trisphosphate formation is liable to thyrotropin-releasing hormone-induced desensitization by a calcium-dependent mechanism. Mol. Cell Endocrinol., *56*:107-114.
Vegesna, R.V., Wu, H.L., Mong, S., and Crooke, S.T. (1988) Staurosporine inhibits protein kinase C and prevents phorbol ester-mediated leukotriene D4 receptor desensitization in RBL-1 cells. Mol. Pharmacol., *33*:537-542.
Winkler, J.D., Sarau, H.M., Foley, J.J., Mong, S., and Crooke, S.T. (1988) Leukotriene B4-induced homologous desensitization of calcium mobilization and phosphoinositide metabolism in U-937 cells. J. Pharmacol. Exp. Ther., *246*:204-210.
Xu, J., and Chuang, D.M. (1987) Muscarinic acetylcholine receptor-mediated phosphoinositide turnover in cultured cerebellar granule cells: desensitization by receptor agonists. J. Pharmacol. Exp. Ther., 242;238-244.

COMPUTER SIMULATION OF PROSTAGLANDIN-REGULATED CYCLIC AMP METABOLISM IN PLATELETS: ACTIVATION AND DESENSITIZATION OF ADENYLATE CYCLASE APPEAR TO BE MEDIATED THROUGH SEPARATE PROSTAGLANDIN RECEPTORS COUPLED TO G_s AND G_i

Barrie Ashby

Thrombosis Research Center and Department of Pharmacology
Temple University Health Sciences Center
Philadelphia, PA 19140

ABSTRACT

We have examined and modeled the kinetics of prostaglandin-regulated cyclic AMP formation in intact human platelets and platelet lysates. In the case of iloprost, a stable analogue of PGI_2, the shape of the time-course of cyclic AMP formation was dependent on prostaglandin concentration. In intact platelets, low concentrations of iloprost gave a rise to a plateau with little decrease in cyclic AMP level over many minutes, while high concentrations gave a rise to a peak followed by a fall to a plateau level lower than that observed at low concentrations of iloprost. In contrast, PGE_1 gave a rapid rise and fall to a plateau at all concentrations, although the initial rate, peak maximum and final steady-state level of cyclic AMP all increased as a saturable function of PGE_1 concentration. The time-dependent fall in cyclic AMP was still evident in the presence of phosphodiesterase inhibitors, indicating that it was due to inhibition of adenylate cyclase, rather than activation of phosphodiesterase activity. Time-dependent inhibition was abolished by treating platelets with agents that impair the function of G_i, indicating that G_i mediates prostaglandin-induced inhibition. We have simulated cyclic AMP metabolism over a wide range of prostaglandin concentration by a model that includes adenylate cyclase activation followed by reversible, slow inhibition of the enzyme mediated by a distinct prostaglandin receptor. The putative inhibitory receptor has an affinity for prostaglandin lower than the stimulatory receptor in the case of iloprost and higher than the stimulatory receptor in the case of PGE_1. The model requires that receptor affinity declines with prostaglandin concentration, which may be related to the existence of high and low affinity forms of receptor that depend on the activation state of the appropriate G protein. Time- and prostaglandin concentration-dependent inhibition of adenylate cyclase was also observed in platelet lysates in the presence of non-hydrolyzable GTP analogues and kinetic curves could be modeled by considering slow transitions among the subunits of G_s and G_i. We conclude that activation and time-dependent inhibition (desensitization) of platelet adenylate cyclase are linked to separate prostaglandin receptors coupled to G_s and G_i; we postulate that desensitization results from a slow change in the equilibrium between protein components of the adenylate cyclase system.

INTRODUCTION

Cyclic AMP metabolism in intact cells involves its formation by adenylate cyclase and removal by various phosphodiesterase activities. In the continuous presence of a stimulatory agonist the time-course of cyclic AMP formation might be expected to rise to a plateau level, representing a steady-state between synthesis and decay of the compound that can be expressed as:

$$[\text{cAMP}] = k_s/k_d(1 - e^{-k_d \cdot t}) \qquad [1]$$

where k_s is the zero-order rate constant for synthesis and k_d is the first-order rate constant for decay of cyclic AMP, assuming that the substrate [ATP] is saturating with respect to adenylate cyclase and the cyclic AMP level remains below the K_m of the phosphodiesterase.

In most cells, however, the time-course of cyclic AMP formation exhibits a rise to a peak followed by a decline to a stable plateau level (Barber et al., 1978), which generally reflects desensitization since a second challenge with the same agonist results in an attenuated response. For this reason, Barber et al. (1978; 1980) modified Equation 1 to include reversible, time-dependent transformation of adenylate cyclase to an inactive form according to the equilibrium:

$$\text{adenylate cyclase: active} \underset{k_n}{\overset{k_m}{\rightleftharpoons}} \text{adenylate cyclase: inactive}$$

where k_m is a rate constant for conversion to the inactive form and k_n is the first order rate constant for reversal of inactivation. The time-course of cyclic AMP formation can then be described by the expression:

$$[\text{cAMP}] = k_s\left[\frac{k_n(1 - e^{-k_d \cdot t})}{(k_m + k_n)k_d} + \frac{k_m[e^{-(k_m + k_n) \cdot t} - e^{-k_d \cdot t}]}{(k_m + k_n)(k_d - k_m - k_n)}\right] \qquad [2]$$

We have shown previously (Ashby, 1988) that in intact platelets challenged with iloprost, a stable analogue of PGI_2, the shape of the time-course of cyclic AMP metabolism varies as a function of prostaglandin concentration. Low concentrations of iloprost give a rise in cyclic AMP level to a steady-state plateau, while high concentrations give higher initial rates of cyclic AMP formation followed by peak and plateau effects similar to those described by Equation 2. To account for the different prostaglandin concentration dependences of activation and inhibition of cyclic AMP formation we proposed that there are separate stimulatory and inhibitory prostaglandin receptors, and we modified Equation 2 to express k_s (the rate constant for cyclic AMP synthesis) and k_m (the rate constant for desensitization) as separate, saturable functions of prostaglandin concentration. By fitting time-courses to Equation 2, we were able to determine EC_{50} values as a function of prostaglandin concentration for both the stimulatory and the inhibitory effects (Ashby, 1988). Time-courses obtained with intact platelets in the presence of phosphodiesterase inhibitors also showed time-dependent inhibition of cyclic AMP formation with a different prostaglandin concentration from activation, indicating that inhibition of adenylate cyclase was responsible for desensitization rather than time-dependent activation of phosphodiesterase activity. In washed platelet lysates, in the presence of phosphodiesterase inhibitors, prostaglandins also induced a time-dependent inhibition of their own rapid activation of adenylate cyclase with a different concentration dependence from activation (Ashby, 1986).

Despite our ability to model time-courses using Equation 2 there were marked discrepancies between the values of the constants used to fit data in the absence and presence of phosphodiesterase inhibitors, suggesting that other factors needed to be considered in order to describe cyclic AMP metabolism. Indeed, the description of cyclic AMP metabolism provided by Equation 2 is only approximate, in part because of the limiting assumptions used in its derivation that a) the intracellular level of cyclic AMP never exceeds the K_m for the phosphodiesterase and b) there is only one platelet phosphodiesterase activity. Neither assumption is true since prostaglandins can induce a rise in cyclic AMP to almost 2% of total adenine nucleotides, corresponding to a concentration of 150 μM, based on an estimate of about 7.5 mM for the concentration of total adenine nucleotides in the cytoplasmic pool of the platelet (Ashby, 1987), while platelets possess at least two phosphodiesterases with Km values of 0.18 and 50 μM. Consequently Equation 1 should be written as:

$$\frac{d[cAMP]}{dt} = k_s - \frac{V_H[cAMP]}{K_H + [cAMP]} - \frac{V_L[cAMP]}{K_L + [cAMP]} \qquad [3]$$

where V_H and V_L, and K_H and K_L are the maximum velocities and K_m values of the high and low K_m phosphodiesterase activities, respectively. Equation 3 cannot be explicitly integrated and, under these conditions, it is impossible to obtain an integrated rate equation that includes desensitization.

To avoid the limitations of explicit integrated rate equations we have examined cyclic AMP metabolism by use of KINSIM (Barshop et al., 1983), a powerful kinetic simulation program that employs numerical integration. Use of KINSIM also permits more flexibility in modeling time-courses. In particular, we were able to describe the prostaglandin concentration-dependence of stimulation and inhibition more precisely by incorporating apparent negative cooperativity of agonist-receptor interactions that is in accord with actual observations of radioligand binding to membranes and intact cells. Apparent negative cooperativity is thought to arise from high and low affinity receptor states whose existence depends on the activation state of the relevant G protein. KINSIM also provides a convenient method of simulating time-courses obtained with broken cell preparations, allowing us to examine mechanisms of prostaglandin-regulated cyclic AMP formation in more detail.

Computer simulation of time-courses by numerical integration Simulations using KINSIM (Barshop et al., 1983) were performed on a MicroVAX computer (Digital Equipment Corporation). KINSIM is written in FORTRAN and can simulate mechanisms with up to forty individual steps. Mechanisms are written as a series of chemical reactions represented either by a dissociation constant for rapid equilibrium steps or by forward and backward rate constants in kinetically determined steps. The computer generates a table of differential equations and performs direct numerical integration to yield the concentration of the designated species as a function of time.

A model for cyclic AMP metabolism in intact platelets Computer simulations involving cyclic AMP metabolism in intact platelets were based on an identical model to that described for the integrated rate expression (Equation 2) except that the Michaelis constant for phosphodiesterase could be set at any value and other constants could be varied with respect to each other in a more flexible fashion. Hence, again the model involves rapid prostaglandin-mediated activation of adenylate cyclase followed by slow, reversible transformation of adenylate cyclase to an inactive form mediated through a distinct inhibitory receptor. The model is indicated by the series of chemical equilibria in Table I. For simplicity, we treated the receptor/G protein/adenylate cyclase complex as a tightly coupled entity represented as a single enzyme species E. The enzyme E has a binding site for substrate ATP (A) as well as two distinct binding sites for prostaglandin (P). The species EA represents the enzyme substrate complex which has negligible activity; binding of P to a stimulatory binding site gives rise to the species EP and subsequently to EPA which is active, generating EP and cyclic AMP, represented as C. P can also bind to a separate, inhibitory binding site to give the species PE. PE can slowly and reversibly convert to PF, generating the inactive enzyme species F. Once formed, cyclic AMP (C) is removed by phosphodiesterase (D) through the enzyme-substrate complex DC and enters a sink of AMP.

Table I. Equilibria describing prostaglandin-regulated cyclic AMP metabolism in intact platelets

	Step	Fig. 1b	Fig. 2b
1	E + A = EA	0.1%	0.1%
2	E + P = EP	0.021 μM	0.017 μM
3	E + P = PE	0.07 μM	0.07 μM
4	EP + A = EPA	0.1%	0.1%
5	EP + P = PEP	5.0 μM	5.0 μM
6	EPA + P = PEPA	5.0 μM	5.0 μM
7	EA + P = PEA	0.07 μM	0.07 μM
8	PE == PF	0.95; 0.12	0.95; 0.16
9	PEP == PFP	0.95; 0.12	0.95; 0.16
10	PEPA == PFPA	0.95; 0.12	0.95; 0.16
11	PEA == PFA	0.95; 0.12	0.95; 0.16
12	EPA == EP + C	$4x10^6$; 0	$2.8x10^6$;0
13	PEPA == PEP + C	$4x10^6$; 0	$2.8x10^6$;0
14	D + C = DC	-	1%
15	DC == D + N	-	$2.2x10^6$; 0

Constants used to simulate curves in Figs. 1b and 2b are indicated in the appropriate columns. Single equal signs signify rapid equilibrium steps; double equal signs signify kinetically determined steps, where the first number refers to the forward rate constant and the second number refers to the reverse rate constant. All rate constants are in units of min^{-1}. Values given in units of % represent cyclic AMP concentrations expressed as a percent of total adenine nucleotides, so that simulations were directly comparable to real data. ATP level was maintained at 100%, simulating the ability of the cell to replenish ATP through cellular metabolism. The concentration of adenylate cyclase (E) was set at 1×10^{-6} μM so that depletion of free A or P by binding to E was negligible; the rate of cyclic AMP formation is the product of E and the rate constant so that the latter was multiplied by 10^6 to compensate for the enzyme concentration. Similarly, the concentration of phosphodiesterase was set at 1×10^{-6} μM.

RESULTS AND DISCUSSION

Measurement and simulation of prostaglandin-regulated cyclic AMP formation in intact platelets. Time-courses of cyclic AMP formation were obtained over a wide range of iloprost concentration in the presence of high concentrations of the phosphodiesterase inhibitor IBMX (Fig.1a). Under these conditions, only adenylate cyclase activity was being measured and both instantaneous and possible time-dependent effects of phosphodiesterase activity were eliminated. Low concentrations of iloprost gave rise to a linear rate, reflecting the zero-order rate of prostaglandin-stimulated cyclic AMP formation at saturating substrate concentration. The initial rate of cyclic AMP formation increased as a saturable function of prostaglandin concentration, reflecting saturation of a stimulatory prostaglandin receptor. However, high concentrations of prostaglandin caused a time-dependent decrease in the rate of cyclic AMP formation to reach a new steady-state rate. The decrease in rate can be attributed to a time-dependent decrease in adenylate cyclase activity since phosphodiesterase activity is inhibited and it can be described as desensitization of adenylate cyclase since a second challenge with prostaglandins results in an attenuated response (Mills, 1982). The extent of inhibition increases with iloprost with a different concentration dependence from stimulation of cyclic AMP formation.

The simplest way to account for the difference in prostaglandin concentration-dependence of stimulation and inhibition is to propose that there are separate stimulatory and inhibitory receptors and the time-courses could be simulated by considering the two-receptor model indicated in Table 1. It was apparent that treating stimulation and inhibition as simple, independent, saturable processes was not sufficient to describe the experimental data in anything other than general terms. More precise fits to the real data (Fig. 1b) were obtained by inserting apparent negative cooperativity into the binding profile of prostaglandin interaction with the stimulatory and inhibitory sites so that the affinity of each receptor for agonist declines with agonist concentration. This could be achieved by varying the apparent dissociation constants of Steps 2 and 7 with respect to those of Steps 5 and 6, according to the values indicated in Table I. In other words, binding of prostaglandin to the stimulatory site appears to weaken binding to the inhibitory site and *vice versa*. The difference in apparent affinity at low and high prostaglandin concentrations was 70-fold. The family of curves that can be obtained by this addition to the model (Fig. 1b) is virtually superimposable on the real data over four orders of magnitude of prostaglandin concentration (3 nM to 30 μM).

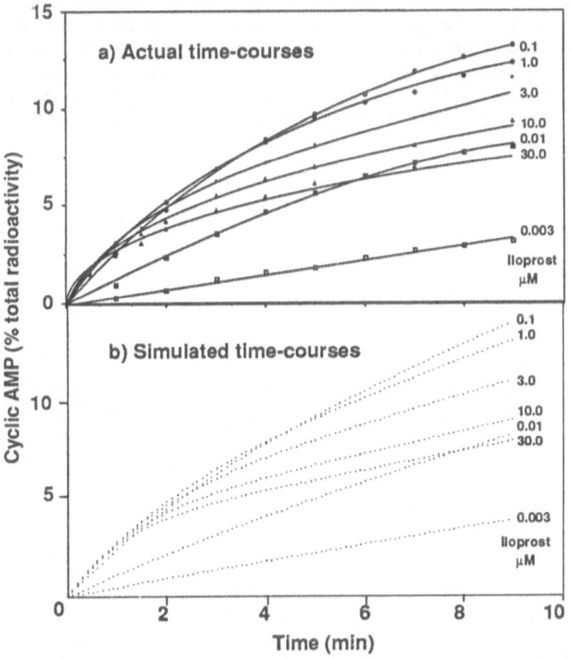

Fig. 1. Actual and simulated time-courses of cyclic AMP formation in intact platelets as a function of iloprost concentration in the presence of 5 mM IBMX. a) Actual time-courses were obtained as follows. Cyclic AMP formation was measured by incubating platelet rich-plasma for 1 h with 2 μCi of [^{3}H]adenine prior to washing. Time-courses were generated at 25°C by addition of prostaglandins to a suspension of radiolabeled, washed platelets, withdrawing 0.5 ml samples at appropriate times into 0.5 ml of stopping solution containing 0.6 M $HClO_4$ and 2% SDS with [^{14}C]cyclic AMP as a recovery standard. The cyclic AMP content of each sample was determined according to Salomon (1979). Results are given as percent of total radioactivity appearing as cyclic AMP. As indicated previously (Ashby, 1988), little cyclic AMP leaked from the cells during the time-course of the experiments reported here. Iloprost was a gift of Berlex Laboratories, Inc., Cedar Knolls, NJ. b) Time-courses were simulated using KINSIM according to the model indicated in Table 1.

At first sight, apparent negative cooperativity appears to be a major addition to the model, implying that perhaps the stimulatory and inhibitory sites interact with one another. However, the apparent negative cooperativity at each receptor can also be treated as independent events and may be explained in terms of current models of agonist/receptor/G protein interaction. Apparent negative cooperativity is generally observed in agonist binding to both stimulatory and inhibitory receptors linked to adenylate cyclase. The explanation for this phenomenon is that the receptor-G protein complex exhibits a high affinity for agonist while activation of the G protein upon binding GTP, dissociates the complex to yield a receptor form with low affinity for agonist (Stadel et al., 1982). It seems reasonable to propose that this mechanism is responsible for the apparent fall in affinity of both stimulatory and inhibitory receptors observed in the current work.

In the absence of phosphodiesterase inhibitors, cyclic AMP is metabolized to AMP and the shape of the time-course changes. Data obtained using iloprost are shown in Fig. 2a. It is possible to model this situation by including phosphodiesterase activity in the simulation. The simulated curves (Fig.2b) clearly contain many of the features of the real data: a) The shape of the time-course varies with prostaglandin concentration; b) At low concentrations of iloprost, the level of cyclic AMP rises to a plateau and shows little decline over 20 min. At higher concentrations of iloprost, the initial rate of cyclic AMP formation is faster than at low concentrations but is followed by a marked decline in cyclic AMP level to a stable plateau; c) Peak and plateau effects increase with prostaglandin concentration to the point that the curves cross.

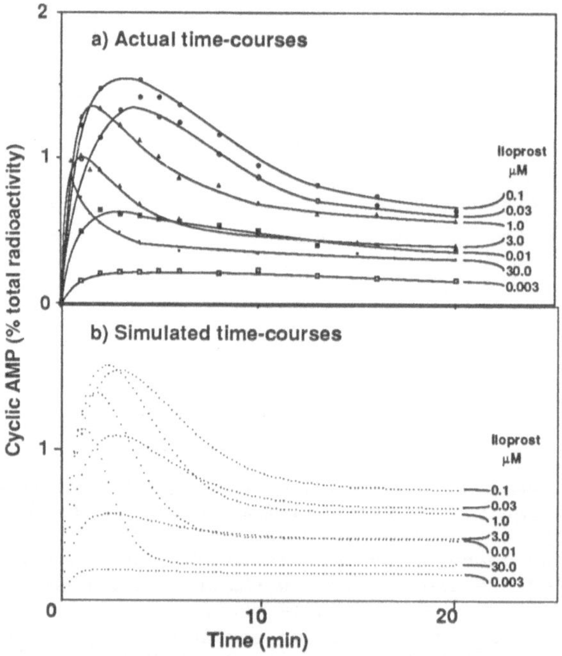

Fig. 2. Actual and simulated time-courses of cyclic AMP formation in intact platelets as a function of iloprost concentration

a) Actual time-courses were obtained as described in the legend to Fig.1.

b) Time-courses were simulated using KINSIM according to the model indicated in Table 1.

Good fits again were obtained by considering apparent negative cooperativity in the agonist binding profile of both the stimulatory and inhibitory receptors. In addition, the K_m of the phosphodiesterase activity could be set at values that approximate those observed in platelets. In this case, the K_m was set at 1% of total platelet nucleotides, corresponding to about 75 µM. Note that we have not attempted to add a second phosphodiesterase activity since we do not know the relative proportions of low and high K_m activities.

We have previously shown that prostaglandin E_1 gives marked peak and plateau effects in cyclic AMP formation in intact platelets. In contrast to iloprost, the shape of the time-course did not change with prostaglandin concentration although the initial rate, peak maximum and final steady-state level all increased as a saturable function of PGE_1 concentration. Analysis of the time-courses in terms of a two-receptor model showed that the EC_{50} and maximal extent of stimulation by PGE_1 were similar to those for iloprost while the EC_{50} for inhibition was lower than that for stimulation (Ashby, 1988). Based on a two-site model, in which saturation of stimulatory and inhibitory receptors by either prostaglandin results in the same final extent of activation and stimulation, it is possible to predict that saturating levels of iloprost should give rise to the same time-course as saturating levels of PGE_1. Hence curves obtained at low concentrations of PGE_1 or iloprost bear no resemblance to each other, while curves obtained at high concentrations are superimposable (data not shown).

<u>Evidence that G_i is involved in prostaglandin-mediated inhibition of adenylate cyclase</u> The presence of a prostaglandin inhibitory receptor implies that inhibition is mediated through the guanine nucleotide-binding regulatory protein G_i. In most cells G_i function can be probed specifically by use of pertussis toxin, however, the toxin is not effective against intact platelets since they apparently lack the appropriate gangliosides for toxin entry. Jakobs and coworkers (Jakobs et al., 1985; Katada et al., 1985; Watanabe et al., 1985) have shown that phorbol ester treatment of platelets leads to protein kinase C mediated phosphorylation of G_i and impairment of its function. We have shown that phorbol ester also abolishes time-dependent inhibition induced by prostaglandins (Ashby, 1988). However, the picture is confused by the fact that phorbol ester treatment also apparently inhibits prostaglandin-stimulated adenylate cyclase activity and may also activate platelet phosphodiesterase activity. Consequently, as an additional probe for G_i involvement in prostaglandin-mediated inhibition, we have treated platelets with N-ethylmaleimide, which is a fairly selective modifier of G_i. Time-dependent inhibition induced by iloprost or PGE_1 was markedly attenuated by pretreatment of platelets with N-ethylmaleimide.

<u>Examination of the mechanism of prostaglandin-mediated inhibition of adenylate cyclase in platelet lysates</u> The kinetics of cyclic AMP formation by washed platelet lysates in response to prostaglandins were similar to those observed with intact platelets in the presence of phosphodiesterase inhibitors. Time-courses obtained in the presence of Gpp(NH)p and various concentrations of iloprost are shown in Fig. 3a. In the absence of iloprost, the time-course showed a lag before reaching a steady-state linear rate after about 10 min. The lag may be attributed to displacement of tightly bound GDP from the α-subunit of G_s, with subsequent dissociation of the αβγ complex. Increasing concentrations of iloprost progressively abolished the lag, presumably by increasing the rate of exchange of GDP for Gpp(NH)p, so that at 0.01 µM iloprost the initial rate was linear. At higher concentrations of iloprost, however, the time-courses showed a time-dependent fall in the rate of cyclic AMP formation to give new, linear steady-state rates. The extent of inhibition was a saturable function of iloprost concentration and, at 30 µM iloprost, the final steady-state rate of cyclic AMP formation was the same as the final rate observed in the absence of iloprost.

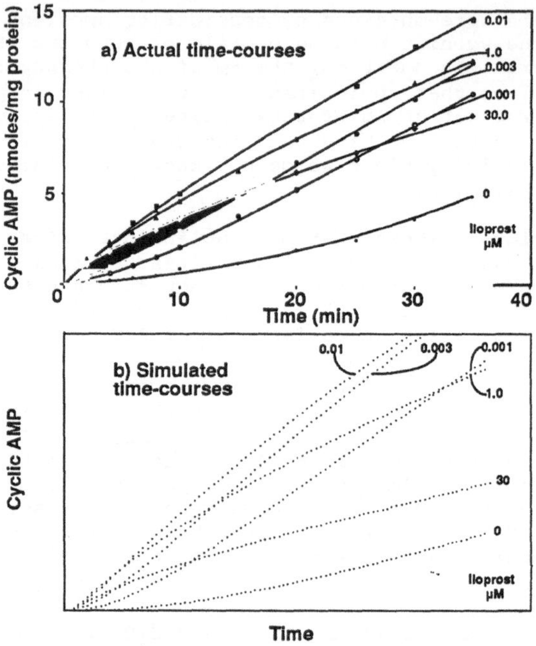

Fig. 3. Actual and simulated time-courses of iloprost-stimulated cyclic AMP formation by platelet lysates in the presence of Gpp(NH)p.

a) Actual time time-courses were generated using washed freeze/thaw lysates of platelets (Ashby, 1986) in an assay mix containing 100 µM [^{3}H]ATP, 10 mM Gpp(NH)p, 10 mM $MgCl_2$, 150 mM NaCl, and various concentrations of iloprost, in 25 mM Hepes, pH 7.4.

b) Time courses were modeled by considering the reactions indicated in the following scheme in which irreversible dissociation of $\alpha_s\beta\gamma$ and $\alpha_i\beta\gamma$ (perhaps rate-limited by GDP-Gpp(NH)p exchange) is induced in a time and iloprost concentration manner.

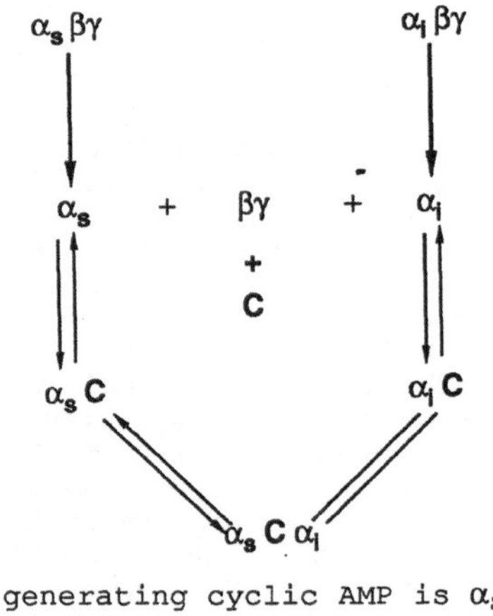

The only active species generating cyclic AMP is $\alpha_s C$, where C is the catalytic unit of adenylate cyclase.

The pattern of curves can be explained by a two-receptor model and it is possible to measure the initial and final steady-state rates of cyclic AMP formation to obtain EC_{50} values for the stimulatory and inhibitory processes

(Ashby, 1986). Furthermore, the ability to observe time-dependent inhibition in broken cell preparations, similar to that observed in the intact cell, allows a more detailed investigation of the molecular mechanism of the inhibition. In this regard, we have found no evidence for the involvement of protein kinases in time-dependent inhibition. Hence, time-dependent inhibition could not be induced by incubating lysates with cyclic AMP dependent protein kinase or partially purified protein kinase C; nor could it be prevented by treatment with inhibitors of these enzymes. Time-dependent inhibition was only observed in the presence of non-hydrolyzable GTP analogues but not in the presence of GTP itself, suggesting the involvement of G proteins. In addition, Katada et al. (1984) have shown that the purified β-subunit of G_i can cause slow inhibition of adenylate cyclase in platelet membranes. In fact, there is plenty of room to explain time- and agonist concentration-dependent inhibition in terms of current models of receptor-G protein-adenylate cyclase interactions. Receptor-mediated effects on G_s and G_i are thought to cause dissociation of both proteins into their α and βγ components. Free α_s binds to and activates the catalytic unit, while inhibition may involve either a direct effect of α_i on the catalytic unit or a mass action effect of βγ in competing with the catalytic unit for α_s (Gilman, 1984). In the absence of agonists, dissociation of the αβγ complex is slow and the process is accelerated by agonists. It is possible that dissociation of the α_s-catalytic unit is also slow. Simulations of a mechanism incorporating some of these slow changes in G protein structure are presented in Fig. 3b and they reproduce the general features of the actual data.

It is clear that the events involved in regulation of G protein interactions are much more complicated than indicated, however, the simulations presented in Fig. 3b suggest that time-dependent activation and inhibition can be explained in terms of rearrangement of the various protein components of the adenylate cyclase system. The availability of purified components means that such models are experimentally testable. We conclude that activation and time-dependent inhibition (desensitization) of platelet adenylate cyclase are linked to separate prostaglandin receptors coupled to G_s and G_i; we postulate that desensitization results from a slow change in the equilibrium between protein components of the adenylate cyclase system.

Acknowledgements - This work was supported by NIH grant HL 36579. I am grateful to Ms. Eve Wernick for technical assistance; to Dr. Carl Frieden for providing a copy of KINSIM; and to Ethan Waldman of the Computer Center at Temple University and Dr. James L. Daniel for help in installing KINSIM.

REFERENCES

Ashby, B., 1986, Kinetic evidence indicating separate stimulatory and inhibitory prostaglandin receptors on platelet membranes. J. Cyc. Nucl. Protein Phos. Res.,11: 291.

Ashby, B., 1987, Nucleotide Metabolism, in Platelet Responses and Metabolism, Vol. II, H. Holmsen, ed. CRC Press, Boca Raton, FL p. 215.

Ashby, B., 1988, Cyclic AMP turnover in response to prostaglandins in intact platelets: evidence for separate stimulatory and inhibitory prostaglandin receptors. Second Messengers and Phosphoproteins, 12: 45.

Barber, R., Clark, R.B., Kelly, L.A., and Butcher, R.W., 1978, A model of desensitization in intact cells. Adv. Cyc. Nucl. Res. 9: 507.

Barber, R., Ray, K.P., and Butcher, R.W., 1980, Turnover of adenosine 3',5'-monophosphate in WI-38 cultured fibroblasts. Biochemistry, 19: 2560.

Barshop, B.A, Wrenn, R.F., and Frieden, C., 1983, Analysis of numerical methods for computer simulation of kinetic processes: development of KINSIM - a flexible, portable system. Analytical Biochem., 130: 134.

Gilman, A.G., 1984, Guanine nucleotide-binding regulatory proteins and dual control of adenylate cyclase. J. Clin. Invest., 73: 1.

Jakobs, K.H., Bauer, S., and Watanabe, Y., 1985, Modulation of adenylate cyclase of human platelets by phorbol ester. Eur. J. Biochem., 151: 425.

Katada, T., Bokoch, G.M., Northup, J.K., Ui, M., and Gilman, A.G., 1984, The inhibitory guanine nucleotide-binding regulatory component of adenylate cyclase. J. Biol. Chem., 259: 3568.

Katada, T., Gilman, A.G., Watanabe, Y., Bauer, S. and Jakobs. K.H., 1985, Protein kinase C phosphorylates the inhibitory guanine-nucleotide-binding regulatory component and apparently suppresses its function in hormonal inhibition of adenylate cyclase. Eur. J. Biochem., 151: 431.

Mills, D.C.B., 1982, The role of cyclic nucleotide in platelets in Handbook of Experimental Pharmacology, Volume 58/II (Kebabian, J.W. and Nathanson, J.A., eds.) p. 732, Springer-Verlag.

Salomon, Y., 1979, Adenylate cyclase assay. Adv. Cyc. Nucl. Res., 10: 35.

Stadel, J.M., De Lean, A., and Lefkowitz, R.J., 1982, Molecular mechanisms of coupling in hormone receptor-adenylate cyclase systems. Adv. Enzymol., 53: 1.

Watanabe, Y., Horn, F., Bauer, S., and Jakobs, K.H., 1985, Protein kinase C interferes with N_i-mediated inhibition of human platelet adenylate cyclase. FEBS Lett., 192: 23.

THE FAMILY OF G PROTEINS

Eva J. Neer*, Yung-Kang Chow*, Suzanne Garen-Fazio,
Thomas Michel*, Carl J. Schmidt*, and Seth Silbert*

Departments of Medicine*, and Neurobiology
Harvard Medical School
Boston, Mass. 02115

MEMBERS OF THE G PROTEIN FAMILY

The guanine nucleotide binding proteins, which couple receptors for light, hormones, neurotransmitters, chemotaxis factors to a variety of intracellular effectors, are a family of heterotrimeric proteins composed of α, ß and γ subunits (reviewed by Gilman, 1987; Lochrie and Simon, 1988; Neer and Clapham, 1988). Each of these subunits consists of a super gene family with many members. The α subunits bind guanine nucleotides and all have an intrinsic GTPase activity which hydrolyzes bound GTP to GDP. This enzymatic activity is central to the function of the G proteins since they are active in the GTP-liganded form and inactive in the GDP-liganded form. Activation of the guanine nucleotide binding proteins by guanosine nucleotide triphosphate causes dissociation of the heterotrimer into α and ßγ subunits. Although the molecule is heterotrimeric, it is functionally a heterodimer because the ß and γ subunits do not dissociate in the native state. The ß and γ peptides can only be separated under strongly denaturing conditions (Neer, unpublished).

At present, there are 12 known types of α subunits and the list is still growing (see Table I). The easiest way to classify the α subunits is according to their ability to be modified by a bacterial toxin. Both cholera toxin and pertussis toxin transfer ADP-ribose from NAD to the α subunit of the G proteins. Although the reaction catalyzed is similar, the site of modification is different for the two toxins. Cholera toxin modifies an arginine at position 201 of the α subunit which stimulates adenylate cyclase (Van Dop et al., 1984). This α subunit also regulates calcium channels (Yatani et al., 1987). Pertussis toxin modifies another class of G proteins whose α subunits, named α_{i-1}, α_{i-2}, α_{i-3} and α_o, are involved in inhibition of adenylyl cyclase, regulation of several ion channels and are probably involved in regulation of phospholipase C activity (reviewed by Gilman, 1987; Neer and Clapham, 1988). The guanine nucleotide binding protein which couples the light-activated receptor molecule in the retina, rhodopsin, to a cyclic GMP phosphodiesterase, can be modified by both bacterial toxins (Abood, et al., 1982; West et al., 1985). In transducin, cys 347 is the site of ADP-ribosylation by

TABLE 1. THE FAMILY OF G PROTEIN α SUBUNITS

MODIFYING TOXIN	SUBSTRATE	ALTERNATIVE SPLICING IN CODING REGION	DISTRIBUTION	TARGET ENZYMES AND PROTEINS
Cholera Toxin	α_s	Yes (4 forms)	Ubiquitous	Adenylyl cyclase Ca^{++} channels Mg^{++} transport
	α_{olf}	No	Olfactory neuroepithelium	Adenylyl cyclase
Pertussis Toxin	α_{i-1}	No	Absent from some cells	Phospholipase C Ion channels Adenylyl cyclase
	α_{i-2}	No	Ubiquitous	
	α_{i-3}	No	Ubiquitous	
	α_o	No	Major neural α subunit	
Both Toxins	α_{t-1}	No	Retinal rods	cGMP phospho-diesterase
	α_{t-2}	No	Retinal cones	cGMP-phospho-diesterase
Neither Toxin (?)	α_z	No	More in brain than other tissues	Unknown

pertussis toxin (West et al., 1985). In addition, there is an α subunit of as yet unknown function (α_z) which may be a substrate for neither toxin (Fong et al., 1988; Matsuoka et al., 1988). A recently identified guanine nucleotide binding protein in olfactory neuroepithelium (Jones and Reed, 1989) is also a substrate for cholera toxin.

So far, three types of ß subunit are known. The first to be cloned was $ß_1$ which is the 36 kDa form which predominates in most tissues (Codina et al., 1986; Fong et al., 1986). It is the only form of ß subunit in the retina. $ß_2$ is a 35 kDa form of the ß subunit which is a minor component in most tissues, and is absent from the retina (Fong et al., 1987; Gao et al., 1989). $ß_3$ is a newly identified ß subunit first reported at the present meeting.[1] $ß_1$ and $ß_2$ are approximately 90% identical to each other. $ß_3$ is more different and is 80% identical to the other two. We have recently identified a third ß subunit in bovine brain. We do not know whether this is the newly reported $ß_3$ or still another form (Neer, unpublished).

Less is known about the γ subunit. The γ subunit from retina is immunologically and structurally distinct from the γ subunit associate d with ßs in nonretinal cells. A nonretinal γ, which is considerably different from the retinal form, has recently been cloned by M. Simon and his colleagues (Gautam et al., 1989). By immunologic criteria, there are probably at least two or three other nonretinal γs whose structure is, as yet, unknown (Evans et al., 1987). Because ß and γ cannot be separated in the native state, nothing is known about the individual functions of ß or γ subunits.

SPECIFICITY OF SUBUNIT FUNCTIONS

The existence of so many similar subunits raises a number of questions about the specificity of their action in coupling receptors to effectors. It is clear that all of the α subunits identified so far are able to function in activating more than one effector system. For example, α_s can activate adenylyl cyclase directly (Bender et al., 1984; May et al., 1985). It also regulates ion channels. Similarly, in the pertussis toxin family, multiple subunits are able to interact with receptors with apparently equal efficacy (for example, see Florio and Sternweis, 1985). Similarly, effectors can be regulated by more than one type of α subunit. For example, the muscarinic-gated K^+ channel in cardiac atria can be activated by all three α_i subtypes and also by α_o (Logothetis et al., 1988; Yatani et al., 1988). In vitro reconstitutions have uniformly shown a lack of specificity of G proteins for receptors or effectors. The lack of specificity is not absolute since, for example, only α_s can activate adenylyl cyclase, but none of the pertussis toxin substrates can do so. Nevertheless, within a particular category either of cholera toxin substrates or pertussis toxin substrates, a great deal of interchange is possible (reviewed by Neer and Clapham, 1988).

1Levine MA, Smallwood PM, Moen P, Helman LJ, Modi W. Identification of a cDNA encoding a third form of the ß subunit of GTP-binding regulatory proteins. Biology of Cellular Transducing Signals, 1989, No, 57, Ninth International Washington Spring Symposium, Washington, DC, May 8-12, 1989.

DISTRIBUTION OF G PROTEIN α SUBUNITS

How does the cell handle this potential for crosstalk? One solution is that taken by the retina where G proteins are strictly segregated into different cells. Thus, the rods contained rhodopsin, transducin-1 and a rod-specific type of cGMP phosphodiesterase (Lerea et al., 1986). In contrast, the cones contained different opsins, their own G protein, transducin-2, and their own cGMP phosphodiesterase. However, the strict segregation of G proteins into different cells is not the rule. Using specific probes for the three human α_i subtypes, we have shown that clonal cell lines are able to express more than one type of α subunits (Kim et al., 1988). Most cell lines express all three of the α_i subtypes, although mRNA levels vary from one cell to another. Among the α_i subtypes, the most variable is α_{i-1} which is not expressed in most cells of myelocytic lineage and is also missing in some other cell types, for example GH_4 cells, a pituitary cell line (Neer, unpublished).

Another example of a G protein with greatly skewed cellular distribution is α_o. This 39 kDa G protein was first described and purified by our laboratory (Neer et al., 1984), and independently and simultaneously purified by Sternweis and Robishaw (1984). The protein is especially abundant in the brain where it makes up approximately 0.5% of brain particulate protein (Huff et al., 1985). In peripheral tissues, the α_o protein is present at one-tenth or less the concentration in the central nervous system. A concentration of 0.5% of particulate protein is already high but, in fact, the concentration in some cells is even higher. An immunocytochemical study performed by Worley et al. (1986), using antibody to α_o generated in our laboratory, showed that α_o was not uniformly distributed throughout the central nervous system, but concentrated in particular regions. In those regions, it was found in the neuropil rather than in cell bodies. If G_o is, in fact, concentrated in particular cells rather than distributed uniformly throughout the central nervous system, then its concentration in those cells where it is present, it would be ten times higher than the average concentration. That would make the protein approximately 5% of particulate protein in some cell types. Analysis by Western blots shows that the amount of $\beta\gamma$ in the central nervous system is at least equal to that of the α subunits, so there seems to be enough $\beta\gamma$ to go around (Huff et al., 1985). It is not clear why a protein whose ostensible function is to amplify a transmembrane signal should exist in such high concentrations in some cells. The observation suggests that perhaps the α_o protein has functions different from those of the signal transducing G proteins, but it is not yet known what such other functions might be.

DEGRADATION RATES OF α_o IN CARDIOCYTES AND GH_4 CELLS

One of the puzzles about α_o is that although it is an extremely abundant protein in the central nervous system, the amount of mRNA is extremely low, making it very difficult to detect (Jones and Reed, 1987; Silbert et al., submitted). Furthermore, the amount of mRNA is similar in cells which express very high amounts of α_o protein such as those in the brain, and cells which express much lower amounts of the protein such as heart or pituitary cells. One possibility for the low levels of mRNA might be that α_o is an extremely stable protein. For this reason, we investigated the degradation rate of α_o in two cell types, GH_4 cells and cardiac cells. Two cell types have approximately equal amounts of mRNA for α_0 and approximately equal amounts of α_0 protein (expressed as α_0 per milligram of total protein). Nevertheless, the degradation rate for α_0 is very different between GH_4 cells and cardiocytes. In the former, the half-time for degradation is approximately 28 hours which is the average degradation rate for most membrane proteins. In contrast, in cardiocytes, the half-time for degradation could not be accurately measured because it was over 72 hours. Since the steady-state levels of α_0 protein are similar in the two cell types but the degradation rates differ by a factor of approximately 3, the rates of synthesis of α_0 protein must be different in the two cells, being higher in GH_4 cells than in cardiocytes. The different rates of synthesis occur despite similar levels of mRNA for α_0 in the two cell types, suggesting that translational control may be one of the mechanisms which regulate levels of α_0 protein (Silbert et al., submitted).

CHARACTERIZATION OF A DROSOPHILA α_0 HOMOLOGUE

Although α_0 has been implicated in the regulation of a variety of different types of ionic channels (Hescheler et al., 1987; Van Dongen et al., 1988), the full range of its function is not yet known. For this reason, it would be extremely useful to have a system which could be manipulated genetically in order to try to generate use which could elucidate the function of this protein. To this end, C. Schmidt screened a genomic library from Drosophila to try to determine whether an analogue of α_0 existed in that organism. He was able to clone first genomic DNA and then a cDNA clone which encoded an α subunit that was 81% identical to the mammalian α_0 protein. The α subunit predicted by this cDNA was more similar to α_0 than it was to any mammalian α_i or α_s, and also more similar to α_0, than it was to another Drosophila α subunit recently described by Provost et al. (1988). Hybridization of a fragment of cDNA from the 3'untranslated region showed that the Drosophila α_0 mRNA was located in the cell bodies of the central nervous system, in the thoracic and abdominal ganglia, and in the ovary. Y.-K. Chow expressed the Drosophila α_0 protein as a fusion protein E. coli and generated polyclonal antibodies to Drosophila α_0. Immunocytochemical studies by S. Garen-Fazio showed that the Drosophila α_0 is located in the neuropil of the central nervous system and the abdominal ganglion just as the mammalian form of the protein is located in the neuropil of certain parts of the mammalian central nervous system. These concordances in sequence and immunocytochemical localization suggests that in Drosophila, as in mammals, the α_0 protein is especially important to neuronal function, and suggests further, that elucidating the functional role of α_0 in Drosphila may provide important clues to the function in higher organisms (C. Schmidt et al., submitted).

FUNCTIONS OF THE $\beta\gamma$ SUBUNIT

The stress so far in this chapter has been on the function of α subunits. However, as noted in the introduction, activation of a G protein leads to dissociation of α from $\beta\gamma$ subunits. This dissociation potentially generates two molecules which could each act as activators of different effector systems. An example of such a bifurcating pathway is the regulation of the cardiac K^+ channel. Approximately 2 years ago,

Drs. Birnbaumer, Brown and their colleagues demonstrated that α subunits could regulate the muscarinic K^+ channel (Codina et al., 1987) while Dr. David Clapham's laboratory, in collaboration with our laboratory, showed that ßγ subunits were able to activate the same channel (Logothetis et al., 1987). Initially, both laboratories interpreted their experiments as showing direct activation of the channel by the α or the ßγ subunits. Subsequent work has shown that activation of the channel, by the ßγ subunit, is not direct, but involves the generation of lipid bound second messengers (Kim et al., 1989). The complexity of action of the ßγ subunit emphasizes the fact that reconstitution of channel regulation in excised patches of cell membrane is not the same as reconstituting pure components, and that the results of experiments in patches cannot prove the direct interaction of any subunit with an effector. Although a patch is very small, it is very large compared to the size of an ion channel and contains many enzymes and proteins. The excised patches can certainly be used to show that no <u>soluble</u> second messengers are required for activation of channels by G protein subunits, but cannot prove that the subunit directly interacts with the channel polypeptide itself.

Subsequent experiments by Logothetis et al. (1988) showed that both α and ßγ subunits are able to activate the K^+ channel in inside out patches, and that different α subunits are able to do so. This was confirmed by Brown and Birnbaumer who showed that all three α_i subtypes were able to activate the K^+ channel with equal potency. The α and ß subunits are not additive in their ability to activate the K^+ channel. This result showed that the two subunits are able to activate the same population of channels in the patch. Although the two subunits are able to activate the same channel, the concentration dependence for the two is quite different, with α subunits acting in the 1-10 pM range while ßγ subunits begin to activate at 200 pmol with maximal activity in the 1-10 nM range. The reason for the difference in concentration dependenc e is not known. It may reflect different ways in which the two subunits interact with the membrane. The ßγ subunit from some tissues is quite hydrophobic and is partially aggregated at the low detergent concentrations which are necessary to prevent disruption of the excised patch. If only unaggregated ßγ subunits can activate the channel, then the actual concentration of effective ßγ subunits is lower than the total concentration.

The observation by Jelsema and Axelrod (1987) that ßγ subunits could activate phospholipase A_2 in the retina, together with observations that arachidonic acid metabolites could activate the muscarinic K^+ channel, led Clapham and his colleagues to test the hypothesis that phospholipase A_2 activation might be the mechanism by which ßγ subunits lead to the activation of K^+ channels (Kim et al., 1989). Antibodies against phospholipase A_2 generated by Dr. Dafna Bar-Sagi, Cold Spring Harbor, block activation of the K^+ channel by ßγ. However, they do not block activation of the channel by acetylcholine. This result is not unexpected since Logothetis et al. (1988) had previously shown that the effects of α and ßγ are not additive, and that the channels are maximally activated by either pathway. Since activation of G proteins by the muscarinic receptor would generate both α and ßγ subunits, blockade of ßγ activity would not be expected to block activation of the channel through the simultaneously produced α subunit. However, it is not known whether the ßγ subunit is actually involved in regulating the channel through the muscarinic receptor or whether it is the means by which other factors modulate K^+ channel activity.

The idea that ßγ subunits may act as positive activators of cellular effector enzymes may turn out to be a general one. For example, Whiteway et al. (1989) have shown by genetic means that ßγ is a positive element in the response of the yeast <u>Saccharomyces cerevisiae</u> to mating pheromones. Deletion of the ß or γ genes blocks the response while disruption of the α subunit is lethal in haploid cells, presumably because the pheromone response pathway, which includes cell cycle arrest, is constitutively expressed. Thus, Whiteway et al. (1989) suggest that the free ßγ subunit may mediate the pheromone response.

CONCLUSION

In vitro experiments alluded to earlier in this chapter suggest that there is considerable potential for crosstalk among the different G protein subunits. Nevertheless, in cells which contain multiple different kinds of G protein α and ßγ subunits, the response to a particular hormone is limited. Therefore, in the cell, a variety of other mechanisms may operate to enhance the specificity of the response. The stoichometry of the various α and ßγ subunits is likely to be important, as well as the relative affinity of different α subunits for ßγ. The stoichometry and relative subunit affinity are especially important if the components of the transmembrane signalling system are able to move freely and independently in the plane of the lipid bilayer. While there is evidence that free diffusion may occur in some cells, there is also evidence for physical or functional compartmentation of receptors, G proteins, and effectors (reviewed by Neer and Clapham, 1988). Interaction with the cytoskeleton or with other cellular proteins could constrain the freedom of motion of the hormone-responsive system, although dissociation of components might still occur within a confined space. If the different α and ßγ subunits cannot exchange, they would not be able to crosstalk. It is likely that different cells use different combinations of these potential mechanisms to ensure the specificity of hormonal responses. The present challenge is to define how G proteins actually function in situ and how the general principles developed through analysis of purified proteins are applied in living cells.

ACKNOWLEDGMENT

The work reported in this paper was supported by NIH grants GM36259 and GM35417 to EJN; a Clinician-Scientist Award from the American Heart Association to TM; NSF grant BNS8508135 to SGF; a Dennison Manufacturing Company Fellowship to Y-KC; and a Pharmacology Training grant (NS07009) Award to CJS.

REFERENCES

Abood, M.E., Hurley, J.B., Pappones, M.C., Bourne, H.R., and Stryer, L., 1982, Functional homology between signal-coupling proteins. Cholera toxin inactivates the GTPase activity of transducin, J Biol Chem., 257:10540.

Bender, J., Wolf, L.G., and Neer, E.J., 1984, Interaction of forskolin with resolved adenylate cyclase components. Adv Cyc Nucl Res., 17:101.

Codina, J., Stengel, D., Woo, S.L.C., and Birnbaumer, L., 1986, ß-Subunits of the human liver G_s/G_i signal-transducing proteins and those of bovine retinal rod cell transducin are identical, FEBS Lett., 207:187.

Codina, J., Yatani, A., Grenet, D., Brown, A.M., and Birnbaumer, L., 1987, The α subunit of the GTP-binding protein G_k opens atrial potassium channels. Science, 238:1288.

Evans, T., Fawzi, A., Fraser, E.D., Brown, M.L., and Northup, J.K., 1987, Purification of a $ß_{35}$ form of the ßγ complex common to G-proteins from human placental membranes, J Biol Chem., 262:176.

Florio, V.A., and Sternweis, P.C., 1985, Reconstitution of resolved muscarinic cholinergic receptors with purified GTP-binding proteins, J Biol Chem., 260:3477.
Fong, H.K.W., Amatruda, T.T., Birren, B.W., and Simon, M.I., 1987, Distinct forms of the ß subunit of GTP-binding regulatory proteins identified by molecular cloning, Proc Natl Acad Sci USA., 84:3792.
Fong, H.K.W., Hurley, J.B., Hopkins, R.S., Miake-Lye, R., Johnson, M.S., Doolittle, R.F., and Simon, M.I., 1986, Repetitive segmental structure of the transducin ß subunit: Homology with the CDC4 gene and identification of related mRNAs, Proc Natl Acad Sci USA., 83:2162.
Fong, H.K.W., Yoshimoto, K.K., Eversole-Cire, P., and Simon, M.I., 1988, Identification of a GTP-binding protein α subunit that lacks an apparent ADP-ribosylation site for pertussis toxin, Proc Natl Acad Sci USA., 85:3066.
Gao, B., Gilman, A.G., and Robishaw, J.D., 1987, A second form of the ß subunit of signal-transducing G proteins, Proc Natl Acad Sci USA., 84:6122.
Gautam, N., Baetscher, M., Aebersold, R., and Simon, M.I., 1989, A G protein gamma subunit shares homology with ras proteins, Science, 244:971.
Heschler, J., Rosenthal, W., Trautwein, W., and Schultz, G., 1987, The GTP-binding protein G_o regulates neuronal calcium channels. Nature (Lond)., 325:445.
Huff, R.M., Axton, J.M., and Neer, E.J., 1985, Physical and immunological characterization of a guanine nucleotide-binding protein purified from bovine cerebral cortex, J Biol Chem., 260:10864.
Jelsema, C.L., and Axelrod, J., 1987, Stimulation of phospholipase A_2 activity in bovine rod outer segments by the ßγ subunits of transducin and its inhibition by the α subunit, Proc Natl Acad Sci USA., 84:3623.
Jones, D.T., and Reed, R.R., 1987, Molecular cloning of five GTP-binding protein cDNA species from rat olfactory neuroepithelium, J Biol Chem., 262:14241.
Jones, D.T., and Reed, R.R., 1989, G_{olf}: An olfactory neuron specific-G protein involved in odorant signal transduction, Science, 244:790.
Kim, D., Lewis, D.L., Graziadei, L., Neer, E.J., Bar-Sagi, D., and Clapham, D.E., 1989, G-protein ßγ-subunits activate the cardiac muscarinic K^+-channel via phospholipase A_2, Nature, 337:557.
Kim, S.Y., Ang, S.-L., Bloch, D., Bloch, K., Kawahara, Y., Lee, R.L., Tolman, C., Seidman, J.G., and Neer, E.J., 1988a, Molecular characterization of cDNA encoding a new human GTP-binding protein, Proc Natl Acad Sci USA., 85:4153.
Kim, S., Ang, S.-L., Bloch, D.B., Bloch, D.D., Kawahara, Y., Tolman, C., Lee, R., Seidman, J.G., and Neer, E.J., 1988b, Identification of cDNA encoding an additional α subunit of a human GTP-binding protein: Expression of three α_i subtypes in human tissues and cell lines, Proc Natl Acad Sci USA., 85:4153.
Lerea, C.L., Somers, D.E., Hurley, J.B., Klock, I.B., and Bunt-Milam, A.M., 1986, Identification of specific transducin α subunits in retinal rod and cone photoreceptors, Science, 234:77.
Lochrie, M.A., and Simon, M.I., 1988, G-protein multiplicity in signal transduction systems, Biochemistry, 27:4957.
Logothetis, D.E., Kim, D., Northup, J.K., Neer, E.J., and Clapham, D.E., 1988, Specificity of action of guanine nucleotide-binding regulatory protein subunits on the cardiac muscarinic K^+ channel, Proc Natl Acad Sci USA., 85:5814.
Logothetis, D.E., Kurachi, Y., Galper, J., Neer, E.J., and Clapham, D.E., 1987, The ßγ subunits of GTP-binding proteins activate the muscarinic K^+ channel in heart, Nature, 325:321.
Matsuoka, M., Itoh, H., Kozasa, T., and Kaziro, Y., 1988, Sequence analysis of cDNA and genomic DNA for a putative pertussis toxin-

insensitive guanine nucleotide-binding regulatory protein α subunit, Proc Natl Acad Sci USA., 85:5384.
May, D.C., Ross, E.M., Gilman, A.G., and Smigel, M.D., 1985, Reconstitution of catecholamine-stimulated adenylate cyclase activity using three purified proteins, J Biol Chem., 260:15829.
Neer, E.J., and Clapham, D.E., 1988, Roles of G-proteins in transmembrane signalling, Nature (Lond)., 333:129.
Neer, E.J., Lok, J.M., and Wolf, L.G., 1984, Purification and properties of the inhibitory guanine nucleotide regulatory unit of brain adenylate cyclase, J Biol Chem., 259:14222.
Provost, N.M., Somers, D.E., and Hurley, J.B., 1988, A Drosophila melanogaster G protein α subunit gene is expressed primarily in embryos and pupae, J Biol Chem., 263:12070.
Schmidt, C.J., Garen-Fazio, S., Chow, Y.-K., and Neer, E.J., 1989, Neuronal expression of a newly identified Drosophila melanogaster G protein α_o subunit, Submitted.
Silbert, S., Michel, T., Lee, R., and Neer, E.J., 1989, Differential degradation rates of the G protein α_o in cultured cardiac and pituitary cells, Submitted.
Sternweis, P.C., and Robishaw, J., 1984, Isolation of two proteins with high affinity for guanine nucleotides from membranes of bovine brain, J Biol Chem., 259:13806.
Van Dongen, A.M., Codina, J., Olate, J., Matera, R., Joho, R., Birnbaumer, L., and Brown, A.M., 1988, Newly identified brain potassium channels gated by the guanine nucleotide binding protein G_o, Science, 242:1433.
Van Dop, C., Tsubokawa, M., Bourne, H.R., and Ramachandran, J., 1984, Amino acid sequence of retinal transducin at the site ADP-ribosylated by cholera toxin, J Biol Chem., 259:696.
West, R.E., Jr., Moss, J., Vaughan, M., Liu, T., and Liu, T.-Y., 1985, Pertussis toxin-catalyzed ADP-ribosylation of transducin, J Biol Chem., 260:14428.
Whiteway, M., Hougan, L., Dignard, D., Thomas, D.Y., Bell, L., Saari, G.C., Grant, F.J., O'Hara, P., and MacKay, V.L., 1989, The STE4 and STE18 genes of yeast encode potential ß and γ subunits of the mating factor receptor-coupled G protein, Cell, 56:467.
Worley, P.F., Baraban, J.M., Van Dop, C., Neer, E.J., and Snyder, S., 1986, G_o, a guanine nucleotide-binding protein: Immunohistochemical localization in rat brain resembles distribution of second messenger systems, Proc Natl Acad Sci USA., 83:4561.
Yatani, A., Codina, J., Imoto, Y., Reeves, J.P., Birnbaumer, L., and Brown, A.M., 1987, A G protein directly regulates mammalian cardiac calcium channels, Science, 238:1288.
Yatani, A., Mattera, R., Codina, J., Graf, R., Okabe, K., Padrell, E., Iyengar, R., Brown, A.M., and Birnbaumer, L., 1988, The G protein-gated atrial K^+ channel is stimulated by three distinct G_i α-subunits, Nature, 336:680.

ROLES OF G PROTEIN SUBUNITS IN COUPLING OF RECEPTORS TO IONIC CHANNELS AND OTHER EFFECTOR SYSTEMS

Lutz Birnbaumer, Atsuko Yatani, Rafael Mattera,
Juan Codina, and Arthur M. Brown

Departments of Cell Biology and Molecular Physiology
and Biophysics; Baylor College of Medicine, Houston, Texas

INTRODUCTION

G proteins play a central role in coupling receptors to effector systems. Like receptors, which are increasing in number rapidly, effectors affected by the activated forms of G proteins are also increasing, most notably through the discovery in 1986-1987 that ionic channels form part of the family of molecules regulated by G proteins. Using cell free systems such as provided by excision of membrane patches from cells and incorporation of plasma membrane vesicles into lipid bilayers, it was shown that ionic channels are indeed regulated by activated G proteins. Some of these channels had long before been predicted to be under the control of G protein coupled receptors by means other than soluble second messengers. The mechanism by which G proteins regulate some of these channels is at the very least "membrane delimited" and independent of any phosphorylation event or of changes in cytoplasmic levels of second messengers such as cAMP, Ca^{2+} or IP_3 and is very likely due to direct interaction of the G protein α subunit and the channel proper (for review see Brown and Birnbaumer, 1988). Our group was prominent in providing some of the initial as well as subsequent supporting data for these conclusions (Brown and Birnbaumer, 1988; Yatani et al., 1987a, 1987b, 1987c, 1988a, 1988b, 1988c, Codina et al., 1987a, 1987b, 1988a, Imoto et al., 1988; Kirsch et al., 1988). Another research group has claimed that ionic channels, specifically the heart muscarinic K^+ channel, now referred to as G_i-gated K^+ channel (see below), may be regulated by receptors not <u>via</u> a specific G protein α subunit, but <u>via</u> the $\beta\gamma$ subunits of G proteins (Logothetis et al., 1987). This has been corrected in part by Kurachi et al. (1989a), as well as by Logothetis et al. (1988), who now recognize that α subunits are active. The issue whether $\beta\gamma$ dimers stimulate K^+ channel activity in inside out membrane patches is not yet settled and, as shown below, is the object of intense research. The recognition that G proteins may affect ionic channels under cell free conditions led us as well as others to investigate this possibility further. By the most recent count (May, 1989) six classes of ionic channels comprising at least 12 separate molecular species, defined by distinct kinetic and pharmacological properties, have been shown to be stimulated or inhibited under cell free conditions by exogenously added G protein α subunits or by activation of a nearby G protein by GTPγS as seen in inside-out membrane patches or after incorporation of membrane vesicles into planar lipid bilayers (summarized in Table I). These ionic channels include channels that are essentially silent unless a G protein is activating them,

Table I. Ionic Channels Regulated Under Cell-Free Conditions by Pure G Protein and/or GTPγS

Channel Type	G Protein	Effect	Tissue/Cell
K_{Gk} 40 pS	G_i (1,2,3)	Stim.	Heart
K_{Gk} 50 pS	G_{i3}	Stim.	GH_3
K_{Gk} 4 types	G_{o1}	Stim.	Hippocampal Neurons
K_{ATP}	G_{i3}	Stim.	RIN, Heart, Skeletal Muscle
$Na/K_{[Amil]}$	G_{i3}	Stim.	Renal Med. Collect.Tubule
$Ca_{[DHP]}$L-type	G_s (all 4)	Stim.	Heart, Skeletal Muscle
$Na_{[TTX]}$	G_s	Inh.	Heart
$K_{Ca[Charyb]}$	? (Iso/GTP)	Stim.	Uterine Smooth Muscle
$Ca_{[Ni]}$T-type	? (GTPγS)	Inh.	Rat DRG's

referred to as G protein-gated ion channels (Yatani et al, 1987a and b; VanDongen et al., 1988), as well as channels that are merely regulated by the G proteins, such as an ATP-sensitive (Parent and Coronado 1988; Ribalet et al., 1989; Kirsch, Birnbaumer and Brown, unpublished); an amiloride sensitive (Light et al., 1989), a Ca^{2+}-activated charybdotoxin-sensitive (Toro et al., 1989) and various voltage-gated (Yatani et al., 1987c; 1988a; Schubert et al., 1989; Dolphin et al., 1988) channels (summarized in Table I).

In addition to the ca. 12 channels thus far shown to be influenced by G proteins in a manner that appears to be direct, i.e., not involving protein phosphrylation or changes in levels of second messengers such as cyclic nucleotides, diacylglycerol or Ca^{2+}, there may still be several more. This is suggested by reports on effects of PTX treatment or of GTPγS and G protein subunit injection on whole cell currents. These effects include the inhibition by GTPγS or PTX-sensitive G_o- and G_i-type G proteins of voltage-gated Ca^{2+} channels in chick (Holz et al., 1986) and rat (Scott and Dolphin, 1986; Ewald et al., 1988, 1989) dorsal root ganglion cells and of similar channels in AtT-20 (Lewis et al., 1986) and NG108-15 (Hescheler et al., 1987) cells, and the stimulation by PTX-sensitive G_i-type G protin of voltage-gated Ca^{2+} channels in adrenal Y1 and GH_3 cells (Hescheler et al., 1988; Rosenthal et al., 1988) (summarized in Table II). However, in these cases the involvement of soluble second messengers in the mediation of the effects of the G protein has yet to be ruled out.

The case of cell free regulation of dihydropyridine (DHP)-sensitive Ca^{2+} channels by G_s is of special interest. It was unexpected for of two reasons: one because DHP-sensitive Ca^{2+} channels had been shown to be stimulated upon phosphorylation by the catalytic unit of cAMP-dependent protein kinase (Kameyama et al., 1985), indicating that nature uses dual pathways to regulate a single function, one fast and membrane delimited, the other slower with a longer life span; the other, because it had been thought that G proteins might be "monogamous", i.e., specific for single effector functions, and here we were faced with proof for multifunctionality in G protein actions.

These advances were all the result of a multi-disciplinary approach to the problem of signal transduction by G proteins which brought together classical biochemistry, sophisticated single channel recordings and modern molecular biology. The background experiments, especially those of Nargeot et al.(1983), Soejima and Noma (1984); Pfaffinger et al. (1985), and Breitwieser and Szabo (1985), which led to the discovery that ionic channels

Table II. Ionic Channels That May Be Under Direct Regulation By G Proteins as Inferred from Whole Cell Recordings

Channel Type	Agonist	G Protein	Effect	Cell	Mimicked by TPA
Ca (N-type?)	Opioid	G_o>G_i	Inh.	NG108-15	?
Ca (L-type)	Ang II	G_i	Stim.	Y1 Adrenal	?
Ca (L-type)	SST	? GTPγS	Inh.	AtT-20	yes
Ca (L-type)	GnRH	G_i/G_o	Stim.	GH_3	?
Ca (L-type)	SST	G_i/G_o	Inh.	GH_3	?
Ca (N-type?)	$GABA_{(B)}$	? GTPγS	Inh.	Rat DRG	no
Ca (?)	NPY	Go>G_i	Inh.	Rat DRG	yes
	Bradyk.	Go=G_i	"	" "	"
Ca (?)	$GABA_{(B)}$	? GTPγS	Inh.	Chick DRG	yes
	NE (αAR)	"	"	" "	"

All the G proteins involved are PTX-sensitive

are effector systems of G proteins akin to adenylyl cyclase and cGMP-phosphodiesterase, and the initial experiments showing effects of purified G proteins and protein subunits on channels in excised membrane patches, were reviewed in Brown and Birnbaumer (1988). The present article will center on some of the more recent results from our laboratories dealing with a pertussis toxin sensitive G_i-type family of G proteins, their effects of ionic channels, our efforts in assigning defined functions to in-dividual G proteins as they are known from biochemical and molecular cloning studies, and some speculations that follow from the results obtained as to why G proteins dissociate and how transduction pathways are set up.

PRIMARY STRUCTURE OF G PROTEINS

The primary αβγ structure of G proteins has been reviewed recently by Lochrie and Simon (1988), as well as by us (Birnbaumer et al., 1989), and are presented by A.G. Gilman in a previous chapter of this book. Key to the understanding of their functioning is of course both their subunit dissociation cyclase superimposed on their GTPase cycle, and their transient -- or perhaps not so transient -- interactions with receptors and effectors. In our view it is clear that the receptor signal is carried to the effector by the α subunit as exemplified in the experiment of Fig. 1. At the time of this writing twelve α subunits, encoded in 9 genes, two β subunits and at least two γ subunits are known. Of these, all the α, the two β and one of the γ subunits have been cloned. β and γ subunits form dimers which may be of two types, $\beta_{36}\gamma$ and $\beta_{35}\gamma$, if a cell expresses only one type of γ subunit, or of four types if it expresses two. To our knowledge, α subunits, when combining with βγ dimers to form holo-G proteins, do not distinguish among βγ dimers (Okabe et al., 1989). This is not to say that all tissues have the same complement of βγ dimers. Quite the contrary, as shown recently the β_{35}/β_{36} ratio differs in human placenta, human erythrocytes, bovine brain and bovine retinal rod cells (Okabe et al., 1989). α subunits bind GTP, hydrolyze GTP, dissociate from the βγ dimer on activation by GTP analogs and or the $NaAlF_4$/GDP, and, with few exceptions are substrates for ADP-ribosylation by cholera and/or pertussis toxin (CTX and PTX). Studies with transducin-α identified an arginine at the approximate center of the molecules as the amino acid ADP-ribosylated by CTX (VanDop et al., 1984) and a cysteine at position -4 from the carboxyl terminal end as the site of ADP-ribosylation by PTX (West et al., 1985). It is worth mentioning that not all G proteins have been either purified or cloned, and that both not all the known G proteins, such as G_{o1} and G_{o2} or some of

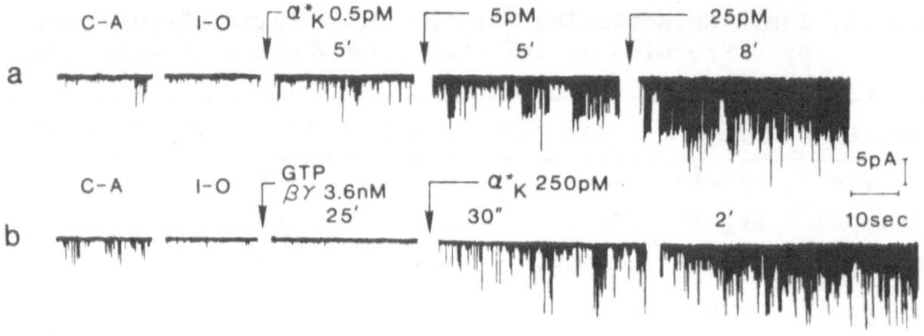

Fig. 1 a. Stimulation of single channel K^+ currents in GH_3 cell membrane patches by increasing concentrations of native GTPγS-activated human erythrocyte α_{i3} (α_k^*). b. Lack of intrinsic stimulatory effects of human erythrocyte $\beta\gamma$ dimer or of a mixture of human erythrocyte GTPγS-activated α_s plus $\beta\gamma$ dimer (G_s^*) in a responsive membrane patch. 500 nM Lubrol PX was present throughout. (Adapted from Codina et al., 1987b).

the G_i's, have unequivocally assigned functions, or that all functionally recognized G proteins, such as the inhibitory G_i of adenylyl cyclase or the stimulatory G_p of phospholipases have been identified.

FUNCTIONAL STUDIES

Combined Use of Natural and Recombinant α Subunits Made in Bacteria to Define Their Functions

The abundance at which different α_s and α_i molecules are expressed varies from tissue to tissue raising the question as to whether functional differences are associated with the structural differences, or whether the G_s's and, respectively, the G_i's should be thought of merely as isoforms. Although the final word on these questions is not yet in, we have during this last year developed the method(s) required to answer them. Thus, we have thus far succeeded in purifying two types of G_i from human erythrocytes (hRBCs) and a third from bovine brain, from which we also purified the two forms of G_o. This allowed us to test for potential differential biological functions. We also expressed biologically active forms of the α subunits in bacteria, designated as recombinant α subunits, so that we could predict/confirm biological functions of various cloned and or purified G protein α subunits.

We have not yet carried out all of the studies. However, we were able to determine first of all that both the natural purified and the recombinant forms of α_{i3}, α_{i1} and α_{i2}, all stimulate K^+ channels (Mattera et al., 1989a; Yatani et al., 1988c). We failed to observe significant differences between type 1, 2 and 3 α_i molecules, even though the recombinant forms all had potencies between 30 and 50 fold lower than their natural counterparts (Yatani et al., 1988c). Thus, with respect to atrial muscarinic K+ channels, G_{i1}, G_{i2} and G_{i3} must be considered iso-G proteins. Studies are currently in progress in atrial membrane patches in which the endogenous G_k has been uncoupled from receptors by treatment with PTX (Yatani et al., 1987b; Brown and Birnbaumer, 1988), to determine whether muscarinic and/or P_2 purinergic receptors exhibit selectivity for interaction with one or another of the G_i proteins.

Second, by testing the effect of recombinant α_s, we were able to determine that the stimulation of Ca^{2+} channels obtained with purified $_h$RBC G_s (Yatani et al., 1987c, 1988a; Imoto et al., 1988) is indeed mediated by G_s as opposed to being due to a contaminant. Further, in collaboration with Michael Graziano and Al Gilman we tested the recombinant forms of three of the possible four splice variants of α_s for their Ca^{2+} channel stimulatory activity and found that they all do so with indistinguishable potency and efficacy (Mattera et al., 1989b). As was the case with recombinant α_i subunits, the recombinant α_s subunits also displayed a 20-fold reduced potency with respect to that of native human erythrocyte α_s. This applied not only to Ca^{2+} channel stimulation, but equally to adenylyl cyclase stimulation. Very likely bacteria fail to carry out a critical postranslational modification that exists only in eukaryotic cells, or alternatively, modify the α subunit in a manner that eukaryotic cells do not.

Gating of Ionic Channels as a Tool to Discover New Roles for G Proteins: Effects of G_o on Neuronal K^+ Channels

One of the properties of the "muscarinic" K^+ channels is that they are essentially silent in the absence of stimulation by an activated G protein (G_k). That is, in the absence of activated G protein their Po is close to zero. The possibility existed that not only G_i proteins regulate K^+ channels but also the structurally closely related G_o. Since nerve tissue is rich in G_o, central nervous system neurons, specifically hippocampal pyramidal cells, were placed into culture and studied for potential presence of both G_i and G_o gated K^+ channels. Although these studies are still in progress (VanDongen, Birnbaumer, Brown, unpublished), the initial findings with highly purified bovine brain GTPγS-activated G_{o1} (G_{o1}^*) were of interest (VanDongen et al., 1988). They identified the existence of several novel G-protein gated, more precisely G_o-gated K^+, channels that are distinct from G_i-gated K^+ channels. Thus, application of purified bovine brain G_{o1}^* to the cytoplasmic aspect of inside-out membrane patches of cultured hippocampal pyramidal cells resulted in appearance of three new types of single channel K^+ currents consistent with the existence of the three non-rectifying of K^+ channels having sizes of 13pS, 40pS and 55pS, respectively, plus an inwardly rectifying K^+ channel with a slope conductance of 40 pS. No such channel activities were observed with hRBC G_i^*-3 or hRBC α_i^*-3. G_{o2}^* or α_{o2}^* have not been tested in this system as yet. In contrast to earlier observations with G_o^* added to guinea pig atrial membrane patches, which showed only marginal effects of GTPγS-activated G_{o1} at 2 nM (Yatani et al., 1987b), the hippocampal K^+ channel is highly sensitive to G_o^* (VanDongen et al., 1988), and significant activation was obtained at 1 pM and half maximal effects were obtained at about 10 pM. The identity of the active G protein in the G_{o1} preparations used was confirmed with recombinant GTPγS-activated α_{o1}. The G_o-gated channels were stimulated in the absence of Ca^{2+} or ATP, in the presence or absence of AMP-P(NH)P, added routinely to inhibit ATP-sensitive 70 pS K^+ channels. Further EGTA did not interfere with the actions of Go_1 or recombinant α_{o1}. Thus, in hippocampal pyramidal cells of the rat, G_{o1} is a G_k, and the K^+ channels gated by it are several and differ from those present in atrial cells in various aspects including G protein specificity. These findings raise the question which if any of the G proteins that gate K^+ channels regulate the other known PTX sensitive effector systems.

$\beta\gamma$ DIMERS INHIBIT K^+ CHANNEL GATING BY G PROTEIN

Logothetis et al. (1987, 1988) reported twice that $\beta\gamma$ dimers stimulate atrial K^+ channel activity. Their finding is not reproduced in our hands. Quite the contrary, when we add $\beta\gamma$ dimers to inside-out membrane patches in which K^+ channels have been stimulated either by GTP only (baseline activity) or by carbachol plus GTP (agonist-stimulated activity), we find consistently inhibition of activity (Fig. 2). On their own, i.e., when added to silent patches in the absence of GTP, $\beta\gamma$ dimers have no effect under our assay conditions (Fig. 1; Codina et al., 1987a; Kirsch et al., 1988; Okabe et al., 1989).

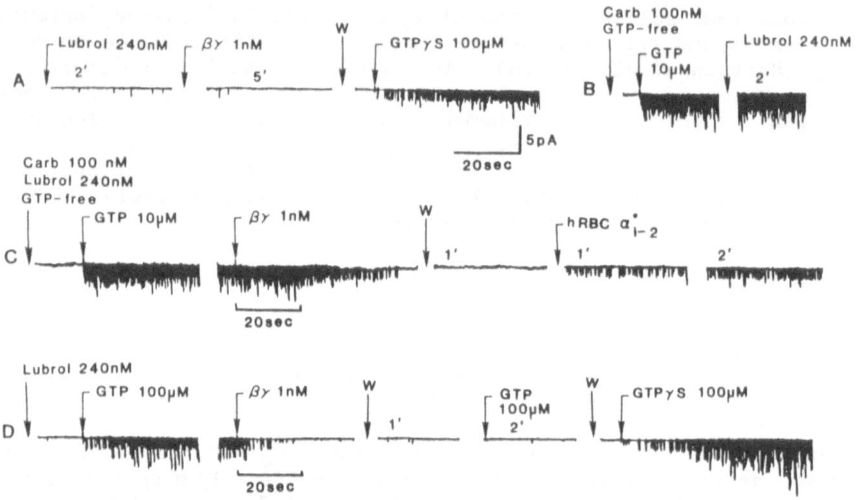

Fig. 2 Inhibition of G protein gated K^+ channel activity by $\beta\gamma$ dimers. Inside-out membrane patches from adult guinea pig atrial cells were exposed to the bathing solution (140 mM KCl, 2 mM $MgCl_2$, 5 mM EGTA, 10 mM Hepes-K, pH 7.4) containing the additives shown on the figure. The pipette solution was identical to the bathing solution and contained 100 nM carbachol when indicated. The composition of bathing solution was changed by a concentration clamp method. Note that: Lubrol PX and/or bovine serum albumin, used to maintain $\beta\gamma$ dimers in suspension do not interfere with stimulation of activity by GTP (D) or GTP plus agonist (B, C), and that $\beta\gamma$ dimers have no effect on their own (A) but inhibit atrial K^+ channel stimulation by the membrane G_k (C, D). Inhibition is faster and is elicited with lower concentrations of $\beta\gamma$ dimers when K^+ channels are operating under baseline conditions (GTP only, experiment D: cessation of activity after 16 sec) than when they are stimulated by agonist (carbachol plus GTP: experiment C: cessation of activity after 50 sec). Numbers above records denote time elapsed in min between solution change and the beginning of the record shown.

Concentration effect studies showed clearly that are more potent in inhibiting agonist independent than agonist stimulated activity, and that this phenomenon applies not only to $\beta\gamma$ dimers suspended in Lubrol-PX, such as those from human placenta, human erythrocytes and bovine brain, but also to $\beta\gamma$ dimers presented to the patches in aqueous media, such as transducin $\beta\gamma$ (Fig. 3).

The fact that $\beta\gamma$ dimers inhibit GTP-dependent activity in the absence of agonist at lower concentrations as compared to inhibition in the presence of agonist was observed previously in purified components reconstituted into phospholipid vesicles (Birnbaumer, 1987), and serves to support our thesis that $\beta\gamma$ dimers act in intact membranes as suppressor of "noise" generated by agonist-unoccupied receptors.

The inhibitory effects of $\beta\gamma$ dimers obtained by us need to be contrasted to stimulatory effects obtained by Clapham, Neer, and their collaborators (Logothetis et al., 1987, 1988; Kim et al., 1989). We do not understand exactly the reasons for the discrepancy. We obtain inhibition also in the absence of Lubrol PX using $\beta\gamma$ dimers from transducin which are water soluble. The claim that $\beta\gamma$ dimers may be acting by stimulation of arachidonic acid formation (Kim et al., 1989), is suspect. In an adjacent report, Kurachi et al. (1989b), describe that arachidonic acid and its

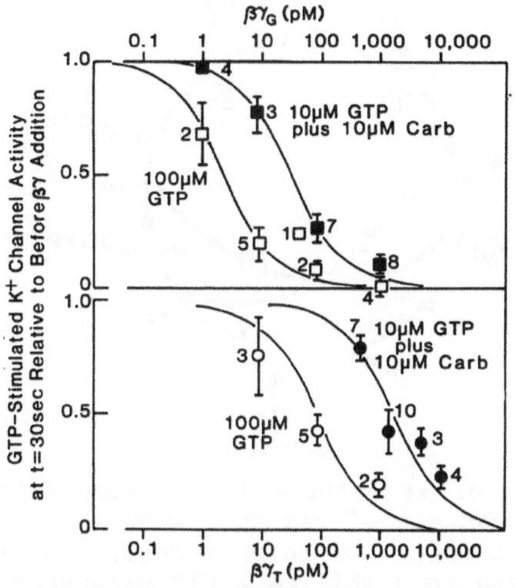

Fig. 3 Effect of agonist (carbachol) on the dose-dependent inhibition by $\beta\gamma$ dimers of GTP-dependent K^+ channel activities in inside-out guinea pig atrial membrane patches by $\beta\gamma$ dimers. When 1 μM carbachol was present in the pipette, GTP was 10 μM; in the absence of carbachol GTP was 100 μM. $\beta\gamma$G: data obtained with $\beta\gamma$ dimers derived from either human erythrocyte, human placenta or bovine brain were pooled. $\beta\gamma_T$: data obtained with $\beta\gamma$ derived from bovine rod outer segments. We thank Dr. Tony evans for the gift of human placental $\beta\gamma$ dimers, and Dr. J.-K. Ho, for their gift of bovine rod outer segment $\beta\gamma$ dimers. (Adapted from Okabe et al., 1989).

metabolites require for their action the presence of GTP and are blocked by GDPβS. Yet the stimulatory effects of $\beta\gamma$ dimers occur in the absence of GTP (Logothetis et al., 1987; 1988; Kim et al., 1989).

CURRENT VIEWS ON HOW SIGNAL TRANSDUCTION BY GTP AND RECEPTORS COMES ABOUT AND WHICH RECEPTOR ACTS ON WHICH G PROTEIN TO REGULATE WHICH EFFECTOR SYSTEM

Taken together the results discussed in the previous sections lead to several conclusions: 1. ionic channels are targets of direct regulation by G proteins as are adenylyl cyclase and the cGMP-specific phosphodiesterase, 2. α subunits and not $\beta\gamma$ dimers are the specificity determinants of signal transduction pathways; 3. several G proteins may have the same function, e.g., stimulation of K^+ channels by three G_i's, and 4. a single G protein may have more than one function, i.e., be multifunctional (e.g., stimulation of adenylyl cyclase and the dihydropyridine sensitive Ca^{2+} channel by a single G_s).

Two independent sets of questions emerge from these findings. The first deals with the subunit dissociation reaction, and asks what type of advantage it confers onto the system by its existence. An answer to these questions can be found upon analyzing in detail the G protein regulatory cycle and the mechanism by which receptors promote G protein activation by GTP.

The second set of questions deals with crosstalk between signal transduction pathways, i.e., whether receptors act on more than one G proteins, and if so which, whether G proteins interact with more than one receptor as well as with more than one effector, and if so, how frequent this is.

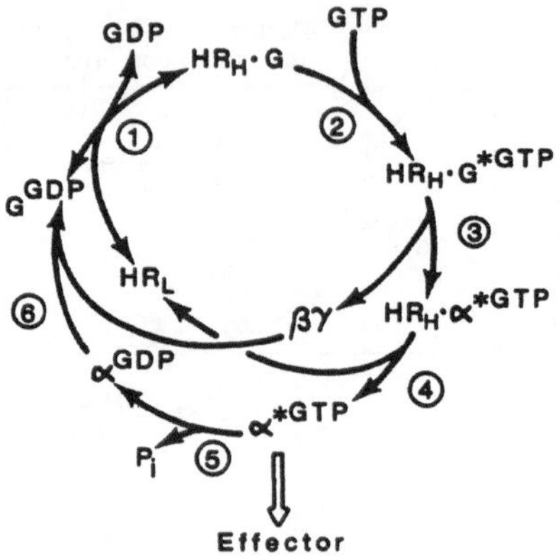

Fig. 4 Integrated view of receptor-mediated catalytic activation of a G protein in the context of the dual subunit dissociation and GTPase cycles of G proteins. The role of receptors is to promote nucleotide exchange and to stabilize a GTP-dependent "activated" form of the G protein and the G protein undergoes a cyclical dissociation-reassociation reaction and oscillates between GDP, nucleotide free and GTP states. The cycles are driven energetically forward by the capacity of the G protein to hydrolyze GTP (Cassel and Selinger, 1978; Godchaux and Zimmerman, 1979; Brandt and Ross, 1986; Cerione et al., 1984), and kinetically by the dissociation of $\beta\gamma$ dimer from the activated receptor-G proteins complex. The receptor has high affinity for agonist (R_H) when associated with the nucleotide-free trimeric $\alpha\beta\gamma$ form of the G protein, and has low affinity for the agonist (R_L) when it is free. Further, the receptor has higher affinity for the trimeric $\alpha\beta\gamma$ form of G than the G-GDP thus accounting for the finding that GDP and GDP analogs promote the R_H to R_L transition. The $\beta\gamma$ dimers are required for the interaction of α with R, and after formation of G-GTP, no G* forms unless it is "aided" by receptor. Thus receptor has an even higher affinity for the G^*GTP state than the nucleotide free state of G. As a consequence receptor dissociation is absolutely dependent on reaction 2 (subunit dissociation). Thermodynamic reasons do not allow R both to stabilize the G^* state and to dissociate from it. Reaction 3 states further that the α*GTP loses its ability to stay associated with receptor and decomposes further into free activated α^*GTP plus free receptor, thus accounting for the fact that under "working" conditions (saturation by both GTP and hormone, and hence sustained regulation of effector) only a small proportion of receptors are found in their high affinity, G-protein associated state. It follows that the G protein cycle is driven forward not only by the GTPase but also, and obligatorily so, by the subunit dissociation reaction. The scheme accounts for the experimental findings a. that receptors act catalytically and therefore need to dissociate from the G protein at one time or another (Tolkovsky and Levitzki (1978); and b. that receptors accelerate the transition from inactive G-GTP$_\gamma$S to active G^*-GTPγS transition and therefore must have higher intrinsic affinity for the activated than the inactive state (Birnbaumer et al., 1980; Iyengar et al., 1980; Iyengar and Birnbaumer, 1982).

Role of Subunit Dissociation: Requirement for Catalytic Action of Receptors

In the second half of the 70's it was demonstrated that receptors act catalytically rather than stoichiometrically to activate adenylyl cyclase (Tolkovsky and Levitzki, 1978) and in 1980 it was found that receptors in

Table III. Intramembrane Roles of G protein $\beta\gamma$ Dimers

Reaction	Product	Role
Reassociation with α^{GDP}	G^{GDP}	Activation by R
Dissociation from $HR.G^{*GTP}$	$HR.\alpha^{*GTP}$	Signal Amplification
Reassociation with $R.\alpha^{*GTP}$	$R.G^{GTP}$	Noise Reduction

addition to promoting GDP/GTP exchange (Cassel and Selinger, 1978; Cassel et al. 1979) also participate in promoting the activation reaction proper of adenylyl cyclase by GTP and its analogs (reviewed in Birnbaumer et al., 1985). These two findings, catalytic action and stabilization of the nucleotide-activated form of adenylyl cyclase, now known to be in fact G_s, are thermodynamically impossible, unless the stabilized form of the G protein undergoes some type of additional spontaneous change that releases it from microscopic reversibility constraints, which happens in all enzymatically catalyzed reactions where the product is chemically different from the starting substrate. Dissociation of $\beta\gamma$ from the activated and stable receptor-G protein complex is such a reaction change. We therefore propose that the role of the dissociation reaction is a requirement without which receptors cannot act catalytically. Both the catalytic nature of receptor mediated activation of G_s and the fact that receptor affects not only GDP release but also the rate at which inactive guanine nucleotide occupied G protein isomerizes from an inactive to an active state have been confirmed in reconstitution experiments with purified receptor and purified G_s (Asano et al., 1984; Brandt and Ross, 1986). Since receptors do not interact with the α subunits except in the context of $\beta\gamma$ (Watkins et al., 1985; Kanaho et al., 1984; Florio and Sternweis, 1985, 1989; Tota et al., 1987), reassociation of the GTPase-deactivated α with $\beta\gamma$ is essential for restimulation by receptors. This then leads to a description of the G protein regulatory cycle under influence of receptor as depicted in Fig. 4.

Taken together, it is thus possible to ascribe three generic functions to $\beta\gamma$ dimers (Table III): 1. activation of α subunit by receptor, for without $\beta\gamma$ receptors do not interact with α; 2. amplification of the receptor signal, for without dissociation receptors cannot act catalytically, and 3. noise reduction, for unoccupied receptors are not silent.

Specificities in Receptor-G Protein-Effector Interaction

Ever since the discoveries in the late 1960's that up to five different hormone receptors can activate in a single adenylyl cyclase system in an isolated membrane (Birnbaumer and Rodbell, 1969) and, in the early and mid 1970's, that receptors can be transferred from one cell to another (Citri and Schramm, 1975) and that there are no species and/or tissue specificity restrictions as to the source of G_s for reconstitution of a hormonally stimulable adenylyl cyclase system in cyc^- membranes (Ross et al., 1978; Kaslow et al., 1979), it has been clear that single G proteins are designed to interact with classes of receptors as opposed to single receptor subtypes. Experimental results mentioned above indicate that one G protein can interact with more than one effector and that several G proteins may regulate a single effector. Finally, Askenazi et al. (1987) showed that single receptors may affect more than one G protein. The important notion that emerges from these considerations is that the wiring diagram describing signal transduction by G proteins needs to be determined individually and separately for each cell or tissue of interest. This includes the determination not only of which receptors are present but also of which G proteins and which effector functions are expressed.

REFERENCES

Asano, T., Pedersen, S. E. Scott, C. W. and Ross, E. M. (1984) "Reconstitution of Catecholamine-Stimulated Binding of Guanosine 5'-O-(3-Thiotriphosphate) to the Stimulatory GTP-Binding Protein of Adenylate Cyclase." Biochemistry 23, 5460-5467.

Ashkenazi, A., Winslow, J. W., Peralta, E. G., Peterson, G. L., Schimerlik, M. I., Capon, D. J. and Ramachandran, J. (1987) "An M2 Muscarinic Receptor Subtype Coupled to Both Adenylyl Cyclase and Phosphoinositide Turnover." Science 238, 672-675.

Birnbaumer, L. (1987) "Which G Protein Subunits are the Active Mediators in Signal Transduction." Trends Pharmacol. Sci. 8, 209-211.

Birnbaumer, L. and Rodbell, M. (1969) "Adenyl Cyclase in Fat Cells. II. Hormone Receptors." J. Biol. Chem. 244, 3477-3482.

Birnbaumer, L., Swartz, T. L., Abramowitz, J., Mintz, P. W. and Iyengar, R. (1980) "Transient and Steady State Kinetics of the Interaction of Nucleotides with the Adenylyl Cyclase System from Rat Liver Plasma Membranes: Interpretation in Terms of a Simple Two-state Model." J. Biol. Chem. 255, 3542-3551.

Birnbaumer, L., Hildebrandt, J. D., Codina, J., Mattera, R., Cerione, R. A., Sunyer, T., Rojas, F. J., Caron, M. G., Lefkowitz, R. J. and Iyengar, R. (1985) "Structural Basis of Adenylate Cyclase Stimulation and Inhibition by Distinct Guanine Nucleotide Regulatory Proteins," In, "Molecular Mechanisms of Signal Transduction," (Cohen, P., and Houslay, M. D., eds.) Elsevier/North Holland Biomedical Press, Amsterdam, pp. 131-182.

Birnbaumer, L., Codina, J., Yatani, A., Mattera, R., Graf, R., Olate, J., Themmen, A.P.N., Liao, C.-F., Sanford, J., Okabe, K., Imoto, Y., Zhou, Z., Abramowitz, J., Suki, W.S., Hamm, H.E., Iyengar, R., Birnbaumer, M. and Brown, A. M. (1989) "Molecular Basis of Regulation of Ionic Channels by G Proteins." Rec. Prog. Horm. Res. 45, 120-208.

Brandt, D. R. and Ross, E. M. (1986) "Catecholamine-stimulated GTPase Cycle. Multiple Sites of Regulation by beta-Adrenergic Receptor and Mg^{2+} Studied in Reconstituted Receptor-G_s Vesicles." J Biol. Chem. 261, 1656-1664.

Breitweiser, G. E. and Szabo, G. (1985) "Uncoupling of Cardiac Muscarinic and beta-Adrenergic Receptors from Ion Channels by a Guanine Nucleotide Analogue." Nature 317, 538-540.

Brown, A. M. and Birnbaumer, L. (1988) "Direct G Protein Gating of Ion Channels" Am. J. Physiol. 23, H401-H410.

Cassel, D. and Selinger, Z. (1978) "Mechanism of Adenylate Cyclase Activation Through the beta-Adrenergic Receptor: Catecholamine-induced Displacement of Bound GDP by GTP." Proc. Natl. Acad. Sci. USA 75, 4155-4159.

Cassel, D., Eckstein, F., Lowe, M. and Selinger, Z. (1979) "Determination of the Turn-off Reaction for the Hormone-Activated Adenylate Cyclase" J. Biol. Chem. 254, 9835-9838.

Cerione, R. A., Codina, J., Benovic, J. L., Lefkowitz, R. J, Birnbaumer, L., Caron, M. G. (1984) "The Mammalian $beta_2$-Adrenergic Receptor: Reconstitution of the Pure Receptor with the Pure Stimulatory Nucleotide Binding Protein (N_s) of the Adenylate Cyclase System." Biochemistry 23, 4519-4525.

Citri, Y. and Schramm, M. (1975) "Resolution, Reconstitution and Kinetics of the Primary Action of Hormone Receptor." Nature 287, 297-300.

Codina, J., Stengel, D., Woo, S. L. C. and Birnbaumer, L. (1986) "Beta Subunits of the Human Liver G_s/G_i Signal Transducing Proteins and Those of Bovine Retinal Rod Cell Transducin are Identical." FEBS Letters 207, 187-192.

Codina, J., Yatani, A., Grenet, D., Brown, A. M. and Birnbaumer, L. (1987a) "The alpha Subunit of the GTP Binding Protein G_k Opens Atrial Potassium Channels." Science 236, 442-445.

Codina, J., Grenet, D., Yatani, A., Birnbaumer, L. and Brown, A. M. (1987b) "Hormonal Regulation of Pituitary GH_3 Cell K^+ Channels by G_k is Mediated by Its alpha Subunit." FEBS Letters 216, 104-106.

Codina, J., Olate, J., Abramowitz, J., Mattera, R., Cook, R. G. and Birnbaumer, L. (1988) "'$Alpha_i$-3 cDNA Encodes he alpha Subunit of G_k, the Stimulatory G Protein of Receptor-regulated K^+ Channels." J. Biol. Chem. 263, 6746-6750.

Ewald, D. A., Sternweis, P. C., and Miller, R. J. (1988) "Guanine Nucleotide-binding Protein G_o-induced Coupling of Neuropeptide Y Receptors to Ca^{2+} Channels in Sensory Neurons." Proc. Natl. Acad. Sci. USA 85, 3633-3637.

Ewald, D.A., Pang, I.-H., Sternweis, P.C. and Miller, R.J. (1989) "Differential G Protein-Mediated Coupling of Neurotransmitter Receptors to Ca^{2+} Channels in Rat Dorsal Root Ganglion Neurons In Vitro." Neuron 2, 1185-1193.

Florio, V. A. and Sternweis, P. C. (1985) "Reconstitution of Resolved Muscarinic Cholinergic Receptors with Purified GTP-binding Proteins." J. Biol. Chem. 260, 3477-3483.

Florio, V.A. and Sternweis, P.C. (1989) "Mechanism of Muscarinic Receptor Action on G_o in Reconstituted Phospholipid Vesicles." J. Biol. Chem. 264, 3909-3915.

Godchaux, W., III, and Zimmerman, W. F. (1979) "Membrane-dependent Guanine Nucleotide Binding and GTPase Activities of Soluble Protein from Bovine Rod Cell Outer Segments." J. Biol. Chem. 254, 7874-7884.

Hescheler, J., Rosenthal, W., Hinsch, K.-D., Wulfern, M., Trautwein, W. and Schultz, G. (1988) "Angiotensin II-induced Stimulation of Voltage-dependent Ca^{2+} Currents in an Adrenal Cortical Cell Line." The EMBO J. 7, 619-624.

Hescheler, J., Rosenthal, W., Trautwein, W. and Schultz, G. (1987) "The GTP-binding Protein, N_o, Regulates Neuronal Calcium Channels." Nature 325, 445-447.

Holz, G. G., Rane, S. G. and Dunlap, K. (1986) "GTP-binding Proteins Mediate Transmitter Inhibition of Voltage-dependent Calcium Channels." Nature 319, 670-672.

Imoto, Y., Yatani, A., Reeves, J. P., Codina, J., Birnbaumer, L. and Brown, A. M. (1988) "α Subunit of G_s Directly Activates Cardiac Calcium Channels in Lipid Bilayers." Amer. J. Physiol. 255, H722-H728.

Iyengar, R., Abramowitz, J., Riser, M. and Birnbaumer, L. (1980) "Hormone Receptor-mediated Stimulation of the Rat Liver Plasma Membrane Adenylyl Cyclase System: Nucleotide Effects and Analysis in Terms of a Two-state Model for the Basic Receptor-affected Enzyme." J. Biol. Chem. 255, 3558-3564.

Iyengar, R. and Birnbaumer, L. (1982) "Hormone Receptors Modulate the Regulatory Component of Adenylyl Cyclases by Reducing its Requirement for Mg Ion and Increasing its Extent of Activation by Guanine Nucleotides." Proc. Natl. Acad. Sci. USA 79, 5179-5183.

Kameyama, M., Hescheler, J., Hofmann, F. and Trautwein, W. (1986) "Modulation of Ca Current During the Phosphorylation Cycle in the Guinea Pig Heart." Pfluegers Arch. 407, 121-128.

Kanaho, Y., Tsai, S.-C., Adamik, R., Hewlett, E. L., Moss, J. and Vaughan, M. (1984) "Rhodopsin-enhanced GTPase Activity of the Inhibitory GTP-binding Protein of Adenylate Cyclase." J. Biol. Chem. 259, 7378-7381.

Kaslow, H. R., Farfel, Z., Johnson, G. L. and Bourne, H. R. (1979) "Adenylate Cyclase Assembled in Vitro: Cholera Toxin Substrates Determine Different Patterns of Regulation by Isoproterenol and Guanosine 5'-triphosphate." Mol. Pharmacol. 15, 472-483.

Kim, D., Lewis, D. L., Graziadei, L., Neer, E. J., Bar-Sagi, D. and Clapham, D. E. (1989) "G Protein $\beta\gamma$-Subunits Activate the Cardiac Muscarinic K^+ Channel via Phospholipase A_2." Nature 337, 557-560.

Kirsch, G., Yatani, A., Codina, J., Birnbaumer, L. and Brown, A. M. (1988) "The alpha Subunit of G_k Activates Atrial K^+ Channels of Chick, Rat and Guinea Pig." Am. J. Physiol. 254 (Heart Circ. Physiol. 23), H1200-H1205.

Kurachi, Y., Ito, H., Sugimoto, T., Katada, T. and Ui., M. (1989a) "Activation of Atrial Muscarinic K^+ Channels by Low Concentrations of $\beta\gamma$ Subunits of G Rat Brain Protein." Pflügers Arch. 413, 325-327.

Kurachi, Y., Ito, H., Sugimoto, T., Shimizu, T., Miki, I., and Ui, M. (1989b) "Arachidonic Acid Metabolites as Intracellular Modulators of the G Protein-gated Cardiac K^+ Channel." Nature 337, 555-557.

Lewis, D. L., Weight, F. F. and Luini, A. (1986) "A Guanine Nucleotide-binding Protein Mediates the Inhibition of Voltage-dependent Calcium Current by Somatostatin in a Pituitary Cell Line." Proc. Natl. Acad. Sci. USA 83, 9035-9039.

Light, D. B., Ausiello, D. and Stanton, B. A. (1989) "A GTP Binding Protein, α_i^*-3, Directly Activates a Sodium-Conducting Ion Channel in a Renal Epithelium." J. Clin. Invest. (in press).

Lochrie, M. A. and Simon, M. I. (1988) "G Protein Multiplicity in Eukaryotic Signal Transduction Systems." Biochemistry 17, 4957-4965.

Logothetis, D. E., Kurachi, Y., Galper, J., Neer, E. J. and Clapham, D. E. (1987) "The $\beta\gamma$ Subunits of GTP-binding Proteins Activate the Muscarinic K^+ Channel in Heart." Nature 325, 321-326.

Logothetis, D. E., Kim, D., Northup, J. K., Neer, E. J. and Clapham, D. E. (1988) "Specificity of Action of Guanine Nucleotide-binding Regulatory Protein Subunits on the Cardiac Muscarinic K^+ Channel." Proc. Natl. Acad. Sci. USA 85, 5814-5818.

Mattera, R., Yatani, A., Kirsch, G. E., Graf, R., Olate, J., Codina, J., Brown, A. M. and Birnbaumer, L. (1989a) "Recombinant α_i-3 Subunit of G Protein Activates G_k-gated K^+ Channels." J. Biol. Chem. 264, 465-471.

Mattera, R., Graziano, M. P., Yatani, A., Zhou, Z., Graf, R., Codina, J., Birnbaumer, L., Gilman, A. G. and Brown, A. M. (1989b) "Individual Splice Variants of the α Subunit of the G Protein G_s Activate both Adenylyl Cyclase and Ca^{2+} Channels." Science 243, 804-807.

Nargeot, J., Nerbonne, J. M., Engels, J. and Lester, H. A. (1983) "Time Course of the Increase in the Myocardial Slow Inward Current After a Photochemically Generated Concentration Jump of Intracellular cAMP." Proc. Natl. Acad. Sci. USA 80, 2395-2399.

Okabe, K., Yatani, A., Evans, T., Ho, K.-J., Codina, J., Birnbaumer, L., Brown, A.M. (1989) "$\beta\gamma$ Dimers of G Proteins: Inhibition of Receptor-mediated activation of Atrial K^+ Channels and Lack of Specificity for Interaction with α Subunits." J. Biol. Chem. (in press)

Parent, L. and Coronado, R. (1989) "Reconstitution of the ATP-sensitive Potassium Channel of Skeletal Muscle. Activation of a G-Protein Dependent Process." J. Gen. Physiol. (in press).

Pfaffinger, P. J., Martin, J. M., Hunter, D. D., Nathanson, N. M. and Hille, B. (1985) "GTP-binding Proteins Couple Cardiac Muscarinic Receptors to a K Channel." Nature 317, 536-538.

Ribalet, B., Ciani, S., and Eddlestone, G.T. (1989) "Modulation of ATP-sensitive K Channels in RINm5F Cells by Phosphorylation and G Proteins." Biophys. J. 55, 587A.

Rosenthal, W., Hescheler, J., Hinsch, K.-D., Spicher, K., Trautwein, W. and Schultz, G. (1988) "Cyclic AMP-independent, Dual Regulation of Voltage-dependent Ca^{2+} Currents by LHRH and Somatostatin in a Pituitary Cell Line." The EMBO J. 7, 1627-1633.

Ross, E.M., Howlett, A.C., Ferguson, K.M. and Gilman, A.G. (1978) "Reconstitution of Hormone-sensitive Adenylate Cyclase Activity with Resolved Components of the Enzyme," J. Biol. Chem. 253, 6401-5412.

Schubert, B., VanDongen, A. M. J., Kirsch, G. E. and Brown, A. M. (1989) "Modulation of Cardiac Na Channels by Beta-adreno Receptors and the G protein G_s." Biophys. J. 55, 229A

Scott, R. H. and Dolphin, A. C. (1986) "Regulation of Calcium Currents by GTP Analogue: Potentiation of (-)-Baclofen-Mediated Inhibition." Neuroscience Letters 69, 59-64.

Dolphin, A.C. , Scott, R.H. and Wooton, J.F. (1988) Proc. Physiol. Soc. Leicester Meeting 4-5 Nov. 1988, p. 19P.

Soejima, M. and Noma, A. (1984) "Mode of Regulation of the ACh-sensitive K-channel by the Muscarinic Receptor in Rabbit Atrial Cells." Pfluegers Arch. 400, 424-431.

Tolkovsky, A. M. and Levitzki, A. (1978) "Mode of Coupling between the β-Adrenergic Receptor and Adenylate Cyclase in Turkey Erythrocytes." Biochemistry 17, 3795-3810.

Tota, M. R., Kahler, K. R. and Schimerlik, M. I. (1987) "Reconstitution of the Purified Porcine Atrial Muscarinic Acetylcholine Receptor with Purified Porcine Atrial Inhibitory Guanine Nucleotide Binding Protein." Biochemistry 26, 8175-8182.

Toro, L., Ramos-Franco, J. and Stephani, E. (1989) "K_{Ca} Channels in Myometrium Can Be Directly Activated by a G-Protein Coupled to a β-Adrenergic Receptor." submitted

Van Dop, C., Tsubokawa, M., Bourne, H. R. and Ramachandran, J. (1984) "Amino Acid Sequence of Retinal Transducin at the Site ADP-ribosylated by Cholera Toxin." J. Biol. Chem. 259, 696-699.

VanDongen,T., Birnbaumer, L., and Brown, A.M., unpublished.

VanDongen, A., Codina, J., Olate, J., Mattera, R., Joho, R., Birnbaumer, L. and Brown, A. M. (1988) "Newly Identified Brain Potassium Channels Gated by the Guanine Nucleotide Binding (G) Protein G_o." Science 242, 1433-1437.

Watkins, P. A., Burns, D. L., Kanaho, Y., Liu, T.-Y., Hewlett, E. L. and Moss, J. (1985) "ADP-ribosylation of Transducin by Pertussis Toxin." J. Biol. Chem. 260, 13478-13482.

West, R. E., Jr., Moss, J. , Vaughan, M., Liu, T. and Liu, T.-Y. (1985) "Pertussis Toxin-catalyzed ADP-ribosylation of Transducin. Cystein 347 is the ADP-ribose acceptor site." J. Biol. Chem. 260, 14428-14430.

Yatani, A., Codina, J., Brown, A. M. and Birnbaumer, L. (1987a) "Direct Activation of Mammalian Atrial Muscarinic Potassium Channels by GTP Regulatory Protein G_k." Science 235, 207-211.

Yatani, A., Codina, J., Sekura, R. D., Birnbaumer, L. and Brown, A. M. (1987b) "Reconstitution of Somatostatin and Muscarinic Receptor Mediated Stimulation of K_+ Channels by Isolated G_k Protein in Clonal Rat Anterior Pituitary Cell Membranes." Mol. Endocrinol. 1, 283-289.

Yatani, A., Codina, J., Imoto, Y., Reeves, J. P., Birnbaumer, L. and Brown, A. M. (1987c) "A G Protein Directly Regulates Mammalian Cardiac Calcium Channels." Science 238, 1288-1292.

Yatani, A., Imoto, Y., Codina, J., Hamilton, S. L., Brown, A. M. and Birnbaumer, L. (1988a) "The Stimulatory G Protein of Adenylyl Cyclase, G_s, Directly Stimulates Dihydropyridine-Sensitive Skeletal Muscle Ca^{2+} Channels. Evidence for Direct Regulation Independent of Phosphorylation by cAMP-Dependent Protein Kinase." J. Biol. Chem. 263, 9887-9895.

Yatani, A., Hamm, H., Codina, J., Mazzoni, M. R., Birnbaumer, L. and Brown, A. M. (1988b) "A Monoclonal Antibody to the α Subunit of G_k Blocks Muscarinic Activation of Atrial K^+ Channels." Science 241, 828-831.

Yatani, A., Mattera, R., Codina, J., Graf, R., Okabe, K., Padrell, E., Iyengar, R., Brown, A. M. and Birnbaumer, L. (1988c) "The G Protein-gated Atrial K^+ Channel is Stimulated by Three Distinct $G_i\alpha$-subunits." Nature 336, 680-682.

Acknowledgements: Supported in part by United States Public Health Service research grants DK-19318, HD-09581, HL-31164 and HL-37044 to Lutz Birnbaumer, HL-39262 to Dr. Arthur M. Brown (Department of Physiology and Molecular Biophysics); a Welch Foundation grant (Q175) and the Baylor College of Medicine Diabetes and Endocrinology Research Center grant DK-27685.

RECEPTOR-MEDIATED ALTERATIONS OF CALCIUM CHANNEL FUNCTION IN THE REGULATION OF NEUROSECRETION

Kathleen Dunlap and George G. Holz

Department of Physiology
Tufts Medical School
Boston, Massachusetts 02111

INTRODUCTION

Modulation of voltage-dependent channel function by neurotransmitters was first described almost twenty years ago for the action of norepinephrine on cardiac muscle (Reuter, 1983; Tsien, 1987). The transmitter-induced increase in calcium channel current observed following application of norepinephrine likely explains the increase in the force of contraction that is produced upon sympathetic nerve stimulation to the heart. These findings were of further interest in that the action of norepinephrine was found to result not from a direct action of the transmitter on the channel but rather from a receptor-mediated activation of adenylate cyclase (Sperelakis, 1985).

Since these initial ground-breaking experiments, it has become clear that calcium channel modulation plays an important role in regulating neuronal as well as cardiac function. In addition, studies of calcium channel modulation in neurons have uncovered certain regulatory mechanisms that are different from those described for heart. Two events, in particular, appear to be closely tied to neuronal channel modulation: 1) the involvement of a pertussis toxin-sensitive GTP-binding protein and 2) the activation of protein kinase C. In this chapter, we will focus on the role played by G proteins in receptor-mediated regulation of calcium channel function and neurosecretion. A review of the role of protein kinase C in the modulatory pathway can be found elsewhere (Rane and Dunlap, 1989).

CALCIUM CHANNEL SUBTYPES

Functionally and pharmacologically distinct types of voltage-dependent calcium channel have been described in a variety of cells. Classification has been made on the basis of single channel conductance, voltage-dependence of activation and inactivation, and pharmacology. It is clear from work published to date that single channel recording techniques are essential in the classification of channel types. Although the time- and voltage-dependence of single channel and macroscopic current recordings have been found to closely correlate for the low threshold, 8 pS channel (T-type) and the high threshold, dihydropyridine-sensitive, 25 pS channel (L-type) in many cell types (Fenwick et al., 1982; Hagiwara and Ohmori, 1983; Carbone and Lux, 1984; Nilius et al., 1985; Fox et al., 1987a,b; Benham et al., 1987; Kostyuk et al., 1988), this has not been the case for other high threshold, intermediate conductance (13 pS) channels described in neurons (N-type) (Fox et al., 1987a,b; Hirning et al.,1988; Kostyuk et al., 1988). Variability in the inactivation kinetics among preparations and discrepancies between the kinetics of ensemble averages of single N-type channel recordings and those of whole cell currents have led to difficulties in understanding the single channel basis of macroscopic currents.

Recent results (Plummer et al., 1989) suggest that single channel recording may be the only means by which the different high threshold channel types can be distinguished and studied in isolation. These investigators report that the inactivation kinetics of N-type channels recorded in superior cervical ganglion neurons depend upon the holding potential. Depolarization to +20 mV from a holding potential of -80 mV produces single channel openings that inactivate with a time constant on the order of 95 ms while depolarization from a holding potential of -30 mV evokes channel openings that do not inactivate during the depolarization. Thus, long-lasting macroscopic currents in these cells cannot be taken as evidence for the presence of L-type channels. Such effects of holding potential on channel gating suggest that, in the absence of pharmacological tools that are highly specific for a single type of calcium channel, studies of channel subtypes can be done with certainty only at the single channel level.

CALCIUM CHANNEL MODULATION IN NEURONS

In light of these problems in distinguishing channel types underlying the macroscopic currents, studies of transmitter-mediated modulation of whole cell currents are difficult to evaluate. The majority of published reports indicate that long-lasting macroscopic calcium currents in neurons are under regulatory control by transmitters and intracellular second messengers. Are these L-type calcium currents (as with cardiac muscle) or are other calcium channel types modulated as well? The answers to these important questions await studies of neuronal calcium channel modulation at the single channel level.

Transmitter-mediated inhibition of long-lasting, high threshold calcium currents has been described in a number of preparations for a wide variety of neurotransmitters (Tsien, 1987; Rane and Dunlap, 1989). Work in our lab has focused on the effects of norepinephrine and GABA on calcium channels in the soma membrane of embryonic chick dorsal root ganglion neurons. These two transmitters, when applied in μM concentrations, produce reversible, dose-dependent decreases in the slowly-inactivating, high threshold calcium current. Norepinephrine and GABA produce their effects through pharmacologically distinguishable receptors (α_2 and $GABA_B$, respectively).

On what calcium channel type do norepinephrine and GABA exert their effects? Previous work characterizing calcium channels in dorsal root ganglion neurons (Nowycky et al., 1985; Fox et al., 1987a,b) would suggest that the L-type channel is likely to be the locus of transmitter action in our studies. These investigators report that, in these cells, only the L-type channel produces non-inactivating current. N-type channels inactivate with a time constant on the order of 80 ms at 0 mV. Results of recent single channel experiments suggest that, indeed, dihydropyridine-sensitive L-type calcium channels in dorsal root ganglion neurons are modulated by norepinephrine (Anderson and Dunlap, 1988) in a manner consistent with the transmitter's action at the macroscopic level. It remains to be determined 1) whether GABA exerts a similar action on L-type channels and 2) whether there is an additional action of the transmitters on N-type calcium channels.

The single channel experiments have been performed on cell-attached patches of membrane. Norepinephrine is applied to the extrapatch membrane. Receptor-mediated alterations in calcium channel function observed under these experimental conditions implies that direct interaction between the transmitter and the calcium channel is not necessary for the modulatory effect. In other words, these results argue strongly for a second messenger mediated transduction pathway that converts transmitter receptor binding ultimately into a change in calcium channel function. This conclusion is supported by experiments on macroscopic currents as well. The effects of norepinephrine and

GABA require the activation of both a GTP-binding protein (Holz et al., 1986) and protein kinase C (Rane and Dunlap, 1986; Rane et al.,1989).

G PROTEIN-MEDIATED MODULATION OF CALCIUM CHANNELS

Effects on calcium currents

Thiophosphate derivatives of GTP and GDP have proven useful in studies of receptor-mediated G protein regulation since the rate of activation and inactivation of G proteins is controlled by the affinity of their α subunits for GTP relative to GDP as well as by the hydrolysis of GTP following its binding to the α subunit. GDP-β-s can prevent G protein activation (by competing with GTP for binding) while GTP-γ-s can prevent G protein inactivation (since the nucleotide can bind to the α subunit but is resistant to hydrolysis).

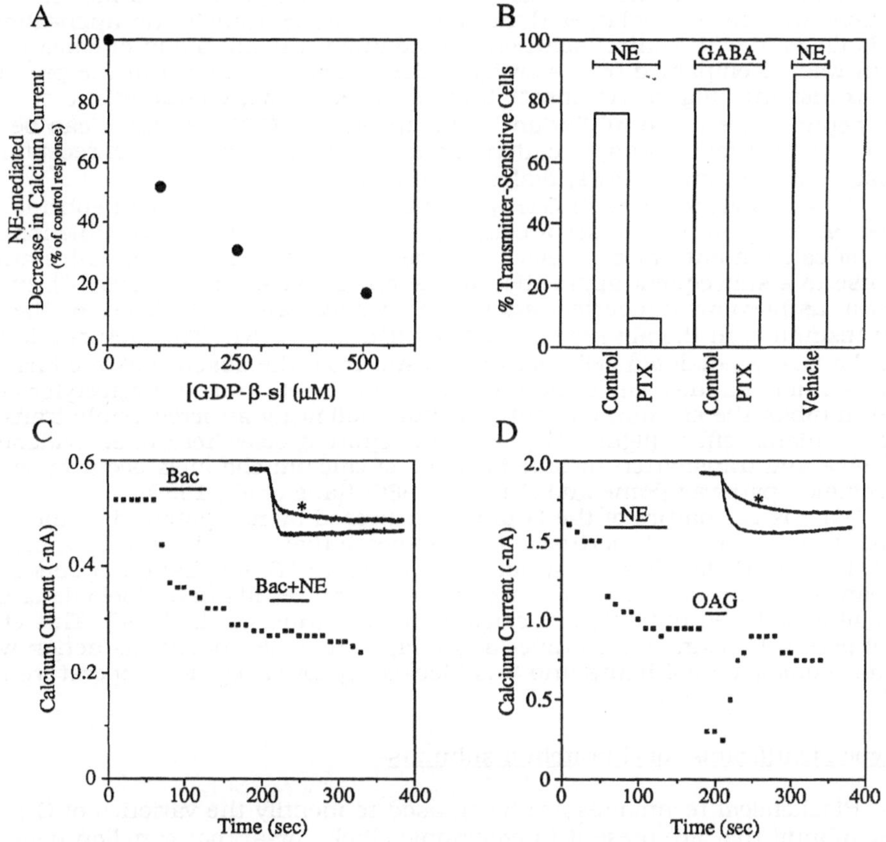

Figure 1. Effects of pertussis toxin (PTX) and thiophosphate derivatives of guanine nucleotides on calcium channel modulation. A) Dose-inhibition curve for the effect of GDP-β-s on the inhibition of calcium current by norepinephrine (NE). B) Histogram illustrating the effect of PTX pretreatment on the sensitivity of dorsal root ganglion neuron calcium currents to NE and GABA (the transmitter tested is noted above the relevant bars of the histogram. 'Vehicle' is ammonium sulfate in which the PTX is stored as a stock solution. C,D) Effects of saturating concentrations of inhibitory transmitters (10 μM NE and baclofen, Bac) and protein kinase C activator (60 μM OAG) on calcium currents before and after (*) application of Bac (C) or NE (D). The inset in C shows traces recorded (±Bac) in a neuron dialyzed without GTP-γ-s, and is included to illustrate that Bac in the absence of GTP-γ-s, produces a slowing in the activation kinetics similar to that observed following NE application to GTP-γ-s dialyzed neurons (D).

Recording macroscopic calcium currents with the whole cell variation of the patch clamp technique offers the advantage of studying the effects of intracellular application of these nucleotides. Inclusion of small molecules in the patch pipette solution allows them to passively diffuse into the cell cytoplasm (dialysis) when intracellular access is obtained. Results of experiments in which dorsal root ganglion neurons were dialyzed with GDP-β-s point to the involvement of a G protein in mediating the effects of the transmitters (Holz et al., 1986). This nucleotide produces a dose-dependent inhibition of the norepinephrine-induced decrease in calcium current (figure 1A).

Dialysis of the cells with GTP-γ-s produces the opposite effect. Concentrations of GTP-γ-s above 100 μM produced an acceleration of the rundown normally observed for calcium currents under whole cell recording conditions. In other words, relatively high concentrations of GTP-γ-s produce irreversible inhibition of calcium current. This calcium channel 'washout' is also accompanied by a slowing of the calcium current activation kinetics. Dialysis of the neurons with 100 μM GTP-γ-s did not produce a significant alteration in the macroscopic calcium current by itself but promoted an irreversible transmitter-mediated inhibition of calcium current. The inhibition produced under these conditions was invariably larger than that produced by transmitter application to cells dialyzed with control solutions. In addition, the inhibition of calcium current was accompanied by a slowing of the activation kinetics in the presence of the transmitter (figure 1D, inset). Such an effect is not characteristic of norepinephrine-mediated inhibition in the absence of GTP-γ-s, but it can be readily observed following application of baclofen (a selective $GABA_B$ receptor agonist), even without GTP-γ-s, dialysis (figure 1C, inset).

These experiments with GTP-γ-s further suggest that norepinephrine and GABA receptors may share a common pool of G protein. Irreversible inactivation of calcium current by either norepinephrine or baclofen eliminates any response to a subsequent application of the other transmitter (figure 1C). In contrast, as previously reported, additive responses can be observed for these two transmitters in the absence of GTP-γ-s (Dunlap, 1981). The lack of additivity in the GTP-γ-s-dialyzed cells does not result from the absence of 'modulatable' calcium channels since the protein kinase C activator, oleoylacetylglycerol (OAG), inhibits the calcium current remaining following an irreversible transmitter-mediated effect (figure 1D). Results reported elsewhere offer evidence that OAG- and transmitter-mediated actions of calcium channels occur by way of a common pathway (Rane and Dunlap, 1986; Rane et al., 1989).

What is the nature of the G protein activated by norepinephrine and GABA? Experiments with pertussis toxin indicate that it is likely to be a member of the G_i or G_o families. As summarized in figure 1B, incubation of dorsal root ganglion neurons in 140 ng/ml pertussis toxin results in a blockade of the transmitter-induced inhibition of calcium current (Holz et al., 1986). This effect of pertussis toxin is both time- and temperature-dependent; its action was virtually complete by 4 hours and was blocked by lowering the temperature to 4°C.

Subtype identification of G protein α subunits

Biochemical techniques are being used to identify the varieties of G protein α subunit that are present in embryonic chick dorsal root ganglion neurons. In these cells pertussis toxin catalyzed the ADP-ribosylation of substrates that migrate on one-dimensional SDS-PAGE gels as a doublet of molecular weight 40-41 kDa (figure 2A). Western blot analysis of these one-dimensional gels, using antisera directed against synthetic peptides corresponding to amino acid sequences predicted for cDNAs for pertussis toxin-sensitive G protein α subunits reveals a minimum of two Gi-like proteins (40 and 41 kDa) and a third G_o-like protein (40 kDa) (Holz et al., 1989). Two dimensional techniques, combining isoelectric focusing (IEF) with SDS-PAGE, demonstrate that this doublet can be resolved as four distinct spots with pIs between 5 and 6 (figure 2B).

Subcellular fractionation of chick cerebral cortical homogenates by sucrose density gradient centrifugation (Whittaker et al., 1964) indicates that these ribosylatable, immunoreactive G protein subunits are present in synaptosomal plasma membrane and synaptic vesicle fractions (Holz, Turner, and Dunlap, 1989). This suggests that these transduction molecules may play a critical role in the control of neurosecretion.

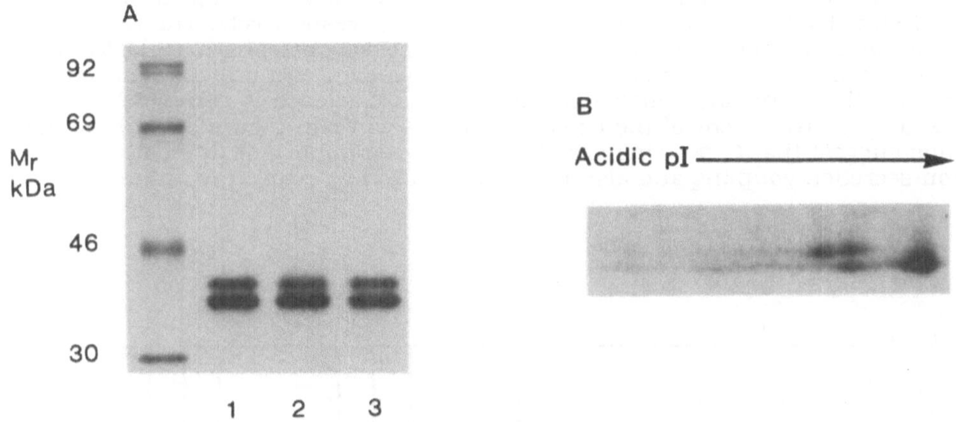

Figure 2. Pertussis toxin (PTX) substrates in sensory neurons. Autoradiograms of one-dimensional SDS-PAGE (A) and two-dimensional IEF/SDS-PAGE (B) gels. A) The left-most lane includes the molecular weight standards. Lanes 1 through 3 were loaded with 40 μg each of Lubrol-solubilized plasma membrane proteins from chick cerebral cortex (1), spinal cord (2), and sensory neurons (3), that were ADP-ribosylated in the presence of 1.4 μg/ml PTX and 1 μCi ^{32}P-NAD. B) On two-dimensional gels, the ribosylated proteins co-migrated with standards of molecular weight 41 kDa (upper two) and 40 kDa (lower two) and exhibited pIs between 5 and 6.

ROLE OF G PROTEINS IN REGULATING NEUROSECRETION

We have employed electrical field stimulation techniques to assess the involvement of G proteins in the regulation of substance P release from dorsal root ganglion neurons *in vitro* (Holz et al., 1988). This method involves the passage of a brief (3 ms) current pulse between two extracellular platinum-iridium electrodes in contact with the bathing solution which surrounds the cultured neurons. If the electrical field generated in the solution is sufficient, the neurons will be depolarized to threshold for action potential generation. Such a technique has been shown to effectively evoke action potentials in virtually every neuron in the dish; the stimulus is innocuous, as evidenced by visual and electrical measurements. Following stimulation at 1 Hz for 1-5 minutes, radioimmunoassay to detect the presence of substance P in the bathing solution shows that the peptide is effectively released with this protocol. Furthermore, the release is calcium-dependent and blocked by the addition of cobalt to the bathing solution.

Treatment of dorsal root ganglion neurons with pertussis toxin (140 ng/ml) for 4-8 hours prior to evoking release has dramatic effects on stimulus-secretion coupling. In experiments in which three phases of stimulation (S1-S3) were used to evoke the secretion of substance P, release of the peptide during S1 was accentuated (ca. 1.3-2 fold) in pertussis toxin-treated cells over that measured in control cultures (figure 3A). During phases S2 and S3, however, release was comparable for toxin-treated and control cells. This finding suggests that in control cultures prior to electrical stimulation (i.e., under steady-state, equilibrium conditions) pertussis toxin-sensitive G proteins may exert a tonic, inhibitory influence on stimulus-secretion coupling, possibly by regulating the size of the releasable pool of substance P. Treatment with pertussis toxin apparently disinhibits this regulatory mechanism.

In addition to its effects on substance P release during S1, pertussis toxin blocks the receptor-mediated modulation of release produced by norepinephrine and GABA (Holz et al., 1989). The effects of the transmitters are summarized in figure 3B. Doses of norepinephrine and GABA that are saturating with respect to their action on calcium current, produce a 60 and 52% inhibition of electrically-stimulated substance P release, respectively. The receptors that mediate this action of the transmitters on release are pharmacologically identical to those that modulate the voltage-dependent calcium channel. Furthermore, the transmitter-mediated modulation of release is completely blocked by pretreatment of the cells with pertussis toxin (figure 3B). Thus, our results suggest that G proteins are likely to play an important role both in excitation-secretion coupling and also in its modulation by neurotransmitters.

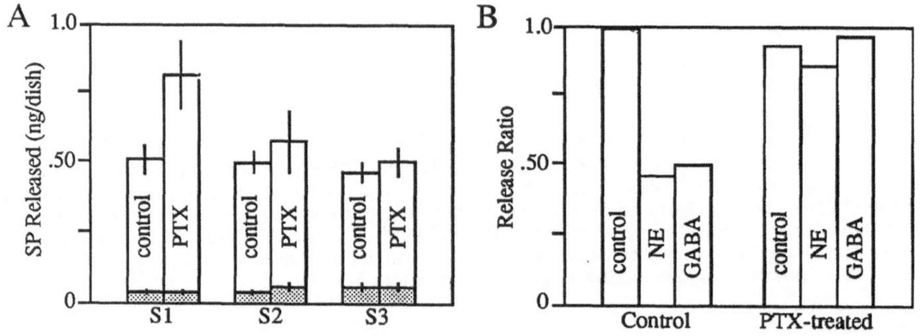

Figure 3. Modulation of substance P release by pertussis toxin (PTX) and neurotransmitters. A) Histogram illustrating the effect of PTX on the three phases of stimulation. Control cultures were incubated in standard saline solution containing ammonium sulfate (the vehicle in which PTX stock solutions were stored). B) Histogram illustrating the inhibition of electrically-stimulated substance P release produced by NE and GABA in control cultures and in those pre-treated with PTX for 4-8 hours. The release ratio was calculated as the amount of substance P released during S2 divided by the average release during phases S1 and S3.

WHICH G PROTEIN IS RESPONSIBLE FOR RECEPTOR-MEDIATED MODULATION OF NEURONAL CALCIUM CHANNELS?

The nature of the G protein that transduces these effects on secretion remains to be determined. Two different approaches have proven useful in addressing this question in other systems. G protein reconstitution experiments involve the inactivation of endogenous G proteins by pretreatment with pertussis toxin. Subsequent dialysis of the toxin-treated cells with purified and activated G protein subunits (or holoenzymes) and testing for a transmitter-mediated inhibition of calcium current has allowed investigators to determine which of the subunits is most efficient at restoring the receptor-mediated inhibition. Such experiments have been performed on neuroblastoma-glioma hybrid cells whose calcium current is inhibited by enkephalin in a pertussis toxin-sensitive manner. Dialysis of toxin-treated cells with α–GTP from Go or with activated Gi restored the enkephalin-induced inhibition of calcium current in the cells. (Hescheler et al. 1987). In these reconstitution experiments, Goα was found to be approximately ten times more potent than Gi. The investigators have, thus, concluded that Go is likely to be the G protein that transduces the effects of enkephalin.

Similar experiments have been performed on rat dorsal root ganglion neuron calcium currents that are inhibited by neuropeptide Y (Ewald et al., 1989). These investigators reported that three purified and activated α subunits (from G_{i1}, G_{i2}, and Go) were tested for their ability to reconstitute neuropeptide Y-mediated inhibition of calcium current in pertussis toxin-treated neurons and that only $G_{o\alpha}$ (at 100 nM) was able to fully restore function. The G_i subunits were ineffective at this concentration.

Conclusions based on the reconstitution experiments are necessarily limited to a comparison of potencies of the subunits tested. As a more direct means of identifying the nature of the endogenous transducing element, a somewhat different approach has been taken using antibodies which are selective for the various G protein subunits and which block G protein function. These antibodies have been injected into cells and tested for their effectiveness in blocking transmitter-mediated calcium current inhibition. For both the dopamine-induced inhibition of calcium current in *Helix* neurons (Harris-Warrick et al., 1988) and the noradrenaline-induced inhibition of calcium current in NG108-15 hybrid cells (McFadzean et al., 1989), antibodies to Go but not those to G_i block the transmitters' effects.

Taken together these data from other systems suggest that the neuron-specific G protein, Go, plays a major role in coupling transmitter receptors to alterations in voltage-dependent calcium channel function. This is significant not only because of its implications for the regulation of neurotransmitter release but also because an understanding of this G protein's functions has heretofore been elusive.

ACKNOWLEDGEMENTS

The authors wish to acknowledge the formal collaboration with Allen Spiegel who has supplied us with the antibodies for the Western blot analysis of G protein subunits. This work was supported by grants from the PHS (NS16483), the American Heart Association (with funds from the Massachusetts Affiliate), and the McKnight Fund for Neuroscience.

REFERENCES

Anderson, C.S. and Dunlap, K., 1988, Single L-type calcium channels in dorsal root ganglion neurons can be modulated by norepinephrine, Soc. for Neurosci. Abstr., 14:644.

Benham, C.D., Hess, P., and Tsien, R.W., 1987, Two types of calcium channels in single smooth muscle cells from rabbit ear artery studied with whole-cell and single-channel recordings, Circulation Research, 61:I-10.

Carbone, E. and Lux, H.D., 1984, A low voltage-activated, fully inactivating Ca channels in vertebrate sensory neurones, Nature, 310:501.

Ewald, D.A., Pang, I-H., Sternweis, P.C., and Miller, R.J., 1989, Differential G protein-mediated coupling of neurotransmitter receptors to calcium channels in rat dorsal root ganglion neurons in vitro, Neuron, 2:1185.

Fenwick, E.M., Marty, A., and Neher, E., 1982, Sodium and calcium channels in bovine chromaffin cells, Journal of Physiology, 331:599.

Fox, A.P., Nowycky, M.C., and Tsien, R.W., 1987a, Kinetic and pharmacological properties distinguishing three types of calcium currents in chick sensory neurones, Journal of Physiology, 394:149.

Fox A.P., Nowycky, M.C., and Tsien, R.W., 1987b, Single-channel recordings of three types of calcium channels in chick sensory neurones, Journal of Physiology, 394:173.

Hagiwara, S. and Ohmori, H., 1983, Studies of single calcium channel currents in rat clonal pituitary cells, Journal of Physiology, 336:649.

Harris-Warrick, R. M., Hammond, C., Paupardin-Tritsch, D., Homburger, V., Rouot, B., Bockaert, J., and Gerschenfeld, H. M., 1988, An α40 subunit of a GTP-binding protein immunologically related to Go mediates a dopamine-induced decrease of Ca^{2+} current in snail neurons, Neuron, 1:27.

Hescheler, J., Rosenthal, W., Trautwein, W., and Schultz, G., 1987, The GTP-binding protein, Go, regulates neuronal calcium channels, Nature, 325:445.

Hirning, L.D., Fox, A.P., McCleskey, E.W., Olivera, B.M., Thayer, S.A., Miller, R.J., and Tsien, R.W., 1987, Dominant role of N-type calcium channels in evoked release of norepinephrine from sympathetic neurons, Science. 239:57.

Holz, G.G., Dunlap, K., and Kream, R.M., 1988, Characterization of the electrically evoked release of substance P from dorsal root ganglion neurons: methods and dihydropyridine sensitivity, Journal of Neuroscience, 8:463.

Holz, G.G., Kream, R.M., Spiegel, A., and Dunlap, K., 1989, G proteins couple α-adrenergic and GABAB receptors to inhibition of peptide secretion from peripheral sensory neurons, Journal of Neuroscience, 9:657.

Holz, G. G., Rane, S. G., and Dunlap, K., 1986, GTP-binding proteins mediate transmitter inhibition of voltage-dependent calcium channels, Nature, 319:670.

Holz, G.G., Turner, T.J., and Dunlap, K., 1989, ADP-ribosylation and immunological characterization of G proteins that mediate presynaptic inhibition, Soc. for Neurosci. Abstr., (In Press).

Kostyuk, P.G., Shuba, Ya.M., and Savchenko, A.N., 1988, Three types of calcium channels in the membrane of mouse sensory neurons, Pflugers Archiv, 411:661.

McFadzean, I., Mullaney, I., Brown, D.A., and Milligan, G., 1989, Antibodies to the GTP-binding protein, Go, antagonise noradrenaline-induced calcium current inhibition in NG108-15 hybrid cells, Neuron, (In Press).

Nilius, B., Hess, P., Lansman, J.B., and Tsien, R.W., 1985, A novel type of cardiac calcium channel in ventricular cells, Nature, 316:443.

Nowycky, M.C., Fox, A.P., and Tsien, R.W., 1985, Three types of neuronal calcium channel with different calcium agonist sensitivity, Nature. 316:440.

Plummer, M.R., Logothetis, D.E., and Hess, P., 1989, Elementary properties and pharmacological sensitivities of calcium channels in mammalian peripheral neurons, Neuron, (In Press).

Rane, S. G., and Dunlap, K., 1986, Kinase C activator 1,2-oleoylacetylglycerol attenuates voltage-dependent calcium current in sensory neurons, Proc. Nat'l. Acad. Sci. USA, 83:184.

Rane, S.G. and Dunlap, K., 1989, G protein and protein kinase C mediated regulation of voltage-dependent calcium channels, in: "G proteins", R. Iyengar and L. Birnbaumer, eds., Academic Press, San Diego.

ANTIBODIES AS PROBES OF G-PROTEIN STRUCTURE AND FUNCTION

Allen M. Spiegel, William F. Simonds, Paul K. Goldsmith, and Cecilia G. Unson*

Molecular Pathophysiology Branch, National Institute of Diabetes, Digestive, and Kidney Diseases, National Institutes of Health Bethesda, Md. 20892, and *Department of Biochemistry, Rockefeller University, New York N. Y. 10021

GENERAL FEATURES OF G-PROTEIN STRUCTURE AND FUNCTION

G-proteins involved in signal transduction are members of a guanine nucleotide-binding protein superfamily that includes cytoskeletal proteins such as tubulin, soluble proteins (initiation and elongation factors involved in protein synthesis), and low molecular weight GTP-binding proteins such as the *ras* p21 protooncogenes and *ras*-related proteins (Spiegel, 1987; Gilman, 1987; Iyengar & Birnbaumer, 1987). Members of the G-protein subset of the GTP-binding protein superfamily share certain general features in common with other members of the GTP-binding protein superfamily: 1) all GTP-binding proteins bind guanine nucleotides with high affinity and specificity, and possess intrinsic GTPase activity that modulates interactions between the GTP-binding protein and other elements; 2) GTP-binding proteins serve as substrates for ADP-ribosylation by bacterial toxins; this covalent modification disrupts normal function.

G-proteins share other features that distinguish them from other GTP-binding proteins. These features include: 1) association with the cytoplasmic surface of the plasma membrane (*ras* p21 and some other low molecular weight GTP-binding proteins are also associated with the cytoplasmic membrane surface); 2) function as receptor-effector couplers; 3) heterotrimeric structure. G-proteins contain α, β, and γ subunits, each distinct gene products. The latter two subunits are tightly, but noncovalently linked in a βγ complex. α subunits bind guanine nucleotide, serve as toxin substrates, confer specificity in receptor-effector coupling, and directly modulate effector activity. Upon activation by GTP, α subunits are thought to dissociate from the $\beta\gamma$ complex. The latter is required for G-protein-receptor interaction, can inhibit G-protein activation by blocking α subunit dissociation, and may in some cases directly regulate effector activity.

SPECIFIC FEATURES OF G-PROTEIN STRUCTURE AND FUNCTION

Molecular cloning provides evidence for a minimum of eight distinct α subunit genes, including G_s, G_o, G_{i1}, G_{i2}, G_{i3}, G_{t1}, G_{t2}, (Spiegel, 1987; Gilman, 1987; Iyengar & Birnbaumer, 1987) and $G_{x(z)}$ (Matsuoka et al., 1988; Fong et al., 1988; Lochrie & Simon, 1988). Further diversity is created by alternative splicing leading to the expression of four forms of G_s (Bray et al., 1986). At least two distinct genes each exist for both β and γ subunits. The expression of certain α subunits is highly restricted, e.g. G_{t1} and G_{t2} found only in photoreceptor rod and cone cells, respectively, whereas others such as G_s are expressed ubiquitously.

The specificity of G-protein interactions with receptors and effectors has been defined in very few cases. Studies involving reconstitution of purified receptors and G-proteins in phospholipid vesicles showed that β-adrenergic receptors couple, in decreasing order of efficiency, to $G_s > G_i >> G_t$, and that for rhodopsin, the selectivity of coupling is $G_t = G_i >> G_s$ (Cerione et al., 1985). Similar studies involving G-protein-effector interaction indicated that only G_s can activate adenylyl cyclase (G_s also appears to stimulate another effector, a Ca^{++} channel), and that G_t uniquely activates retinal cGMP phosphodiesterase (see Roof et al., 1985, for example). The endogenous G-proteins coupled to most other receptors and effectors, however, remain to be identified. Many G-protein-coupled receptors (D_2-dopaminergic for example, Senogles et al., 1987), and a variety of effectors, including adenylyl cyclase (inhibition), certain Ca^{++} (inhibition) and K^+ (stimulation) channels, and phospholipase C in cell types such as neutrophils (e.g. f-Met-Leu-Phe receptor) are regulated by one or more pertussis toxin-sensitive G-proteins (Spiegel, 1987; Gilman, 1987; Iyengar & Birnbaumer, 1987). Since, not only both forms of G_t, but also G_{i1}, G_{i2}, G_{i3}, and G_o are pertussis toxin-sensitive G-proteins, demonstration of an effect of pertussis toxin on receptor or effector regulation does not uniquely identify the relevant endogenous G-protein. In most cells, phospholipase C is regulated by a pertussis toxin-insensitive G-protein. $G_{z(x)}$ may play this role, but definitive evidence is lacking. Phospholipase A_2 activity may also be regulated by one or more G-proteins, perhaps by the $\beta\gamma$ complex. The latter has also been suggested to stimulate a K^+ channel, but recent evidence suggests this is an indirect effect (Kim et al., 1989).

PEPTIDE ANTIBODIES AS PROBES OF G-PROTEIN STRUCTURE AND FUNCTION

We have generated numerous antisera against synthetic peptides corresponding to the predicted amino acid sequence of G-protein subunits. Such antisera have proved very useful in identifying the putative proteins encoded by cloned cDNAs (Goldsmith et al., 1987, 1988b), in discriminating between closely related G-proteins (Goldsmith et al., 1988a), and in quantitating and localizing G-proteins. For example, three antisera, LD, LE, and SQ were raised against an internal decapeptide (residues 159-168) of α subunits of G_{i1}, G_{i2}, and G_{i3}, respectively. These proved capable of discriminating between α subunits of 41 and 40kDa purified from brain, and of 41kDa purified from HL-60 cells, and served to identify these as the products of G_{i1}, G_{i2}, and G_{i3}, respectively (Goldsmith et al., 1988a and 1988b).

One approach theoretically capable of identifying endogenous G-proteins coupling to particular receptors and effectors involves the use of specific antibodies that can bind to native G-proteins and affect their function. Antibodies directed against the carboxy-terminus of G_α subunits hold particular promise in this regard. Several lines of evidence point to the importance of the extreme carboxy-terminus in G-protein-receptor coupling. These include the locus of the receptor-uncoupling mutation (*unc*) in S49 mouse lymphoma cells (6th residue from carboxy-terminus, Sullivan et al., 1987), and the ability of pertussis-toxin catalyzed ADP-ribosylation of a cysteine residue (4th from carboxy-terminus) to uncouple G-proteins from receptors (West et al., 1985).

To study G-protein coupling with this approach, antibodies with well-defined specificity are essential. We had previously raised antisera AS/6 and 7, and GO/1 (referred to henceforth as AS and GO, see Table I) against synthetic decapeptides corresponding to the carboxy-termini of α subunits of G_t and G_o, respectively. We characterized the specificity of these antisera by performing immunoblots of several distinct purified G-proteins. With this approach, we found that AS reacts, not only with G_t purified from rod outer segments (presumably G_{t1}), but also with α_{41} and α_{40} proteins (presumably G_{i1} and G_{i2}, respectively) purified from brain. We also showed that AS does not react with purified brain α_{39} (G_o), but does react, albeit weakly, with an α_{41} protein purified from HL-60 cells and tentatively identified as G_{i3} (Goldsmith et al., 1987 and 1988a). GO antiserum was shown to react with brain α_{39} (G_o), but not with brain α_{41} (G_{i1}) or α_{40} (G_{i2}); reactivity against G_{i3} had not been tested (Goldsmith et al., 1988b).

In order to generate antisera capable of recognizing each of the known G_α subunits, we synthesized additional decapeptides corresponding to the carboxy-termini of G_s, $G_{x(z)}$, and G_{i3}, and used these to immunize rabbits (Table I).

We assessed the reactivity and specificity of these antibodies against proteins encompassing the immunogenic peptide sequences by performing immunoblots and immunoprecipitation of bacterial lysates containing unique, defined G_{α} subunits expressed by recombinant DNA techniques (kindly provided by J. Codina, Baylor College of Medicine, Houston, Texas). On immunoblot, RM and QN antisera are absolutely specific for G_s and G_x, respectively. G_s and G_x, moreover, are not reactive with GO, EC, or AS antisera. AS antiserum reacts strongly and equivalently with G_{i1} and G_{i2}, shows weak reactivity against G_{i3}, and no reactivity against G_o. GO and EC antisera react best with G_o and G_{i3}, respectively, but they display very substantial reciprocal

Table I. Carboxy-terminal Decapeptides and Corresponding Antisera

peptide	sequence[a]	antiserum[b]	G-α subunit[c]
RM	RMHLRQYELL	RM	G_s
QN	QNNLKYIGLC	QN	$G_{x(z)}$
GO	ANNLRGCGLY	GO	G_o
EC	KNNLKECGLY	EC	G_{i3}
KE	KENLKDCGLF	AS	G_{t1}, G_{t2}
KN	KNNLKDCGLF	-	G_{i1}, G_{i2}

[a]Single letter amino acid code.
[b]Antiserum raised by immunization with designated peptide.
[c]Designated peptide represents carboxy-terminus of corresponding G-α subunit (Lochrie & Simon, 1988).

cross-reactivity. The somewhat unexpected pattern of reactivity of AS, GO, and EC reflects the importance of the carboxy-terminal residue, phenylalanine vs. tyrosine, in determining antigenicity (Table I). The identical pattern of reactivity was observed in immunoprecipitation experiments which also indicated that the antisera can recognize native G-proteins (Fig. 1).

Using purified proteins reconstituted into phospholipid vesicles, we found that affinity-purified AS antibodies specifically block rhodopsin-G_t (transducin) interaction, but do not block interaction between activated G_t and its effector, cGMP phosphodiesterase (Cerione et al., 1988). This approach was then extended to G-proteins *in situ*, i.e. in native membranes. RM antibodies were shown on immunoblots to recognize the multiple forms of G_s-α derived from alternative splicing. Affinity-purified RM antibodies block

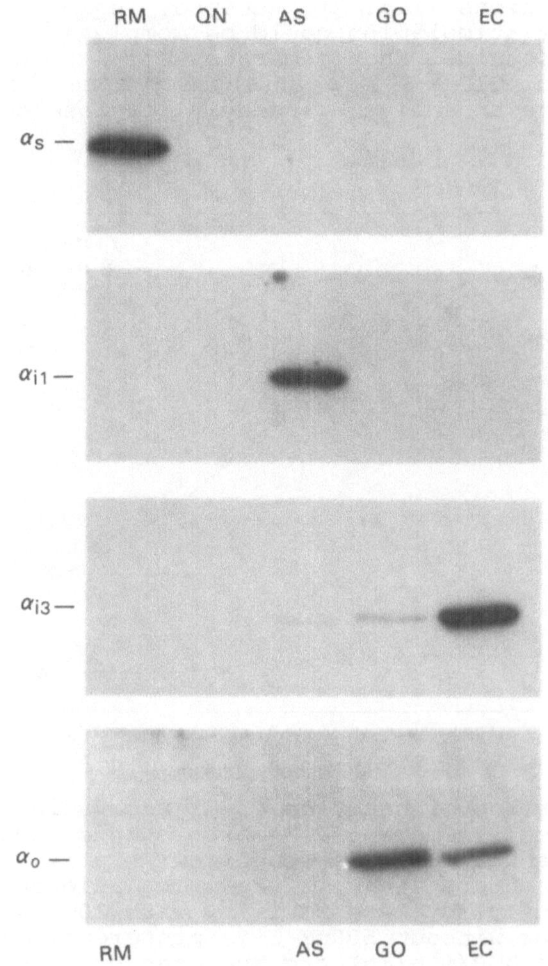

Figure 1. Analysis of specificity of G-protein antisera by immunoprecipitation of G-α subunits expressed in E. Coli. Soluble fractions of E. Coli lysates containing recombinant α subunits of G_s, G_{i1}, G_{i3}, and G_o, were incubated with affinity-purified antibodies (20ug/ml final concentration), and immunoprecipitation performed with protein-A sepharose. For each α subunit, five equivalent aliquots were incubated with each of the indicated antibodies. α subunit content of the washed, immunoprecipitated pellets were determined by immunoblot analysis, using the appropriate antibody for each subunit and iodinated protein-A for detection of bound first antibody. The positions of the immunoreactive G_α subunits are indicated.

receptor-mediated adenylyl cyclase stimulation in S49 mouse lymphoma cell membranes, but inhibit fluoride stimulation (which bypasses the receptor to activate G_s) only at much higher concentrations (Fig. 2). Antibody inhibition of adenylyl cyclase stimulation could be completely prevented by the cognate peptide. When membranes are preactivated (with fluoride or GTP-γ-S), RM antibodies specifically immunoprecipitate an active G_s-adenylyl cyclase complex (Fig. 3).

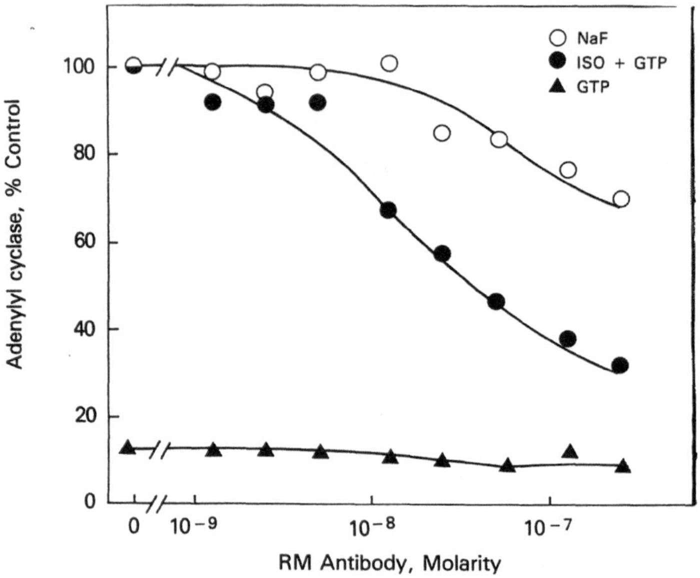

Figure 2: RM antibody inhibition of G_s activation in S49 mouse lymphoma cell membranes. Membranes were incubated with increasing concentrations (as shown) of RM affinity-purified antibody for two hours at 4°, after which aliquots were assayed for adenylyl cyclase activity with 10mM NaF, or with 100uM GTP with or without 500uM l-isoproterenol (as shown). Values (in pmol cAMP/min/mg) are the mean of triplicate determinations and are expressed as percent of control (84 for NaF and 60 for isoproterenol).

These results have major implications for our understanding of G-protein structure and function, and for the elucidation of the specificity of receptor-effector coupling by G-proteins. The ability of carboxy-terminal decapeptide antibodies to block receptor-G-protein interaction emphasizes the importance of this domain of the α subunit in receptor interaction. The ability of antibodies to interact with the membrane-bound G-protein implies that this region is exposed at the cytoplasmic surface, and is consistent with the ability of pertussis toxin to covalently modify a cysteine within the carboxy-

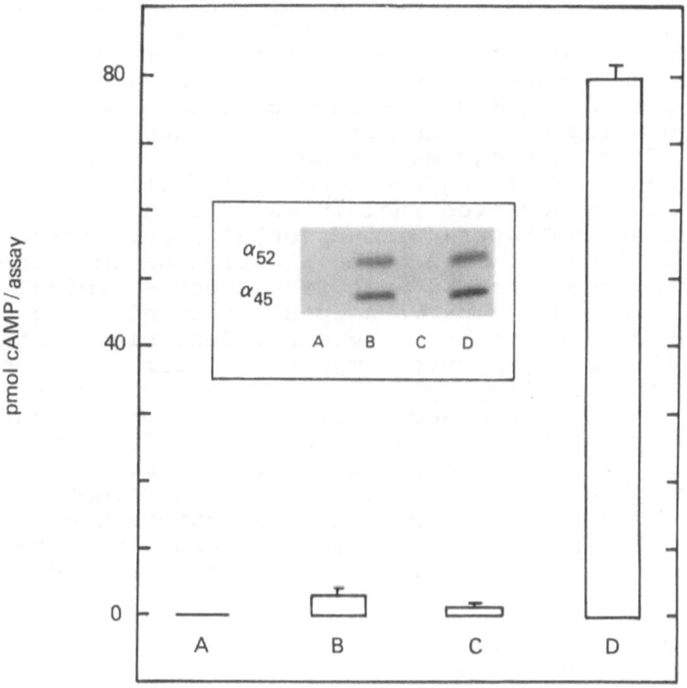

Figure 3: Immunoprecipitation of adenylyl cyclase activity and G_s by RM antibody. Detergent extracts were prepared from bovine brain membranes with (C and D) or without (A and B) preactivation by 100uM GTP-γ-S for 20min at 30°. Extracts (3mg/ml) were incubated with normal rabbit immunoglobulin (A and C) or RM antibody (B and D) for four hours at 4°. Immunoprecipitation was performed with Protein A Staph. Aureus cells (Pansorbin). The immunoprecipitated pellet was washed, resuspended in detergent buffer, and adenylyl cyclase activity determined. The activity of 20uL aliquots from each immunoprecipitate is shown. Inset: parallel immunoblot analysis of α_s content in immunoprecipitates. The positions of the 52 and 45kDa forms of α_s are indicated.

terminal decapeptide in the membrane-bound G-protein. Antibody binding to the carboxy-terminus, however, does not result in global inhibition of G-protein function. Indeed, RM antibody immunoprecipitation of a G-protein-effector complex implies that the carboxy-terminal decapeptide of the α subunit is _not_ critically involved in effector interaction.

The ability of carboxy-terminal decapeptide antibodies to block receptor-G-protein interaction in native membranes should be useful in defining the specificity of these interactions. The immunoblot studies described above indicate that, at a minimum, one should be able to distinguish coupling to G_{i1} and/or G_{i2} from coupling to G_{i3} and/or G_o. Studies in progress already indicate the utility of this approach in defining the specificity of α_2-adrenergic receptor-mediated inhibition of adenylyl cyclase in human platelet membranes (Simonds et al., unpublished observations). Likewise, immunoprecipitation of activated G-protein-effector complexes by specific carboxy-terminal peptide antibodies should prove helpful in identifying effectors linked to specific G-proteins. Recombinant DNA techniques will clearly provide important information on receptor-effector coupling by G-proteins, e.g. α subunits of G_{i1}, G_{i2}, and G_{i3} expressed in E. Coli all proved capable of stimulating a K^+ channel (Yatani et al., 1988), but such approaches can only indicate which G-proteins are potentially capable of coupling to individual receptors and effectors. The use of specific antibodies offers the advantage of identifying the endogenous G-protein(s) that actually does couple to individual receptors and effectors.

ACKNOWLEDGMENTS

We thank J. Codina, Baylor College of Medicine, for providing G-proteins expressed in E. Coli, and R. Cerione, Cornell University, for collaborative experiments.

REFERENCES

Bray, P., Carter, A., Simons, C., Guo, V., Puckett, C., Kamholz, J., Spiegel, A., and Nirenberg, M., 1986, Human cDNA clones for four species of G-α_s signal transduction protein, _Proc. Natl. Acad. Sci. U. S. A._ 83:8893.

Cerione, R.A., Staniszewski, C., Benovic, J.L., Lefkowitz, R.J., Caron, M.G., Gierschik, P., Somers, R., Spiegel, A.M., Codina, J., and Birnbaumer, L., 1985, Specificity of the Functional Interactions of the beta-Adrenergic Receptor and Rhodopsin with Guanine Nucleotide Regulatory Proteins Reconstituted in Phospholipid Vesicles, _J. Biol. Chem._, 260:1493.

Cerione, R.A., Spencer K., Rajaram R., Unson, C., Goldsmith, P., and Spiegel, A., 1988, An Antibody Directed against the Carboxyl-terminal Decapeptide of the α Subunit of the Retinal GTP-binding Protein, Transducin, _J. Biol. Chem._, 263:9345.

Fong, H.K.W., Yoshimoto, K.K., Eversole-Cire, P., and Simon, M.I., 1988, Identification of a GTP-binding protein

alpha subunit that lacks an apparent ADP-ribosylation site for pertussis toxin, Proc. Natl. Acad. Sci. U.S.A., 85:3066.
Gilman, A., 1987, G Proteins: Transducers of receptor-generated signals, *Ann. Rev. Biochem.*, 56:615.
Goldsmith, P., Gierschik, P., Milligan, G., Unson, C.G., Vinitsky, R., Malech, H., and Spiegel, A., 1987, Antibodies Directed against Synthetic Peptides Distinguish between GTP-binding Proteins in Neutrophil and Brain, *J. Biol. Chem.*, 262:14683.
Goldsmith, P., Backlund, Jr., P.S., Rossiter, K., Carter, A., Milligan, G., Unson, C.G., and Spiegel, A., 1988a, Purification of Heterotrimeric GTP-Binding Proteins from Brain: Identification of a Novel Form of Go, *Biochemistry*, 27:7085.
Goldsmith, P., Rossiter, K., Carter, A., Simonds, W., Unson, C.G., Vinitsky, R., and Spiegel, A., 1988b, Identification of the GTP-binding Protein Encoded by Gi3 Complementary DNA, *J. Biol. Chem.*, 263:6476.
Iyengar, R., and Birnbaumer, L., 1987, Signal transduction by G-proteins, ISI Atlas of Science: Pharmacology, 1:213.
Kim, D., Lewis, D.L., Graziadei, L., Neer, E.J., Bar-Sagi, D., and Clapham, D.E., 1989, G-protein βγ-subunits activate the cardiac muscarinic K^+-channel via phospholipase A_2, *Nature*, 337:557.
Lochrie, M.A. and Simon, M.I., 1988, G Protein multiplicity in eukaryotic signal transduction systems, *Biochemistry*, 27:4957.
Matsuoka, M., Itoh, H., Tohru, K., and Kaziro, Y., 1988, Sequence analysis of cDNA and genomic DNA for a putative pertussis toxin-insensitive guanine nucleotide-binding regulatory protein α subunit, *Proc. Natl. Acad. Sci. U. S. A.*, 85:5384.
Roof, D.J., Applebury, M.L., and Sternweis, P.C., 1985, Relationships within the family of GTP-binding proteins isolated from bovine central nervous system, *J. Biol. Chem.*, 260:16242.
Senogles, S.E., Benovic, J.L., Amlaiky, N., Unson, C., Milligan, G., Vinitsky, R., Spiegel, A.M., and Caron, M.G., 1987, The D2-dopamine receptor of anterior pituitary is functionally associated with a pertussis toxin-sensitive guanine nucleotide binding protein, *J. Biol. Chem.*, 262:4860.
Spiegel, A., 1987, Signal transduction by guanine nucleotide binding proteins, *Mol. Cell. Endocrinol.*, 49:1-16.
Sullivan, K.A., Miller, R.T., Masters, S.B., Beiderman, B., Heideman, W., and Bourne, H.R., 1987, Identification of receptor contact site involved in receptor-G protein coupling, *Nature*, 330:758.
West, R.E.Jr., Moss, J., Vaughan, M., Liu, T., and Liu, T-Y., 1985, Pertussis toxin-catalyzed ADP-ribosylation of transducin: Cysteine 347 is the ADP-ribose acceptor site, *J. Biol. Chem.*, 260:14428.
Yatani, A., Mattera, R., Codina, J., Graf, R., Okabe, K., Padrell, E., Iyengar, R., Brown, A.M., and Birnbaumer, L., 1988b, The G protein-gated atrial K^+ channel is stimulated by three distinct G_i α-subunits, *Nature*, 336:680.

TRANSDUCIN : A MODEL OF BIOLOGICAL COUPLING ENZYMES

Yee-Kin Ho

Department of Biological Chemistry
University of Illinois at Chicago
Chicago, Illinois 60612

I. Introduction

In order to survive, living organisms must be able to respond to changes in their environment. By coordinating a diversity of cellular processes, the responses range from chemotaxis, cell differentiation and proliferation, sporulation and germination, to changes inside the cell, such as the modulation of enzyme activities, activation of protein kinases, and expression of new gene products. The network of cellular coupling processes is too complicated and diverse to be discussed in details. However, based on their basic functions, one can dissect the biological coupling processes into four categories: (1) Mechanical Coupling, (2) Signal Transduction, (3) Informational Transfer and (4) Energy Coupling. In spite of the diversity of the functions involved, recent studies have shown that all coupling processes are mediated by nucleotide binding proteins. For examples, mechanical coupling regulates cell motility which is mediated by the actions of actinomyosin ATPase, dynein ATPase, kinesin ATPase and tubulin. In signal transduction pathways, a host of GTP-binding proteins, the G-proteins, are now known to couple cell surface receptors to cytosolic effector enzymes. In the informational transfer processes such as DNA replication and recombination, RNA transcription and splicing, ATP-binding proteins including the rec A protein, DNA polymerase the rho transcription termination factor, the DNA gyrase play an essential role in controlling the coupling processes. In protein synthesis, GTP-binding proteins such as the elongation, iniation and termination factors carry out individual coupling step during translation. Finally, in energy transduction, the F_oF_1 ATPase catalyzes the reaction that transforms energy stored in the ion gradient to the chemical form of ATP. Most of these coupling processes are essential for cell survival. It is of interest to understand their regulatory mechanism. Do all these biological coupling enzymes share similar mechanisms in regulating various cellular processes ? And, how are these coupling processes and the coupling enzymes evolved? In this chapter, the retinal signal coupling G-protein, transducin and mechanical coupling proteins, actinomyosin and elongation factor EF-Tu and Ts will be used as examples of biological coupling enzymes for critical comparison which focuses on the mode of coupling actions, their reaction dynamics and possible evolutionary links among these biological coupling enzymes.

II. The Cycles of Biological Coupling Enzymes

A. The Coupling Cycle of Transducin

In vertebrate rod photoreceptor cells, the visual excitation process involves the activation of a cGMP enzyme cascade (For a review, see Liebman, *et al.*, 1987). Transducin (T) mediates the activation signal from photolyzed rhodospsin (R*) to the cGMP phosphodiesterase (PDE). The activation of PDE leads to reduction of the cytosolic cGMP concentration which in turn, regulates the Na^+ permeability of the plasma membrane by controlling a cGMP-sensitive channel. By this mechanism, a photochemical signal generated by the absorption of a photon by rhodopsin is transduced into a neural signal recognizable by other cells in the retina.

Transducin is a heterotrimer composed of three polypeptides: T_α (39 kDa), T_β (37 kDa) and T_γ (8.5 kDa). T_α contains a single guanine nucleotide binding site and is the activator of PDE. The GDP-bound form of T_α has a high affinity for $T_{\beta\gamma}$ and R*, and a low affinity for PDE. $T_{\beta\gamma}$ functions as a modulator which enhances the binding of T_α to R*. R*/ $T_{\beta\gamma}$ catalyzes the nucleotide exchange. Upon exchanging bound GDP for GTP, T_α-GTP dissociates from $T_{\beta\gamma}$ and R*. This dissociation allows the free R* to recycle and bind another transducin molecule. The T_α-GTP, which is no longer interacting with R* and $T_{\beta\gamma}$, has a very tight binding for GTP and acquires a high affinity for PDE. This interaction activates the latent PDE which is terminated when bound GTP is spontaneously hydrolyzed to GDP by the intrinsic GTPase activity associated with T_α (Fung, 1983). The basic mechanism of signal transduction utilized by transducin is a subunit dissociation and association cycle regulated via GTP binding and hydrolysis. By switching between GDP and GTP, transducin can exist in two conformations which differentiate between the binding to the receptor and effector. The transducin coupling cycle is depicted in Figure 1.

B. The Coupling Cycle of Actinomysin

The actinomyosin ATPase in muscle contraction carries out many cycles of sequential association and dissociation between the actin and the myosin filaments (Hibberd and Trentham, 1986). In the resting stage, myosin heavy chain contains a tightly bound ADP. The interaction between actin and myosin facilitates the exchange of bound ADP for ATP on the myosin site. The binding of ATP triggers a conformational change of myosin heavy chain, which leads to the dissociation of the myosin heavy chain from the actin filament. After the hydrolysis of the bound ATP, the myosin-ADP, reassociates with another actin subunit along the actin filament for the next cycle of reaction. In this system, the myosin heavy chain is the activator equivalent to T_α. The actin, which facilitates the ATP exchange, is the modulator. Nature uses a similar mechanism in dynein ATPase and kinesin ATPase to control movement of cilia, flagella. Here tubulin rather than actin acts as the modulator and facilitates the ATP exchange reaction.

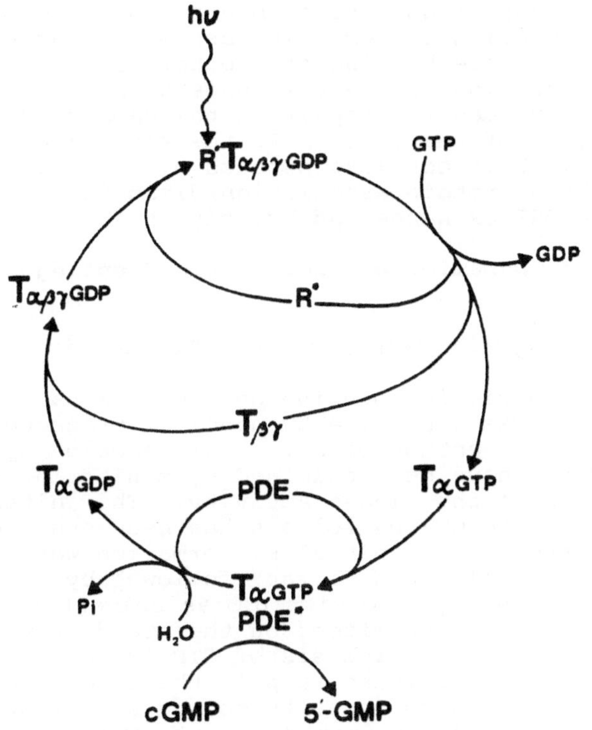

Figure 1. The Transducin Coupling Cycle in Visual Excitation

C. The Coupling Cycle of Elongation Factors Tu-Ts

Elongation factor couples the amino-acyl-tRNA to the ribosome for the confirmation of codon on the mRNA which results into the incorporation of a specific amino-acid to the nacent polypeptide chain (Kaziro, 1978). The latent form of EF-Tu contains tightly bound GDP and is associated with EF-Ts. The Tu-TS association facilitates the exchange reaction in which GTP is incorporated into EF-Tu. EF-Tu is equivalent to T_α of transducin and Ts functions as the modulator. The Tu-GTP complex dissociates from Ts and forms a complex with its effector, the amino-acyl tRNA molecule, which then binds to the ribosome to allow the coupling between the amino-acyl-tRNA and the mRNA to occur. After the hydrolysis of the bound GTP, Tu-GDP dissociates from the tRNA and reassociates with Ts. The Tu-GDP-Ts complex is then ready for another activation cycle.

D. The Coupling Cycle of Rec A Protein

Homologous recombination of DNA represents a major coupling event in biological systems. The rec A protein facilitates these processes by forming an assembly for the binding of the two DNA molecules (Kowalczykowski, 1987). Rec A protein contains an ATP binding site and is the activator. The assembly of rec A proteins is again regulated by the exchange and hydrolysis of ATP. Dissociated rec A protein contains tightly bound ADP in its latent form. The complex of single strand DNA (ssDNA) with ssDNA binding protein catalyzes the exchange of ATP for bound ADP. The resulting rec A-ATP complex dissociates from the ssDNA region and polymerizes around the double-stranded DNA, the effector of this coupling system. The assembled rec A filament provides the site for the binding of another double-stranded DNA and permits the screening of the existence of homologous sequences that are needed for repairing the damaged DNA. After the hydrolysis of the bound ATP, the rec A-ADP dissociates from the filament. This coupling process by rec A protein is again generated by a protein association/dissociation cycle regulated via the ATP exchange and hydrolysis.

III. The Reaction Dynamics of Biological Coupling Enzymes

A. The GTP Hydrolysis Mechanism of Transducin

Purified transducin in solution does not exchange its bound nucleotide or hydrolyze GTP. The turnover of transducin requires the catalytic action of R*. The reaction dynamics of the transducin cycle have been examined by monitoring the pre-steady kinetics of the GTPase activity. The initial rate of the Pi formation due to GTP hydrolysis has been studied using a acid quenching method. The initial Pi formation was found to be biphasic with an initial rapid "burst" followed by a slow steady state rate. Substituting H_2O with D_2O as solvent slows down the initial burst but has no effect on the steady rate. This observation suggests that hydrolysis of GTP is not rate-limiting. The rate-limiting step is the release of the hydrolysis product, Pi which constitutes the steady state rate. The accumulation of T_α-GDP·Pi should occur. The rate of Pi release from the T_α-GDP·Pi complex was found to have a half life of ~20 sec which is similar to the rate of PDE deactivation due to the turnover of GTP. This result suggests that T_α-GDP·Pi has an active conformation which is fully capable of activating the effector enzyme. The overall reaction is shwon below (Ting and Ho, 1989).

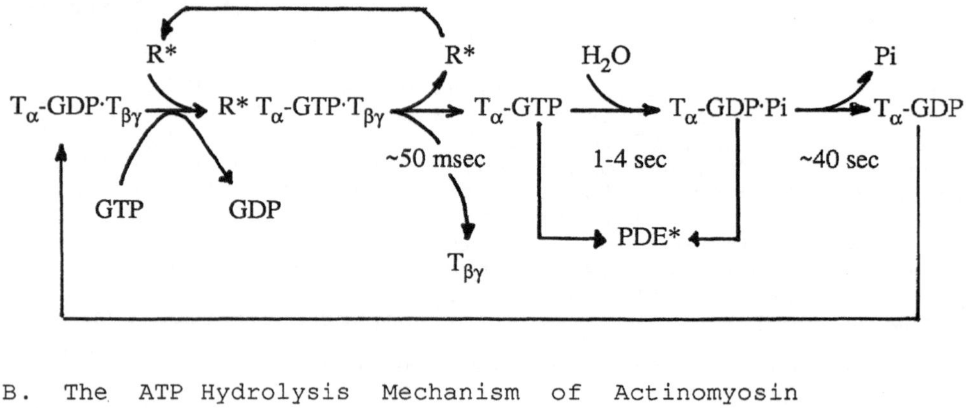

B. The ATP Hydrolysis Mechanism of Actinomyosin

Homologies among these coupling cycles suggest that they may share similar characteristics in their GTP/ATP hydrolysis mechanism. The reaction dynamics of several systems such as actin-myosin, tubulin, kinesin, dynein, EF-Tu, have been elucidated. The reaction mechanism of myosin ATPase and its relationship to muscle concentration can be used as an example. As described in the following scheme, the binding of ATP to myosin is facilitated via the interaction of actin. The hydrolysis of the bound ATP is fast, but the release of the hydrolysis products, ADP and Pi is slow and represents the rate-limiting step of the actin-myosin coupling cycle. As a result, the myosin-ADP.Pi intermediate is accumulated. Myosin-ADP.Pi has an active conformation which remains dissociated from actin (Lymn and Taylor, 1971).

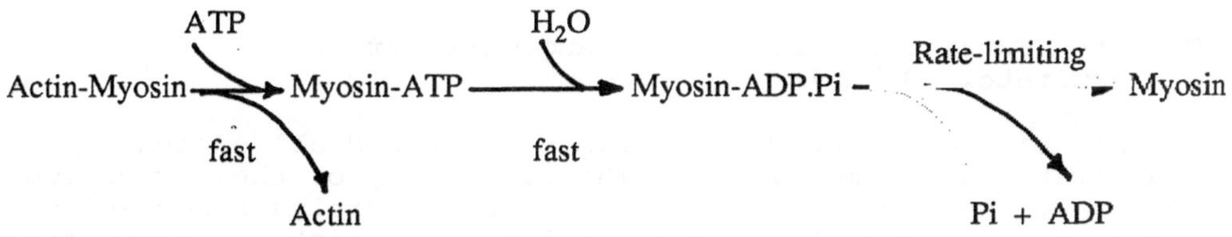

Besides actinomyosin, tubulin, kinesin, dynein and EF-Tu, all share the similar dynamics that the release of the hydrolytic product, Pi, is rate limiting. The similarity among all the coupling enzymes is striking.

C. Common Features of Nucleotide Binding Sites

Several features common to the nucleotide binding sites among the biological coupling enzymes are noteworthy.

1. The hydrolysis of ATP or GTP by the activator is via a direct hydrolysis mechanism without the formation of a phosphorylated intermediate. This feature distinguishes the transduction enzymes from the ion translocation Na^+/K^+ ATPase and Ca^{2+} ATPase. In general, the biological coupling enzymes are less sensitive to vanadate inhibition which is specific for those ATPase with a phosphorylated intermediates in their hydrolysis mechanisms.

2. The bound nucleotides in the coupling enzymes serve as a conformational control trigger the coupling processes. The hydrolysis of the bound nucleotide (ATP or GTP) resets the system to its latent state and is not for energy requirement in the coupling processes. All of the coupling enzymes can be activated by non-hydrolyzable analogues of ATP or GTP. Moreover, the reaction dynamics of the coupling enzymes indicate that an active intermediate of activator-NDP.Pi is accumulated. Phosphate analogues, such as AlF_4^- and BeF_4^- are capable of activating the coupling enzymes in the presence of either GDP or ADP.

3. The coupling enzymes described above are all sensitive to sulfhydryl modification especially after the removal of the tightly bound nucleotide. The relationship of the sulfhydryl groups to the coupling mechanism of the enzymes remains to be established.

4. Based on the x-ray crystal structure of EF-Tu and the ras-p21 protein, the coupling enzymes have a separate Mg^{2+} binding site close to the nucleotide site. The Mg^{2+} may form a mono-dentate ligand with the terminal phosphate of the bound GTP. As a result, the coupling enzymes show no selectivity on

the stereoisomers of Cr(III) β,γ-bidentate ATP or GTP complexes. This feature has been demonstrated in transducin and actinomysin and is likely to be shared by other coupling enzymes (Frey, et al., 1988). This property distinguishes the nucleotide binding site of the coupling enzymes to that of kinases which ultilize the Mg^{2+}-ATP bidentate complex as substrate for the phosphoryl transfer reaction. The kinases generally select either the Δ or the Λ isomers of the Cr(III)ATP complexes as substrate.

IV. The Functional Domain Arrangement of the Activator Subunit

All biological coupling enzymes discussed above consist of an activator and a modulator. The similarity of their coupling cycles and the homology of the nucleotide hydrolysis mechanism suggest that they might have an evolutionary link. Judging from the diversity of receptors, effectors and modulators in different coupling systems, such an evolutionary link should exist only on the nucleotide binding activators. It is possible that the development of the biological coupling processes originated from a nucleotide binding protein, which existed in two different conformations as GTP or GDP, is bound to the site. The existence of these two conformations was later utilized to couple information between the receptor and effector. By splicing in different receptor and effector interacting domains to this nucleotide binding protein, a battery of biological enzymes coupled to a variety of cellular processes could be generated. If this is the case, the arrangement of their functional domains with respect to the binding of modulator/receptor, nucleotide and the effector on a linear peptide map of the activator may exhibit common order. A comparison of the functional peptide maps of the T_{α} subunit of transducin, EF-Tu and myosin heavy chain is shown in Figure 2. In general, the nucleotide binding site consists of two regions, one near the amino-terminus for interacting with the γ-phosphate and another near the carboxyl-terminus for interacting with the purine moiety of the nucleotide. In between these two regions lies the proposed effector binding sites for the proteins. Examples of this arrangement are the putative PDE binding domain of T_{α}, the tRNA binding site of EF-Tu. The amino- and carboxyl-terminal peptide generally contains the modulator/receptor binding site. This seemingly uniform arrangement of functional domains may be related to the evolution of these coupling enzymes.

V. Concluding Remarks

Important progress has been made in the characterization of the regulatory role of transducin in visual excitation. The structure and function of transducin has been reviewed extensively (Ho, et al., 1989) and will not be repeated in this chapter. Closely related to transducin in structure and function are a host of GTP-binding regulatory proteins that mediate diverse signal transduction processes. The GTP-binding protein superfamily haw also been highlighted in several excellent reviews (Stryer & Bourne, 1986; Gilman, 1987) and will not be presented here. Instead, a comparison of the coupling cycle, the reaction dynamics and the structural homology of transducin to that of other biolgoical coupling enzymes is presented. A survey of the literature suggests that most biological coupling processes are mediated by nucleotide binding

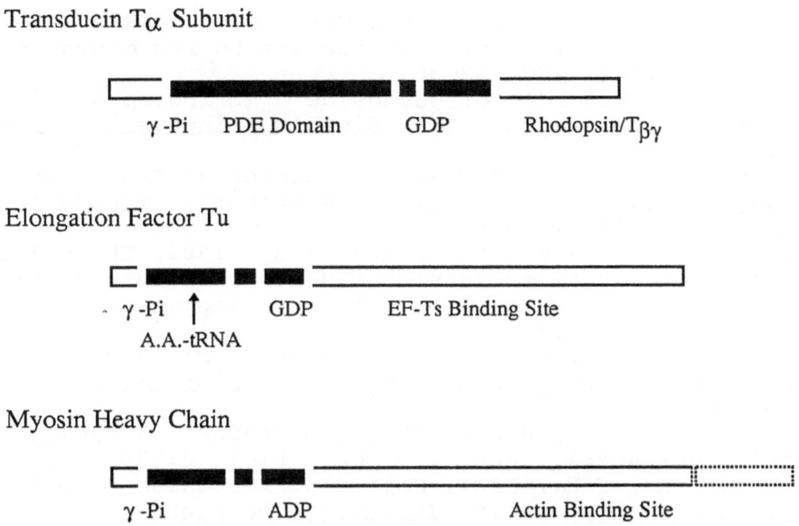

Figure 2 . The Functional Domain Arrangements of Some Coupling Enzymes. Relative locations of various functional domains are based on structural analyses of the proteins (Hingorani and Ho, 1987; Jurnack, 1985; Warrick and Spudich, 1987)

protein. Their coupling actions can be formulated in a similar manner as transducin in the visual excitation cycle. The homology existed is not only limited to GTP-binding proteins but extends to all ATP-binding coupling enzymes. The existence of tightly bound nucleotides and their exchange and hydrolysis cycles may represent a common motif in controlling multi-subunit enzymes in cellular regulation.

Reference

Frey, S.E., Hingorani, V.N., Su-Tsai, S.-M., and Ho, Y.-K., 1988, Chromium (III) β,γ-bidentate guanine nucleotide complexes as probes of the GTP-activated cGMP cascade of retinal rod outer segments. Biochemistry 27: 8209-8218.

Fung, B.K.-K., 1983, Characterization of transducin from bovine retinal rod outer segments. Separation and reconstitution of the subunits. J. Biol. Chem. 25: 10495-10502.

Gilman, A.G., 1987, The G proteins. Ann. Rev. Biochem. 56: 615-649.

Hibberd, M.G. and Trentham, D.R., 1986, Relationship between chemical and mechanical events during muscular contraction. Ann. Rev. Biophys. Biophys. Chem. 15:119-161.

Hingorani, V.N., and Ho, Y.-K., 1987, A structural model for the α-subunit of transducin: Implications of its role as a molecular switch in the visual signal transduction mechanism. FEBS Lett. 220: 15-22.

Ho, Y-K., Hingorani, V.N., Navon, S.E. and Fung, B. K-k., 1989, Transducin : A signalling switch regulated by guanine nucleotides. Current Topics in Cellular Regulation. Vol 30, in press.

Jurnack, F., 1985, Structure of the GDP domain of EF-Tu and location of the amino acids homologous to ras oncogene proteins. Science 230: 32-36.
Kaziro, Y., 1978, The role of guanosine 5'-triphosphate in polypeptide chain elongation. Biochim. Biophys. Acta. 505: 95-127.
Kowalczykowski, T., 1987, Mechanistic aspect of DNA strand exchange activity of E. coli rec A protein. Trends in Biochem. Sci. 12: 141-145.
Liebman, P.A., Park, K.R., and Dratz, E.A., 1987, The molecular mechanism of visual excitation and its relation to the structure and composition of rod outer segment. Ann. Rev. Physiol. 49: 765-791.
Lymn, R.W. and Taylor, E.W., 1971, Mechanism of Adenosine triphosphate hydrolysis by actomyosin. Biochemistry 10:4617-4624.
Stryer, L., and Bourne, H.R., 1986, G proteins : A family of signal transducers. Ann. Rev. Cell Biol. 2: 391-419.
Ting, T.D. and Ho, Y-K., 1989, Molecular mechanism of GTP hydrolysis by transducin. Investigative Ophthalmology and Visual Science 30: No 3. Supplement. Abs. p. 172.
Warrick H.M. and Spudich, J.A., 1987, Myosin structure and function in cell motility. Ann. Rev. Cell Biol. 3: 379-421

CO-LOCALIZATION BY IMMUNOFLUORESCENCE OF THE α SUBUNIT(S) OF G_i WITH CYTOPLASMIC STRUCTURES

Jean M. Lewis†, Marilyn J. Woolkalis†, George L. Gerton‡ and David R. Manning†

†Department of Pharmacology and ‡Department of Obstetrics and Gynecology, University of Pennsylvania School of Medicine Philadelphia, Pennsylvania 19104-6084

INTRODUCTION

GTP-binding proteins established in signal transduction (G proteins) are heterotrimers consisting of α, β, and γ subunits. From the initial discovery G_s, the family of G proteins has expanded in both number and scope of action. Arrangements of G proteins within the plasma membrane and among subcellular structures, however, remain largely uncharacterized. Compartmentation within the plasma membrane may indeed provide the means by which agonists can operate selectively on particular enzymes or channels (Brass et al., 1988; Neer et al., 1988). G proteins have also been detected within the cytosol (Bokoch et al., 1988; Rotrosen et al., 1988; Spicher et al., 1988) and in granule membranes (Toutant et al., 1987; Rotrosen et al., 1988). While cytoplasmic populations of G proteins may simply represent internal stores bound for recruitment to the plasma membrane, there is evidence that they may perform more active roles critical to normal cell function. The existence of guanine nucleotide-sensitivity, for example, has been demonstrated in transport of protein within the Golgi apparatus (Melançon et al., 1987), degranulation (Cockroft et al., 1987; Lindau and Nusse,1987), release of Ca^{++} from endoplasmic reticulum (Nasmith and Grinstein, 1987), and perhaps translocation of nascent proteins (Audigier et al., 1988; Connolly and Gilmore, 1986; Hoffman and Gilmore, 1988). In parallel to these findings, a recent study has shown that a 41-kDa, $G_{i\alpha}$-like protein co-fractionates with rough endoplasmic reticulum (Audigier et al, 1988). Clearly, knowledge of the distributions of G proteins within the cell will be critical to our understanding of G protein function.

Previous studies examining distributions of G proteins have relied chiefly on protocols of cell fractionation. Interpretations of data, however, are often complicated by issues of purity and yield. Studies of G proteins employing immunocytochemistry have also emerged, but these have focused either on specialized G proteins such as G_o (Brabet et al., 1988; Worley et al., 1986) and G_t (Grunwald et al., 1986; Lerea et al., 1986), or on specialized cells and tissues (Garty et al., 1988; Cortes et al., 1988). In our studies, we have examined the subcellular distribution of $G_{i\alpha}$. This subunit has far-ranging activities, has been cited to exist in association with cytoplasmic structures, and is ubiquitous. Distribution was assessed by immunofluorescence microscopy, and the cells employed were those maintained routinely in monolayer culture.

RESULTS

Demonstration of antibody specificity is the cornerstone of any immunocytochemical study. This issue was approached here in two ways. First, we employed a panel of rabbit antisera comprising two classes: those generated with a combination of purified G_i and G_o, and those raised with KLH-conjugated peptides that correspond to regions of primary structure present within one or more forms of $G_{i\alpha}$ (Figure 1). Antisera obtained from mice were also employed. By using antisera differing in specificities for epitopes within $G_{i\alpha}$, the potential for problems arising through cross-reactivities of any one antiserum with otherwise unrelated proteins was minimized. Second, we established the ability of each antiserum to recognize through Western blotting $G_{i\alpha}$ as the predominant reactant(s) in low- and high-speed pellets obtained from cell homogenates. Recognition of $G_{i\alpha}$ with antiserum 8729, the most commonly used in immunofluorescence, was virtually exclusive. No reactivity with proteins in high-speed supernatants was evident. Other controls for specificity, used within the context of immunofluorescence itself, are described below.

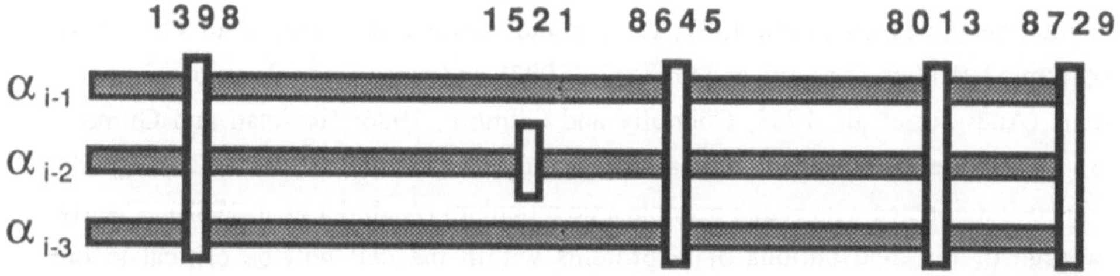

Figure 1. **Schematic depiction of epitopes recognized by peptide-directed antibodies.** Antisera were generated in rabbits by multiple-site intradermal injection of synthetic peptides conjugated to keyhole limpet hemocyanin with glutaraldehyde or m-maleimidobenzoyl-N-hydroxysuccinimide ester. All conjugates were injected in Freund's complete adjuvant. The sequences of conjugated peptides are: CGAGESGKSTIVK (antiserum 1398), CLERIAQSDYI (antiserum 1521), CFEDLNKRKDTKE (antiserum 8013), CFDVGGQRSERKK (antiserum 8645), and KNNLKDCGLF (antiserum 8729).

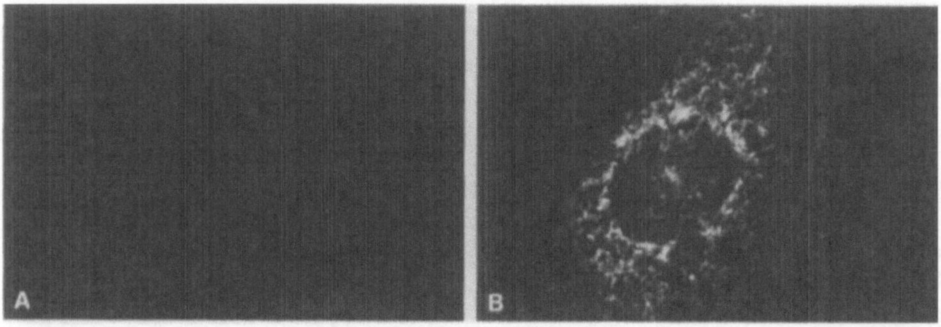

Figure 2. **Visualization of $G_{i\alpha}$ within Swiss 3T3 cells.** Primary antibodies were those of (A) preimmune serum and (B) antiserum 8729, both diluted to 1:100. Secondary antibodies in both (A) and (B) were fluorescein-conjugated goat anti-rabbit IgG (Cappell Labs, Malvern, PA.). Samples were mounted in Fluoromount (Fisher) containing 2.5% DABCO (Sigma). Cells were viewed through a Zeiss photomicroscope III equipped for epifluorescence, using a 100X objective, and photographed with Kodak T-max 400 film.

Cells were prepared for indirect immunofluorescence microscopy in the following manner. First, cells were cultured on coverslips for at least 24 hours to allow recovery from trypsinization. Following washes in PBS, the cells typically were fixed in 4% formaldehyde in 100 mM PIPES buffer, pH 6.8, containing 2.0 mM $MgCl_2$ and 2.0 mM EGTA for 20 minutes at room temperature. Cells were permeabilized for 10 minutes in 0.2% Triton X-100, and non-specific sites of protein binding were blocked with 10% normal goat serum. Incubations with primary and secondary antibodies were conducted for 45 minutes at 37 °C. Primary antibodies were those of any one of the antisera described above, diluted with PBS 1:50 - 1:500. Secondary antibodies were either fluorescein- or rhodamine-conjugated goat anti-rabbit IgG.

The pattern of immunofluorescence obtained for mouse 3T3 fibroblasts with antiserum 8729 is depicted in Figure 2, and consisted of two components. The first, and less distinctive, was that associated with the plasma membrane. This is evident as a diffuse fluorescence especially prominent at the cell periphery. The second, and more striking, was that associated with an array of tubule-like structures, perinuclear and extending outwards towards the edges of the cell. These correspond to phase-dense structures (not shown), and possibly are composed of vesicles or organelles arranged in linear arrays. That these structures are cytoplasmic in nature - and do not simply represent imperfections in the cell surface - was confirmed by confocal microscopy (not shown).

The results obtained with 8729 were corroborated with 1521 and 8645 (not shown), and with 8013 and an antiserum (5296) generated with a mixture of purified G_i and G_o (Figure 3). While differences in titer and in pre-immune background were observed, each of these antisera labeled the tube-like structures and plasma membrane in a manner characteristic of antiserum 8729. In contrast, antiserum 1398 provided no appreciable fluorescence. This was possibly attributable to occlusion of the epitope upon fixation.

The pattern of immunofluorescence obtained for 3T3 cells was also evident in other

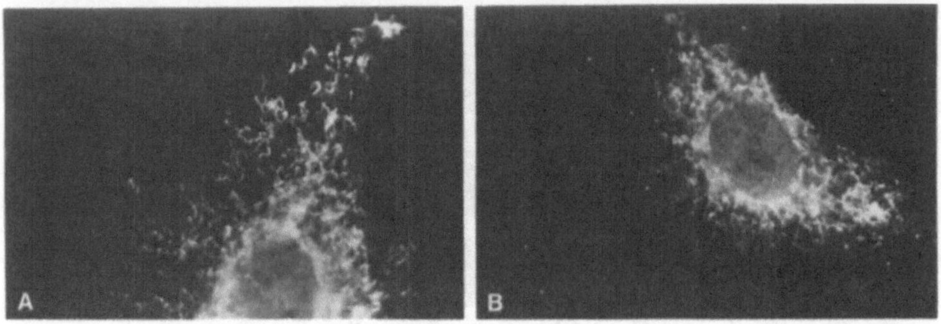

Figure 3. **Visualization of $G_{i\alpha}$ within Swiss 3T3 cells by two other antisera.** Primary antibodies were those of (A) antiserum 8013, and (B) antiserum 5296, both diluted 1:100. Samples were viewed as described in the legend to Figure 2.

types of cells. These included C6 glioma cells and primary cultures of human umbilical vein endothelial cells (Figure 4). While each type of cell has a characteristic morphology, the features of fluorescence are essentially the same: intense illumination of cytoplasmic tube-like structures, coupled with a diffuse highlighting of the plasma membrane.

The fact that a number of antisera designed to recognize different determinants within $G_{i\alpha}$ all label cytoplasmic structures strongly argues that the protein visualized by immunofluorescence is indeed $G_{i\alpha}$. Various types of controls helped to confirm this conclusion. First, the ability of alternate fixations, such as acetone or methanol at -20 °C, to produce cells that exhibited essentially the same pattern of fluorescence showed that the observed distribution of proteins was not an artifact of fixation. Second, pre-immune sera, tested over a wide range of concentrations, did not elicit this pattern of fluorescence. Finally, preincubation of at least antiserum 8729 with the immunizing peptide completely abolished the immunofluorescent staining pattern; irrelevant peptides were without effect.

CONCLUSIONS

The finding that a large proportion of $G_{i\alpha}$ exists at any given time on cytoplasmic structures may provide insights into processing or function of this subunit. One possibility is that the $G_{i\alpha}$ associated with cytoplasmic structures represents newly synthesized protein being transported to the plasma membrane in a constitutive fashion. If so, the observed distribution of subunit between internal structures and plasma membrane would imply that such transport occurs slowly relative to degradation of the subunit following insertion into the latter. Transport from the endoplasmic reticulum to the Golgi apparatus is most often considered to be the rate-limiting step in biosynthetic protein transport (Pfeffer and Rothman, 1987). Audigier et al. (1988) indeed found the majority of a 41-kDa substrate(s) for pertussis toxin to co-fractionate with rough endoplasmic reticulum . Still, retention at the endoplasmic reticulum of proteins destined for export to the plasma membrane is felt to reflect partly the time required for assumption of proper tertiary or quaternary structure.

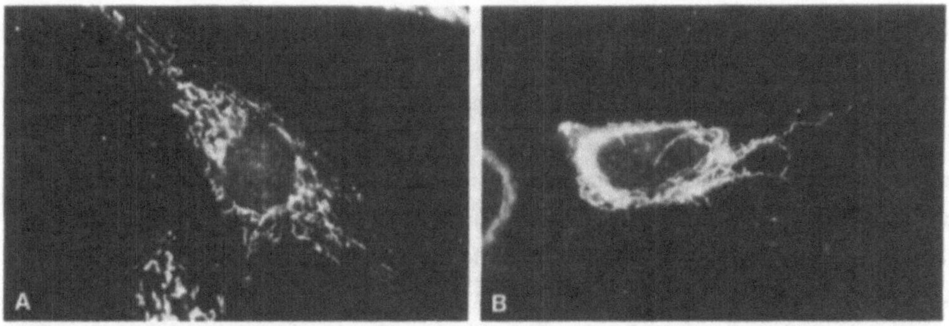

Figure 4. **Visualization of $G_{i\alpha}$ within two other types of cells.** Immunofluorescence patterns were examined for (A) C6 glioma cells, and (B) Human umbilical vein endothelial cells. Primary antibodies were those of 8729, diluted 1:100. Samples were viewed as described in the legend to Figure 2.

The competency of the 41-kDa protein to undergo ADP-ribosylation indicates, by the requirement for combination of $G_{i\alpha}$ with $\beta\gamma$, that assumption of proper structure has occurred. In protocols of double-labeling with antibodies specific for the Golgi apparatus, we find no evidence for $G_{i\alpha}$ existing in significant quantities within this organelle. At this point, we cannot preclude the possibility that $G_{i\alpha}$ exists in either the transreticular network or within vesicles, perhaps arranged in tandem, leading directly to the plasma membrane.

The $G_{i\alpha}$ visualized could also be that awaiting recruitment to the plasma membrane in response to regulatory signals, such as in the case of insulin receptors or glucose transporters. Brief pulses of insulin, administered within a rat hepatocyte perfusion system, have been shown to cause the re-insertion of internalized insulin receptors into the plasma membrane (Goodner et al., 1988). Binding of insulin to cells has also been shown to shift the balance of type II insulin-like growth factor receptors from an intracellular pool to the plasma membrane of adipose cells (Wardzala et al., 1984). Insulin binding also mediates an increase in plasma membrane-associated glucose transporters, in parallel to a decrease in the amount of Golgi-associated glucose transporters, in rat adipocytes (Wang, 1985). Possibly, one consequence of activating those receptors associated with G proteins at the plasma membrane may be recruitment of new G proteins from intracellular stores.

Alternatively, we may be observing G proteins at work, performing regulatory or other functions at the organelle level. A role for G proteins in events such as translocation of proteins from ribosomes to endoplasmic reticulum membranes (Connolly and Gilmore, 1986; Hoffman and Gilmore, 1988), vesicular traffic between the endoplasmic reticulum and the Golgi apparatus (Beckers and Balch, 1989), or in exocytosis from mast cells (Cockcroft el al., 1987), has been suggested by the existence of a GTPγS-sensitive step in each of these phenomena. In this regard, the functions of $G_{i\alpha}$ would conceivably be integrated with those of yeast *ras*-like gene products (Mr~21,000), whose roles in different aspects of protein transport in yeast are now emerging (Segev et al., 1988; Nakano et al., 1988).

Most significantly, our results demonstrate the existence of a population of $G_{i\alpha}$ that is almost certainly unavailable to interact with receptors for agonists at the plasma membrane. This has an immediate impact on studies involving quantitation. For example, changes observed in amounts of G_{α} or $G_{\beta\gamma}$ within the context of differentiation (Watkins et al., 1987) or pathophysiological processes (Woolkalis et al., 1986; Gawler et al., 1987) must now be reconciled with the proximity of these changes to receptors and targets. The results presented here, upon extension, may also help to explain the paradox posed by Ransnas and Insel, who determined that, while the amount of G_s should be the limiting factor in stimulation of adenylyl cyclase, it exists in a 100-fold 'excess' over β-adrenergic receptors in S49 mouse lymphoma cells. Finally, $G_{i\alpha}$ has been cited with other G_{α} subunits to be a substrate for co- and post-translational modifications. The occurence of these modifications in relation to spatially distinct pools of $G_{i\alpha}$ must now be examined.

The relative distribution of the different subtypes of $G_{i\alpha}$ within both the plasma membrane and the cytoplasmic structures remains to be investigated. Of interest will be identification of the cytoplasmic structures, ascertainment of functional ramifications, and extension of protocols to other G proteins.

ACKNOWLEDGMENTS

We would like to extend our thanks to John Murray for providing confocal analysis of immunofluorescence. This work was supported by National Institutes of Health grant GM34781.

REFERENCES

Audigier, Y., Nigam, S.K., and Blobel, G., 1988, Identification of a G protein in rough endoplasmic reticulum of canine pancreas, J. Biol. Chem., 263:16352.

Beckers, C.J.M., and Balch, W.E., 1989, Calcium and GTP: Essential components in vesicular trafficking between the endoplasmic reticulum and golgi apparatus, J. Cell Biol., 108:1245.

Bokoch, G.M., Bickford, K., and Bohl, B.P., 1988, Subcellular localization and quantitation of the major neutrophil pertussis toxin substrate, G_n, J. Cell Biol., 106:1927.

Brabet, P., Dumuis, A., Sebben, M., Pantaloni, C., Bockaert, J., and Homburger, V., 1988, Immunocytochemical localization of the guanine nucleotide-binding protein G_o in primary cultures of neuronal and glial cells, J. Neuroscience, 8:701.

Brass, L.F., Woolkalis, M.J., and Manning, D.R., 1988, Interactions in platelets between G proteins and the agonists that stimulate phospholipase C and inhibit adenylyl cyclase, J. Biol. Chem., 263:5348.

Cockroft, S., Howell, T.W., and Gomperts, B.D., 1987, Two G-proteins act in series to control stimulus-secretion coupling in mast cells: use of neomycin to distinguish between G-proteins controlling polyphosphoinositide phosphodiesterase and exocytosis, J. Cell Biol., 105:2745.

Connelley, T., and Gilmore, R., 1986, Formation of a functional ribosome-membrane junction during translocation requires the participation of a GTP-binding protein, J. Cell Biol., 103:2253.

Cortes, R., Hokfelt, T., Schalling, M., Goldstein, M., Goldsmith, P., Spiegel, A., Unson, C., and Walsh, J., 1988, Antiserum raised against residues 159-168 of the guanine nucleotide-binding protein $G_{i3\text{-}\alpha}$ reacts with ependymal cells and some neurons in the rat brain containing cholecystokinin- or cholecystokinin- and tyrosine 3-hydroxylase-like immunoreactivities, Proc. Natl. Acad. Sci., U.S.A., 85:9351.

Garty, N., Galiani, D., Aharonheim, A., Ho, Y.-K., Phillips, D.M., Dekel, N., and Salomon, Y., 1988, G-proteins in mammalian gametes: an immunocytochemical study, J. Cell Science, 91:21.

Gawler, D., Milligan, G., Spiegel, A.M., Unson, C.G., and Houslay, M.D., 1987, Abolition of the expression of inhibitory guanine nucleotide regulatory protein G_i activity in diabetes, Nature, 327:229.

Goodner, C., Sweet, I.R., and Harrison, H.C., 1988, Rapid reduction and return of surface insulin receptors after exposure to brief pulses of insulin in perfused rat hepatocytes, Diabetes, 37:1316.

Grunwald, G., Gierschik, P., Nirenberg, M., and Spiegel, A., 1986, Detection of α-transducin in retinal rods but not cones, Science, 231:856.

Hoffman, K., and Gilmore, R., 1988, Guanosine triphosphate promotes the post-translational integration of opsin into the endoplasmic reticulum membrane, J. Biol. Chem., 263:4381.

Lerea, C.L., Somers, D.E., Hurley, J.B., Block, I.B., and Bunt-Milam, A.H., 1986, Identification of specific transducin α subunits in retinal rod and cone photoreceptors, Science, 234:77.

Lindau , M., and Nusse, O., 1987, Pertussis toxin does not affect the time course of exocytosis in mast cells stimulated by intracellular application of GTP-γ-S, FEBS Letts., 222:317.

Melançon, P., Glick, B.S., Malhotra, V., Weidman, P.J., Serafini, T., Gleason, M.L., Orci, L., and Rothman, J.E., 1987, Involvement of GTP-binding "G" proteins in transport through the Golgi stack, Cell, 51:1053.

Nakano, A., Brada, D., and Schekman, R., 1988, A membrane glycoprotein, Sec12p, required for protein transport from the endoplasmic reticulum to the golgi apparatus in yeast, J. Cell Biol., 107:851.

Nasmith, P., and Grinstein, S., 1987, Phorbol ester-induced changes in cytoplasmic Ca^{2+} in human neutrophils, J. Biol. Chem., 262:13558.

Neer, E.J., and Clapham, D.E., 1988, Roles of G protein subunits in transmembrane signalling, Nature, 333:129.

Pfeffer, S.R., and Rothman, J.E., 1987, Biosynthetic protein transport and sorting by the endoplasmic reticulum and Golgi, Ann. Rev. Biochem., 56:829.

Ransnas, L.A., and Insel, P.A., 1988, Quantitation of the guanine nucleotide binding regulatory protein G_s in S49 cell membranes using antipeptide antibodies to α_s, J. Biol. Chem., 263:9482.

Rotrosen, D., Gallin, J.I., Spiegel, A.M., and Malech, H.L., 1988, Subcellular localization of $G_{i\alpha}$ in human neutrophils, J. Biol. Chem., 263:10958.

Segev, N., Mulholland, J., and Botstein, D., 1988, The yeast GTP-binding YPT1 protein and a mammalian counterpart are associated with the secretion machinery, Cell, 52:915.

Spicher, K., Hinsch, K.D., Gausepohl, H., Frank, R., Rosenthal, W., and Schulz, G., 1988, Immunochemical detection of the alpha-subunit of the G-protein, G_z, in membranes and cytosols of mammalian cells, Biochem. Biophys. Res. Commun., 157:883.

Toutant, M., Aunis, D., Bockaert, J., Homburger, V., and Rouot, B., 1987, Presence of three pertussis toxin substrates and the $G_{o\alpha}$ immunoreactivity in both plasma and granule membranes of chromaffin cells, FEBS Lett., 215:339.

Wang, C., 1985, Insulin-stimulated glucose uptake in rat diaphragm during postnatal development: Lack of correlation with the number of insulin receptors and of intracellular glucose transporters, Proc. Natl. Acad. Sci., U.S.A., 82:3621.

Wardzala, L.J., Simpson, I.A., Rechler, M.M., and Cushman, S.W., 1984, Potential mechanism of stimulatory action of insulin on insulin-like growth factor II binding to the isolated rat adipose cell: apparent redistribution of receptors recycling between a large intracellular pool and the plasma membrane, J.Biol.Chem., 259:8378.

Watkins, D.C., Northrup, J.K., and Malbon, C.C., 1987, Regulation of G-proteins in differentiation, J. Biol. Chem., 262:10651.

Woolkalis, M.J., Nakada, M.T., and Manning, D.R., 1986, Alterations in components of adenylate cyclase associated with transformation of chicken embryo fibroblasts by Rous sarcoma virus, J. Biol. Chem., 261:3408.

Worley, P.F., Baraban, J.M., Van Dop, C., and Neer, E.J., 1986, G_o, a guanine nucleotide-binding protein: Immunohistochemical localization in rat brain resembles distribution of second messenger systems, Proc. Natl. Acad. Sci., U.S.A., 83:4561.

'CROSSTALK' AND THE REGULATION OF REPATOCYTE ADENYLATE CYCLASE ACTIVITY: DESENSITIZATION, G_1 AND CYCLIC AMP PHOSPHODIESTERASE REGULATION

Miles D. Houslay
Molecular Pharmacology Group
Department of Biochemistry, University of Glasgow
Glasgow G12 8QQ, Scotland, UK

Introduction

Challenge of a variety of different cell types by receptor agonists capable of stimulating adenylate cyclase leads to a transient increase in the intracellular concentrations of cyclic AMP. In hepatocytes, glucagon achieves just this (fig.1). We show that such an event is determined by the initial rapid activation of adenylate cyclase which is then followed by the activation of a specific cyclic AMP phosphodiesterase and desensitization of adenylate cyclase. These events identify 'interplay' or 'cross-talk' occurring between two distinct cellular signalling systems.

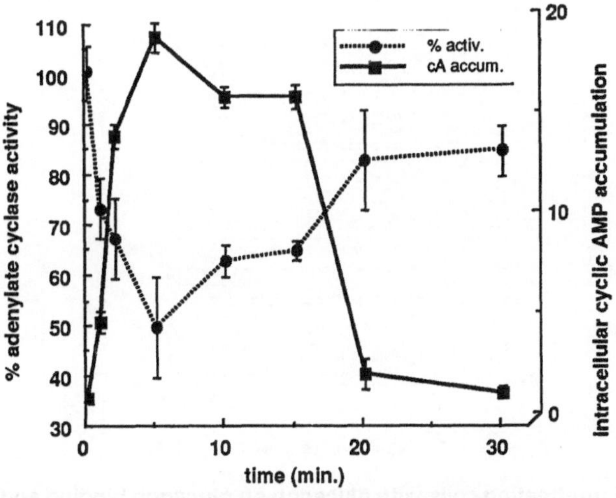

Figure 1. Timecourse of glucagon-elicited cyclic AMP accumulation and desensitization in intact hepatocytes. Hepatocytes were incubated with glucagon (1nM) and samples taken for both the determination of intracellular [cAMP] and for assay of glucagon-stimulated adenylate cyclase activity in a washed membrane preparation.

Changes in the regulation of adenylate cyclase are shown to occur at both the stimulatory and inhibitory points of control of adenylate cyclase activity and these normal physiological controls are further identified as being affected in pathophysiological states such as is seen in insulin resistance found in diabetes and obese conditions.

Desensitization of adenylate cyclase results from the uncoupling of the glucagon receptor from G_s

Challenge of intact hepatocytes by glucagon leads [Heyworth & Houslay, 1983] to a transient increase in the intracellular accumulation of cyclic AMP (fig. 1). This occurred whether or not inhibitors of cyclic AMP phosphodiesterase were present and was not accompanied by any marked efflux of cyclic AMP from the cells. The lesion must therefore lie primarily at the site of synthesis of cyclic AMP i.e. desensitization of adenylate cyclase has occurred.

In order to identify any changes in adenylate cyclase activity, hepatocytes were first challenged with glucagon for various times, harvested, disrupted and a washed membrane fraction prepared for the assay of adenylate cyclase. Using such a procedure we showed [Heyworth & Houslay, 1983] that the glucagon-stimulated adenylate cyclase activity of these membranes was markedly reduced compared with that seen in membranes from control cells. This desensitization was time-dependent, with a $t_{0.5}$ of 1.5-2min and dose-dependent, with an EC_{50} of $4x10^{-10}$M glucagon.

The reduced glucagon-stimulated adenylate cyclase activity (fig. 2) could have resulted from lesions occurring at either the catalytic unit of adenylate cyclase (assessed by measuring either forskolin-stimulated or basal activity); at

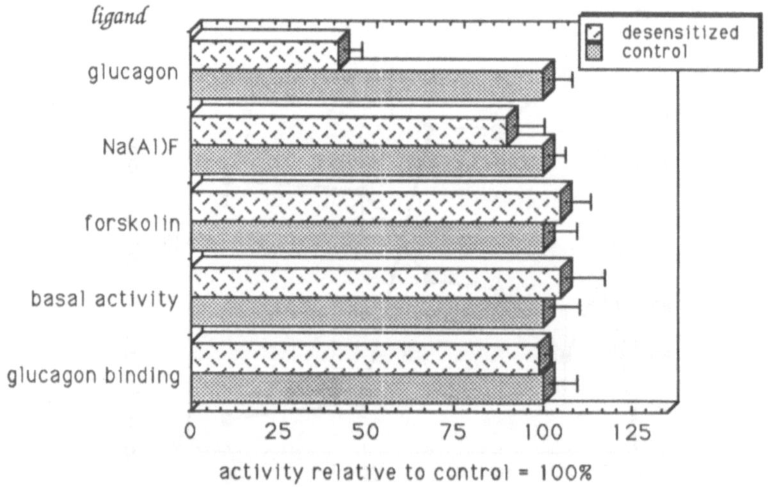

Figure 2. Action of pre-treating cells with glucagon on glucagon binding and the expression of adenylate cyclase activity in a washed membrane fraction. Cells were treated with 1nM-glucagon for 5min. before a washed membrane fraction was prepared for the assay of adenylate cyclase activity under basal and ligand-stimulated conditions. I^{125}-Glucagon binding is shown for this fraction, with identical results for the amount of cell surface glucagon receptors.

the coupling of G_s to adenylate cyclase (assessed by measuring Na(Al)F-stimulated activity); at the coupling of the glucagon receptor to G_s and could also result from either loss of receptor binding or receptor internalization. Evaluation of these various possibilities by, for example, assaying adenylate cyclase activity, when stimulated by different ligands in isolated membranes from glucagon pre-treated cells (fig. 2), showed quite clearly that receptors are present in the plasma membrane and but that they are unable to couple appropriately to G_s and thus to stimulate adenylate cyclase [Heyworth & Houslay, 1983]. The lesion is thus due to a modification of the coupling interface between the glucagon receptor and G_s; this could result from a modification of either G_s or the glucagon receptor itself (fig. 3).

Desensitization of adenylate cyclase is a cyclic AMP-independent process

Desensitization cannot be mimicked by the addition of either dibutyryl cyclic AMP or 8-bromo cyclic AMP to intact hepatocytes under conditions where these ligands activate A-kinase. Furthermore, the half-time for the onset of desensitization is similar in cells where the cyclic AMP phosphodiesterase inhibitor IBMX was present and cyclic AMP concentrations were some 5-fold greater than in cells which were not treated with this agent ($t_{0.5}$ of 1.5-2min).

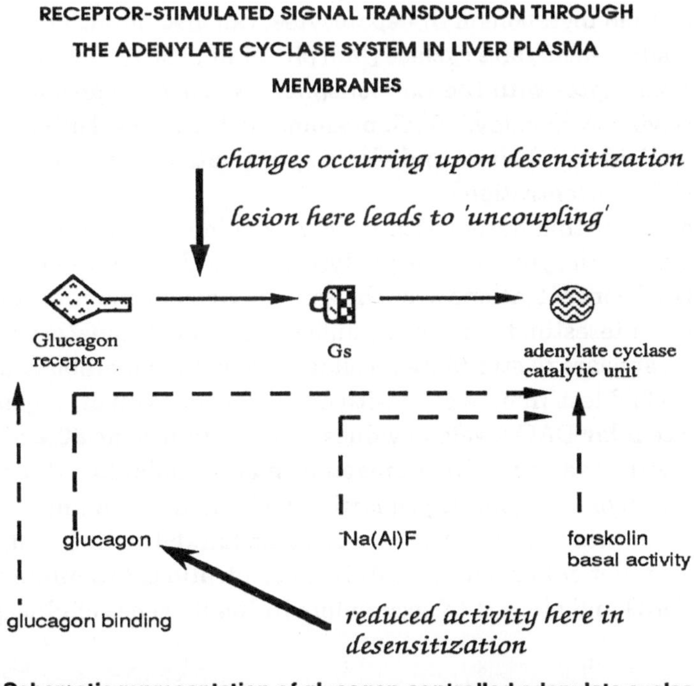

Figure 3. Schematic representation of glucagon-controlled adenylate cyclase in hepatocytes, identifying the site of the lesion found in desensitization. The scheme shows the various assays used to define the site of the lesion. This is located at the coupling interface between G_s and the glucagon receptor.

These data, coupled with the fact that the EC_{50} values for desensitization were an order of magnitude lower than those noted for glucagon's ability to both activate adenylate cyclase and to increase the intracellular concentration of cyclic AMP, indicated that desensitization was mediated by a cyclic AMP-independent process.

The proof for this was provided [Murphy et al., 1987] using an analogue of glucagon called TH-glucagon ([1-*N*-α-trinitrophenylhistidine, 12-homoarginine] glucagon). This derivative, whilst capable of eliciting desensitization in intact hepatocytes at concentrations as low as 10^{-10}M, was incapable of either activating adenylate cyclase or increasing the intracellular concentration of cyclic AMP even at concentrations of up to 10^{-5}M THG.

So how does desensitization ensue? The first clue we had to a molecular mechanism was provided by our observations [Heyworth et al., 1984] that treatment of intact hepatocytes with the tumour-promoting phorbol ester TPA appeared to mimic the desensitization process. As this compound serves to activate protein kinase C, by mimicking the action of the natural agonist diacyl glycerol (DAG), we set out to determine whether glucagon elicited desensitization through such a route. Indeed, we [Wakelam et al., 1986] and others [Whipps et al., 1987] have now been able to show that both glucagon and TH-glucagon elicit a small stimulation of inositol phospholipid metabolism, through which we hypothesized diacyl glycerol would be produced and desensitization could occur. Consistent with this idea were our observations that treatment of hepatocytes with either vasopressin or angiotensin, both of which stimulate inositol phospholipid metabolism in hepatocytes, can also mimic glucagon's ability to desensitize adenylate cyclase [Murphy et al., 1987]. Furthermore, treatment of hepatocytes with the Ca^{2+} ionophore A23187 can also mimic desensitization [Irvine & Houslay, 1988], presumably because it both stimulates inositol phospholipid metabolism and also may activate C-kinase by increasing intracellular Ca^{2+} concentrations.

We can now show that desensitization can be mimicked by treatment of intact hepatocytes with synthetic diacyl glycerols and that glucagon increases intracellular DAG concentrations over the same period of time that it elicits desensitization. Interestingly, however, under maximal stimulatory conditions, whilst glucagon appears to stimulate inositol phospholipid metabolism to levels which are 10-20 fold less than is seen with either vasopressin or angiotensin it increases intracellular DAG levels to values which attain some 50% of that recorded using either vasopressin or angiotensin as stimulants. This may indicate that glucagon stimulates, primarily, the breakdown of another phospholipid to elevate DAG levels, perhaps phosphatidylcholine. Alternatively, it could be that enhanced cyclic AMP concentrations attenuate the breakdown of DAG and allow it to accumulate in the presence of glucagon.

Resensitization of adenylate cyclase is attenuated by elevated cyclic AMP concentrations

We noted that desensitization of adenylate cyclase was maximal after 2-5 min, whereupon a slow resensitization of the system followed on. This resensitization process appears susceptible to inhibition by elevation of cyclic AMP concentrations to supra-physiological levels. This can be achieved using cyclic AMP phosphodiesterase inhibitors such as IBMX and Ro-2074 [Murphy & Houslay, 1988]. The molecular basis of this requires elucidation.

Desensitization of adenylate cyclase is blocked by treatment with pertussis toxin whose action in doing this does not appear to involve G_i and does not involve an effect upon inositol phospholipid metabolism

Hepatocytes contain a functional G_i whose activity can be identified by the ability of either high concentrations of GTP or low concentrations of its non-hydrolysable analogue guanylyl 5'-imidodiphosphate (p[NH]ppG) to inhibit forskolin-stimulated adenylate cyclase activity [Gawler et al., 1987; Heyworth et al., 1983b]. In intact cells, G_i appears to be coupled to P_2-purinergic receptors [Okajima et al., 1987]. Our analysis of G_i forms in hepatocytes, using specific anti-peptide antibodies, shows that both G_i-2 and G_i-3, but not G_i-1, are expressed in hepatocytes. We can confirm this using Northern blotting techniques with specific 34-mer oligonucleotide probes.

Treatment of intact hepatocytes with pertussis toxin causes the ADP-ribosylation of these two forms of G_i and the inactivation of functional G_i as assessed by the three criteria stated above [Gawler et al., 1987; Heyworth et al., 1983; Heyworth et al., 1985]. We do not know unequivocally whether one or both of these forms of G_i serves to mediate inhibition of adenylate cyclase. However, as stated below, we have some evidence which suggests that this action may be mediated by G_i-2.

We have observed that treatment of intact hepatocytes with pertussis toxin prevents the glucagon-mediated desensitization of adenylate cyclase [Heyworth et al., 1983; Murphy et al., 1989]. This could be because desensitization was caused by the constitutive activation of G_i (by C-kinase). We, however, believe this to be unlikely as such an action might be expected to lead to the decreased functioning of both forskolin-stimulated and basal adenylate cyclase activity: which does not occur. Furthermore, we have been able to demonstrate (i) that G_i is inactivated by C-kinase action(see below) [Pyne et al., 1989]; (ii) that desensitization occurs in hepatocytes from diabetic animals which do not exhibit a functional G_i [Murphy et al., 1989] and (iii) that desensitization occurs just as well in hepatocytes from immature animals which show levels of G_i which are some 50% of those seen in mature animals [Murphy et al., 1989]. All of which militate against an involvement of G_i in the desensitization process.

The mechanism whereby pertussis toxin exerts its inhibitory effect on desensitization is thus unclear. Certainly, it is not restricted to the action of glucagon as desensitization of this system elicited by vasopressin and angiotensin is blocked similarly [Murphy et al., 1989], indicating that the site of action is at a point in the mechanism which is common to all of these hormones. However, this is not due to any action of pertussis toxin on the stimulation of inositol phospholipid metabolism, achieved by either vasopressin, angiotensin or glucagon, under conditions where this toxin causes the ADP-ribosylation of G_i-2 and G_i-3 in hepatocytes; an experiment which also shows that not only is the stimulation of inositol phospholipid metabolism not affected by pertussis toxin in hepatocytes but that G_i-2 and G_i-3 cannot serve as coupling G-proteins to phospholipase C in hepatocytes. It is possible then that pertussis toxin could exert its action on this system by inhibiting a species of C-kinase involved in mediating the desensitization process.

In some instances it has been shown that this toxin can exert biological effects by virtue of its B-(binding) subunit, however, treatment of hepatocytes with the isolated B-subunit failed to elicit any inhibition of desensitization [Murphy et al., 1989].

Inactivation and phosphorylation of G_i-2 by treatment of intact hepatocytes with agents which activate C-kinase

Treatment of intact hepatocytes with the phorbol ester TPA gives rise to the functional inactivation of G_i. This is accompanied by the phosphorylation of the α-subunit of this G-protein [Pyne et al., 1989], when immunoprecipitated with the antiserum AS7. We now know that this antiserum can immunoprecipitate G_i-1 and G_i-2 but not G_i-3. As liver does not express G_i-1 then the phosphorylated species immunoprecipitated must be the α-subunit of G_i-2. We have subsequently performed experiments with an anti-peptide antiserum which is specific for G_i-3 and failed to show any phosphorylation of the α-subunit of this form of G_i. Such experiments suggest that the major form of G_i which exerts inhibitory effects upon adenylate cyclase in hepatocytes is in fact G_i-2.

Similarly, challenge of intact hepatocytes with either vasopressin, angiotensin, glucagon or TH-glucagon leads to the inactivation and phosphorylation of G_i (fig. 4). It thus seems likely that these affects are mediated by C-kinase action. However, we have noted that treatment of intact hepatocytes with dibutyryl cyclic AMP can also give rise to an increase in the phosphorylation state of G_i. This may indicate that A-kinase can also phosphorylate this G-protein. However, from observations of the protein sequence, there is no obvious A-kinase phosphorylation site consensus sequence. Thus it is possible that dibutyryl cyclic AMP may not be exerting this action directly through A-kinase. We are currently addressing these problems.

Inactive G_i is characteristic of insulin-resistant states

In 1983 we first mooted [Houslay & Heyworth, 1983] the idea that insulin might exert certain of its actions upon target cells by modifying the functioning of members of the G-protein family and some evidence has arisen in support of

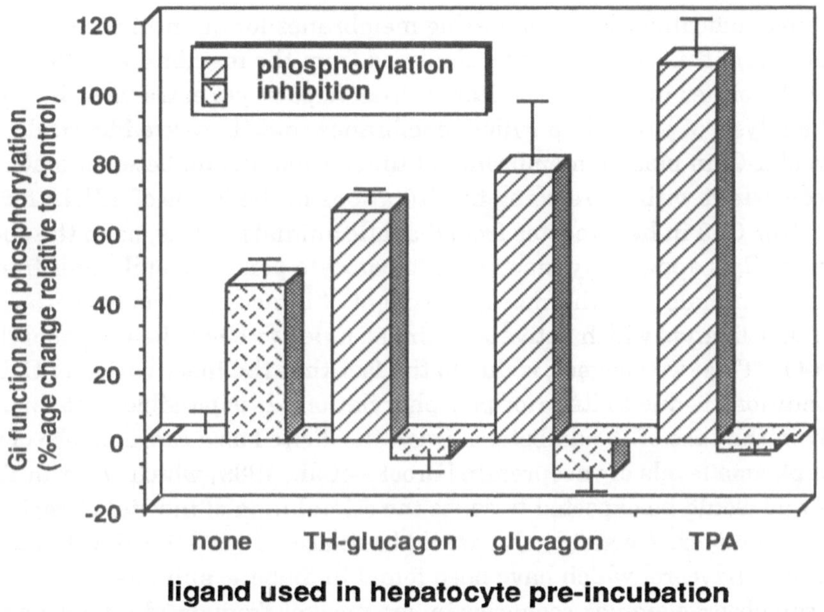

Figure 4. Phosphorylation and inactivation of G_i in intact hepatocytes. Cells were pre-incubated with ligands prior to either the preparation of a washed membrane fraction for the determination of functional G_i or specific immunoprecipitation of ^{32}P-loaded cell extracts for assessment of phosphorylation of α-G_i.

this concept [Houslay & Siddle, 1989]. However, our recent studies have focussed on the functioning of hepatocyte G_i in insulin-resistant states found in animal models of diabetes and obesity [Gawler et al., 1987; Houslay et al., 1989].

Young, obese Zucker rats at 8-10 weeks of age show acute insulin resistance, hyperinsulinaemia but are not diabetic, i.e. are normoglycaemic. In this model system of insulin-resistant obesity, we observed that hepatocytes from lean but not obese Zucker rats expressed functional G_i activity [Houslay et al., 1989]. Using both anti-peptide antibodies and pertussis toxin-catalysed ADP-ribosylation to quantify the α-subunit of G_i it appears, however, that both the lean and obese animals expressed identical amounts of α-G_i. It thus seems that G_i is functionally inactive in the membranes from the obese animals. We believe that the reason for this lack of functional G_i is due to its phosphorylation. Evidence which indicates that this is the case comes from our ability to phosphorylate the α-subunit of G_i in hepatocytes from the lean but not the obese animals.

We have performed a similar study of G_i function in rats made diabetic with either streptozotocin or alloxan [Gawler et al., 1987]. In these cases, experimental diabetes elicited insulin-resistance in the liver and a loss of functional G_i. We originally attributed this to a loss of α-G_i protein in the plasma membranes from these animals [Gawler et al., 1987]. However, our recent studies have indicated that in preparing membranes for quantification of G_i, the α-subunit of this G-protein was displaced from the membranes. We can now show by either immunoprecipitation from hepatocyte extracts or by Western blot analysis of partially purified membranes that there are identical amounts of α-G_i in plasma membranes of the normal and diabetic animals. This is consistent with there being no alterations in the levels of mRNA for either G_i-2 or G_i-3 in hepatocytes from diabetic animals. Thus, as in the obese zucker rats, G_i is apparently inactive in hepatocyte plasma membranes from the diabetic animals. Furthermore, treatment of intact hepatocytes with either TPA or other ligands which activate C-kinase failed to elicit the phosphorylation of α-G_i. This, we suggest, is due to the fact that the inactivation of G_i in these conditions is due to its prior phosphorylation. It is possible that this, certainly in the case of the diabetic animals, is elicited as a consequence of the elevated plasma levels of vasopressin [Brooks et al., 1989] which occur in these animals and would be expected to cause the stimulation of inositol phospholipid metabolism in liver. Certainly this would be in accord with the elevated tissue levels of diacylglycerol which have been found in diabetic animals.

We can observe similar scenarios in hepatocytes from db^+/db^+ mice and ob^+/ob^+ mice, indicating that insulin resistance is associated with a loss of functional G_i. we do not construe from this that insulin might exert actions through G, but rather that activation of C-kinase may be an underlying feature of insulin resistance which happens to modify both G_i and the insulin receptor in an inhibitory fashion. Interestingly, however, glucagon desensitization occurs in hepatocytes from diabetic animals and in obese zucker rats. This suggests that one can dissociate the pathways which control G_i phosphorylation and inactivation from those which control desensitization and resensitization. If the pathway leading to inactivation is the same, i.e. C-kinase activation, then it suggests that the action of phosphatases, which presumably act as one of the points controlling reversal, are different. It also implies that even in the diabetic and obese animals C-kinase cannot be maximally activated as desensitization can occur.

Activation of a high affinity, cyclic AMP-specific phosphodiesterase from hepatocytes by A-kinase-mediated phosphorylation

We have undertaken a characterization of all of the various cyclic nucleotide phosphodiesterases in hepatocytes and evaluated their potential for rapid control by the action of glucagon and insulin. Thus plasma membranes possess at least two high affinity enzymes [Pyne et al., 1986; Heyworth et al., 1983a], one of which is activated by insulin [Heyworth et al., 1983a; Houslay, 1986]; mitochondria express two species [Cercek & Houslay, 1983]; smooth endoplasmic reticulum express one species [Cercek et al., 1983]; rough endoplasmic reticulum expresses three species; lysosomes and nuclei express none and there is another species associated with an as yet undefined intracellular vesicle fraction, which we have called the 'dense vesicle' phosphodiesterase [Heyworth et al., 1983a]. It is this latter species which can be activated by both insulin and by glucagon, albeit through distinct routes [Heyworth et al., 1984]. Indeed, in that study [Heyworth et al., 1984] we also showed that at least a third cyclic AMP-specific phosphodiesterase must be activated by insulin through a rapidly reversible process. Recently [Pyne & Houslay, 1988], we have suggested that this could be due to the effect of an, as yet undefined, insulin-produced soluble 'second messenger' or 'mediator' substance acting upon one or both of the cyclic GMP-activated enzymes found in the soluble and plasma membrane fractions. We can also identify five distinct cyclic nucleotide phosphodiesterase forms in the cytosol fraction [Lavan et al., 1989], although we have no evidence [Heyworth et al., 1983a] which suggests that any of these, save the cyclic GMP-activated enzyme [Heyworth et al., 1984] may have activities which are under rapid hormonal regulation by hormones.

It is quite clear, however, that glucagon causes the rapid activation of the so-called 'dense vesicle' cyclic AMP phosphodiesterase and that this process is triggered by the increase in hepatocyte intracellular cyclic AMP that glucagon achieves [Heyworth et al., 1983a]. Thus activation can be mimicked by treating hepatocytes with either forskolin, cholera toxin, 8-bromo cyclic AMP ordibutyryl cyclic AMP [Heyworth et al., 1983a]. On this basis we have suggested [Heyworth et al., 1983a] that this enzyme may be a substrate for phosphorylation and activation by A-kinase. We have now purified this enzyme to apparent homogeneity [Pyne et al., 1987a] and prepared antisera to it which do not appear to cross-react with any other phosphodiesterase species [Pyne et al., 1987a; Pyne et al., 1987b]. Using these antisera together with one against the peripheral, insulin-activated plasma membrane enzyme, we have shown both enzymes are differentially distributed in rat tissues. We have also employed specific antisera against the 'dense vesicle' cyclic AMP phosphodiesterase to demonstrate, in both hepatocytes [Kilgour et al., 1989] and adipocytes [Anderson et al., 1989], that A-kinase-treatment of a membrane fraction containing this enzyme leads to its activation and stoichiometric phosphorylation.

Simulation of cyclic AMP metabolism using the known kinetic properties of all the enzymes involved indicates that activation of the 'dense vesicle' cyclic AMP phosphodiesterase has a major influence on the glucagon-stimulated accumulation of cyclic AMP in hepatocytes. The nature of this appears to be essentially in damping the magnitude of the rise in cyclic AMP attained after challenge of hepatocytes with glucagon.

Concluding remarks

Signal transduction systems in intact hepatocytes appear to involve complex co-ordinated interactions between the various signalling systems present there. Our studies clearly show that the signalling systems for glucagon (major action on cyclic AMP), vasopressin/angiotensin (inositol phospholipid) and insulin (tyrosyl kinase) do not function in isolation but, rather, the 'quality' of signals passing through these systems is modulated by the flux passing through the other interacting systems. These interacting systems may form a 'network' which allows the hepatocyte to adapt and learn from various distinct environmental stimuli caused by different hormonal challenges which arise under particular physiological conditions. Thus they would form a system which could respond appropriately under a variety of different conditions. The presise 'wiring' and functioning of these networks in normal and pathological conditions demands further study.

Acknowledgements

Work in the authors laboratory was supported from the MRC, AFRC, British Diabetic Association, Lipha Pharmaceuticals and the CMRF.

References

Anderson, N.G., Kilgour, E. & Houslay, M.D. (1989) Biochem. J. (in press). Subcellular localisation and hormone sensitivity of adipocyte phosphodiesterases

Brooks, D.P., Nutting, D.F., Crofton, J.T. & Share, L. (1989) Diabetes 38, 54-57.
Elevated plasma vasopressin levels in diabetic rats.

Cercek, B. & Houslay, M.D. (1982) Biochem. J. 207, 123-132.
Submitochondrial localisation and disposition of cyclic AMP-phosphodiesterases.

Cercek, B., Wilson, S.R. & Houslay, M.D. (1983) Biochem. J. 213, 89-97.
Heterogeneity of cyclic AMP-phosphodiesterases in endoplasmiv reticulum.

Gawler, D., Milligan, G., Spiegel, A.M., Unson, C.G. & Houslay, M.D. (1987) Nature (Lond.) 327, 229-232.
Abolition of the expression of Gi in diabetes.

Heyworth, C.M. & Houslay, M.D. (1983) Biochem. J. 214, 93-98.
Challenge of ehpatocytes with glucagon triggers the rapid modulation of adenylate cyclase activity in membranes.

Heyworth, C.M., Wallace, A.V. & Houslay, M.D. (1983a) Biochem. J. 214, 99–110
Insulin and glucagon regulate the activation of two cyclic AMP-phosphodiesterases in hepatocytes.

Heyworth, C.M., Hanski, E. & Houslay, M.D. (1983b) Biochem. J. 222, 189-194.
IAP blocks glucagon desensitization in intact hepatocytes

Heyworth, C.M., Whetton, A.D., Kinsella, A.R. & Houslay, M.D. (1984a) FEBS Lett. 170, 38-42.
TPA inhibits adenylate cyclase activity.

Heyworth, C.M., Wallace, A.V., Wilson, S.R. & Houslay, M.D. (1984b) Biochem. J. 222, 183-187.
Insulin-stimulated cyclic AMP-phosphodiesterases

Heyworth, C.M., Whetton, A.D., Wong, S., Martin, B.R. & Houslay, M.D. (1985) Biochem. J. 228, 593-603.

Insulin inhibits the cholera toxin catalysed ADP-ribosylation of a 25kDa protein.
Houslay, M.D. (1986) Biochem. Soc. Trans 14, 183-193.
Insulin, glucagon and the receptor-mediated control of cyclic AMP metabolism.
Houslay, M.D. & Heyworth, C.M. (1983) Trends Biochem. Sci.8, 449-457.
Insulin; in search of a mechanism.
Houslay, M.D. & Siddle, K. (1989) Brit. Med. Bull. 45, 264-284.
Molecular basis of insulin receptor function.,
Houslay, M.D., Gawler, D., Milligan, G. & Wilson, A. (1988) Cellular Signalling 1, 9-22.
Multiple defects in the G-protein system of liver plasma membranes from Zucker rats.
Irvine, F.J. & Houslay, M.D. (1988) Biochem. Pharmacol. 37, 2773-2779.
Effects of treatment of intact hepatocytes with A23187 on cAMP metabolism
Kilgour, E., Anderson, N.G. & Houslay, M.D. (1989) Biochem. J. 260, 27-36.
Activation and phosphorylation of the 'dense-vesicle' cyclic AMP phosphodiesterase
Lavan, B.E., Lakey, T. & Houslay, M.D. (1989) Biochem. Pharmacol. (in press)
Resolution of soluble cyclic nucleotide phosphodiesterase isoenzymes from rat liver
Murphy, G.J. & Houslay, M.D. (1988) Biochem. J. 249, 543-547.
Resensitization of adenylate cyclase.
Murphy, G.J., Hruby, V.J., Trivedi, D., Wakelam, M.J.O. & Houslay, M.D. (1987) Biochem. J. 243, 39-46.
Rapid desensitization of adenylate cyclase is a cyclic AMP-independent event.
Murphy, G.J., Gawler, D., Milligan, G., Wakelam, M.J.O., Pyne, N.J. & Houslay, M.D. (1989) Biochem. J. 259, 191-197.
Glucagon desensitization does not involve the inhibitory G-protein Gi.
Okajima, F., Tokumitsu, Y., Kondo, Y. & Ui, M. (1987) J. Biol. Chem. 262, 13483-13490.
P2-purinergic receptor-mediated events in hepatocytes
Pyne, N.J., Cooper, M.E. & Houslay, M.D. (1987a) Biochem.J. 242, 33-42.
Insulin- and glucagon-stimulated 'dense-vesicle' phosphodiesterase.
Pyne, N.J., Anderson, N., Lavan, B.E., Milligan, G., Nimmo, H.G. & Houslay, M.D. (1987b) Biochem. J. 248, 897-901.
Tissue distribution of cyclic AMP-phosphodiesterases in rat
Pyne, N.J., Cooper, M.E. & Houslay, M.D. (1986) Biochem.J. 234, 325-334.
Cytosolic and particulate forms of the cGMP-stimulated cyclic AMP-phosphodiesterase.
Pyne, N.J. & Houslay, M.D. (1988) Biochem. Biophys. Res. Commun. 156, 290-296.
An insulin 'mediator' preparation stimulates the cGMP-stimulated cyclic AMP-phosphodiesterase.
Pyne, N.J., Murphy, G.J., Milligan, G. & Houslay, M.D. (1989) FEBS Lett.
Treatment of intact hepatocytes with either glucagon or TPA causes the phosphorylation and functional inactivation of Gi.

Wakelam, M.J.O., Murphy, G.J., Hruby, V.J. & Houslay, M.D. (1986) Nature (Lond.) 323, 68-71.
Activation of two signal transduction systems in hepatocytes by glucagon
Whipps, D.E., Armstron, A.E., Pryor, H.J. & Halestrap, A.P. (1987) Bio chem. J. 241, 835-845.
Glucagon effects on inositol phospholipd metabolism.

ACTIVATION OF HUMAN NEUTROPHILS BY GRANYLOCYTE-MACROPHAGE COLONY-STIMULATING FACTOR: ROLE OF GUANINE-NUCLEOTIDE BINDING PROTEINS

R.I. Sha'afi, J. Gomez-Cambronero, M. Yamazaki, M. Durstin, T.F.P. Molski, and C.-K. Huang*

Departments of Physiology and Pathology*
University of Connecticut Health Center
Farmington, Connecticut, 06032, U.S.A.

INTRODUCTION

The human hormone granulocyte-macrophage colony-stimulating factor (GM-CSF), which is released by activated T lymphocytes, endothelium and fibroblasts, is an important stimulus for the proliferation of erythroid and myelomonocytic stem cells in vitro (Golde and Gasson, 1988). GM-CSF is also an important in vitro regulator of granulopoiesis and neutrophil function. It induces significant leukocytosis and shortens the period of neutropenia following bone marrow transplantation, and in patients with AIDS, it increases the number of neutrophils, eosinophils and monocytes (Groopman et. al., 1987; Donahue et. al., 1986; Metcalf, 1985; Golde and Gasson, 1988). The addition of GM-CSF to mature human neutrophils primes these cells to subsequent stimulation by chemotactic factors such as fMet-Leu-Phe and thus it plays an important role in the host defense (Weisbart et. al., 1987; Dahinden et. al., 1988, Sullivan et. al., 1987; Naccache et. al., 1988; Weisbart, et. al., 1986; Mege et. al., 1989). The responses which are primed include oxidative burst, phagocytosis, arachidonic acid release and others. In addition to the priming of neutrophils, the addition of GM-CSF to human neutrophils produces on its own some responses (Coffey et. al., 1988; Gomez-Cambronero et. al., 1989; McColl et. al., 1989; Mege et. al., 1989; Nathan, 1989). These include stimulation of guanylate cyclase activity, a reduction in adenylate cyclase activity and phosphorylation of several proteins, superoxide production (when cells are on surface), Na^+-influx, the induction of c-fos mRNA, cationic protein synthesis, phosphate uptake and other responses. GM-CSF modulates the number and the affinity of fMet-Leu-Phe receptors (Weisbart et. al., 1986).

Although it is generally agreed that GM-CSF produces its effects through a plasma membrane associated receptors, the exact molecular weight and the identity of these receptors are not known. Based on cross-linking with radioactive ligand [^{125}I] GM-CSF studies the most recent value for the undegraded GM-CSF is 84 kDa (Dipersio et. al., 1988).

In spite of its importance, the mechanism of GM-CSF action is totally unknown. Very recently, the human genes for several growth factors have been cloned and expressed in transfected cell culture systems, providing for the first time relatively large quantities of pure, human-active factors for study (Golde and Gasson, 1988). Now it is possible to investigate in detail the mechanism of GM-CSF action, and in particular the roles of guanine-nucleotide regulatory proteins, protein kinase C, cyclic nucleotides, cellular organelles, phosphoinositides pathways, and lipid mediators in mediating GM-CSF actions.

POTENTIATION OF SUPEROXIDE PRODUCTION BY GM-CSF

These studies were undertaken to examine two questions. First, does GM-CSF potentiate superoxide production in human neutrophils stimulated by all agonists? Second, does the potentiation require intact neutrophils or can it be produced in cytoplasts? In order to answer the first question, superoxide production in control and GM-CSF treated human neutrophils stimulated by several stimuli were measured. The results of these studies clearly show that GM-CSF potentiates superoxide production produced by fMet-Leu-Phe and PAF but not by PMA, opsonized zymosan or unopsonized zymosan. The magnitude of potentiation is greatest in the case of fMet-Leu-Phe. It was also found that although the protein kinase C inhibitor H-7 inhibits the superoxide production by PMA, it has no effect on the potentiation by GM-CSF of the fMet-Leu-Phe stimulated superoxide generation.

In order to answer the second question, we have measured superoxide production in cytoplasts stimulated by fMet-Leu-Phe and PMA. In these studies two sets of experiments were carried out. In the first set, cytoplasts were first prepared and then incubated with GM-CSF for 1.5 hours before they were stimulated. The results of these studies show that GM-CSF is unable to potentiate superoxide generation produced by fMet-Leu-Phe. The cytoplasts are capable of generating a large amount of superoxide when stimulated with PMA. In the second set of experiments, the cells were incubated with GM-CSF for 1.5 hours prior to the preparation of cytoplasts. The data show that when stimulated with fMet-Leu-Phe, cytoplasts prepared from GM-CSF treated neutrophils generated higher amounts of superoxide than that produced by cytoplasts prepared from control cells.

GM-CSF does not potentiate superoxide production by cytoplasts stimulated with fMet-Leu-Phe. This is not due to the inability of cytoplasts to synthesize proteins since cyclohexamide added to intact neutrophils for 10 min prior to the addition of GM-CSF does not inhibit the potentiation by GM-CSF. This lack of stimulation by GM-CSF is most likely due to the loss of the granules, the nucleus and other intracellular compartments, and it is not due to the experimental procedure (cytochalasin B, Ficoll). The finding that intact cells treated identically as those cells to be enucleated except for the ultracentrifugation step showed the same pattern of superoxide production as untreated cells supports this view. In addition, the observation that cytoplasts prepared from GM-CSF treated human neutrophils generate more superoxide than those isolated from normal cells when stimulated with fMet-Leu-Phe is consistent with this view. This potentiation by GM-CSF is much smaller than that observed for intact neutrophils. In cytoplasts prepared from GM-CSF-treated cells the enhancement is small (60%) whereas in intact cells it is large (>280%). The inability of GM-CSF to potentiate oxidative burst in cytoplasts suggests the importance of the granules in the oxidative burst in the action of GM-CSF. It will be interesting to see if other actions produced by GM-CSF depend on the granules, nucleus or both.

IS THERE A ROLE FOR GUANINE-NUCLEOTIDE REGULATORY PROTEINS IN GM-CSF ACTION?

In spite of its importance, the mechanism of GM-CSF action is totally unknown. In this regard, one main unanswered question is the role of any of the known guanine nucleotide regulatory proteins in the effects of GM-CSF on neutrophils. To ascertain this point it is necessary to use a stimulus whose effect is potentiated by GM-CSF, and whose action is not inhibited by any of the known toxins that catalyze the ADP-ribosylation of one or more G-proteins and/or to identify a change that can be elicited by GM-CSF alone. To investigate the role of G-proteins in GM-CSF action we made use of pertussis toxin which catalyzes the ADP-ribosylation of the γ-subunit of G_c (Sha'afi and Molski, 1988) cholera toxin which ribosylates the γ-subunit of G_i (Gilman, 1987) and botulinum D toxin which has been shown recently to catalyze the ADP-ribosylation of a 22 kDa protein in neutrophils (Mege et. al., 1989). Two types of experiments were carried out. The first dealt with the priming action of GM-CSF, and the second dealt with the effect of GM-CSF alone on various biochemical changes.

Cell Priming by GM-CSF

The effects of various toxins on superoxide generation in control and GM-CSF-treated neutrophils stimulated with fMet-Leu-Phe was studied (Table I). These data clearly show that neither cholera nor botulimun D toxin affects superoxide generation in control or GM-CSF treated human neutrophils.

The addition of platelet-activating factor (PAF) to neutrophils causes a rise in the intracellular free calcium concentration, and this rise is not greatly ihibited by pertussis toxin (Naccache et. al., 1985; Verghese, et. al., 1987; McColl et. al., 1989). GM-CSF potentiates PAF-induced calcium rise, and this potentiation is inhibited when the cells were incubated for 1 hour with pertussis toxin (0.5 µg/ml) before the addition of GM-CSF (McColl et. al., 1989; Mege et. al., 1989).

Phosphate Uptake and GM-CSF

The effect of the addition of GM-CSF on phosphate uptake was investigated. These studies clearly show that phosphate uptake is significantly ($p < 0.001$) stimulated by GM-CSF. The effect of GM-CSF was evident only after the cells were treated with GM-CSF for 15 min. The stimulation by GM-CSF is dose-dependent. Significant increases can be seen at concentrations > 27 pM. Most of the increase in the counts in GM-CSF-treated cells is associated with the TCA-precipitate. After 20 minutes of incubation with the radioactive phosphate, the total radioactivity in GM-CSF-treated cells was 1.95 ± 0.15 relative to control cells, and the counts in TCA-precipitate was 1.80 ± 0.1 relative to control.

The effects of GM-CSF on phosphate uptake in control and toxin-treated human neutrophils were also examined. These studies show that the stimulation by GM-CSF is significantly ($p < 0.05$) inhibited in pertussis toxin treated cells but not in cholera or botulinum D toxin treated neutrophils (Gomez-Cambronero et. al., 1989).

Na^+/H^+ Antiport and GM-CSF

The effects of GM-CSF on the basal and stimulated ^{22}Na-influx in human neutrophils stimulated with fMet-Leu-Phe have been studied. In these experiments, the effect of GM-CSF on cells treated with pertussis toxin was also examined. The results are shown in Figure 1. The data

Table 1. Effect of GM-CSF on Superoxide Production in Control and Toxin-Treated Neutrophils Stimulated with fMet-Leu-Phe.

Exp. Condition[b]	Superoxide Production ($nmol/10^7$ cells, 2min)[a]	
	Control	+ GM-CSF
Normal cells	17.1 ± 4	66 ± 8
Pertussis toxin	3.5 ± 4	8.5 ± 4
Cholera toxin	15.0 ± 6	60 ± 10
Botulinum D toxin	20.3 ± 4	58 ± 5

[a]The cells were stimulated with fMet-Leu-Phe (5 nM) for 2 minutes. Each value represents the mean ± SEM of five experiments.

[b]The cells were incubated with botulinum D toxin (0.5 μg/ml), pertussis toxin (0.5 μg/ml) and cholera (2 μg/ml) for 1 hour and then with GM-CSF (100 pM) for an additional 90 minutes.

reveal several points. (1) The addition of GM-CSF causes an increase in ^{22}Na-influx, and the action of GM-CSF on ^{22}Na-influx is very rapid. (2) The GM-CSF induced increase is inhibited by amiloride. (3) The fMet-Leu-Phe stimulated ^{22}Na-influx is increased in GM-CSF-treated cells. (4) Both increases are statistically ($p<0.001$) less in pertussis toxin treated cells.

The effect of GM-CSF on the intracellular pH in control and stimulated human neutrophils was also examined and the results are shown in Figure 2. The data show three points. (1) The addition of GM-CSF increases the intracellular pH, and this effect can be seen as early as 5 min after its addition. (2) This increase is totally abolished by amiloride, and is diminished in a statistically significant manner ($p <0.001$) in pertussis toxin treated neutrophils. (3) The observed increase in intracellular pH found in neutrophils stimulated with either PMA or fMet-Leu-Phe is inhibited in GM-CSF-treated cells. This indicates that the activation of this antiport by GM-CSF inhibits this system to further stimulation.

Proto-oncogenes c-fos and c-myc and GM-CSF

The addition of GM-CSF to human neutrophils has been shown to stimulate the expression of surface antigens as well as the induction of the synthesis of mRNA of the proto-oncogenes c-fos and c-myc (Lopez et. al., 1986; Arnaout et. al., 1986; Colotta et. al., 1987; McColl et. al., 1989). Preincubation of human neutrophils with pertussis toxin significantly inhibited the induction of c-fos mRNA produced by GM-CSF (McColl et. al., 1989).

GM-CSF and GTPase Activity

The basal and stimulated GTPase activity in membrane preparation isolated by nitrogen cavitation from control and GM-CSF-treated cells were determined. The basal activity is significantly ($p<0.05$, paired sample t-test) higher (27%) in GM-CSF cells (45 min). Pertussis toxin reduced significantly the increase in the basal and stimulated activity in GM-CSF-treated cells as well as the GTPase activities in control cells ($p<0.001$ for both fMet-Leu-Phe and PAF) stimulated by PAF or fMet-Leu-Phe. These effects of GM-CSF on the basal and stimulated

GTPase activity can be shown to be statistically significant only when GM-CSF is added to the intact cells prior to the preparation of the membrane. GM-CSF added after the membranes were prepared showed small increases in both the basal and stimulated GTPase activities, but the increases were within experimental errors. This lack of effect may be due to the method used to prepare the membrane.

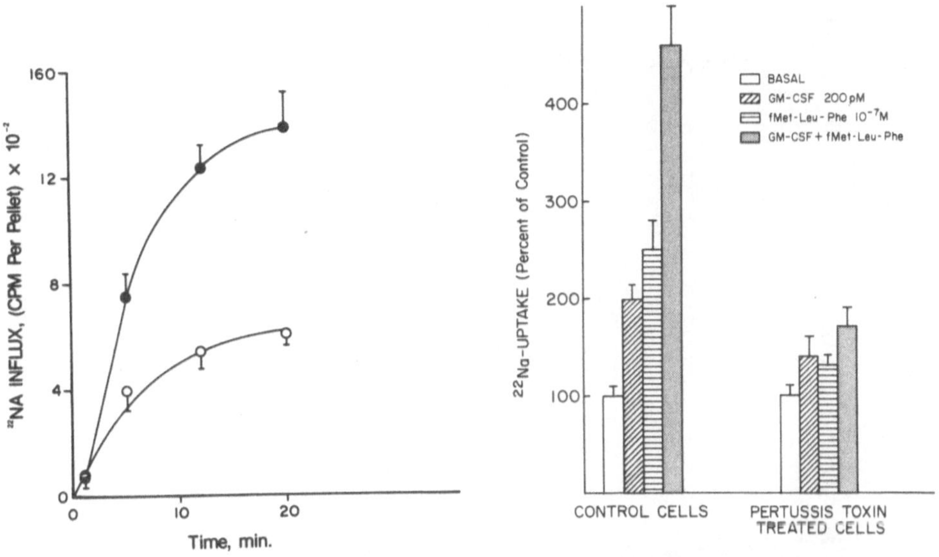

Figure 1. (Left panel) The effect of GM-CSF on the time course of ^{22}Na-influx in human neutrophils. GM-CSF (200 pM) and the radioactive sodium were added at time zero. Each point represents the mean ± SEM of at least three experiments.
(Right panel) Effect of pertussis toxin on ^{22}Na-influx in human neutrophils stimulated with GM-CSF, fMet-Leu-Phe and GM-CSF and fMet-Leu-Phe. The cells were incubated with 0.5 µg/ml pertussis toxin for 45 minutes.

GM-CSF and Tyrosine Phosphorylation

The addition of GM-CSF to human neutrophils causes an increase in the tyrosine phosphorylation levels of at least five different proteins whose molecular weights are 118 kDa, 92 kDa, 78 kDa, 54 kDa and 40 kDa (Figure 3). The increase in tyrosine phosphorylation levels in these proteins are significantly reduced in pertussis toxin treated cells. It is tempting to speculate that one of these proteins is probably the receptor for GM-CSF. Based on cross-linking with radioactive ligand ^{125}I-GM-CSF studies, the most recent value for the undegraded GM-CSF receptor is 84 kDa. Because of its low molecular weight, the GM-CSF receptor probably does not have intracellular tyrosine kinase domain. If so, GM-CSF receptor must be coupled to some intracellular tyrosine kinases. Nonreceptor tyrosine kinases have been found in mammalian cells (Bishop, 1983; Morla et. al., 1988). The possible coupling of

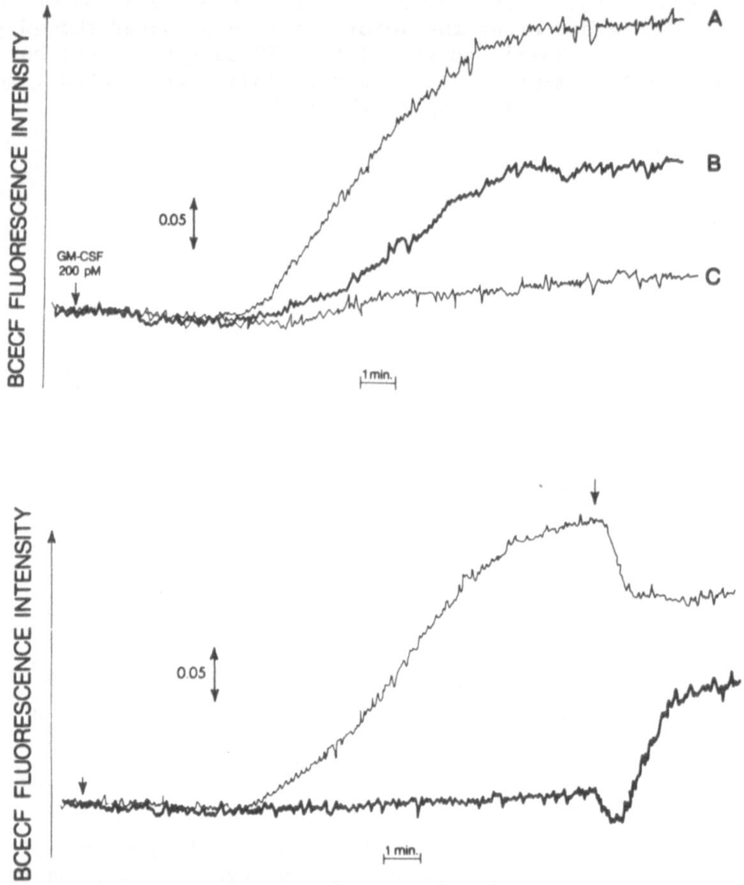

Figure 2. (Top panel) Effect of GM-CSF on the intracellular pH in control (A) pertussis toxin (B) and amiloride (C) treated human neutrophils. The arrow represents the time of the addition of GM-CSF (200 pM). The cells were treated with 1mM amiloride for 5 min or pertussis toxin (0.5 μg/ml) for 45 min before the addition of GM-CSF. The tracings are taken from a single experiment that was carried out at least three separate times with essentially identical results. Upward deflection represents an increase in intracellular pH.
(Bottom panel) Effect of GM-CSF-induced increase in intracellular pH on the action of the subsequent addition of PMA. The first arrow represents the time of addition of GM-CSF (200 pM), and the second arrow represents the time of addition of PMA (0.05 μg/ml). The lower figure represents control cells stimulated with PMA, and the upper figure represents cells stimulated first with GM-CSF and then with PMA.

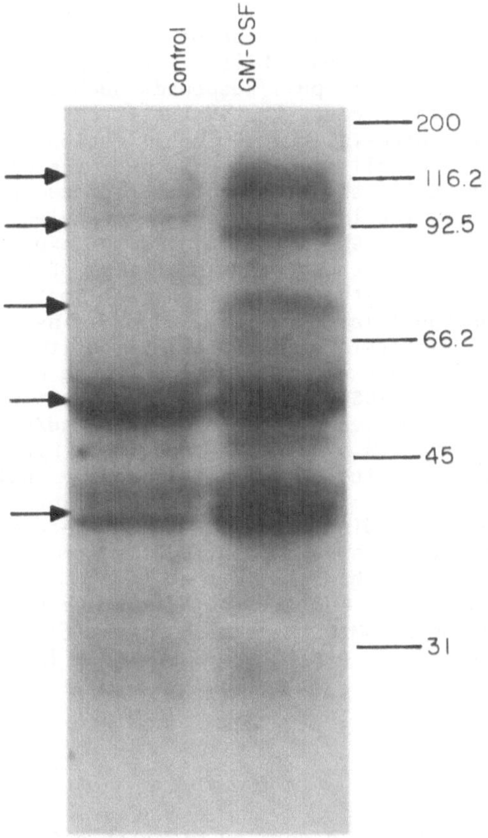

Figure 3. Autoradiograph showing the patterns of immunoblotting of proteins from control and GM-CSF stimulated human neutrophils. Unlabelled cells were stimulated with GM-CSF (200 pM) for 10 min at 37°C. Proteins were analyzed by SDS-PAGE and followed by immunoblotting with antiphosphotyrosine antibodies. GM-CSF increases the tyrosine phosphorylation levels in five proteins whose molecular weights are 118 kDa, 92 kDa, 78 kDa, 54 kDa and 40 kDa. The autoradiograph represents a single experiment that was carried out at least three separate times with similar results.

the GM-CSF receptor to intracellular kinases could be achieved by several mechanisms (Morla et. al., 1988). First, the kinase may be a component of the receptor complex. The activation of the kinases could be by a protein-protein interaction or mediated through a guanine-nucleotide regulatory protein. Second, the occupancy of the GM-CSF receptor could stimulate the production of a diffusible second messenger. Third, the binding of GM-CSF to its receptor could alter the location or the conformation of the substrate proteins so they can become phosphorylated.

SUMMARY

Several important conclusions can be drawn from the studies of GM-CSF effects on mature human neutophils. First, preincubation of neutrophils with GM-CSF potentiates neutrophil responses such as phagocytosis, oxidative burst, arachidonic acid release, the synthesis of platelet-activating factor, leukotriene synthesis, calcium mobilization and other responses produced by some but not all agonists. Second, the potentiation by GM-CSF requires intact cells, and it cannot be reproduced in cytoplasts. This strongly suggests the importance of the granules in the priming effects of GM-CSF. Third, GM-CSF stimulates the expression of surface antigens, the induction of the synthesis of mRNA of the proto-oncogenes c-fos and c-myc, the activity of the Na^+/H^+ antiport, phosphate uptake and the tyrosine phosphorylation of several proteins. Fourth, GM-CSF does not cause the hydrolysis of phosphatidylinositol 4,5-bisphosphate suggesting that it does not activate phospholipase C. However, under appropriate conditions GM-CSF stimulates the production of diacylglycerol indicating that it may activate phospholipase D and/or PLC that activates other phospholipids such as phosphatidylcholine. Fifth, all of the effects of GM-CSF that have been studied are significantly inhibited by pertussis toxin but not by cholera toxin or botulinum D toxin. The effect of pertussis toxin is distal to the binding of GM-CSF to its receptors. This suggests that the GM-CSF receptors in human neutrophils are coupled to a G-protein. Whether this G-protein is the same as that which mediates the hydrolysis of phosphoinositides is not known. The 40 kDa and 54 kDa which are phosphorylated on tyrosine by GM-CSF are most likely the $Gi_{\alpha 2}$ and a protein kinase respectively. The protein tyrosine kinase is probably the product of the gene hck.

ACKNOWLEDGEMENTS

This work is supported in part by NIH grants GM-37694, AI-24935, and AI-20943. C-K. Huang is a recepient of the American Heart Association Established Investigator Award. We thank Dr. John Casnellie for his help on the use of antiphosphotyrosine antibodies. J. Gomez-Cambronero is a recipient of a "Consejo Superior de Investigaciones Cienetificas" fellowship.

REFERENCES

Arnaout, M. A. Wang, E. A., Clark, S. C., and Sieff, C. A., 1986, Human recombinant granulocyte-macrophage colony-stimulating factor increases cell-to-cell adhesion and surface expression of adhesion-promoting surface glycoproteins on mature granulocytes. J. Clin. Invest., 78: 597.

Bishop, J.M., 1983, Cellular oncogenes and retroviruses. Annu. Rev. Biochem., 52: 301.

Coffey, R. G., Davis, J. S., and Djeu, J. Y., 1988, Stimulation of guanylate cyclase activity and reduction of adenylate cyclase activity by granulocyte-macrophage colony-stimulating factor in human blood neutrophils. J. Immunol. 140: 2695.

Colotta, F., Wang, J. M., Polentaruti, N., and Mantovani, A., 1987, Expression of c-fos protooncogene in normal human peripheral blood granulocytes. J. Exp. Med., 165: 1224.

Dahinden, C. A., Zingg, J., Maly, F. E., and DeWeck, A. L., 1988, Leukotriene production in human neutrophils primed by recombinant human granulo-cyte macrophage colony-stimulating factor and stimulated

with the complement component CSa and FMLP as second signals. J. Exp. Med., 167: 1281.
Dipersio, J., Billing, P., Kufman, S., Egresady, P., Williams, R. E., and Gasson, J. C., 1988, Characterization of the human granulocyte-macrophage colony-stimulating factor (GM-CSF) receptor, J. Biol. Chem., 263: 1834.
Donahue, R. E., Karlsson, S., Clark, S., Anderson, W. F., and Nienhuis, A. W., 1986, Recombinant human GM-CSF accelerates neutrophil and platelet recovery following autologous bone marrow transplantation. Blood, 68: 281a.
Gilman, A. G., 1987, G-proteins: Transducers of receptor-generated signals, Ann. Rev. Biochem., 56: 615.
Golde, D. W., and Gasson, J. C., 1988, Hormones that stimulate the growth of blood cells, Scientific American, 259: 62.
Gomez-Cambronero, J., Yamazaki, M., Metwally, F., Molski, T. F. P., Bonak, V. A., Huang, C.-K., Becker, E. L., and Sha'afi, R. I., 1989, Granulocyte-macrophage colony-stimulating factor and human neutrophils: Role of guanine-nucleotide regulatory proteins, Proc. Natl. Acad. Sci. U. S. A., in press.
Groopman, J. E., Mitsuyasu, R. T., Deleo, M. J., Oette, D. H., and Golde, D. W., 1987, Effect of recombinant human granulocyte-macrophage colony-stimulation factor on myelopoiesis in the acquired immunodeficiency syndrome. N. Engl. J. Med., 317: 593.
Lopez, A., Williamson, J., Gamble, J., Begley, G., Harlan, J., Klebanoff, S., Waltersdorph, A., Wong, G. G., Clark, S. C., and Vadas, M., 1986, Recombinant human granulocyte-macrophage colony-stimulating factor stimulates in vitro mature human neutrophil and eosinophil function,surface receptor expression and survival, J. Clin. Invest., 78: 1220.
McColl, S. R., Kresi, C., DiPersio, J. F., Borgeat, P., and Naccache, P. H., 1989, Involvement of guanine nucleotide binding proteins in neutrophil activation and priming by GM-CSF. Blood, 73: 588.
Mege, J. L., Gomez-Cambronero, J., Molski, T. F. P., Becker, E. L., and Sha'afi, R. I., 1989, Effect of granulocyte-macrophage colonly-stimulating factor on superoxide produciton in cytoplasts and intact human neutrophils: Role of protein kinase C and G-proteins. J. Leuk. Biol., In press.
Metcalf, D., 1985, The granulocyte-macrophage colony-stimulating factors. Science, 229: 16.
Morla, A. O., Schreurs, J., Miyajima, A., and Wang, J. Y. J., 1988, Hematopoietic growth factors activate the tyrosine phosphorylation of distinct sets of proteins in interleukin-3-dependent murine cell line, Mol. Cell. Biol., 8: 2214.
Naccache, P. H., Molski, M. M., Volpi, M., Becker, E. L., and Sha'afi, R. I., 1985, Unique inhibitory profile of platelet activating factor induced calcium mobilization, polyphosphoinositide turnover and granule enzyme secretion in rabbit neutrophils towards pertussis toxin and phorbol esters. Biochem. Biophys. Res. Commun., 130: 677.
Naccache, P. H., Faucher, N., Borgeat, P., J. C., and Depersio, J. F.,1988, Granulocyte-macrophage colony-stimulating factor modulates the excitation-response coupling sequence in human neutrophils, J. Immunol.,140: 3541.
Nathan, C.F., 1989, Respiratory burst in adherent human neutophils: Triggering by colony-stimulating factors CSF-GM and CSF-G., Blood, 73: 301.
Sha'afi, R. I. and Molski, T. F. P., 1988, Activation of the neutrophils, Progress in Allergy, 42: 1.
Sullivan, R., Griffin, J. D., Simons, E. R., Schafer, A. I., Meshulam, T., Fredette, J. P., Maas, A. K., Gadenne, A. S., Leavitt, J. L.,

and Melnick, D. A., 1987, Effects of recombinant human granulocyte and macrophage colony-stimulating factors on signal transduction pathways in humangranulocytes. J. Immunol., 139: 3422.
Verghese, M. W., Charles, L., Jakoi, L., Dillon, S. B., and Snyderman, R., 1987, Role of a guanine nucleotide regulatory protein in the activation of phospholipase C by different chemoattractants. J. Immunol., 138: 4374.
Weisbart, R. H., Golde, D. W., and Gasson, J. C., 1986, Biosynthetic human GM-CSF modulates the number and affinity of neutrophil fMet-Leu-Phe receptors, J. Immunol., 137: 3584.
Weisbart, R. H., Kwan, L., Golde, D. W., and Gasson, J. C., 1987, Human GM-CSF primes neutrophils for enhanced oxidative metabolism in response to the major physiologic chemoattractants, Blood, 69: 18.

CYTOSKELETAL PARTICIPATION IN THE SIGNAL TRANSDUCTION PROCESS:

TUBULIN G PROTEIN INTERACTIONS IN THE REGULATION OF ADENYLATE CYCLASE

Kun Yan and Mark M. Rasenick

Department of Physiology and Biophysics and the Committee on Neuroscience, University of Illinois College of Medicine m/c 901 PO Box 6998, Chicago, IL 60680 USA

Receptors for a variety of hormones or neurotransmitters are coupled to their intracellular effectors via GTP-binding [G] proteins. Although initially discovered as proteins which conferred stimulation [Gs] or inhibition [Gi] upon adenylate cyclase. Since their initial discovery, these proteins have been assigned new roles at an alarming rate. Further, molecular genetic techniques have demonstrated that the signal transducing G proteins [for this review defined as those G proteins which are substrates for ADP ribosylation by pertussis and/or cholera toxins] consist of a family of closely related proteins which are the products of several genes (or alternate splicing of a single gene). Currently, it appears that G proteins may be involved in neutrophil activation as well as the activation of phospholipase C and A2. Further, some G proteins may act as second messengers, via direct effects on ion channels. The purpose of this chapter is to provide a scheme whereby components of the cell thought to be primarily structural, might participate in the signal transduction process.

Roles for G protein subunits in the transduction of cellular signals

The known signal transducing G proteins have a heterotrimeric structure consisting of α, β, and γ subunits. Subsequent to binding of GTP by the α subunit, β and γ subunits appear to dissociate (in vitro). Reconstitution experiments have demonstrated the ability of the β γ subunits to turn off the activated G protein α, subunit. No evidence exists that this mechanism regulates adenylate cyclase within the cell, however. In fact, several schemes for the regulation of G protein mediated second messenger systems have been proposed (see figure 1). In addition to βγ acting as a switch, it has been suggested that βγ might inhibit calmodulin sensitive cellular processes (Katada et.al., 1987) or activate phospholipase a_2. The possibility that Giα exerts a direct effect upon the adenylate cyclase catalyst has been proposed. Additionally, the possibility that α subunits interact with one another has been suggested (Hatta et., al. 1986; Marbach et al., 1988). Some of these schemes for G protein mediated effects on the adenylate cyclase system are seen in figure 1.

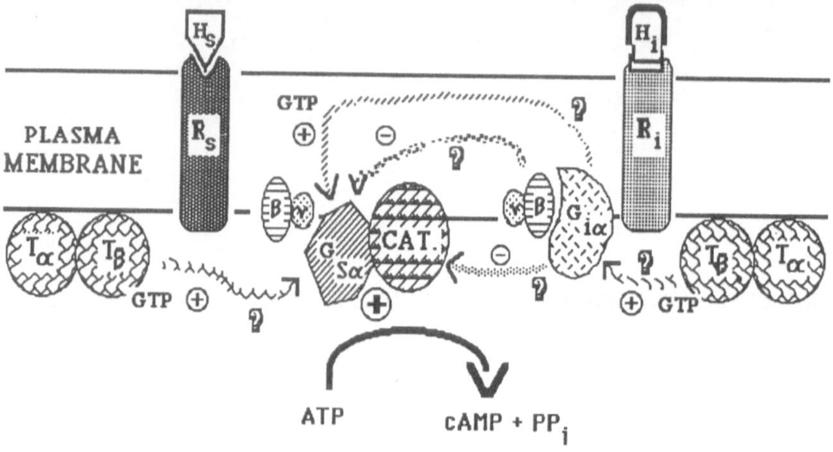

Figure 1 Regulation of adenylate cyclase.

In this schematic, identified components of the adenylate cyclase system and their possible roles in the regulation of that system are illustrated. The text below describes coupling among components of the adenylate cyclase system and this cartoon shows that the binding of an hormone of neurotransmitter (Hs or Hi) to the specific receptors mediating the stimulation (Rs) or inhibition (Ri) of adenylate cyclase cause the activation of the G proteins involved in those processes (Gs or Gi respectively). Subsequent to the activation of those G proteins (and the binding of GTP by those proteins), a complex series of interactions could take place. It does appear that the β and γ subunits can dissociate from the α subunit subsequent to the activation of α, but a variety of actions are possible for these subunits. These actions are described in the text above and below. This laboratory has suggested that GTP (or analogs of GTP) can be transferred from Giα to Gsα and might be one locus for intracellular activation of adenylate cyclase or for activation of adenylate cyclase by other second messenger systems. Other possible participants in the regulation of adenylate cyclase are membrane associated tubulin dimers (Tα Tβ). This laboratory has observed the stable inhibition of synaptic membrane adenylate cyclase by tubulin polymerized with GppNHp and has suggested that the mechanism for this inhibition resides in the transfer of GTP from tubulin to Giα. In C6 cells, tubulin stimulates adenylate cyclase in a receptor-independent fashion and it is suggested that the mechanism for this involved transfer of GTP from tubulin to Gsα.

Receptor-effector coupling in the signal transduction process

Receptor-effector coupling refers to the process whereby an agonist-occupied (activated) receptor activates a G protein and the α subunit of that G protein engages one of the intracellular effectors described above. It has been argued that receptors may be uncoupled from G proteins in the absence of agonist, and that the presence of agonist brings the proteins together. Certain kinetic data as well as data from experiments in which the mobility of membrane proteins is increased favor such hypotheses (see Levitski, 1987). Other theories favor the idea of a "precoupled" receptor-G protein complex, which represents the high affinity state of agonist binding. Solubilization of receptor-G protein complexes (in the absence of agonist) as well as target size analysis favor the latter hypothesis (see Rodbell, 1985).

This laboratory has suggested that the receptor coupled to a G protein constrains that G protein (specifically Gsα) from interacting with the catalytic subunit of adenylate cyclase and activating that enzyme. Experiments with permeable C6 cells [in which tight receptor-G protein coupling is maintained - Rasenick and Kaplan, 1986] and a comparison to the loosely coupled synaptic membrane system in which the G protein "antagonist" GDPβS serves as a partial agonist (Rasenick et. al., 1989a) have led to this theory. Consistent with the above hypothesis, Woon et. al. (1989) have constructed a chimeric G protein [containing the first 356 amino acid residues of Gsα and the final 36 residues of Giα2 at the carboxy terminal] which they have expressed in CHO cells. This protein was unable to couple to the receptor, but the ability of this Gα chimera to stimulate adenylate cyclase was quite high (in fact it may have engaged in constitutive activation of that enzyme). Thus, if a G protein is freed from the constraints of receptor coupling, it would be expected to interact more successfully with the appropriate intracellular effector molecule.

Participation of cytoskeletal components in the process of cellular signal transduction

Additional evidence that certain cellular components constrain a favored Gs-catalytic interaction comes from a variety of studies which suggest that a component of the cytoskeleton (i.e., microtubules) acts to provide such constraint and that microtubule disrupting agents alleviate said constraint, subsequently enhancing adenylate cyclase stimulation by increasing Gs-Catalytic moiety interaction. (see Zor, 1983 and Rasenick et al, 1985 for reviews).

More recent work from this laboratory has demonstrated that, in rat cerebral cortex synaptic membranes, tubulin dimers, rather than microtubules, inhibited adenylate cyclase. Such inhibition (see figure 2b) resulted after incubation of tubulin with synaptic membranes, and was stable to repeated washing of the membranes (Rasenick and Wang, 1988). Tubulin, in these experiments, polymerized with GppNHp [tubulin-GppNHp] had no direct effects upon resolved adenylate cyclase catalyst and required bound GTP (or GTP analog) to exert its effects. When tubulin was polymerized with the hydrolysis-resistant GTP photoaffinity analog, azidoanilido GTP [[^{32}P] AAGTP] and tubulin-AAGTP was incubated with synaptic membranes, AAGTP which had been bound to tubulin appeared on Giα. [In these experiments, membranes were incubated with AAGTP-tubulin, washed and resuspended before the UV exposure which made AAGTP binding covalent.] No nucleotide was released from tubulin into the medium and no nucleotide was released from tubulin in the absence of an available acceptor site (i.e. Giα) on the synaptic membrane. In analogy with earlier results from our laboratory (Hatta et. al., 1986) which suggested that there is a direct transfer of nucleotide between Giα and Gsα, we suggested that tubulin formed an association with Giα and transferred nucleotide directly to that protein.

Tubulin as a G protein

Tubulin shares many features with "traditional" signal transducing G proteins. Tubulin consists of an αβ heterodimer with a molecular weight of approximately 110 KDa. It has at least two GTP (guanine nucleotide) binding sites, one non-exchangeable at the α subunit, and the other exchangeable at the β subunit. Tubulin dimers can undergo polymerization and depolymerization in a Mg^{2+}- and temperature-dependent manner. Unlike the "conventional" G proteins, where GTP appears to facilitate the separation of the heterodimer, the presence of GTP enhances the polymerization process, and hydrolysis of GTP to GDP by what appears to be a tubulin intrinsic GTPase, occurs subsequent to microtubule polymerization {the relationship between polymerization and GTP hydrolysis is not yet clear}. As with the "conventional" G proteins, tubulin is also a substrate for pertussis and cholera toxin catalysed ADP-ribosylation (Amir-salzman et.al., 1978; Lim et. al., 1986). Additionally, α and β tubulin appear to possess the 4 consensus GTP binding sequences (Halliday, 1984) presented in the "conventional" G proteins, except that the third sequence is inverted (Sternlitch, et. al., 1987).

Association of tubulin with "traditional" G proteins

In addition to the data of Rasenick and Wang (1988) described above, several earlier reports gave evidence of the possibility that tubulin might associate with G proteins. Crude microtubule preparations have been demonstrated to contain intrinsic adenylate cyclase activity (Margolis and Wilson, 1979, and Wolff and Cook, 1985). Sahyoun (1981) showed that there is a direct interaction between the catalytic moiety, Gsα, and the Triton X-100-insoluble cytoskeleton isolated from erythrocytes. Release of Gsα from rat brain membranes after colchicine and vinblastin treatment, and the binding of Gsα to a tubulin-Sepharose affinity column have been demonstrated by Rasenick et. al. (1984; 1985). Furthermore, binding of tubulin to a column containing immobilized Gi/Go was reported by Higashi and Ishibashi (1985).

The direct effects of tubulin and the transfer of GTP analogs from tubulin to synaptic membrane G proteins probably require formation of a complex which consists of tubulin and at least one G protein. As indicated above, significant evidence exists which suggests that tubulin (or microtubules) form complexes with members of the adenylate cyclase cascade. Further evidence for tubulin-G protein complex formation is provided in this manuscript. In addition, tubulin interactions with the receptor-coupled C6 system are compared to those in the essentially uncoupled rat cerebral cortex synaptic membrane. It is hoped that such a comparison can help to elucidate the role of tubulin and other elements of the cytoskeleton in the modification of cellular signal transduction.

C6 glioma adenylate cyclase: effects of tubulin

Incubation of C6 glioma membranes with tubulin-GppNHp results in a stimulation of the adenylate cyclase. This stimulation is dose-dependent and saturable with an EC_{50} of approximately 10^{-8} M tubulin-GppNHp (Fig. 2b). In contrast, under the same experimental condition (1 mM MgCl2, 23°C), tubulin-GppNHp inhibits adenylate cyclase on synaptic membranes (Fig. 2a). The same concentrations of IgG were added to both C6 glioma and synaptic membranes and were ineffective in modulating adenylate cyclase (data not shown). Additionally, tubulin-GppNHp does not affect the resolved catalytic unit of the adenylate cyclase prepared from solubilized C6 membranes.

The adenylate cyclase assay in saponin permeable cells was developed in this laboratory (Rasenick and Kaplan 1986), and the permeable cells are tightly coupled to the β adrenergic receptor. When pre-confluent monolayers of C6 glioma cells were treated with saponin solution for two minutes and incubated with GppNHp or tubulin-GppNHp (in the presence or absence of isoproterenol) , tubulin-GppNHp, unlike GppNHp, was able to stimulate adenylate cyclase activity in C6 glioma cells in the absence of the agonist (Fig.3). This result suggests that tubulin-GppNHp is able to bypass the receptor and stimulate the stimulatory G protein in permeable cells. In contrast, GppNHp has an absolute requirement for the presence of agonist to activate the G protein (Gs).

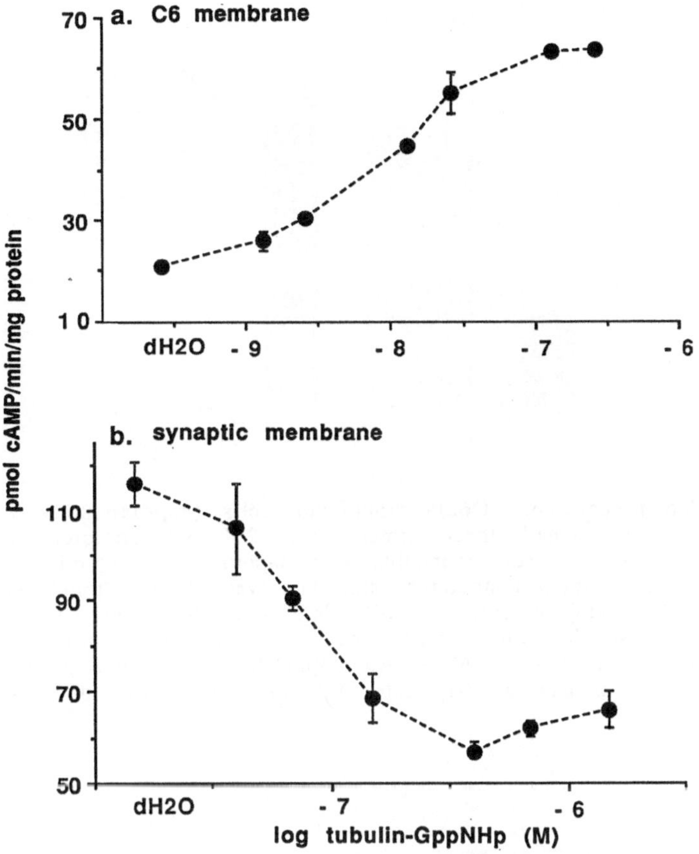

Fig. 2 Effects of tubulin-GppNHp preincubation on adenylate cyclase.

C6 glioma membranes (prepared as described in Rasenick and Kaplan (1986) and rat cerebral cortex synaptic membranes (prepared as described in Rasenick et.al.,1981) were thawed and suspended in a buffer system containing 20 mM Hepes (pH 7.5), 1 mM $MgCl_2$, 0.3 mM PMSF and 1mM DDT. The suspension was kept on ice. Then 25 ml of a reaction cocktail (consisting of 15 mM Hepes, 0.05 mM ATP, α-[^{32}P] ATP (5×10^5 cpm/tube), 1 mM $MgCl_2$, 1 mM dithiothreitol, 0.05 mM cyclic AMP, 60 mM NaCl, 0.25 mg/ml of BSA, 0.5 mM isobutylmethylxanthine and a creatine phosphokinase nucleotide triphosphate regenerating system), Tubulin was prepared according to the method of Shelanski (1973) and further purified by P6-DG desalting and phosphocellulose columns to eliminate the free guanine nucleotides and microtubule associated proteins. 10 ml of tubulin-GppNHp (tubulin polymerized with GppNHp; 0.8 mol bound/mol tubulin) at the indicated concentrations and 40 ml of ddH2O were added. The reaction was initiated by adding 25 ml of C6 glioma or rat brain membranes (15-20 mg of protein/tube). The experimental temperature was 23° C and the reaction was stopped with SDS after 10 minutes incubation. The [^{32}P]-cyclic AMP formed was purified by Dowex - alumina chromatography and quantitated by scintilation counting.

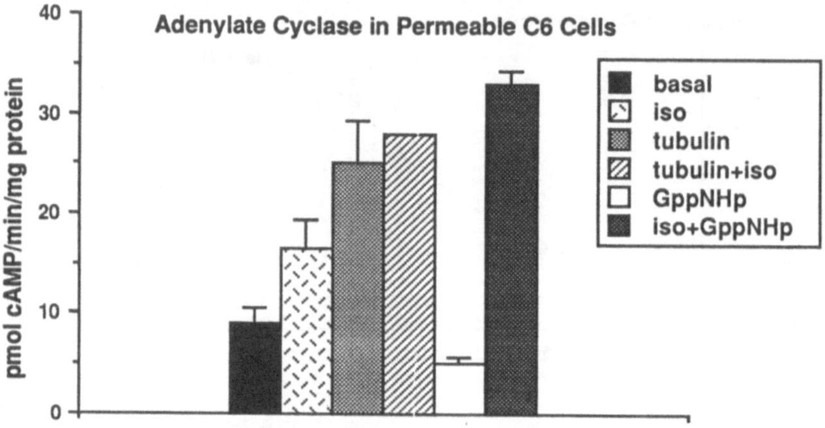

Fig. 3 The preconfluent C62B monolayer cells (approximately 250,000 cells/well) were washed three times with 200 ..ml complete Locke's solution. They were made permeable with saponin (100 mg/ml) treatment for 2.5 minutes at room temperature and then washed three times with KG (potassium glutamate) buffer. Following this, they were assayed for adenylate cyclase by the method of Rasenick and Kaplan (1986). Isoproterenol is present at 1 μM as was GppNHp or tubulin-GppNHp. The values represent means of triplicate determinations from one of three experiments.

Tubulin - G protein specific binding

Based on the above data as well as that from synaptic membranes, we suggested that there is a direct interaction between tubulin and G-protein(s). To demonstrate this possibility, western blot and [^{125}I] tubulin overlay experiments were performed. Western blots indicated that 125-I-tubulin bound to a variety of proteins in C6 membranes. Considering many of these [^{125}I]-tubulin-bound proteins shared electrophoretic mobilities and antigenic determinants with [^{32}P]-AAGTP labeled G proteins, it was likely that G proteins bound tubulin. The failure of the 125-I-ovalbumin to bind to any of the proteins in C6 membranes under the same experimental condition indicated that the interaction between tubulin and G proteins was specific.

Selective binding of tubulin to Gsα and Giα1

To determine precisely which G protein subtypes bind to tubulin, hybridization of 125-I-tubulin with purified and recombinant Ga subtypes was performed (Fig.4). figure 3a shows that [^{125}I]-tubulin binds to αs and αi (from brain) with a similar affinity and that affinity is much greater than the affinity of [^{125}I] tubulin for transducin α. The selectivity of 125-I-tubulin binding to αi subtypes is even more surprising. 125-I-tubulin binds to αi1 with a significantly higher affinity than it does to αi2, αi3 and αo. The binding affinity [Kd ~ 130 nM] of tubulin to Gsα and Gi1α is comparable to the subunit dissociation constant for tubulin dimers (Sackett et al,1989).

Tubulin Binding to Gα Subtypes

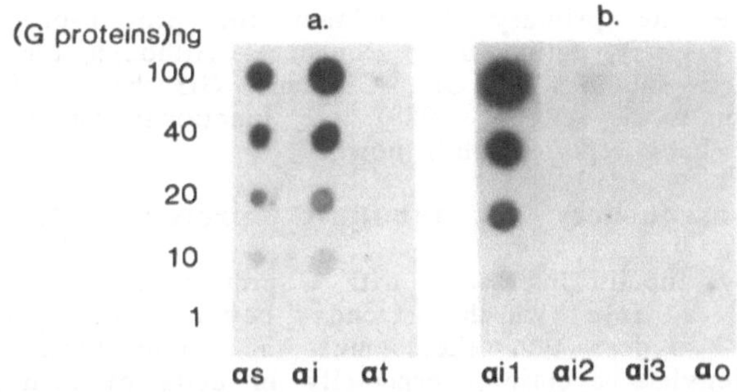

Fig. 4 Purified G-protein a-subunits (as and ai3 from A. Gilman and M. Graziano, αi2 from T. Katada and Y. Kaziro, , αi , αo and heterotrimeric Gi from E. Neer and α transducin from H. Hamm) were dotted on a nitrocellulose sheet in the dilution series indicated. After drying, the nitrocellulose sheet was incubated with blocking buffer containing 40 mg/ml BSA for 2 hours. The same sheet was then incubated with the same buffer plus 125-I-tubulin at room temperature for 2 hours. The radioautograph of the dried nitrocellulose paper is shown here. 125-I-ovalbumin does not bind to any of these G-proteins (Wang et. al., submitted).

Tubulin effects on receptor-G protein coupling

It appears that tubulin-GppNHp is able to modulate adenylate cyclase activity in both C6 and synaptic membranes under the same experimental conditions. However the effects of tubulin-GppNHp in these two systems are opposite. One possible explanation for this discrepancy is the difference in the content of G protein subtypes in these two cell types. It is known that Giα1 is the predominant Giα protein in brain tissue (Brann et. al.); whereas the mRNA of Giα1 has not been detected in the C6 glioma cell line (Lochrie and Simon, 1988). The results from our hybridization studies reveal that tubulin binds to G proteins selectively. Therefore the lack of inhibitory effect of tubulin in C6 glioma membranes is probably due to the lack of Gi1 in that cell line.

As indicated above, the receptor may exert a constraint on the stimulatory G proteins (Rasenick et. al, 1989; Woon et. al.,1989), thereby preventing the guanine nucleotide (GTP or GTP analog) activation of Gs in the absence of agonist. Nevertheless, tubulin-GppNHp is able to bypass this "constraint" and promotes the activation of the stimulatory G protein by itself. This results implies that tubulin-GppNHp may act at the receptor-G protein interface and regulate the adenylate cyclase, in concert with the external signals, as an internal monitor.

The selectivity of tubulin binding to different G protein subtypes, especially Gi subtypes is somewhat surprising. In terms of the amino acid sequence homologies, there is about 90% homology among Gi subtypes (Itoh H., et. al., 1988); 80% among the pertussis toxin substrates (Gi1, Gi2, Gi3, Go and Gt) (Itoh, H., et. al., 1986); and only 45% between Gs and others. However the secondary structures predicted among Gi subtypes are distinct (Itoh H., et. al., 1988). Despite the higher homology among Giα subtypes and lower homology between Gsα and others, tubulin binds quite well to Gs and Gi1, but not as well as to Gi2 and Gi3. This result suggests that the conformational homology of G proteins may be the primary determinant for protein-protein interaction, rather than the primary amino acid sequences. Although the antibody study has shown that α subunits of Gi1, Gi2 and Gi3 have different antigenic determinants (Mumby S., et al., 1988), the functional difference among Giα subtypes has not been reported until now.

Physiologic consequences of tubulin-G protein interaction

Presumably, tubulin interaction with G proteins in C6 cells and synaptic membranes has a role which extends beyond the phenomenological. Specifically, we suggest that there must be some form of intracellular regulation of adenylate cyclase , especially in cells of neural origin. The rationale for this suggestion resides in the fact that in the brain. the pool of neurotransmitter (in terms of amount) which is available for the stimulation of adenylate cyclase is far smaller than the pool of neurotransmitter capable of inhibiting that enzyme. Thus, some additional mechanism must account for the activation of adenylate cyclase. We have suggested that tubulin interaction with G proteins may be subjected to regulation by levels of intracellular Ca^{2+} and that neurotransmitters which elevate intracellular Ca^{2+} might modulate adenylate cyclase by increasing interaction between membrane tubulin and membrane associated G proteins (see Rasenick and Wang, 1988 and Rasenick at al, 1989b). Although exogenously added tubulin appears to associate more readily with Giα (in synaptic membranes), the situation might be considerably more complex inside the neuron, especially due to restrictions imposed by the membrane cytoarchitecture.

It is precisely this cytoarchitecture, of which microtubules form an essential element, that may choreograph the response to a variety of cellular signals. It is suggested that microtubules act not only as a component of the cytoskeleton, but also as a source of tubulin dimers which, in turn, modify G protein mediation of adenylate cyclase. Tubulin-G protein association may provide a mechanism whereby different second messenger systems interact. The specifics of such interaction as well as the significance of tubulin-G protein association have yet to be established. Nonetheless the implications of such interactions as a source of regulatory nuance for cellular signal transduction are intriguing.

Acknowledgment

The authors thank Nan Wang for helpful discussions and Narda Coranado for growth and maintainence of cells. The work reported in this manuscript has been supported by grants from the U.S. Public Health Service (MH39595) and the National Science Foundation (BNS 87-19758). MMR is the reciepient of a Research Scientist Development Award from the National Institute of Mental Health.

References

Amir-Zaltsman, Y., Ezra, Z., Scherson, T., Littauer, U., and Salomon, Y., ADP-ribosylation of microtubule proteins as catalyzed by cholera toxin. EMBO Journal 1(2):181-186, 1982

Brann, M. R., Collins, R. M., and Spiegel, A., Localization of mRNAs encoding the α-subunits of signal-transducing G-proteins with rat brain and among peripheral tissues. FEBS Lett. 222: 191-198, 1987

Carlier, M. F., Didry, D., and Pantaloni, D., Microtubule elongation and guanosine 5'-triphosphate hydrolysis. Role of guanine nucleotides in microtubule dynamics. Biochemistry, 26: 4428-4437, 1987

Halliday, K., Regional homology in GTP-binding proto-oncogen products and elongation factors. J. Cyclic Nucleotide Protein Phosphor. Res. 9: 435-446, 1984

Hatta, S., Marcus M.M., and Rasenick, M. M., Exchange of guanine nucleotide between GTP-binding proteins which regulate neuronal adenylate cyclase. Proc. Natl. Acad. Sci. USA, 83: 5439-5443, 1986

Higashi, K., and Ishibashi, S., Specific binding of tubulin to a guanine nucleotide-binding regulatory component of adenylate cyclase. Biochem. Biophysic. Res. Comm. 132: 193-197, 1985

Itoh, H., Toyama, R., Kozasa, T., Tsukamoto, T., Matsuoka M., and Kaziro, K., Presence of three distinct molecular species of Gi protein α subunit. J. Biol. C.hem. 263:6656-6664, 1988

Itoh, H., Kozasa, T., Nagata, S., Nakamura, S., Katada, T., Ui, M., Iwai, S., Ohtsuka, E., Kawasaki, H., Suzuki, K., Molecular cloning and sequence determination of cDNAs for α subunits of the guanine nucleotide-binding proteins: Gs, Gl and Go from rat brain. Proc. Natl. Acad. Sci. USA. 83: 3776, 1986

Katada, T., Kusakabe, K., Oinuma, M., and Ui, M., A novel mechanism for the inhibition of adenylate cyclase via inhibitory GTP-binding proteins. J. Biol. Chem., 262: 11897-11900, 1987

Levitzki, A., Regulation of hormone-sensitive adenylate cyclase. Trends. in Pharmac. Sci., 8:299-303, 1987

Lim, L., Sekura R., and Kaslow, H. J., Adenine nucleotides directly stimulate pertussis toxin. J. Biol. Chem., 260: 2585-2588, 1985

Lochrie, M. A., and Simon, M. I., G protein multiplicity in eukaryotic signal transduction systems. Biochemistry, 27 (14):4957-4965, 1988

Marbach, I., Shiloach J., and Levitzki, A., Gi affects the agonist-binding properties of ß-adrenoreceptors in the presence of Gs. Eur. J. Biochem. 172, 239, 1988

Margolis, R., and Wilson, L., Regulation of the microtubule steady state in vitro by ATP.. Cell, 18: 673-679, 1979

Mumby, S., Pang, H-I., Gilman A. G., and Sternweis, P. C., Chromatographic resolution and immunologic identification of the α40 and α41 subunits of guanine nucleotide-binding regulatory proteins from bovine brain. J. Biol. Chem., 263:2020, 1988

Peralta, E. G., Ashkenazi, A., Winslow, J. W., Ramachandran, J., and Capon, J. D., Differential regulation of PI hydrolysis and adenylyl cyclase by muscarinic receptor subtypes. Nature, 334: 434-437, 1988

Rasenick, M. M., Stein P. J., and Bitensky, M. W., Evidence that regulatory subunit of adenylate cyclase interacts with cytoskeletal components. Nature, 294:560-562, 1981

Rasenick, M. M., Wheeler, G. L., Bitensky, M. W., Kosack, C. M., Malina, R. L., and Stein, P. J., Photoaffinity identification of colchicine solubilized regulatory subunit from rat brain adenylate cyclase. J. Neurochem. 43: 1447-1454, 1984

Rasenick, M. M., O'Callahan, C. M., Moore C. A., and Kaplan, R. S., GTP-binding proteins which regulate normal adenylate cyclase interact with microtubule proteins. in "Microtubules and microtubule inhibitors", edited by DeBrabander, M. and DeMey, J. pp313-323, Elsevier (Amsterdam) 1985

Rasenick, M. M., and Kaplan, R. S., Guanine nucleotide activation of adenylate cyclase in saponin permeabilized glioma cells. FEBS. 207:296-301, 1986

Rasenick, M. M., and Wang, N., Exchange of guanine nucleotides between tubulin and GTP-Binding proteins that regulate adenylate cyclase: cytoskeletal modification of neuronal signal transduction. J. of Neurochem., 51:301-311, 1988

Rasenick, M. M., Hughes, J. M., and Wang, N., Guanosine -5'-O-thiodiphosphate functions as a partial agonist for the receptor-independent stimulation of neural adenylate cyclase. Brain research, 488: 105-113, 1989 a

Rasenick, M. M., Yan, K., and Wang, N., Tubulin as a G protein? in "The guanine nucleotide binding proteins", edited by Bosch, L., Kraal. B. and Parmeggani, A. Plenum, Amsterdam. (in press) 1989 b

Rodbell, M., Programmable messengers: a new theory of hormone action. Trends Biochem. Sic., 10:461-464, 1985

Sackett, D.L., Zimmerman, D.A. and Wolff, J. Tubulin dimer dissociation and accessibility. Biochemistry, 26:2662-2667 (1989)

Sahyoun, N., LeVine, H., Davis, III J., Hebdon, G., and Catrecasas, P., Molecular complexes involved in the regulation of adenylate cyclase. Proc. Natl. Acad. Sci. USA, , 78:6158-6162, 1981

Shelanski, M., Gaskin F., and Cantor, C., Microtubule assembly in the absence of added nucleotides. Proc. Natl. Acad. Sci. USA, 70:765-768, 1973

Sternlicht, H., Yaffe M., and Farr, G., A model of the nucleotide-binding site of tubulin. FEBS Lett., 214:223-235, 1987

Wang, N., Yan, K., and Rasenick, M. M., Tubulin binds specifically to the signal transducing proteins, Gαs and Giα1. Submitted for publication, 1989

Wolff, J., and Cook, G. H., Microtubule-associated adenylate cyclase. Biochem. Biophysica Acta 884: 34-43, 1985

Woon, C. W., Soparkar, S., Heasley, L., and Johnson, G. L., Expression of a Gas/Gai chimera that constitutively activates cyclic AMP synthesis. J. Biol. Chem. , 264: 5687-5693, 1989

Zor, U., Role of cytoskeletal organization in the regulation of adenylate cyclase-cyclic adenosine monophosphate by hormones. Endocrine Rev., 4:1-21, 1983

INITIATION AND MODULATION OF SIGNAL OUTPUT FROM THE T CELL ANTIGEN RECEPTOR

Robert T. Abraham, James A. Augustine, Janis W. Schlager,
Thomas J. Barna, and Paul J. Leibson

Department of Immunology
Mayo Clinic/Foundation
Rochester, MN

INTRODUCTION

T lymphocytes recognize, and are triggered by, antigenic determinants displayed by accessory cells in association with major histocompatibility complex molecules. The T cell receptor for antigen is a multimeric structure consisting of a polymorphic, antigen-binding heterodimer (T_i) noncovalently associated with a series of at least 5 invariant polypeptides termed CD3 (Samelson et al., 1985; Baniyash et al., 1988). The highly polymorphic nature of the ligand binding site expressed by clonally distributed T_i heterodimers ensures the clonal nature of the T cell activation response to specific antigens. Under appropriate conditions, certain soluble stimuli (phytohemagglutinin [PHA], anti-T_i-CD3 antibodies) can mimic the physiologic ligand and activate T cells. Thus, the T_i-CD3 receptor functions both as a variable recognition element for antigens and as a ligand-activated signal transducer that couples the antigenic stimulus to the T cell activation response. Characteristic activation responses triggered in resting T cells include cell cycle entry, T-cell growth factor receptor expression, and the release of lymphokines, including interleukin 2 (IL2). The subsequent binding of IL2 and other lymphokines to receptors expressed by activated T cells provides the requisite stimuli for cell-cycle progression, proliferation, and differentiation. In this article, we will discuss data related to the biochemical mechanisms involved in the coupling of T_i-CD3 receptor stimulation to signal output and the natures of the second messengers utilized by this receptor for signal transmission.

SIGNAL OUTPUT FROM THE T CELL ANTIGEN RECEPTOR

Numerous studies on the molecular mechanism of T cell activation have documented that T_i-CD3 receptor triggering results in rapid increases in phosphoinositide hydrolysis and intracellular Ca^{2+} concentration ($[Ca^{2+}]_i$) (Weiss et al., 1986). Therefore, the T_i-CD3 receptor, like many other Ca^{2+}-mobilizing receptors, stimulates a polyphosphoinositide-specific phospholipase C (PLC) upon interaction with stimulatory ligands. Receptor-mediated phosphoinositide hydrolysis liberates at least 2 second messengers, inositol trisphosphate and 1,2-diacylglycerol, that initiate bifurcating signal transduction pathways involving $[Ca^{2+}]_i$ elevation and protein kinase C (pkC) activation, respectively. Both Ca^{2+} mobilization and pkC activation have been implicated as critical early signals for the induction of distal T cell activation responses. Indeed, Ca^{2+} ionophores and phorbol esters, which elevate $[Ca^{2+}]_i$ and activate pkC directly, can activate T cells to express lymphokine genes and to proliferate, independent of cell surface receptor stimulation. However, more recent evidence suggests that the association between phosphoinositide hydrolysis and lymphokine gene expression in T cells may not always be obligatory (Sussman et al., 1988; Abraham et al., 1988). Thus, the T_i-CD3 receptor appears functionally coupled to multiple signal transduction pathways in T cells. Signal multiplicity from the T_i-CD3 receptor becomes highly tenable when one considers the multi-subunit organization of the receptor and the highly pleiotropic nature of the T cell activation response.

The first evidence of T_i-CD3 receptor coupling to signal generators other than PLC was provided by the demonstration that receptor perturbation by either specific antigen or mitogenic lectins resulted in the multi-site phosphorylation of the CD3-ζ polypeptide at tyrosyl residues (Samelson et al., 1986; Klausner et al., 1987; Baniyash et al., 1988). The results indicated that the T_i-CD3 receptor regulated a tyrosine protein kinase which was, based on the known structure of the T_i-CD3 complex subunits, not an integral component of the receptor itself. Although the exact nature, mechanism, and functional significance of the tyrosine kinase(s) activated during T_i-CD3 receptor stimulation remain unclear, it is tempting to speculate that non-transmembrane tyrosine kinases, exemplified by the *src* proto-oncogene family, might be involved. Indirect evidence for this proposal is provided by the recent demonstration that the CD4 molecule expressed by helper T cells co-immunoprecipitates with and regulates the activity of the *src* family proto-oncogene product, $p56^{lck}$ (Veillette et al., 1989). Intriguingly, crosslinkage of CD4 molecules stimulated both activation of the $p56^{lck}$ tyrosine kinase and CD3-ζ chain hyperphosphorylation. Thus, the T_i-CD3 receptor appears functionally coupled to at least two signal amplification mechanisms in T cells, one involving the activation of phosphoinositide-specific PLC, and the other mediated through the modulation of one or more protein tyrosine kinases.

SIGNAL INITIATION FROM THE T_i-CD3 RECEPTOR

The studies described above demonstrated that T_i-CD3 receptor stimulation results in signal output via activation of PLC and tyrosine kinase activities extrinsic to the receptor itself. However, the mechanisms involved in coupling of the T_i-CD3 receptor to these signal-generating enzymes remain undefined. The general paradigm for receptor-mediated activation of heterologous signal amplifiers, including adenylyl cyclase or PLC, interposes a guanine nucleotide-binding (G) protein between the receptor and the amplifier. In certain other receptor systems (the fmet-leu-phe receptor on neutrophils is prototypical), GTP or non-hydrolyzable analogs such as guanosine-5'-[γ-thio]triphosphate (GTPγS) synergistically stimulate phosphoinositide hydrolysis in the presence of receptor agonists (Smith et al., 1986). Moreover, receptor coupling to PLC activity is variably sensitive to inhibition by pertussis toxin. In neutrophils, pertussis toxin-dependent inhibition of fmet-leu-phe receptor function correlates with ADP-ribosylation of a 40 kDa membrane protein thought to be homologous to a $G_i\alpha$ subunit (Polakis et al., 1988). In contrast, the T_i-CD3 receptor on T cells has not proven amenable to the standard manipulations used to demonstrate receptor-G protein interactions in other systems. In intact cells, pertussis toxin treatment has no effect on T_i-CD3 receptor-dependent phosphoinositide hydrolysis, Ca^{2+} mobilization, or lymphokine gene expression. The finding that pertussis toxin, in certain instances, actually *stimulates* these responses, was explained by the binding of the pertussis toxin B subunit to the T_i-CD3 complex or a related stimulatory receptor on T cells (Rosoff et al., 1987). A distinct bacterial toxin, cholera toxin (CTX) also modifies receptor-linked G protein α subunits in T cells. Although the known substrate specificity of CTX has traditionally been restricted to the $G_s\alpha$ subunit and transducin, recent studies have demonstrated that CTX modifies additional G proteins in certain cell types (Owens et al., 1985). The seminal observation by Imboden and co-workers (Imboden et al., 1986) that CTX pretreatment specifically blocked T_i-CD3 receptor-PLC coupling in the human T cell leukemic line, Jurkat, strongly suggested that a G protein participated directly in signal initiation by the T_i-CD3 receptor. As will be discussed below, we have confirmed and extended these results, and, although the story is far from complete, these studies should provide significant new information regarding the mechanism of signal initiation by the T_i-CD3 receptor.

Effect of PMA on phosphoinositide-dependent signaling by the T_i-CD3 receptor

The demonstration that T_i-CD3 receptor-mediated Ca^{2+} mobilization was sensitive to CTX prompted further examination of the role of G proteins in signal initiation by the T_i-CD3 receptor. The classical approach to the study of receptor-G protein interactions involves the exposure of cell-free systems to non-hydrolyzable GTP analogs, such as GTPγS, in the absence or presence of receptor agonists. Although numerous studies with permeabilized T cells or T cell plasma membrane preparations appear in the literature, few actually demonstrate the regulation of phosphoinositide hydrolysis in T cells by G proteins. This paucity of data most likely reflects the fact that the cell-free systems often used are unable to mount receptor-mediated biochemical responses. Recent studies with T cells permeabilized by Mg^{2+}-free ATP or S. aureus α-toxin have demonstrated that GTPγS alone can trigger biochemical responses (i.e., exocytosis of cytolytic granules, induction of ornithine decarboxylase activity) reminiscent of those elicited after T_i-CD3 receptor crosslinkage (Mustelin et al., 1986; Schrezenmeier et al., 1988). Our experiments have employed

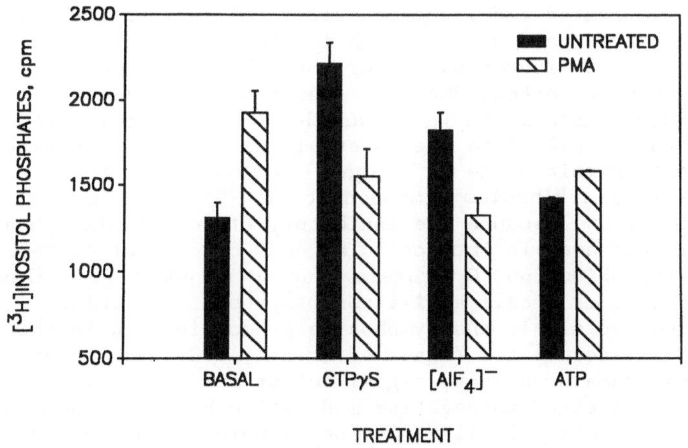

Fig. 1. Effect of PMA on inositol phosphate release from LBRM-33 cell membranes. [^{3}H]inositol-labeled LBRM-33 cells were untreated or stimulated with 20 ng/ml PMA for 10 min. Isolated cell membranes were incubated at 37°C for 20 min in buffer only (basal); GTPγS (10 μM); $[AlF_4]^-$, (10 mM NaF plus 10 μM $AlCl_3$), or 10 μM ATP. Data represent mean cpm [^{3}H]inositol phosphates +/- SD.

both crude membrane fractions and permeabilized T cells for analyses of the regulation of G protein-stimulated phosphoinositide turnover by agents that modulate signal output from the T cell antigen receptor in intact cells. In an earlier study (Abraham et al., 1988) we demonstrated that acute exposure to phorbol myristate acetate (PMA) blocked T_i-CD3 receptor-mediated phosphoinositide hydrolysis and $[Ca^{2+}]_i$ increases in the murine T cell lymphoma line, LBRM-33. Kinetic analysis suggested that the inhibition of T_i-CD3 receptor-mediated signal output by PMA involved 2 distinct mechanisms: acute uncoupling of the receptor from PLC activity followed by down-regulation of T_i-CD3 receptor expression. The early receptor-desensitization phase was not explained by decreases in T_i-CD3 receptor expression or by gross changes in PLC activity toward exogenous phosphoinositides. Subsequently, we were interested in determining whether the uncoupling effect of PMA was exerted at the level of a G protein linked to PLC in LBRM-33 cells. In preliminary experiments, we demonstrated that GTPγS, but not GTP, GDP, or GDPβS, stimulated the release of radiolabeled inositol phosphates (IPs) from crude membrane fractions isolated from $[^3H]$inositol-prelabeled LBRM-33 cells. These results indicated that a G protein-modulated PLC activity was, in fact, detectable in LBRM-33 cell plasma membranes. To determine whether acute PMA treatment altered G protein-PLC coupling in LBRM-33 cells, intact cells were treated with PMA prior to isolation of plasma membranes. As shown in Fig. 1, PMA pretreatment elevated the basal level of IP release relative to that obtained with membranes from untreated cells. However, the GTPγS-stimulated IP release observed in untreated LBRM-33 cells was completely inhibited by pretreatment with TPA. In fact, GTPγS inhibited IP generation by membranes from TPA-treated cells. Similar results were obtained with fluoroaluminate ion ($[AlF_4]^-$), a nonspecific activator of G proteins. Thus, these studies suggested that the inhibitory effect of TPA on signal output from the T_i-CD3 receptor in LBRM-33 cells was due, at least in part, to a functional alteration in G protein regulation of PLC activity. With these preliminary studies as a background, we proceeded to examine a distinct T cell tumor line in which G protein involvement had clearly been implicated in T_i-CD3 receptor-PLC coupling.

The human leukemic T cell line, Jurkat, has been one of the most intensively investigated model systems for T cell activation. Jurkat cells are induced to express and secrete high levels of IL2 as a consequence of synergistic signals delivered by PHA or anti-T_i-CD3 antibodies plus the pkC-activating phorbol ester, PMA. Earlier studies implicated the T_i-CD3 receptor-dependent increase in $[Ca^{2+}]_i$ as the primary signal for IL2 production by this T cell line (Weiss et al., 1986). As described above, CTX pretreatment inhibited the Ca^{2+}-signal in Jurkat cells stimulated with anti-T_i-CD3 receptor antibodies (Imboden et al., 1986). This study was especially noteworthy, because the inhibitory effect of CTX appeared independent of increases in intracellular cAMP concentration resulting from activation of the ubiquitous G_s protein coupled to adenylyl cyclase. The sensitivity of various T cell activation responses, including phosphoinositide hydrolysis and lymphokine production, to inhibition by cAMP has been well-documented (Klausner et al., 1987; Farrar et al., 1987). For reasons that remain unclear, signal output from T_i-CD3 receptors on Jurkat cells is resistant to negative modulation by cAMP-dependent mechanisms. Thus, this cell line provides a unique opportunity to study the role of CTX-sensitive G proteins in signal transmission from the T_i-CD3 receptor.

Effect of CTX on phosphoinositide-dependent signaling by the T_i-CD3 receptor

In initial experiments, we confirmed earlier findings (Imboden et al., 1986) that CTX blocked both inositol phosphate (IP) generation and

Ca^{2+} mobilization induced by PHA or anti-CD3 antibodies. The effect of CTX on IP release from [^{3}H]inositol-labeled Jurkat cells was both concentration- and time-dependent. T_i-CD3 receptor-dependent IP release was remarkably sensitive to inhibition by CTX. After 2 hrs of pretreatment, the CTX concentration required for half-maximal inhibition (IC_{50}) of the response to PHA or anti-CD3 antibodies was approximately 16 ng/ml or less then 5 ng/ml CTX, respectively. Expression of the maximal inhibitory effect of CTX required at least one hour of exposure to the toxin prior to cellular stimulation. The latter result suggests that a time-dependent process, putatively ADP-ribosylation, underlies the disruption of T_i-CD3 receptor-PLC coupling by CTX. In contrast, the purified B subunit of CTX, which binds to cells but lacks ADP-ribosylating activity, was inactive in all of these assays. Moreover, CTX pretreatment (2 hrs) had no effect on the density of T_i-CD3 receptors expressed by Jurkat cells, indicating that the observed receptor desensitization was not a consequence of receptor down-regulation. Finally, it was crucial to determine whether cAMP elevation mimicked the effect of CTX on second messenger generation by the T_i-CD3 complex on Jurkat cells. The results of these studies were conclusive: under no circumstances were intracellular cAMP elevations by forskolin, 8-bromo-cAMP, or dibutyryl cAMP capable of inhibiting PHA- or anti-CD3 antibody-stimulated IP release in Jurkat cells. In parallel studies, untreated or CTX-pretreated Jurkat cells were loaded with the Ca^{2+}-sensitive dye, indo-1, and changes in $[Ca^{2+}]_i$ induced by T_i-CD3 receptor stimulation were measured by flow cytofluorimetry. The results of these $[Ca^{2+}]_i$ measurements directly reflected the IP release experiments. CTX abrogated T_i-CD3 receptor-dependent $[Ca^{2+}]_i$ increases, whereas the other cAMP agonists had virtually no effect on this response. The logical conclusion was that CTX disrupted T_i-CD3 receptor-PLC coupling in a fashion independent of intracellular cAMP increases and, presumably, G_s-mediated adenylyl cyclase activation.

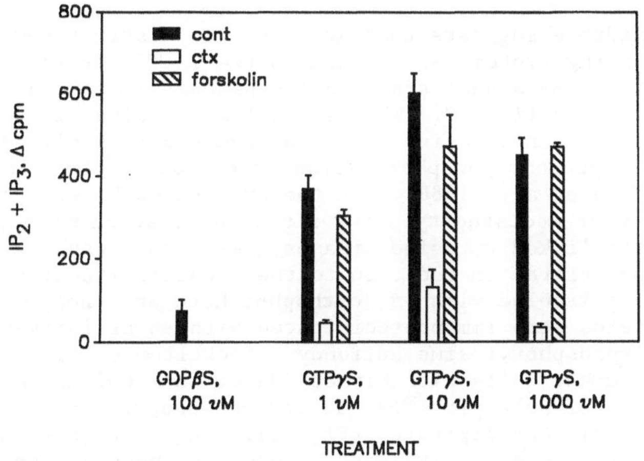

Fig. 2. Effect of CTX on phosphoinositide hydrolysis in permeabilized Jurkat cells. Intact Jurkat cells were pretreated with medium (cont), CTX (200 ng/ml), or forskolin (30 μM) for 2 hrs. After electro-permeabilization, cells were stimulated with the indicated GDP or GTP analogs for 10 min at 37°C. Data represent mean cpm ([^{3}H]IP_2 + [^{3}H]IP_3) +/- SD.

The inhibitory effect of CTX on T_i-CD3 receptor-PLC coupling might be exerted at any site in the putative receptor-transducer-PLC triad. The specific catalytic activity of CTX toward certain G proteins suggested that CTX might act at the level of a G protein involved in T_i-CD3 receptor-PLC coupling. To test this possibility, Jurkat cells were permeabilized to permit experimental control of Ca^{2+} and guanine nucleotide concentrations, parameters known to regulate phosphoinositide-specific PLC activity. Electric field-induced permeabilization was chosen because this procedure results in minimal disruption of the functional coupling among membrane receptors, signal-transducing enzymes, and phospholipid substrates. Electrically-permeabilized, [^{3}H]inositol-prelabeled Jurkat cells were shown to release both IP_3 and IP_2 during stimulation with the non-hydrolyzable GTP analog, GTPγS. These studies indicated that a G protein-regulated PLC activity exists in Jurkat cells; however, to date, we have been unable to demonstrate a modulatory effect of T_i-CD3 crosslinkage on this biochemical response. Nonetheless, the availability of permeabilized Jurkat cells provided the opportunity to evaluate the direct effects of CTX pretreatment on G protein-dependent PLC activation. As shown in Fig. 2, GTPγS stimulated a concentration-dependent increase in IP generation by untreated, permeabilized Jurkat cells. The concentration-response curve was consistently biphasic, with stimulation apparent at GTPγS concentrations as low as 0.5 μM, a maximal effect at 10μM GTPγS, and a sub-maximal response in the presence of 1 mM GTPγS. As expected for a G protein-mediated response, the GDP analog, GDPβS, did not stimulate significant IP generation in the permeabilized cells. In contrast, treatment with CTX for 2 hrs prior to cell permeabilization markedly inhibited the GTPγS-dependent release of IPs. Once again, the inhibitory effect of CTX on second messenger release was not mimicked by direct activation of adenylyl cyclase with forskolin. Thus, these experiments demonstrate that CTX pretreatment blocks the ability of GTPγS to regulate phosphoinositide-specific PLC activity in Jurkat cells and further implicate the involvement of a CTX-sensitive G protein in T_i-CD3 receptor-PLC coupling.

Effect of CTX on T_i-CD3 Receptor-dependent Tyrosine Kinase Activation

Current evidence suggests that at least two distinct second messenger-generating systems are linked to the T_i-CD3 receptor. As discussed earlier, one signal transduction pathway is initiated by phosphoinositide-specific PLC. The second limb involves the activation of an unidentified protein tyrosine kinase(s) and was initially defined by the T_i-CD3 receptor-dependent phosphorylation of the CD3-ζ chain at tyrosine residues (Baniyash et al., 1988). In the present work, we hypothesized that T_i-CD3 receptor-mediated tyrosine kinase activation might, by analogy to known receptor-linked tyrosine kinases, serve to phosphorylate other intracellular substrates in addition to the receptor itself. Therefore, Jurkat cells were labeled with [^{32}P]orthophosphate and phosphotyrosine-containing proteins were immunoprecipitated with an affinity-purified monoclonal anti-phosphotyrosine antibody (Frackelton et al., 1983). Stimulation of Jurkat cells with either PHA or anti-CD3 antibodies resulted in the appearance of multiple [^{32}P]-labeled phosphoproteins in the anti-phosphotyrosine immunoprecipitates (Fig. 3). In addition to numerous hyperphosphorylation events, at least 5 novel phosphoproteins (indicated by arrows) were detected in immunoprecipitates from PHA or anti-CD3 antibody-stimulated cells. Protein tyrosine phosphorylation was evident within 2 min of PHA exposure, suggesting that protein tyrosine kinase activation was a proximal response to T_i-CD3 receptor perturbation. Although the functional roles of these tyrosine phosphorylation events are presently undefined, these studies further support the proposal that protein tyrosine kinase activation participates in transmembrane signaling by the T_i-CD3 receptor. Moreover, the inability of PMA to mimic the consequences of T_i-

CD3 receptor stimulation indicated that tyrosine kinase activation was not secondary to activation of pkC. Subsequently, experiments were performed to determine whether protein tyrosine phosphorylation stimulated by T_i-CD3 receptor crosslinkage was sensitive to inhibition by CTX. The results in Fig. 3 demonstrate that this was indeed the case: CTX pretreatment uncoupled T_i-CD3 receptor activation from early increases in protein tyrosine phosphorylation. In contrast, direct elevation of intracellular cAMP with dibutyryl cAMP failed to mimic the inhibitory activity of CTX on tyrosine kinase activation. These studies demonstrate that bifurcating signal transduction pathways coupled to the T_i-CD3 receptor exhibit common sensitivity to inhibition by CTX in Jurkat cells. Thus, a CTX-sensitive G protein may play a fundamental role in signal relay from T_i-CD3 receptors to both PLC and tyrosine kinase activities in T lymphocytes.

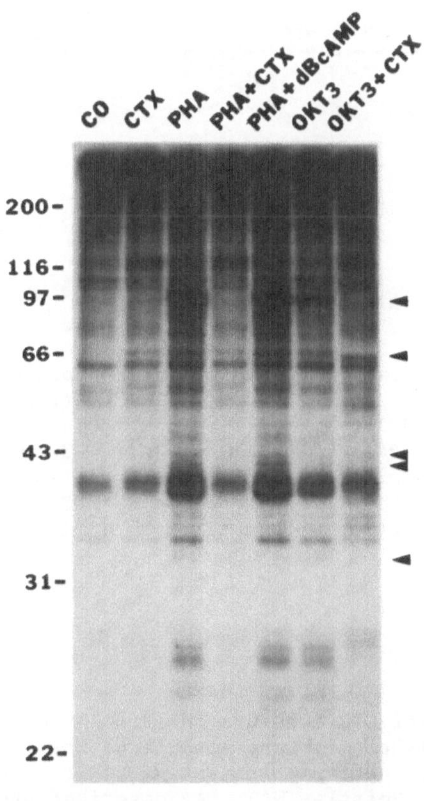

Fig. 3. Effect of CTX on stimulus-dependent protein tyrosine phosphorylation in Jurkat cells. Prior to stimulation, [^{32}P] orthophosphate-labeled cells were pretreated with medium only, CTX (1 μg/ml) for 4 hrs, or with dibutyryl cAMP (dBcAMP, 1 mM) for 30 min. Subsequently, cells were stimulated for 30 min with medium only (CO), PHA (10 μg/ml) or anti-CD3 antibodies (OKT3, 2 μg/ml) for 30 min. Phosphoproteins were immunoprecipitated with monoclonal anti-phosphotyrosine antibodies. After SDS-PAGE in a 10% gel, [^{32}P]-labeled phosphoproteins were visualized by autoradiography. Numbers on left represent molecular mass standards. Arrows on right designate novel phosphorylation events induced by PHA or OKT3 antibody stimulation.

A prominent candidate for the receptor-regulated tyrosine kinase activity in T cells is the lymphocyte-specific *src* family kinase, $p56^{lck}$. Recent studies demonstrated that T_i-CD3 receptor stimulation in T cells provokes a rapid structural modification of $p56^{lck}$ (Veillette et al., 1988; Marth et al., 1989). Receptor-dependent $p56^{lck}$ modification was detected on $p56^{lck}$ immunoblots by the loss of native protein and the concomitant appearance of multiple immunoreactive species (collectively termed $p59^{lck}$) characterized by decreased mobilities during SDS-PAGE. The mechanism underlying the $p56^{lck}$ to $p59^{lck}$ conversion is uncertain, although activation of a pkC-like serine-threonine kinase appears to initiate the mobility shift. The rapid modification of $p56^{lck}$ following antigen receptor perturbation is, nonetheless, consistent with the presumptive view that $p56^{lck}$ is somehow involved in the lymphocyte activation sequence. To

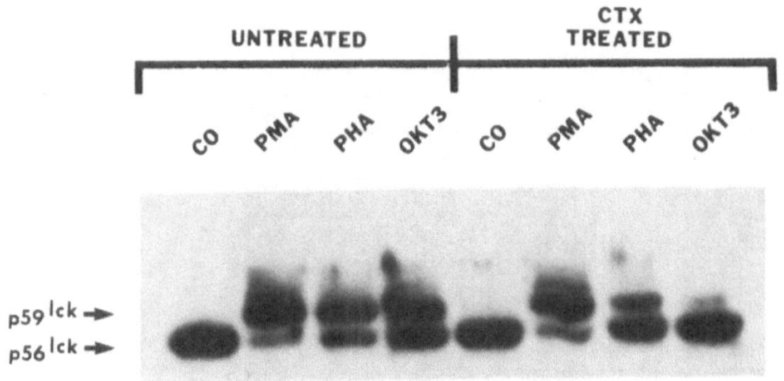

Fig. 4. Effect of CTX on stimulus-dependent $p56^{lck}$ modification. Untreated or CTX (200 ng/ml, 2 hrs) - pretreated Jurkat cells were stimulated with PMA (20 ng/ml), PHA (10 μg/ml) or anti-CD3 antibodies (OKT3, 2 μg/ml) for 30 min. Detergent-soluble membrane proteins were resolved by SDS-PAGE followed by immunoblotting with anti-$p56^{lck}$ antibodies.

determine the effect of CTX exposure on stimulus-dependent $p56^{lck}$ modification, we analyzed membrane-associated forms of $p56^{lck}$ by immunoblotting with an anti-peptide antiserum provided by Dr. R. Perlmutter (Univ. of Washington, Seattle, WA). As described previously (Perlmutter et al., 1988), PMA elicited a dramatic conversion of $p56^{lck}$ to $p59^{lck}$ in Jurkat cells (Fig. 4). Furthermore, both PHA and anti-CD3 antibody stimulation induced substantial $p59^{lck}$ generation, demonstrating a direct correlation between T_i-CD3 receptor stimulation and $p56^{lck}$ modification in Jurkat cells. CTX pretreatment strongly, but not completely, inhibited formation of $p59^{lck}$ induced by PHA or anti-CD3 antibody stimulation. In contrast, CTX exposure failed to inhibit $p56^{lck}$ to $p59^{lck}$ conversion induced by the T_i-CD3 receptor-independent stimulus, PMA. Thus, CTX selectively disrupts the coupling between the T_i-CD3 receptor and the $p56^{lck}$ tyrosine kinase, suggesting that the proposed functional linkage between the T_i-CD3 receptor and $p56^{lck}$ also involves a CTX-sensitive G protein in Jurkat cells.

CONCLUSION

In recent years, significant progress has been made toward defining the structure and function of the T_i-CD3 receptor expressed by T lymphocytes. At the same time, these studies have served to highlight the extraordinary complexity of the receptor itself and the mechanisms by which it regulates the T cell activation state. To rationalize the teleologic basis for such complexity, one need only consider the diversity of cellular responses, collectively termed "activation", regulated by this receptor: expression of cell-cycle related genes (*fos*, *myc*), expression of lymphokine and lymphokine receptor genes, secretion of lymphokines and cytotoxic proteins, growth, differentiation, and immunologic "memory". These processes involve stimulus-response lag times ranging from seconds or minutes to days or even months. The multimeric structure of the T_i-CD3 receptor may, therefore, reflect its multi-functional character.

Major questions regarding the mechanisms of signal initiation and signal output by the T_i-CD3 receptor have been raised by the studies summarized above. First, does a G protein intermediate play an obligatory role in signal transduction from the T_i-CD3 receptor? If so, how does this putative G protein communicate with a receptor that bears no structural homology to the "seven transmembrane-spanning segment" structural motif characteristic of other G protein-coupled receptors? The ability of CTX to desensitize T_i-CD3 receptors in Jurkat cells provides a powerful model system for further investigations of the roles of G proteins in signal transmission from T_i-CD3 receptors. A second area of active investigation focuses on the mechanisms and biological consequences of protein tyrosine kinase activation by the T_i-CD3 receptor. Again, the present studies suggest that the receptor regulates an extrinsic tyrosine kinase activity through a CTX-sensitive mechanism. The nature of the protein tyrosine kinase regulated by the T_i-CD3 receptor remains unclear, although the *src*, *fyn*, and *lck* gene products are viable candidates (Perlmutter et al., 1988). At present, $p56^{lck}$ has evoked the most intense experimental interest, primarily due to the recent demonstration that $p56^{lck}$ is functionally linked to the CD4 molecule expressed by helper T cells (Veillette et al., 1989). Although existing data are far from definitive, it is tempting to speculate that the lymphocyte-specific tyrosine kinase, $p56^{lck}$, exists to perform a lymphocyte-specific function: signal transduction from antigen receptors. The third and perhaps most controversial area relates to the specific roles of phosphoinositide-derived second messengers in signal transduction during T cell activation. A growing body of data support the argument that, depending on the nature of the co-stimuli provided, certain T cell subtypes may exhibit partial to complete independence of T_i-CD3 receptor-mediated phosphoinositide hydrolysis and Ca^{2+} mobilization. As the answers to these and other questions begin to unfold, the coming years should bring extensive revisions of current paradigms for the receptor mechanisms and transmembrane signals that regulate T lymphocyte activation.

REFERENCES

Abraham, R. T., Ho, S. N., Barna, T. J., Rusovick, K. M., and McKean, D. J., 1988, Inhibition of T-cell antigen receptor-mediated transmembrane signaling by protein kinase C activation, Mol. Cell. Biol., 8:5448.

Baniyash, M., Garcia-Morales, P., Bonifacino, J. S., Samelson, L. E., and Klausner, R. D., 1988, Disulfide linkage of the zeta and eta chains of the T cell receptor: Possible identification of two structural classes of receptors, J. Biol. Chem., 263:9874.

Baniyash, M., Garcia-Morales, P., Luong, E., Samelson, L. E., and Klausner, R. D., 1988, The T cell antigen receptor ζ chain is tyrosine phosphorylated upon activation, J. Biol. Chem., 263:18225.
Farrar, W. L., Evans, S. W., Rapp, U. R., and Cleveland, J. L., 1987, Effects of anti-proliferative cyclic AMP on interleukin 2-stimulated gene expression, J. Immunol., 139:2075.
Frackelton, A. R., Ross, A. H., and Eisen, H. N., 1983, Characterization and use of monoclonal antibodies for isolation of phosphotyrosyl proteins from retrovirus-transformed cells and growth factor-stimulated cells, Mol. Cell. Biol., 3:1343.
Imboden, J. B., Shoback, D. M., Pattison, G., and Stobo, J. D., 1986, Cholera toxin inhibits the T-cell antigen receptor-mediated increases in inositol trisphosphate and cytoplasmic free calcium, Proc. Natl. Acad. Sci. USA, 83:5673.
Klausner, R. D., O'Shea, J. J., Luong, H., Ross, P., Bluestone, J. A. and Samelson, L. E., 1987, T cell receptor tyrosine phosphorylation. Variable coupling for different activating ligands, J. Biol. Chem., 262:12654.
Marth, J. D., Lewis, D. B., Cooke, M. P., Mellins, E. D., Gearn, M. E. Samelson, L. E., Wilson, C. B., Miller, A. D., and Perlmutter, R. M., 1989, Lymphocyte activation provokes modification of a lymphocyte-specific protein tyrosine kinase ($p56^{lck}$), J. Immunol. 142:2430.
Mustelin, T., Poso, H., and Andersson, L. C., 1986, Role of G-proteins in T cell activation: non-hydrolyzable GTP analogues induce early ornithine decarboxylase activity in human T lymphocytes, EMBO J., 5:3287.
Owens, J. R., Frame, L. T., Ui, M., and Cooper, D. M. F., 1985, Cholera toxin ADP-ribosylates the islet-activating protein substrate in adipocyte membranes and alters its function, J. Biol. Chem. 260:15946.
Perlmutter, R. M., Marth, J. D., Ziegler, S. F., Garvin, A. M., Pawar, S. Cooke, M. P., and Abraham, K. M., 1988, Specialized protein tyrosine kinase proto-oncogenes in hematopoietic cells, Biochem. Biophys. Acta, 948:245.
Polakis, P. G., Uhing, R. J., and Snyderman, R., 1988, The formylpeptide chemoattractant receptor copurifies with a GTP-binding protein containing a distinct 40-kDa pertussis toxin substrate, J. Biol. Chem., 263:4969.
Rosoff, P. M., Walker, R., and Winberry, L., 1987, Pertussis toxin triggers rapid second messenger production in human T lymphocytes, J. Immunol., 139:2419.
Samelson, L. E., Harford, J. B., and Klausner, R. D., 1985, Identification of the components of the murine T cell antigen receptor complex, Cell, 43:223.
Samelson, L. E., Patel, M. D., Weissman, A. M., Harford, J. B., and Klausner, R. D., 1986, Antigen activation of murine T cells induces tyrosine phosphorylation of a polypeptide associated with the T cell antigen receptor, Cell, 46:1083.
Schrezenmeier, H., Ahnert-Hilger, G., and Fleischer, B., 1988, A T cell receptor-associated GTP-binding protein triggers T cell receptor-mediated granule exocytosis in cytotoxic T lymphocytes, J. Immunol., 141:3785.
Smith, C. D., Cox, C. C., and Snyderman, R., 1986, Receptor-coupled activation of phosphoinositide-specific phospholipase C by an N protein, Science, 232:97.
Sussman, J. J., Mercep, M., Saito, T., Germain, R. N., Bonvini, E., and Ashwell, J. D., 1988, Dissociation of phosphoinositide hydrolysis and Ca^{2+} fluxes from the biological responses of a T-cell hybridoma, Nature, 334:625.

Veillette, A., Bookman, M. A., Horak, E. M., Samelson, L. E., and Bolen, J. B., 1989, Signal transduction through the CD4 receptor involves the activation of the internal membrane tyrosine-protein kinase $p56^{lck}$, Nature, 338:257.
Veillette, A., Horak, I. D., Horak, E. M., Bookman, M. A. and Bolen, J. B. 1988, Alterations of the lymphocyte-specific protein tyrosine kinase ($p56^{lck}$) during T-cell activation, Mol. Cell. Biol., 8:4353.
Weiss, A., Imboden, J., Hardy, K., Manger, B., Terhorst, C. and Stobo, J., 1986, The role of the T3/antigen receptor complex in T-cell activation, in: "Annual Review of Immunology," W. E. Paul, C. G. Fathman, and H. Metzger, eds., Annual Reviews Inc., Palo Alto, CA.

SELECTIVE PHOSPHORYLATION OF A G PROTEIN WITHIN PERMEABILIZED AND INTACT PLATELETS CAUSED BY ACTIVATORS OF PROTEIN KINASE C

Kenneth E. Carlson†, Lawrence F. Brass‡, and David R. Manning†

†Department of Pharmacology and ‡Departments of Medicine and Pathology at the University of Pennsylvania School of Medicine, 36th and Hamilton Walk, Philadelphia, PA 19104-6084

INTRODUCTION

Phosphorylation of G_i by the phospholipid- and Ca^{2+}-dependent enzyme protein kinase C has been proposed as a mechanism by which phorbol esters and certain agonists either attenuate inhibition of adenylyl cyclase or impair stimulation of phospholipase C (Jakobs, et al., 1985; Orellana et al., 1987; Smith et al., 1987; Pyne et al., 1989). Support for the phosphorylation of $G_{i\alpha}$ is provided by the occurrence of this modification for one or more forms of $G_{i\alpha}$ in solution with partially purified protein kinase C (Katada et al., 1985). Indeed, phosphorylation of the purified subunit(s) can also be accomplished with cAMP-dependent and insulin receptor kinases (O'Brien et al., 1987; Krupinski et al., 1988; Watanabe et al., 1988). The occurrence of phosphorylation within the cell itself, however, has not been fully documented, nor have functional correlates of phosphorylation been established. In unstimulated rat hepatocytes incubated with $[^{32}P]H_3PO_4$, a 41-kDa phosphoprotein can be immunoprecipitated with a $G_{i\alpha}$-directed antiserum (Rothenberg and Kahn, 1988; Pyne et al., 1989). Treatment of these cells with phorbol 12-myristate 13-acetate (PMA) results in a modest (i.e., 70%) increase in incorporated phosphate, while treatment with insulin has no effect. Incorporation of radiolabel into proteins having isoelectric or immunologic properties similar to $G_{i\alpha}$ has also been reported for PMA-treated human platelets (Halenda et al., 1986; Crouch and Lapetina, 1988). The identities of these proteins as $G_{i\alpha}$, however, have not been rigorously established.

In this study, we sought to investigate the phosphorylation of $G_{i\alpha}$ in response to activators of protein kinase C within permeabilized and intact human platelets. The choice of platelets was based upon precedent for the actions of PMA on pathways of transduction conceivably mediated by G_i, as well as the suitability of platelets to experimental manipulation. Our strategy was to develop a number of antibodies capable of recognizing the various forms of $G_{i\alpha}$, and to use these to immunoprecipitate the subunit(s) from PMA-treated platelets containing [^{32}P]ATP. The results obtained provide no evidence for the phosphorylation of $G_{i\alpha}$. Instead, data point to the phosphorylation of a protein of limited immunological cross-reactivity. This protein is related to the novel G protein subunit $G_{z\alpha}$.

RESULTS

Antisera were generated in rabbits by the injection of keyhole limpet hemocyanin-conjugated synthetic peptides corresponding to different regions of primary structure within one or more forms of $G_{i\alpha}$ (Figure 1; $G_{z\alpha}$ is discussed below). Three of the peptides – those used in the development of antisera 1398, 8730, and 8645 – represent sequences held in common among all three forms, with conservative differences having little impact on immunoreactivity. The peptide used in the development of antiserum 1521 was unique to $G_{i\alpha 2}$. Antibodies in all four antisera recognized both purified $G_{i\alpha}$ and a 40–41-kDa protein(s) predominantly within membranes of human platelets by Western blotting. These, together with protein A-Sepharose, achieved precipitation of up to 65% of the protein(s) from platelets solubilized in SDS (Carlson et al., 1989).

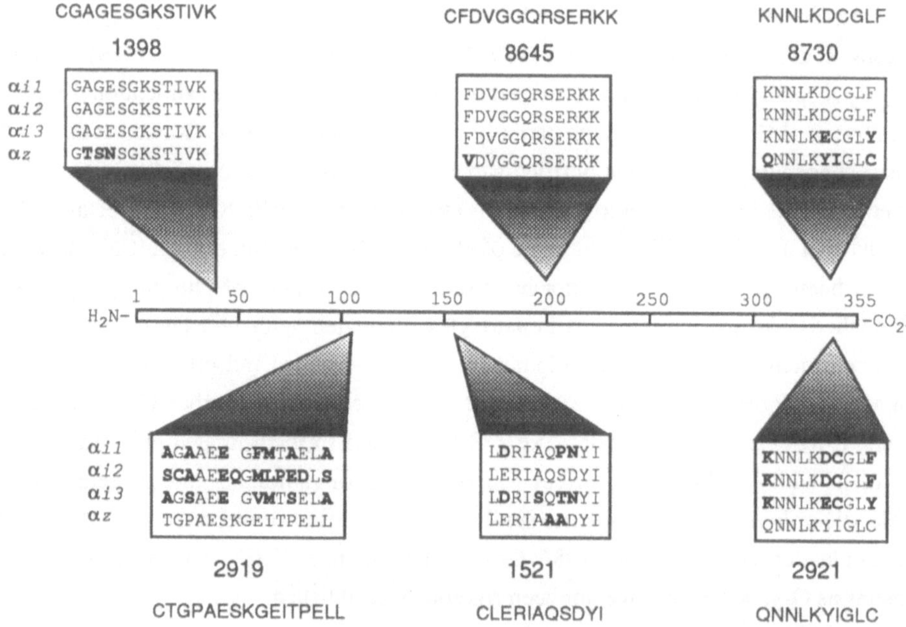

Figure 1. **Comparison of the amino acid sequences of peptide antigens and G_α subunits.** The amino acid sequences of the synthetic peptide antigens and of the corresponding regions present in $G_{i\alpha 1}$, $G_{i\alpha 2}$, $G_{i\alpha 3}$, and $G_{z\alpha}$ are portrayed according to their position within a composite G_α subunit. Differences in amino acids between the antigens and those in G_α subunits are printed in bold type. The sequences for $G_{i\alpha 1}$, $G_{i\alpha 2}$, and $G_{i\alpha 3}$ (inferred from rat cDNA) were taken from Jones and Reed (1987), that for $G_{z\alpha}$ (inferred from human cDNA) was taken from Fong et al. (1988).

The pool of ATP within platelets was radiolabeled by one of two methods: incubation of saponin-permeabilized platelets with [γ-^{32}P]ATP or incubation of intact platelets with [^{32}P]H_3PO_4. Platelets were then incubated in the presence or absence of 100 nM PMA for 1 to 60 minutes at 30 °C, after which solubilization was achieved by boiling in 0.5% SDS and 1 mM DTT. Figure 2 shows autoradiograms following SDS-PAGE of precipitates obtained with each of the antisera. A prominent band in precipitates obtained from PMA-treated platelets with either immune or preimmune serum corresponds to P47, the major substrate for protein kinase C in platelets. Only a small fraction of P47 is present, however, and its appearance is the result of carry-over or nonspecific adsorption. A phosphoprotein having a mobility identical to $G_{i\alpha}$ was also noted, but surprisingly a selectivity in the precipitation of this protein was evident. Specifically, antisera 1398 and 8730 did not achieve precipitation in a manner distinguishable from preimmune sera. The precipitation of the 40-kDa phosphoprotein by antisera 8645 and 1521, however, was clearly discerned, with antiserum 8645 several-fold more effective than 1521. Only small amounts of incorporated radiolabel were noted in the absence of PMA. Identical results were obtained for intact platelets incubated with [^{32}P]H_3PO_4. Phosphorylation of the 40-

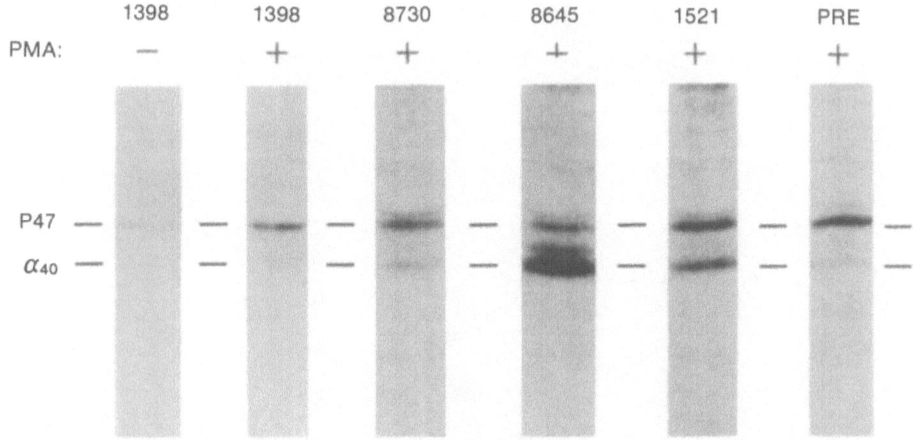

Figure 2. **Immunoprecipitates obtained from [^{32}P]ATP-labeled platelets.** Washed human platelets were incubated with 15 ug/ml saponin and 50 uCi carrier free [γ-^{32}P]ATP without (-) or with (+) 100 nM PMA at 30 °C for 3 min. Samples were then immediately combined with a buffer containing SDS, DTT, protease and phosphatase inhibitors and heated at 100 °C for 3 min. Following incubation of the solubilized material in RIPA buffer (Sefton et al., 1978), antisera or preimmune antiserum (PRE) were added at dilutions of 1:8 (antiserum 1521), 1:20 (antisera 8645 and 8730), or 1:100 (antiserum 1398 and preimmune). Following overnight incubations at 4 °C, protein A-Sepharose was added, and precipitates were isolated by centrifugation. Shown are autoradiograms of SDS-polyacrylamide gels to which equivalent amounts of precipitates were applied. The mobility of the major form of $G_{i\alpha}$ in platelets, referred to as 'α_{40}', was determined in separate experiments involving immunoblotting with antiserum 1398 prior to autoradiography. P47 is a prominent, non-specific contaminant of precipitates prepared from PMA-treated cells.

kDa protein was rapid, evident within 1 min following addition of PMA and reaching a maximum by 3 min, and in intact cells was stable with time. Phosphorylation was also achieved for both permeabilized and intact cells using 10 nM thrombin, a potent activator of phosphoinositide hydrolysis in platelets, in place of PMA.

Evidence that the phosphoprotein precipitated with antiserum 8645 is not a substrate for the pertussis toxin islet-activating factor (PTX) is presented in Figure 3. Immunoprecipitations were undertaken in a sequential fashion for extracts derived either from platelets that had been incubated with PTX and [^{32}P]NAD (Fig. 3A) or incubated in PMA and [γ-^{32}P]ATP (Fig. 3B). Immunoprecipitations were performed identically for the two sets of platelets: precipitates were obtained from extracts with antisera 1398 plus 8730 (left

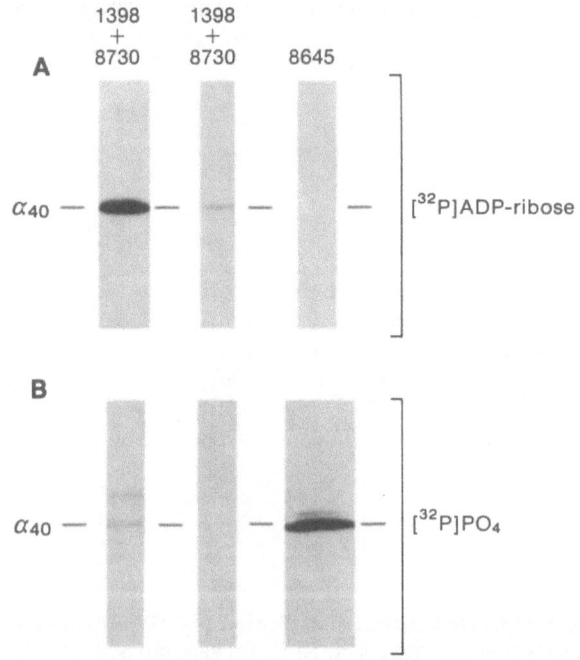

Figure 3. **Immunoprecipitates obtained from [^{32}P]NAD- and [^{32}P]ATP-labeled platelets through sequential incubation with antisera.** Washed platelets were incubated with either (A) saponin, 50 uM [^{32}P]NAD and 10 ug/ml PTX at 30 °C for 60 min or (B) saponin, [γ-^{32}P]ATP, and 100 nM PMA at 30 °C for 3 min. Incubations were terminated as described in Figure 2. Following solubilization in RIPA buffer, precipitates from both sets of incubated platelets were prepared with a combination of antisera 1398 (1:50) and 8730 (1:20; left lane). The resulting supernatants were combined with additional 8730 and 1398 antisera, and a second set of precipitates were collected (middle lane). The resulting supernatants were combined with antiserum 8645 (1:20), and a final set of precipitates were collected (right lane). Shown are autoradiograms of SDS-polyacrylamide gels to which equivalent amounts of precipitates were applied.

lanes), from the resulting supernatant fractions by addition of 1398 plus 8730 again (middle lanes), and from the final set of supernatants by addition of 8645 (right lanes). The majority of [^{32}P]ADP-ribosylated protein was found to be precipitated upon the initial addition of 1398 plus 8730, with a much smaller quantity of labeled protein precipitated upon reiteration. No [^{32}P]ADP-ribosylated protein was evident upon subsequent precipitation with 8645 (antiserum 8645, however, if added prior to precipitation with other antisera precipitates about one-half of the radiolabeled protein evident for 1398 plus 8730). The distribution of radiolabel using [^{32}P]NAD was in direct contrast to that observed for immunoprecipitates obtained from platelets incubated in PMA and [γ-^{32}P]ATP. In this instance, precipitates obtained with 1398 plus 8730 did not contain radiolabeled protein (at least in excess of that achieved with preimmune sera), but that obtained with 8645 did. Thus, while 1398 plus 8730 preempt precipitation by 8645 of [^{32}P]ADP-ribosylated protein, they have no effect on precipitation of the phosphoprotein by this antiserum.

Failure of antisera 1398 and 8730 to precipitate the phosphoprotein and data indicating the protein to be refractory to ADP-ribosylation led us to suspect that the phosphoprotein was not $G_{i\alpha}$. Recognition of the phosphorylated protein by the two other $G_{i\alpha}$-directed antisera together with an apparent molecular weight of 40,000, however, suggested structural similarities. The likely identity of the phosphoprotein became clearer with the discovery of $G_{z\alpha}$ (or $G_{x\alpha}$), identified through analyses of complementary DNA (from human retina and rat C6 glioma cells) and human genomic DNA (Fong et al., 1988; Matsuoka et al., 1988). The deduced primary structure of $G_{z\alpha}$ indicates a protein with a molecular weight of 40,879, which exhibits 67% homology to members of the $G_{i\alpha}$ family; the consensus sequence for ADP-ribosylation is not evident. Of interest to us was the potential for antiserum 8645, at least, to cross-react with $G_{z\alpha}$. Specifically, the peptide to which 8645 was generated differed from a sequence present within $G_{z\alpha}$ by only one residue (N-terminal cysteine aside; Figure 1). Differences in relation to other peptides occurred at two (1521), three (1398), or four (8730) residues, rendering cross-reactivities in these instances less predictable. By Western blotting using recombinant protein, cross-reactivities for two of the antisera were indeed evident. Specifically, recombinant $G_{z\alpha}$ was recognized easily by 8645, much less well by 1521, and not at all by 1398 and 8730 (P. Casey and A. Gilman, personal communication). The phosphorylation of a protein resembling $G_{z\alpha}$ in PMA- or thrombin-treated platelets, and recognition of this protein through cross-reactivities with 8645 and 1521, therefore, formed an attractive hypothesis.

To directly test this proposition, two antisera – 2919 and 2921 – were generated against peptide sequences unique to $G_{z\alpha}$ (Figure 1). These antisera had the expected specificities upon immunotransfer blotting: recognition of a 40-kDa protein in platelet membranes, but not $G_{i\alpha}$ purified from bovine brain or rabbit liver. Importantly, both antisera precipitated a 40-kDa phosphoprotein from PMA-treated platelets (Figure 4). Antiserum 2919 was several-fold more effective at precipitating the phosphoprotein than

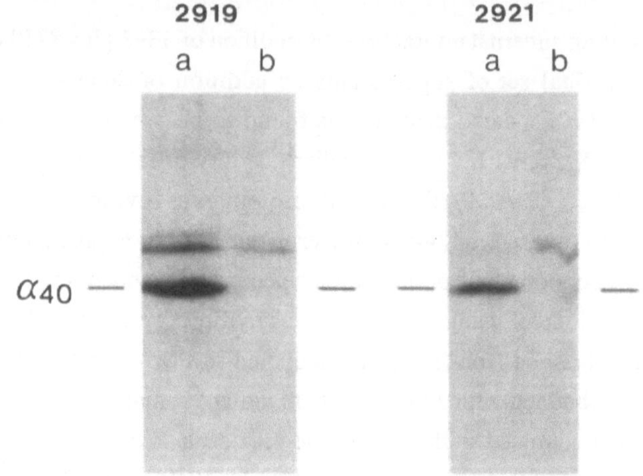

Figure 4. **Immunoprecipitates obtained from [^{32}P]ATP-labeled platelets with antisera 2919 and 2921.** Washed platelets were incubated with saponin, [γ-^{32}P]ATP, and 100 nM PMA at 30 °C for 3 min. Following solubilzation in RIPA buffer, precipitates were prepared with a 1:8 dilution of antiserum 2919 or 2921 (Lanes a) or with respective preimmune sera (Lanes b). Shown are autoradiograms of SDS-polyacrylamide gels to which equivalent amounts of precipitates were applied.

either antiserum 8645 or 2921. Both of the $G_{z\alpha}$-directed antisera were also capable of precipitating the phosphoprotein from thrombin-treated platelets. In an experiment not shown (Carlson et al., 1989), precipitation of the phosphoprotein by antisera 8645, 1521, or 2921 was preempted by prior precipitation with antiserum 2919, indicating that all four antisera are recognizing the same phosphoprotein(s). Thus, the existence and selective phosphorylation of a $G_{z\alpha}$-like protein within platelets was demonstrated.

SUMMARY AND CONCLUSIONS

A G_{α} subunit is phosphorylated upon treatment of intact or permeabilized human platelets with agents promoting activation of protein kinase C. The phosphorylated subunit does not appear to be any of the known forms of $G_{i\alpha}$. Rather, it is related to, if not identical with, the newly described subunit $G_{z\alpha}$. These conclusions are based on the following: 1) two antisera with specificities unique to $G_{i\alpha}$ (1398 and 8730) fail to precipitate phosphoproteins from PMA- or thrombin-treated platelets; 2) two antisera with specificities for both $G_{i\alpha}$ and $G_{z\alpha}$ (8645 and 1521) precipitate a 40-kDa phosphoprotein from PMA- and thrombin-treated platelets; 3) two antisera with specificities unique to $G_{z\alpha}$ (2919 and 2921) precipitate a 40-kDa phosphoprotein from PMA- and thrombin-treated platelets; 4) the 40-kDa protein subject to phosphorylation appears not to be a substrate for ADP-ribosylation by PTX; and 5) precipitation of the 40-kDa phosphoprotein with a $G_{z\alpha}$-selective antiserum (2919) preempts precipitation of this protein with the other antisera.

These findings have several important implications. With respect to the effects of phorbol esters on hormonal inhibition of adenylyl cyclase, the G protein(s) that mediate inhibition – i.e., one or more forms of $G_{i\alpha}$ – are substrate(s) for PTX. The lack of evidence for phosphorylation of G_i and the data supporting phosphorylation of a G_α subunit distinct from a PTX-substrate therefore indicate that the effects of PMA on hormonal inhibition in platelets may not be exerted at the level of G_i. The same may be true for the effects of PMA on PTX-sensitive pathways of phospholipase C stimulation. The functions of G_z at this time are unknown. Because $G_{z\alpha}$ lacks the consensus site for ADP-ribosylation by PTX, it is a leading candidate for mediating activation of phospholipase C through PTX-insensitive pathways. Activation of protein kinase C and consequent phosphorylation of the $G_{z\alpha}$-like protein, therefore, may be an important mechanism for regulation of signal transduction through this particular pathway.

Phosphorylation of $G_{z\alpha}$ within intact or permeabilized platelets occurs selectively, with no evidence for the phosphorylation of $G_{i\alpha}$. Thus, $G_{z\alpha}$ is emerging as a protein with several potentially unique features. These include possible differences in the ability to serve as a substrate for bacterial toxins (Fong et al., 1988; Matsuoka et al., 1988), to recognize or hydrolyze GTP (Fong et al., 1988; Matsuoka et al., 1988), and to undergo phosphorylation by protein kinase C. The extent to which these features are related should prove to be interesting.

ACKNOWLEDGEMENTS

This work was supported by National Institutes of Health Grants CA39712 and HL40387 and by the American Heart Association. KEC was supported in part through the Predoctoral Training Program in Cardiovascular Research HL07502, and LFB is an Established Investigator of the American Heart Association. The authors are indebted to Drs. Patrick Casey and Alfred Gilman for assays of antibody specificity using recombinant proteins.

REFERENCES

Carlson, K.E., Brass, L.F., and Manning, D.R., 1989, Thrombin and phorbol esters cause the selective phosphorylation of a G protein other than Gi in human platelets, *J. Biol. Chem.*, in press.

Crouch, M.F., and Lapetina, E.G., 1988, A role for Gi in control of thrombin receptor-phosphopholipase C coupling in human platelets, J. Biol. Chem., 263:3363-3371.

Fong, H.K.W., Yoshimoto, K.K., Eversole-Cire, P., and Simon, M.I., 1988, Identification of a GTP-binding protein a subunit that lacks an apparent ADP-ribosylation site for pertussis toxin, *Proc. Natl. Acad. Sci. USA*, 85:3066-3070.

Halenda, S.P., Volpi, M., Zavoico, G.B., Sha'afi, R.I., and Feinstein, M.B., 1986, Effects of thrombin, phorbol myristate acetate, and prostaglandin D_2 on 40–41 kDa protein that is ADP ribosylated by pertussis toxin in platelets, *FEBS Lett.*, 204: 341-346.

Jakobs, K.H., Bauer, S., and Watanabe, Y., 1985, Modulation of adenylate cyclase of

human platelets by phorbol ester: Impairment of the hormone-sensitive inhibitory pathway, *Eur. J. Biochem.*, 151:425-430.
Jones, D.T., and Reed, R.R., 1987, Molecular cloning of five GTP-binding cDNA species from rat olfactory neuroepithelium, *J. Biol. Chem.*, 262:14241-14249.
Katada, T., Gilman, A.G., Watanabe, Y., Bauer, S., and Jakobs, K.H., 1985, Protein kinase C phosphorylates the inhibitory guanine-nucleotide-binding regulatory component and apparently suppresses its function in hormonal inhibition of adenylate cyclase, *Eur. J. Biochem.*, 151:431-437.
Krupinski, J., Rajaram, R., Lakonishok, M., Benovic, J.L., and Cerione, R.A., 1988, Insulin-dependent phosphorylation of GTP-binding proteins in phospholipid vesicles, *J. Biol. Chem.*, 263:12333-12341.
Matsuoka, M., Itoh, H., Kozasa, T., and Kaziro, Y., 1988, Sequence analysis of cDNA and genomic DNA for a putative pertussis toxin-insensitive guanine nucleotide-binding regulatory protein a subunit, *Proc. Natl. Acad. Sci. USA*, 85:5384-5388.
O'Brien, R.M., Houslay, M.D., Milligan, G., and Siddle, K., 1987, The insulin receptor tyrosyl kinase phosphorylates holomeric forms of the guanine nucleotide regulatory proteins Gi and Go, *FEBS Lett.*, 212:281-288.
Orellana, S., Solski, P.A., and Brown, J.H., 1987, Guanosine 5'-O-(triotriphosphate)-dependent inositol triphosphate formation in membranes is inhibited by phorbol ester and protein kinase C, *J. Biol. Chem.*, 262:1638-1643.
Pyne, N.J., Murphy, G.J., Milligan, G., and Houslay, M.D., 1989, Treatment of intact hepatocytes with either the phorbol ester TPA or glucagon elicits the phosphorylation and functional inactivation of the inhibitory guanine nucleotide regulatory protein Gi, *FEBS Lett.*, 243:77-82.
Rothenberg, P.L., and Kahn, C.R., 1988, Insulin inhibits pertussis toxin-catalyzed ADP-ribosylation of G-proteins: Evidence for a novel interaction between insulin receptors and G-proteins, *J. Biol. Chem.*, 263:15546-15552.
Sefton, B.M., Beemon, K., and Hunter, T., 1978, Comparison of the expression of the src gene of rous sarcoma virus in vitro and in vivo, *J. Virol.*, 28:957-971.
Smith, C.D., Uhing, R.J., and Snyderman, R., 1987, Nucleotide regulatory protein-mediated activation of phospholipase C in human polymorphonuclear leukocytes is disrupted by phorbol esters, *J. Biol. Chem.*, 262:6121-6127.
Watanabe, Y., Imaizumi, T., Misaki, N., Iwakura, K., and Yoshida, H., 1988, Effects of phosphorylation of inhibitory GTP-binding protein by cyclic AMP-dependent protein kinase on its ADP-ribosylation by pertussis toxin, islet-activating protein, *FEBS Lett.*, 236:372-374.

G PROTEINS AND TOXIN-CATALYZED ADP-RIBOSYLATION

Martha Vaughan and Joel Moss

Laboratory of Cellular Metabolism
National Heart, Lung, and Blood Institute
National Institutes of Health
Bethesda, Maryland

INTRODUCTION

Cholera, a devastating diarrheal disease characterized by abnormalities in fluid and electrolyte flux, results in part from the action of cholera toxin (choleragen), a secretory product of Vibrio cholerae, on the intestinal mucosa (Carpenter, 1980; Kelly, 1986). Cholera toxin exerts its effects by catalyzing the ADP-ribosylation of $G_{s\alpha}$, the α subunit of the heterotrimeric stimulatory guanine nucleotide-regulatory protein that activates the adenylyl cyclase catalytic unit and is involved in the modulation of ion channels (Birnbaumer et al., 1987; Casey and Gilman, 1988; Moss and Vaughan, 1988). Activation of adenylyl cyclase, as well as other toxin-catalyzed reactions, is enhanced by membrane and soluble factors that have been identified in preparations from numerous animal cells and tissues (Gill, 1976; Enomoto and Gill, 1980; Nakaya et al., 1980; Le Vine, III and Cuatrecasas, 1981; Pinkett and Anderson, 1982; Schleifer et al., 1982; Gill and Meren, 1983; Kahn and Gilman, 1984; Kahn and Gilman, 1986; Tsai et al., 1987, 1988; Gill and Coburn, 1987). A 21 kDa protein that stimulated the toxin-catalyzed ADP-ribosylation of $G_{s\alpha}$ was purified from liver membranes and termed ADP-ribosylation factor or ARF (Kahn and Gilman, 1984). A similar ARF protein from bovine brain membranes was shown to bind guanine nucleotides in the presence of dimyristoyl phosphatidylcholine and NaCl (Kahn and Gilman, 1986). It was proposed that ARF acted by complexing with $G_{s\alpha}$, thereby enhancing its ability to serve as a toxin substrate (Kahn and Gilman, 1984).

PURIFICATION AND IMMUNOLOGICAL CHARACTERIZATION OF ARF

ARF was extensively purified from membrane and soluble fractions of bovine brain (Tsai et al., 1987, 1988c). Two soluble (sARF-I and sARF-II) and one membrane (mARF) species were isolated. The soluble forms differed in their elution from carboxymethyl cellulose and in mobility on SDS-polyacrylamide gel electrophoresis. Antiserum prepared against sARF-II reacted with both soluble and membrane species from bovine brain (Tsai et al., 1988b). Immunoreactive ∿20 kDa proteins were identified in a variety of tissues from several species (e.g., rat, frog, mouse, chicken)(Tsai et al., 1988b; Kahn et al., 1988).

A similar distribution of ARF-like proteins was found using antibodies against mARF or using an ADP-ribosylation assay to quantify ARF activity (Tsai et al., 1988b; Kahn et al., 1988). The anti-sARF-II antibodies did not react significantly with the guanine nucleotide-binding proteins such as $G_{s\alpha}$ and $G_{i\alpha}$, which participate in the stimulation and inhibition, respectively, of adenylyl cyclase, $G_{o\alpha}$, which may be involved in the regulation of ion flux, and $G_{t\alpha}$, which regulates the visual excitation pathway (Tsai et al., 1988b). Although ARF proteins are widely distributed, the highest levels are found in brain and neural tissues where the larger, sARF-II-like species appears to predominate. In other tissues, the smaller form of ARF, which has a mobility similar to that of sARF-I, is present.

MECHANISM OF ACTION OF ARF

To define the mechanism by which ARF activates cholera toxin-catalyzed ADP-ribosylation of $G_{s\alpha}$, advantage as taken of the fact that the toxin catalyzes a number of reactions that reflect its ability to activate the ribosyl-nicotinamide bond of NAD and to use water or simple guanidino compounds (e.g., arginine, agmatine) as ADP-ribose acceptors. Since these reactions do not require the presence of $G_{s\alpha}$, they can be used to determine whether ARF activation of the toxin has a G_s requirement. Both the $G_{s\alpha}$-dependent and $G_{s\alpha}$-independent ADP-ribosyltransferase activities of cholera toxin were enhanced by mARF, sARF I, and sARF II (Tsai et al., 1987, 1988c; Noda et al, in press). In all instances, the stimulatory effect of ARF required the presence of GTP or a nonhydrolyzable analogue such as GTPγS; GDP and GDPβS were inactive, as were adenine nucleotides (Table I). ARF activated the cholera toxin A subunit or the reduced and alkylated CTA_1 fragment, in which the single cysteine near the carboxyl terminus is modified (Tsai et al., 1987, 1988c; Noda et al., in press). This is consistent with the view that the function of ARF is not to promote release of the active catalytic unit of the toxin but rather to interact with this enzyme and modify its activity.

Maximal binding of guanine nucleotides by sARF-II required the presence of both dimyristoyl phosphatidylcholine and cholate (Bobak et al., 1988). In contrast to the published findings with mARF, high concentrations of NaCl were not required (Kahn and Gilman, 1986; Bobak et al., 1988). In the presence of dimyristoyl phosphatidylcholine and cholate, maximal binding of guanine nucleotides was not observed for several hours; the binding appeared to be specific for guanine nucleotides (GTP, GDP and GTPγS); adenine nucleotides were inactive (Bobak et al., 1988).

Lineweaver-Burk analysis of the effect of ARF on the NAD:agmatine ADP-ribosyltransferase activity of cholera toxin revealed that ARF served as an allosteric activator of the toxin; it decreased the K_ms for both NAD and agmatine, with little effect on the V_{max} (Noda et al., in press). Activation of cholera toxin by ARF was enhanced by SDS, cholate or DMPC/cholate (Noda et al., in press; Bobak et al., 1988). In the presence of SDS, ARF caused a larger decrease in K_ms and significantly increased V_{max}. The effects of SDS on ARF activity were observed at GTP concentrations in the micromolar range; stimulation by DMPC/cholate required nanomolar GTP concentrations (Noda et al., in press; Bobak et al., 1988). These data are compatible with the results of the binding studies which show high affinity binding in DMPC/cholate but not in SDS.

Table I. Effect of Nucleotides on NAD:agmatine ADP-ribosyltransferase Activity of Choleragen A Subunit in the Presence of sARF I or sARF II

Nucleotide added	[carbonyl-^{14}C]nicotinamide released	
	sARF I	sARF II
30 μM	nmol·μg A subunit^{-1}·h^{-1}	
None	2.10	1.22
GTP	5.48	2.12
GTPγS	4.70	1.91
Gpp(NH)p	4.70	1.73
ATP	2.15	1.26
App(NH)p	2.00	1.11
GDP	2.00	1.16
GDPβS	1.89	1.11

Assays contained choleragen A (1 μg), sARF I (2.8 μg), or sARF II (2.4 μg) with nucleotide as indicated. Data are from Tsai et al., 1988c.

All observations were consistent with the conclusion that ARF interacts directly with and activates the toxin in a GTP-dependent action. In support of this proposal is the finding that, in the presence of SDS and GTPγS, ARF and toxin form tightly associated aggregates that can be isolated by gel permeation chromatography (Tsai et al., 1988a). The ARF-toxin aggregates are considerably more active at catalyzing the auto-ADP-ribosylation of CTA_1, than in catalyzing the ADP-ribosylation of $G_{s\alpha}$. Only a fraction of the purified sARF II used in these experiments was capable of complexing with the toxin, suggesting that these preparations are in some way heterogeneous, despite their appearance on SDS-polyacrylamide gels. Although some of the ARF was incapable of tightly complexing with CTA_1, this ARF, purified by gel permeation chromatography, was still capable of activating the toxin-catalyzed ADP-ribosylation of $G_{s\alpha}$ and agmatine.

To define further the effect of ARF on cholera toxin, advantage was taken of the fact that CTA_1 contains a single cysteine residue near the carboxy terminus. As noted above, prior studies had shown that alkylation of this sulfhydryl did not interfere with stimulation by ARF of the NAD:agmatine ADP-ribosyltransferase activity of the toxin. CTA was reduced and reacted with sulfosuccinimidyl 6-(biotinamido) hexanoate; the biotinylated CTA_1 was then immobilized on avidin-agarose (Noda et al., 1988). Since immobilized CTA_1 catalyzed the ARF-stimulated ADP-ribosylation of agmatine, it appears that the allosteric site for ARF and the catalytic site are not located near the carboxy end of the protein. Although the immobilized CTA_1 catalyzed the ADP-ribosylation of agmatine, consistent with the presence of an intact ADP-ribose acceptor site, its ability to modify $G_{s\alpha}$ was greatly reduced. Perhaps immobilization of the toxin restricts access or in some other way makes the site unsuitable for the larger acceptor molecule (Noda et al., 1988).

As ARF both binds guanine nucleotides and serves to activate the cholera toxin-catalyzed ADP-ribosylation of a guanine nucleotide-binding protein, it appears that a guanine nucleotide-binding protein cascade may be involved in the activation of adenylyl cyclase by cholera toxin (Fig. 1). ARF, with GTP bound, can interact with CTA_1, after it is released from its latent state in the CTA subunit by thiols such as dithiothreitol. CTA_1 is activated by ARF (GTP) and catalyzes the ADP-ribosylation of $G_{s\alpha}$. ADP-ribosyl-$G_{s\alpha}$ containing bound GTP serves to activate the catalytic unit of adenylyl cyclase. It is possible that the stimulatory effects of DMPC/cholate or SDS result from their ability to mimic a membrane-like environment that facilitates functional interaction of ARF with the toxin.

To begin to investigate the physiological role of ARF, which presumably has nothing to do with cholera toxin, purified sARF-II was injected into Xenopus oocytes which were incubated with either progesterone or insulin to induce maturation (Bahnson et al., 1988).

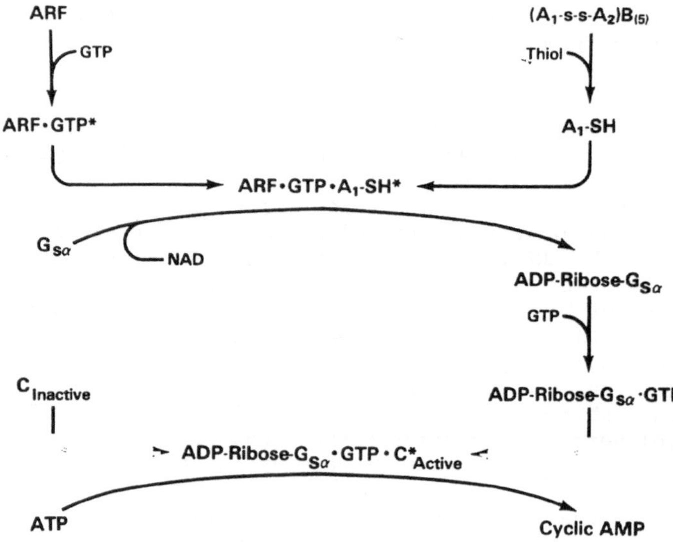

Fig. 1. Activation of adenylyl cyclase by choleragen involves a guanine nucleotide binding-protein cascade. Details are discussed in the text. Figure is reprinted from Tsai et al., 1988c.

ARF inhibited maturation of both progesterone- and insulin-stimulated oocytes. The effect was significantly enhanced by concomitant injection of GTPγS. For progesterone-stimulated maturation, inhibition by ARF was only observed at suboptimal concentrations of the steroid. It was maximal 2 to 8 hrs after ARF injection and then decreased, although ARF immunoreactivity, significantly greater than that found in uninjected oocytes, was still present at later times. This inhibitory action of ARF is in contrast to the stimulatory effect of ras p21 on oocyte maturation (Birchmeier et al., 1985). Perhaps these are representative of two different types of guanine nucleotide-binding proteins that have opposing effects on the maturation process.

STRUCTURAL CHARACTERIZATION OF ARF

Two ARF cDNA clones, ARF-1 and ARF-2, have been isolated from bovine adrenal and retinal libraries, respectively (Sewell and Kahn, 1988; Price et al., 1988). Nucleotide and deduced amino acid sequences of the clones were 80% and 96% identical. Comparison of the deduced amino acid sequence of the bovine retinal clone ARF-2B to the sequences of CNBr peptides from sARF-II revealed that there was identity in 58 of 60 amino acids. The sequences of the CNBr peptides were, however, completely identical to that deduced from the bovine adrenal ARF-1 clone.

The deduced amino acid sequences of ARF are similar to sequences of the G protein α-subunits and to the ras-like proteins in regions believed to be involved in guanine nucleotide-binding and hydrolysis (Price et al., 1988) (Fig. 2). The GTP hydrolysis site is near the amino terminus of these proteins. Based on crystallographic studies of ras and EF-Tu (De Vos et al., 1988; Jurnak, 1985), it appears that four regions dispersed in the linear sequence are brought into proximity in the folded protein to form the guanine nucleotide-binding site.

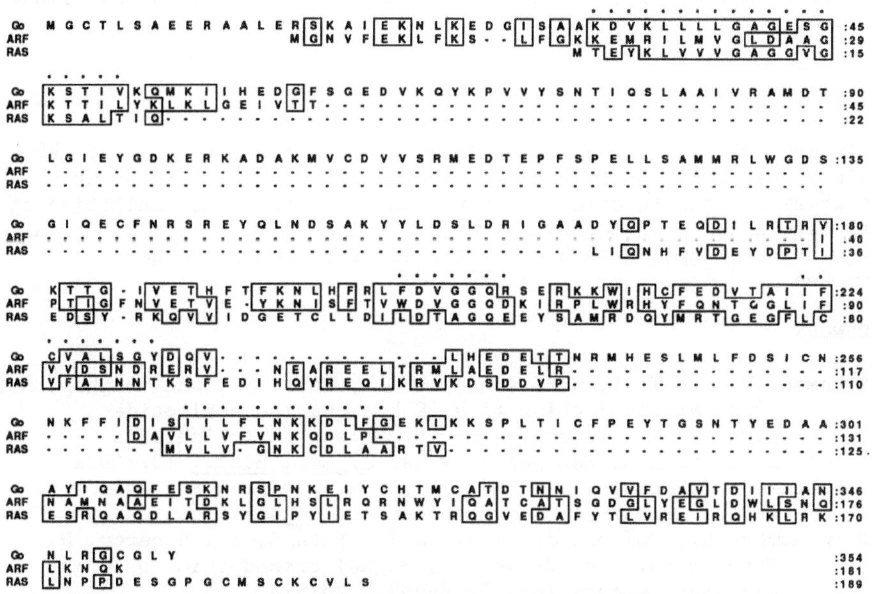

Fig. 2. Comparison of deduced amino acid sequences of bovine G_{oa} (G_o), λARF2B (ARF) and human c-Ha-ras-encoded p21 (RAS). Gaps, indicated by hyphens, were introduced to obtain maximal similarity. Asterisks above the sequences mark regions presumed to be involved in guanine nucleotide binding and GTP hydrolysis. Identical amino acids are boxed with conservative substitutions. Categories of conservative substitutions are as follows: cysteine (C); serine (S), threonine (T), proline (P), alanine (A), and glycine (G); asparagine (N), aspartic acid (D), glutamic acid (E), and glutamine (Q); histidine (H), arginine (R), and lysine (K); methionine (M), isoleucine (I), leucine (L), and valine (V); phenylalanine (F), tryptophan (W), and tyrosine (Y). Figure is reprinted from Price et al., 1988.

Outside of these four regions, the sequences of the different types of GTP-binding proteins differ from each other and from the ARF proteins.

The distribution of mRNA for ARF-2B and ARF-1 was determined by hybridization of the ARF-2B cDNA and oligonucleotide probes specific for either ARF-1 or ARF-2B with poly (A)+ RNA (Tsuchiya et al., 1988). Each of the clones hybridized with a single mRNA species. The ARF-2B and ARF-1 mRNAs of approximately 2.1 and 1.7 kb, respectively, were found in bovine brain, lung, heart, spleen, liver and kidney. Using the ARF-2B cDNA probe, ARF-related mRNA was detected in poly (A)+ preparations from mouse, rat, rabbit and human brains. With restriction fragments derived from the ARF-2B clone as probes, it was determined that conservation of sequence across species is greater in the coding than in the 3'-untranslated regions.

The ARF-2B cDNA was used to isolate a human brain ARF cDNA (Bobak et al., 1989). Its nucleotide and deduced amino acid sequences are greater than 90% and 80% identical, respectively, to those of the two bovine clones, and contain the regions believed to be responsible for guanine nucleotide-binding and GTP hydrolysis. The human cDNA hybridized with human poly (A)+ RNA of ∿3.7 kb, consistent with that representing a third form of ARF, different from both bovine ARF-1 and ARF-2B.

CONCLUSIONS

It appears that there are at least three different types of ARF proteins. Although these proteins and mRNAs are widely distributed, their physiological function(s) is unclear. Whether the inhibition of oocyte maturation produced by injection of ARF reflects a role for ARF in regulation of cell growth or multiplication remains to be determined.

REFERENCES

1. Bahnson, T. D., Tsai, S.-C., Adamik, R., Moss, J., and Vaughan, M., 1988, Microinjection of a 19 kDa guanine nucleotide-binding protein activator of cholera toxin inhibits maturation of Xenopus oocytes, J. Cell Biol., 107:493a.
2. Birchmeier, C., Broek, D., and Wigler, M., 1985, RAS proteins can induce meiosis in Xenopus oocytes, Cell, 43:615.
3. Birnbaumer, L., Codina, J., Mattera R., Yatani, A., Scherer, N., Toro, M.-J., and Brown, M., Signal transduction by G proteins, Kidney Int. 32 Suppl., 23:S14.
4. Bobak, D. A., Bliziotes, M. M., Noda, M., Tsai, S.-C., Moss, M., and Vaughan, M., 1988, Mechanism of activation of adenylate cyclase by cholera toxin: stimulation by a 19 kDa protein that exhibits both low and high affinity guanine nucleotide-binding, Clin. Res., 36:578A.
5. Bobak, D. A., Nightingale, M. S., Murtagh, J. J., Price, S. R., Moss, J., and Vaughan, M., 1989, Guanine nucleotide-binding proteins that stimulate cholera toxin ADP-ribosyltransferase activity: isolation of a human brain ADP-ribosylation factor cDNA, Clin. Res., 37:561A.
6. Carpenter, C. C. J., 1980, Clinical and pathophysiologic features of diarrhea caused by Vibrio cholerae and Escherichia coli, in: "Secretory Diarrhea," M. Field, J.S. Fordtran, S. G. Schultz, eds., American Physiological Society, Bethesda.
7. Casey, P. J., and Gilman, A G., 1988, G protein involvement in receptor-effector coupling, J. Biol. Chem., 263:2577.

8. De Vos, A. M., Tong, L., Milburn, M. V., Matias, P. M., Jancarik, J., Noguchi, S., Nishimura, S., Miura, K., Ohtsuka, E., and Kim, S. H., 1988, Three-dimensional structure of an oncogene protein: catalytic domain of human c-H-ras p21, Science, 239:888.
9. Enomoto, E., and Gill, D. M., 1980, Cholera toxin activation of adenylate cyclase. Roles of nucleoside triphosphates and a macromolecular factor in the ADP ribosylation of the GTP-dependent regulatory component, J. Biol. Chem., 255:1252.
10. Gill, D. M., 1976, Multiple roles of erythrocyte supernatant in the activation of adenylate cyclase by Vibrio cholerae toxin in vitro, J. Infect. Dis., 133:S55.
11. Gill, D. M., and Coburn, J., 1987, ADP-ribosylation by cholera toxin: Functional analysis of a cellular system that stimulates the enzymic activity of cholera toxin fragment A_1, Biochemistry, 26:6364.
12. Gill, D. M., and Meren, R., 1983, A second guanyl nucleotide-binding site associated with adenylate cyclase. Distinct nucleotides activate adenylate cyclase and permit ADP-ribosylation by cholera toxin, J. Biol. Chem., 258:11908.
13. Jurnak, F., 1985, Structure of the GDP domain of EF-Tu and location of the amino acids homologous to ras oncogene proteins, Science, 230:32.
14. Kahn, R. A., and Gilman, A. G., 1984, Purification of a protein cofactor required for ADP-ribosylation of the stimulatory regulatory component of adenylate cyclase by cholera toxin, J. Biol. Chem., 259:6228.
15. Kahn, R. A., and Gilman, A. G., 1986, The protein cofactor necessary for ADP-ribosylation of G_s by cholera toxin is itself a GTP binding protein, J. Biol Chem., 261:7906.
16. Kahn, R. A., Goddard, C., and Newkirk, N., 1988, Chemical and immunological characterization of the 21-kDa ADP-ribosylation factor of adenylate cyclase, J. Biol. Chem., 263:8282.
17. Kelly, M. T., 1986, Cholera: a worldwide perspective, Ped. Infect. Dis. 5:5101.
18. Le Vine, III, H., and Cuatrecasas, P., Activation of pigeon erythrocyte adenylate cyclase by cholera toxin. Partial purification of an essential macromolecular factor from horse erythrocyte cytosol, Biochem. Biophys. Acta, 672:248.
19. Moss, J., and Vaughan, M., 1988, ADP-ribosylation of guanyl nucleotide-binding regulatory proteins by bacterial toxins, Adv. Enzymol., 61:303.
20. Nakaya, S., Moss, J., and Vaughan, M., 1980, Effects of nucleoside triphosphates on choleragen-activated brain adenylate cyclase, Biochemistry, 19:4871.
21. Noda, M., Tsai, S.-C., Adamik, R., Bobak, D. A., Bliziotes, M. M., Moss, J., and Vaughan, M., in press, Activation of cholera toxin by soluble guanine nucleotide-binding proteins, 24th U.S.-Japan Joint Cholera Conference.
22. Noda, M., Tsai, S.-C., Adamik, R., Bobak, D., Moss, J., Vaughan, M., 1988, Thiol-dependent activation by 19 kDa guanine nucleotide-dependent proteins of alkylated choleragen A_1 protein and an immobilized, biotinylated toxin complex, J. Cell Biol., 107:419a.
23. Pinkett, M. O., and Anderson, W. B., 1982, Plasma membrane-associated component(s) that confer(s) cholera toxin sensitivity to adenylate cyclase, Biochim. Biophys. Acta, 714:337.
24. Price, S. R., Nightingale, M., Tsai, S.-C., Williamson, K. C., Adamik, R., Chen, H.-C., Moss, J., and Vaughan, M., Guanine

nucleotide-binding proteins that enhance choleragen ADP-ribosyltransferase activity: Nucleotide and deduced amino acid sequence of an ADP-ribosylation factor cDNA, Proc. Natl. Acad. Sci. USA, 85:5488.

25. Schleifer, L. S., Kahn, R. A., Hanski, E., Northup, J. K., Sternweis, P. C., and Gilman, A. G., 1982, Requirements for cholera toxin-dependent ADP-ribosylation of the purified regulatory component of adenylate cyclase, J. Biol. Chem., 257:20.
26. Sewell, J. L., and Kahn, R. A., 1988, Sequences of the bovine and yeast ADP-ribosylation factor and comparison to other GTP-binding proteins, Proc. Natl. Acad. Sci. USA, 85:4620.
27. Tsai, S.-C., Adamik, R., Moss, J., and Vaughan, M., 1988, Guanine nucleotide-dependent complex formation between choleragen (cholera toxin) A_1 subunit and bovine brain ADP-ribosylation factor, FASEB J., 2:6092.
28. Tsai, S.-C., Adamik, R., Williamson, K., Moss, J., and Vaughan, M., 1988b, Immunological and functional identification of the guanine nucleotide-dependent protein activator of cholera toxin, J. Cell Biology, 107:709a.
29. Tsai, S.-C., Noda, M., Adamik, R., Chang, P. P., Chen, H.-C., Moss, J., and Vaughan, M., 1988c, Stimulation of choleragen enzymic activities by GTP and two soluble proteins purified from bovine brain, J. Biol. Chem., 263:1768.
30. Tsai, S.-C., Noda, M., Adamik, R., Moss, J., and Vaughan, M., 1987, Enhancement of choleragen ADP-ribosyltransferase activities by guanyl nucleotides and a 19 kDa membrane protein, Proc. Natl. Acad. Sci. USA, 84:5139.
31. Tsuchiya, M., Price, S. R., Nightingale, M., Moss, J., and Vaughan, M., 1989, Identification of multiple mRNAs for ADP-ribosylation factors (ARFs) in bovine tissues: evidence for more than one ARF gene, J. Cell Biol., 107:709a.

ACKNOWLEDGEMENT

We thank Dr. S.-C. Tsai, Dr. M. Noda, Dr. D. Bobak, Dr. S. R. Price, Mr. R. Adamik, and Mrs. Maria Nightingale for their major contributions to studies of the ARF proteins and Mrs. Barbara Mihalko for expert secretarial assistance.

ABBREVIATIONS: ARF, ADP-ribosylation factor; mARF and sARF, membrane and soluble ARF, respectively; sARF-I and sARF-II, ARF proteins purified from bovine brain supernatant (Tsai et al., 1988c); DMPC, dimyristoyl phosphatidylcholine; CTA_1 and CTA_2, A_1 and A_2 proteins of cholera toxin; GTPγS, guanosine-5'-O-(3-thiotriphosphate); GDPβS, guanosine-5'-O-(2-thiodiphosphate); G_s, stimulatory guanine nucleotide-binding protein of the adenylyl cyclase system; $G_{s\alpha}$, α subunit of G_s; SDS, sodium dodecyl sulfate.

NOVEL ENDOGENOUS ADP-RIBOSYLATIONS OF PROTEINS IN TRANSFORMED CELLS : Effects of cellular transformation and differentiation

Harjit S. Banga and Maurice B. Feinstein

Department of Pharmacology, University of Connecticut Health Center, Farmington, CT 06032 U.S.A.

INTRODUCTION

ADP-ribosylation is a post-translational modification of proteins that utilizes NAD[1] as a donor for the modification group (Ueda and Hayaishi, 1985). The reaction involves enzymatic transfer of the ADP-ribose moiety of NAD to specific acceptor amino acids on the target protein; nicotinamide being the released product of this reaction. ADP-ribosylation reactions are broadly classified into mono- and poly(ADP-ribosyl)ation on the basis of ADP-ribose chain length. However, there are other differences between the two groups. Mono(ADP-ribosyl)ations generally occur on the side-chain nitrogen atom of the acceptor amino acid, and depending upon this acceptor, can be classified as types A (arginine), C (cysteine) and D (diphthamide) respectively (Moss et al., 1980; Tanuma et al., 1988; Sayhan et al., 1986). On the contrary, poly(ADP-ribosyl)ation starts on the carboxyl groups and forms o-glycosidic linkages on glutamic acid residues within the target protein. In some cases (e.g. in histone H1), COOH-terminal lysine may also be the site for poly(ADP-ribosyl)ation (Ogata et al., 1980). The o-glycosides formed in poly(ADP-ribosyl)ation reactions are usually sensitive to treatment with neutral NH_2OH whereas the N-glycosidic linkages in mono(ADP-ribosyl)ated proteins are largely resistant to NH_2OH (Ueda and Hayaishi, 1985). This forms the basis for a convenient, though less definitive, distinction between these two broad classes of ADP-ribosylation.

ADP-ribosylation by Bacterial Toxins

Many bacterial toxins possess ADP-ribosyltransferase activity that modifies target/acceptor proteins in eukaryotic cells. Three types of mono(ADP-ribosyl)ations are known at present.

1. Cholera toxin mediates the mono(ADP-ribosyl)ation of an arginine residue (type A) of guanine nucleotide binding protein

[1] Abbreviations used : NAD, nicotinamide adenine dinucleotide; NH_2OH, hydroxylamine.

Gs in the adenylyl cyclase system (Katada and Ui, 1982). A similar bond is formed in actin by botulinum toxin C2 (Aktories et al., 1988).

2. Pertussis toxin, on the other hand, modulates the adenylyl cyclase system by mono(ADP-ribosyl)ating a cysteine residue (type C) of the guanine nucleotide binding protein Gi (Hisa et al., 1985). Pertussis toxin substrate(s) also play a role in the activation of phospholipase C in response to stimuli in some, but not all, cell types.

3. Diphtheria toxin and Pseudomonas exotoxin A utilize diphthamide residue (type D) in elongation factor 2 (EF-2) as an acceptor of ADP-ribose and prevent its translocase function in protein sysnthesis (Van Ness et al., 1980).

There is now evidence to indicate that a fourth type of mono(ADP-ribosyl)ation may occur in botulinum ADP-ribosyltransferase C3 mediated modification of platelet cytosolic proteins (Aktories et al., 1988).

ADP-ribosyltransferases in eukaryotic cells

ADP-ribosyltransferases have been purified and characterized from eukaryotic cells as well (Moss et al., 1980; Tanuma et al., 1988; Sayhan et al., 1986; Yost and Moss, 1983; Tanigawa et al., 1984). Interestingly, these enzymes may modify the same acceptors as used by bacterial toxins. It has been recognized that the diphthamide residue of elongation factor-2 is mono(ADP-ribosyl)ated endogenously in polyoma virus transformed baby hamster kidney cells (Fendrick and Iglewski, 1989), and exogenously by a beef liver enzyme (Iglewski et al., 1984). An ADP-ribosyltransferase from turkey erythrocytes catalyzes ADP-ribosylation of arginine (type A), and activates adenylyl cyclase in a manner similar to cholera toxin (Moss and Vaughan, 1978). More recently, yet another ADP-ribosyl-transferase (type C) has been purified from human erythrocytes that, like pertussis toxin, ADP-ribosylates the inhibitory guanine nucleotide binding protein (Gi) of the adenylyl cyclase system (Tanuma et al., 1988). It therefore seems possible that the endogenous ADP-ribosyltranseferases may be involved in the regulation of transmembrane signalling. There is now growing evidence to suggest that endogenous ADP-ribosylation may be modulated by extracellular signals. An increase in ADP-ribosylation of a 40 kDa protein was first reported in bovine thyroid membranes stimulated with thyrotropin (Vitti et al., 1982), but more recently extracellular calcium (Duncan et al., 1988), glucocorticoids (Tanuma et al., 1983), tumor necrosis factor (Agarwal et al., 1988), and agents that stimulate guanylate cyclase (Brune and Lapetina, 1989) have also been shown to influence endogenous ADP-ribosylation. However, unlike the bacterial toxins, the physiological role of these cellular enzymes has not been clearly demonstrated.

Poly(ADP-ribosyl)ation

Mono(ADP-ribosyl)ated proteins usually predominate by far (>10 fold) over poly(ADP-ribosyl)ated ones in mammalian cells, particularly in the extranuclear compartment (Vaughan and Moss, 1982). However, the mono(ADP-ribosyl)ated proteins do not represent the initiation sites for elongation of (ADP-ribose) side chain, and are not the products of degradation of poly(ADP-ribosyl)ated proteins. They probably have different fates and functions than the poly(ADP-ribosyl)ated proteins (Hilz, 1981). Poly(ADP-ribosyl)ation of histones and other nuclear proteins by chromatin-associated poly(ADP-ribosyl) synthetase was first demonstrated in 1968 (Nishizuka et al., 1968). Subsequently, similar activity was also detected in ribosomes of HeLa cells (Roberts et al., 1975) and mitochondria of rat liver (Kun et al., 1975) and testis (Burzio et al., 1981). A single enzyme appears to mediate initiation (transfer of ADP-ribose from NAD to a protein acceptor to form mono(ADP-ribosyl) protein), elongation (addition of more ADP-ribose residues to mono(ADP-ribosyl)ated protein) and branching in the poly(ADP-ribose) chain. The three reactions thus generate oligo- or poly(ADP-ribosyl)-proteins in which the polymer may be linear or branched (Ueda and Hayaishi, 1985).

The biological significance of several mono(ADP-ribosyl)ation reactions are better understood than that of poly(ADP-ribosyl)ations. However, rapidly accumulating data suggests a role for poly(ADP-ribosyl)ation in several cellular processes such as cell proliferation, DNA synthesis and repair, regulation of conformation of chromatin and chromosomal condensation, cell differentiation and possibly in malignant transformation (Hayaishi and Ueda, 1982; Althaus et al., 1985; Shall, 1988). It has been demonstrated in a number of cases that there is an obligatory requirement for endogenous poly(ADP-ribosyl)ating activity during cell differentiation. Similar association between poly(ADP-ribosyl)ation and cellular transformation has also been observed. Nuclei isolated from SV40 transformed cells show two- to ten-fold higher poly(ADP-ribose) synthetase activity over untransformed cells (Miwa et al., 1977). The enzyme activity is significantly higher in leukemic cells than in normal leukocytes (Ueda and Hayaishi, 1982). Furthermore, hormones that influence DNA synthesis and cell replication affect poly(ADP-ribose) synthetase activity (Muira et al., 1972; Kitamura et al., 1979).

Novel ADP-ribosylations in transformed cells

In vitro ADP-ribosylation of a 39 KDa protein by an endogenous activity in saponin-permeabilized human platelets incubated with [^{32}P]NAD was reported by Banga et al. (1988). This protein is different from the 41 kDa pertussis toxin substrate (unpublished observations) and its ADP-ribosylation is enhanced by increasing concentrations of potassium phosphate (Molina Y Vedia et al., 1988), sodium nitroprusside (Brune and Lapetina, 1989) and some platelet agonists[2]. However, the physiological function of this ADP-ribosylation reaction is not known. We extended these studies to several other normal cells and tissues including human and mouse T cells, human B cells, cells from fetal rat calvarea and a variety tissues from adult rat. In all these cases, we found ADP-ribosylation of a 39 kDa protein when cell/tissue homogenates were incubated with [^{32}P]NAD in vitro (not shown).

In striking contrast to the normal leukocytes and lymphocytes, endogenous ADP-ribosylation was observed in several polypeptides in homogenates of HL-60 cells (Fig. 1). The HL-60 cell line was originally derived from peripheral blood leukocytes of a patient with acute promyelocytic

[2] Banga and Feinstein, unpublished observations.

leukemia, and consists predominantly of neutrophilic promyelocytes that produce subcutaneous myeloid tumors in nude mice. The expression of multiple ADP-ribosylations in HL-60 cells, but not in any of the normal blood cells studied, led us to hypothesize that the endogenous ADP-ribosylation observed in cell homogenates was closely related to the state of cellular transformation or differentiation. Further evidence for this hypothesis has been obtained in tumors from two other cell types. Endogenous ADP-ribosylations of membrane proteins in vitro were found in GH_3 (rat pituitary tumor) and ROS 17/2.8 and ROS 25/1 (rat osteosarcoma) cell homogenates. The profile of [^{32}P]-labelled ADP-ribosylated proteins on SDS-PAGE of GH_3 and ROS 17/2.8 cell homogenates are compared with that of HL-60 cells in figure 1. Some degree of variability in the extent of ADP-ribosylation of proteins exists, but the molecular weights of all the proteins modified in these cells are essentially similar.

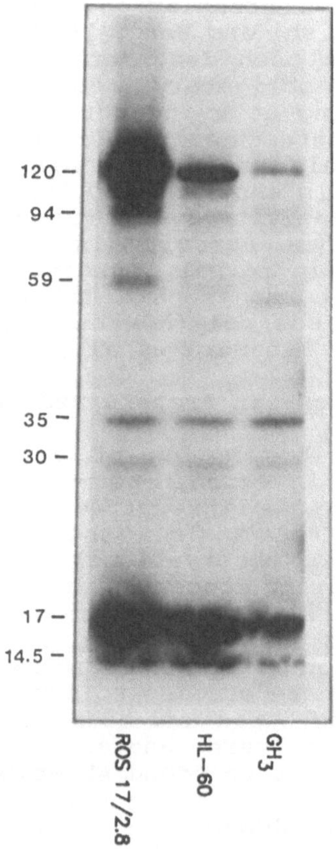

Fig. 1. ADP-ribosylation of multiple proteins in vitro by endogenous enzymatic activity in homogenates of ROS 17/2.8, HL-60 and GH_3 cells. Cell homogenates were incubated with [^{32}P]NAD and the TCA-precipitated material was analyzed by SDS-polyacrylamide electrophoresis on a 5-15% gradient gel. The ADP-ribosylated bands were detected by autoradiography.

Viral Transformation of Normal Cells

In order to study the relationship between cellular transformation and multiple ADP-ribosylations, we shifted our attention from spontaneously transformed cells to normal cells that can be virally transformed in vitro. Cell homogenates from fetal rat bone (calvarea) cells transformed by polyoma virus[3] exhibited multiple ADP-ribosylations of the type seen in spontaneous tumor cells. As pointed out earlier, cell homogenates from normal rat calvarea did not express this reaction in vitro. Similarly, transformation of human B lymphocytes with Epstein-Barr virus was also associated with the expression of in vitro ADP-ribosylation of multiple proteins (Fig. 2).

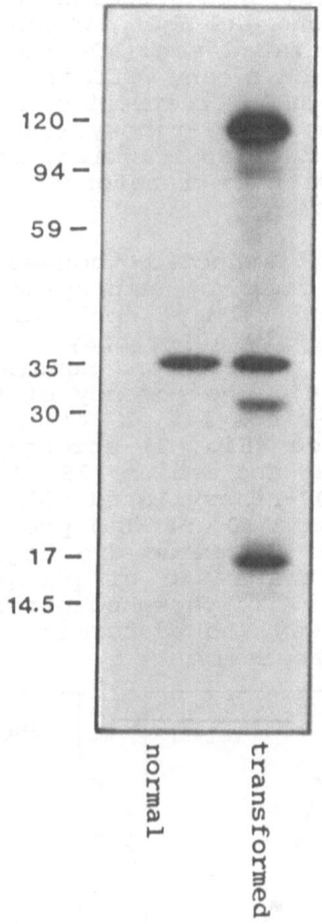

Fig. 2. In vitro ADP-ribosylation of proteins in normal and Epstein-Barr virus-transformed human B lymphocytes.

[3] Kindly provided by Dr. R. Majeska, Mount Sinai Hospital, New York

Chemical Distinction between Mono- and Poly(ADP-ribosyl)ation

Poly(ADP-ribosyl)ation being an o-glycosidic modification is susceptible to hydrolysis by neutral hydroxylamine (NH_2OH) treatment and can usually be readily differentiated from N-glycosidic mono(ADP-ribosyl)ation which is NH_2OH-resistant. However, it should be pointed out that ADP-ribosylated arginine residues show some degree of sensitivity towards NH_2OH treatment (Tanuma et al., 1988), and 10-30% of poly(ADP-ribosyl)ated proteins may not be sensitive to such treatment (Wielckens et al., 1981). Whether this represents polymerized ADP-ribose residues linked by a different bond or an experimental artifact arising due to insolubility of very long chain and branched structures of poly(ADP-ribose) (Miwa et al., 1979), is not known. In HL-60 cells, the [^{32}P] label incorporated into proteins upon incubation of homogenates with [^{32}P]NAD was completely removed by treatment with snake venom phosphodiesterase, thereby confirming that the proteins were ADP-ribosylated (Fensdrick and Iglewski, 1989). A significant fraction of ADP-ribosylated proteins was found to be sensitive to NH_2OH treatment, whereas $HgCl_2$, which removes the label from pertussis toxin-meditated mono(ADP-ribosyl)ation of cysteine residues, did not have any effect in HL-60 cells. Their sensitivity towards NH_2OH treatment combined with the fact that most of the endogenous ADP-ribosylating activity appeared to be localized in the nuclear fraction, led us to speculate that the endogenous activity in transformed cells may be a poly(ADP-ribosyl)ation reaction.

An additional distinction between poly- and mono(ADP-ribosyl)ation was made by employing benzamide, a potent inhibitor of poly(ADP-ribosyl) polymerase that can virtually completely inhibit poly(ADP-ribosyl)ation at concentrations that have minimal effect on mono(ADP-ribosyl)transferase (Rankin et al., 1989). The potency of benzamide for in vitro inhibition of ADP-ribosylation of HL-60 proteins in the presence of benzamide (Fig. 3) strongly suggest that atleast the low (17,14.5 kDa) and medium (39, 30 kDa) molecular weight proteins are poly(ADP-ribosyl)ated (IC_{50} for benzamide = 40 uM). The results with high (120, 94 kDa) proteins are not conclusive since the IC_{50}'s for benzamide are higher; i.e. 1 mM. More direct studies on the ratio of phosphoribosyl-AMP and AMP formed after snake venom phosphodiesterase treatment of each protein are underway to determine the nature of ADP-ribosylation more precisely.

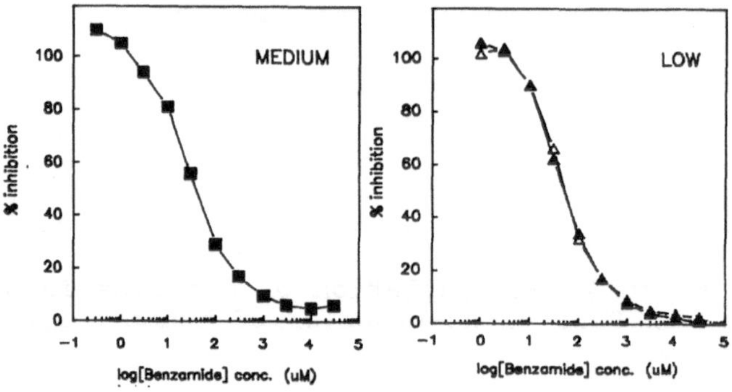

Fig. 3. Effect of benzamide on in vitro inhibition of ADP-ribosylation of proteins in HL-60 cell homogenate.

Chemically Induced Cellular Differentiation of HL-60 Cells Alters ADP-ribosylation

HL-60 cells can be induced to differentiate into cells which have acquired some of the morphological, cytochemical and functional properties of mature granulocytes when cultured in the presence of dimethylsulfoxide (DMSO) or retinoic acid (RA) (Collins et al., 1978; Breitman et al., 1980). In the presence of phorbol esters (PMA) or 1,25-dihydroxyvitamin D_3 (1,25 $(OH)_2$-D_3), they differentiate into monocytes/macrophages (Rovera et al., 1979; Miyaura et al., 1981; Mangelsdorf et al., 1984). Accompanying such differentiation are changes in several biochemical correlates, and over 90% of the cells become capable of producing superoxide anion and carrying out functional phagocytosis. We have found that differentiation of HL-60 cells into granulocytes by treating them with DMSO (1.3%) or into monocytes, induced by PMA (50 nM) treatment, is associated with a striking decrease in the in vitro ADP-ribosylation of these proteins (Fig. 4). It is not known, however, whether these effects are due to changes in the content, or state of ADP-ribosylation of substrate proteins

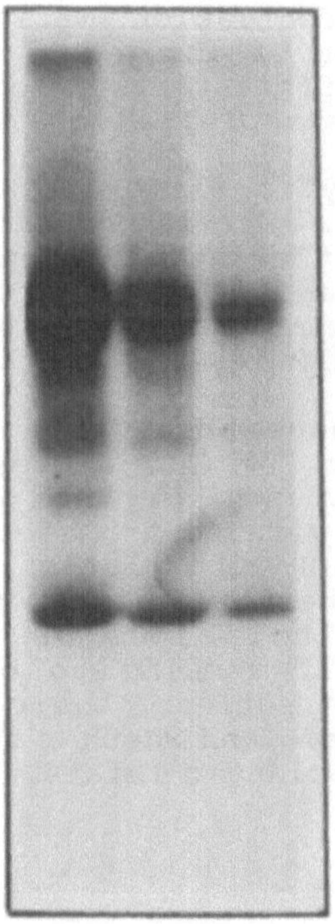

Fig. 4. Changes in the in vitro ADP-ribosylation of proteins in HL-60 cells induced to differentiate either into granulocytes (DMSO) or monocytes (PMA).

and/or the activity or content of ADP-ribosyltransferase itself. Some of the possible mechanisms are considered below:

1. Differentiation may be associated with a decrease in the ADP-ribosyltransferase/polymerase activity in HL-60 cells. Such a decrease in poly(ADP-ribosyl)ation has been reported during differentiation of several cell types (Ueda and Hayaishi, 1985), including a preliminary report on HL-60 cells (Kanai et al., 1982). However, an increase in poly(ADP-ribosyl) polymerase upon differentiation of murine erythroleukemia cells (Rastle and Swetly, 1978), and an early decrease followed by later increase during differentiation of mouse preadipocytes (Pekala et al., 1981) have also been reported.

2. Treatment with differentiating agents may enhance endogenous ADP-ribosylation of target proteins that are involved in the biochemical changes that underlie differentiation. The turnover of these ADP-ribosylated sites may be very slow to maintain a differentiated phenotype. Thus these proteins would not incorporate any additional [^{32}P] label from [^{32}P]NAD when assayed in vitro for ADP-ribosylation subsequent to differentiation. An eight-fold increase in the amount of endogenous poly(ADP-ribose) in HL-60 cells treated with DMSO for 5 days (Kanai et al., 1982) supports such a possibility.

3. Yet another possibility is that diminished ADP-ribosylation is due to a decrease in the expression of substrate proteins during the course of differentiation. Malignant transformations are mediated by expression of oncogenes (Alonso et al., 1988) and agents that lead to differentiation modulate the expression of these oncogenes. For example, a decline in the expression of c-myc is followed by the induction of c-fos and c-fms transcripts during monocytic differentiation of HL-60 cells (Sariban et al., 1985). It is possible that some ADP-ribosylated proteins, or the enzymes themselves, are products of these oncogenes. The possible association between oncogenes, ADP-ribosylation and cellular transformation or differentiation offers intriguing possibilities for further investigation.

In conclusion, we have observed marked changes in the capacity of proteins to be ADP-ribosylated in vitro by endogenous enzymatic activity in the presence of [^{32}P]NAD. Several tumor cells and virally-transformed cells exhibit a high capacity to undergo ADP-ribosylation, in contrast to their normal counterparts. Furthermore, differentiation of one tumor cell line (HL-60) leads to a loss of these in vitro ADP-ribosylation reactions. It remains to be determined if the changes in tumor cells reflect their state of existing endogenous ADP-ribosylation, and whether the reactions observed are related to the neoplastic state, or their state of differentiation.

REFERENCES

Agarwal, S., Drysdale, B. and Shin, H. S., 1988, Tumor necrosis factor-mediated cytotoxicity involves ADP-ribosylation, J. Immunol., 140:4187.

Aktories, K., Just, I. and Rosenthal, W., 1988, Different types of ADP-ribose protein bonds formed by botulinum C2 toxin, botulinum ADP-ribosyltransferase C3 and pertussis toxin. Biochem. Biophys. Res. Commun., 156:361.

Alonso, T., Morgan, R. O., Marvizon, J. C., Zabrl, H. and Santos, E., 1988, Malignant transformation by ras and other oncogenes produces common alterations in inositol phospholipid signaling pathways, Proc. Natl. Acad. Sci. USA, 85:4271.
Althaus, F. R., Hilz, H. and Shall, S., 1985, "ADP-ribosylation of proteins", Springer Verlag, Berlin.
Banga, H. S., Walker, R. K., Winberry, L. K. and Rittenhouse, S. E., 1988, Platelet adenylate cyclase and phospholipase C are affected differentially by ADP-ribosylation, Biochem. J., 252:297.
Breitman, T. R., Selonick, S. E. and Collins, S. J., 1980, Induction of differentiation of human promyelocytic leukemia cell line HL-60) by retinoic acid, Proc. Natl. Acad. Sci. USA, 77:2936.
Brune, B. and Lapetina, E. G., 1989, Activation of cytosolic ADP-ribosyl transferase by nitric oxide-generating agents, J. Biol. Chem., 264:8455.
Burzio, L. O., Saez, L. and Cornejo, R., 1981, Poly(ADP-ribose)synthetase activity in rat testis mitochondria. Biochem. Biophys. Res. Commun., 103:369.
Collins, S. J., Ruscetti, F. W., Gallagher, R. E. and Gallo, R. C., 1978, Terminal differentiation of human promyelocytic leukemia cells induced by dimethyl sulfoxide and other polar compounds, Proc. Natl. Acad. Sci. USA, 75:2458.
Duncan, M. R., Rankin, P. R., King, R. L., Jacobson, M. K. and Dell'Orco, R. T., 1988, Stimulation of mono(ADP-ribosyl)-ation by reduced extracellular calcium levels in human fibroblasts, J. Cell.Physiol., 134:161.
Fendrick, J. L. and Iglewski, W. J., 1989, Endogenous ADP-ribosylation of elongation factor-2 in polyoma virus-transformed baby hamster kidney cells, Proc. Natl. Acad. Sci. USA, 86:554.
Hayaishi, O. and Ueda, K., 1982, "ADP-ribosylation reactions: biology and medicine", Academic Press, New York.
Hilz, H., 1981, ADP-ribosylation of proteins - a multifunctional process, Hoppe-Seyler's Z. Physiol. Chem., 362:1415.
Hisa, J. A., Tsai, S. C., Adamik, R., Yost, D. A., Hewlett, E. L. and Moss, J., 1985, Amino acid-specific ADP-ribosylation. Sensitivity to hydroxylamine of [cysteine(ADP-ribose)]protein and [arginine(ADP-ribose)]protein linkages, J. Biol. Chem., 260:16187
Iglewski,W.J., Lee,H. and Muller,P., 1984, ADP-ribosyl transferase from beef liver which ADP-ribosylates elongation factor-2, FEBS Lett., 173:113.
Kanai, M., Miwa, M., Kondo, T., Tanaka, Y., Nakayasu, M. and Sugimura, T., 1982, Involvement of poly(ADP-ribose) metabolism in induction of differentiation of HL-60 promyelocytic leukemia cells, Biochem. Biophys. Res. Commun., 105:404.
Katada, T. and Ui, M., 1982, Direct modification of the membrane adenylate cyclase system by islet-activating protein due to ADP-ribosylation of a membrane protein, Proc. Natl. Acad. Sci. USA, 79:3129.
Kitamura, A., Tanigawa, Y., Yamamoto, T., Kawamura, M., Doi, S. and Shimoyama, Y., 1979, Glucocorticoid induced reduction of poly(ADP-ribose)synthetase in nuclei from chick embryo liver, Biochem. Biophys. Res. Commun., 87:725.

Kun, E., Zimber, P. H., Chang, A. Y., Puschendorf, B. and Grunicke, H., 1975, Macromolecular enzymatic product of NAD+ in liver mitochondria, Proc. Natl. Acad. Sci. USA, 72:1436.
Mangelsdorf, D. J., Koeffler, P. H., Donaldson, C. A., Pike, J. W. and Jaussler, M. R., 1984, 1,25-Dihydroxyvitamin D3-induced differentiation in a human promyelocytic leukemia cell line (HL-60): receptor-mediated maturation to macrophage-like cells, J. Cell Biol., 98:391.
Miwa, M., Oda, K., Segawa, K., Tanaka, M., Irie, S., Yamaguchi, N., Kuchino, T., Shiroki, K., Shimojo, H., Sakura, H., Matsushima, T. and Sugimura, T., 1977, Cell density dependent increase in chromatin-associated ADP-ribosyl transferase activity in simian virus 40-transformed cells, Arch. Biochem. Biophys., 181:313-321.
Miwa, M., Saikawa, N., Yamaizumi, Z., Nishimura, S. and Sugimura, T., 1979, Structure of poly(adenosine diphosphate ribose): Identification of 2'-[1"-ribosyl-2"-(or 3"-)(1"'-ribosyl)]adenosine-5',5",5"'-tris(phosphate) as a branch linkage, Proc. Natl. Acad. Sci. USA, 76:595.
Miyaura, C., Abe, E., Kuribayashi, T., Tanaka, H., Konno, K., Nishii, Y. and Suda, T., 1981, 1,25-Dihydroxyvitamin D3 induces differentiation of human myeloid leukemia cells, Biochem. Biophys. Res. Commun., 102:937.
Molina Y Vedia, L., Nolan, R. D. and Lapetina, E. G., 1988, Endogenous ADP-ribosylation in human platelets, Biochem.Biophys. Res. Commun., 157:1323.
Moss, J. and Vaughan, M., 1978, Isolation of an avian erythrocyte protein possessing ADP-ribosyltransferase activity and capable of activating adenylate cyclase, Proc.Natl. Acad. Sci. USA, 75:3621.
Moss, J., Stanley, S. J. and Watkins, P. A., 1980, Isolation and properties of an NAD- and guanidine-dependent ADP-ribosyl transferase from turkey erythrocytes, J. Biol. Chem., 255:5838.
Muira, S., Burzio, L. and Koide, S. S., 1972, Studies on deoxyribosenucleic acid synthesis in uteri of immature mouse, Hormone Metab. Res., 4:273.
Nishizuka, Y., Ueda, K., Honjo, T. and Hayaishi, O., 1968, Enzymatic adenosine diphosphate ribosylation of histone and polyadenosine diphosphate ribose synthesis in rat liver nuclei, J. Biol. Chem., 243:3765.
Ogata, N., Ueda, K., Kagamiyama, H. and Hayaishi, O., 1980, ADP-ribosylation of histone H1. Identification of glutamic acid residues 2,14 and the carboxy terminal lysine residues as modification sites, J. Biol. Chem., 255:7616.
Pekala, P. H., Lane, M. D., Watkins, P. A. and Moss, J., 1981, On the mechanism of preadipocyte differentiation. Masking of poly(ADP-ribose)synthetase activity during differentiation of 3T3-L1 preadipocytes, J. Biol. Chem., 256:4871.
Rankin, P. W., Jacobson, E. L., Benjamin, R. C., Moss, J. and Jacobson, M. K., 1989, Quantitative studies of inhibitors of ADP-ribosylation in vitro and in vivo, J. Biol. Chem., 264:4312.
Rastle, E. and Swetly, P., 1978, Expression of poly(adenosine diphosphate ribose)polymerase activity in erythroleukemic mouse cells during cell cycle and erythropoeitic differentiation, J. Biol. Chem., 253:4333.
Roberts, J. H., Stark, P., Giri, C. P. and Smulson, M., 1975, Cytoplasmic poly(ADP-ribose)polymerase during the HeLa

cell cycle, Arch. Biochem. Biophys., 171:305.
Rovera,G., Santoli,D. and Damski,C. (1979) Human promyelocytic leukemia cells in culture differentiate into macrophage-like cells when treated with a phorbol diester, Proc. Natl. Acad. Sci. USA, 76:2779.
Sariban, E., Mitchell, T. and Kufe, D., 1985, Expression of the c-fms pro-oncogene during human monocytic differentiation, Nature (Lond.), 316:64.
Sayhan, O., Ozdemirli, M., Nurten, R. and Bermek, E., 1986, On the nature of cellular ADP-ribosyltransferase from rat liver specific for elongation factor 2, Biochem. Biophys. Res. Commun., 139:1210.
Shall, S., 1988, ADP-ribosylation of proteins: A ubiquitous cellular control mechanism, Adv. Exp. Med. Biol., 231:597.
Tanigawa, Y., Tsuchiya, M., Imai, Y. and Shimoyama, M., 1984, ADP-ribosyltransferase from hen liver nuclei, J. Biol. Chem., 259:2022.
Tanuma, S., Johnson, L. D. and Johnson, G. S., 1983, ADP-ribosylation of chromosomal proteins and mouse mammary tumor virus gene expression, J. Biol. Chem., 258:15371.
Tanuma, S., Kawashima, K. and Endo, H., 1988, Eukaryotic mono(ADP-ribosyl)transferase that ADP-ribosylates GTP-binding regulatory Gi protein, J. Biol. Chem.,263:5485.
Ueda, K. and Hayaishi, O., 1985, ADP-ribosylation, Ann. Rev. Biochem., 54:73.
Ueda, K. and Hayaishi, O., 1982, Leukemia and cancer, in: "ADP-ribosylation reactions: biology and medicine," O. Hayaishi and K. Ueda, eds., Academic Press, New York.
Van Ness, B. G., Howard, J. B. and Bodley, J. W., 1980, ADP-ribosylation of elongation factor 2 by diphtheria toxin, J. Biol. Chem., 255:10710.
Vaughan, M. and Moss, J., 1981, Mono(ADP-ribosyl)transferases and their effects on cellular metabolism, Curr. Top. Cell. Regul., 20:205.
Vitti, P., De Wolf, M. J. S., Acquaviva, A. M., Epstein, M. and Kohn, L. D., 1982, Thyrotropin stimulation of the ADP-ribosyltransferase activity of bovine thyroid membranes, Proc. Natl. Acad. Sci. USA, 79:1525-1529.
Wielckens, K., Bredehorst, R., Adamietz, P. and Hilz, H., 1981, Protein-bound polymeric and monomeric ADP-ribose residues in hepatic tissues, Eur. J. Biochem., 117:69.
Yost, D. A. and Moss, J., 1983, Amino acid specific ADP-ribosylation. Evidence for two distinct NAD-arginine ADP-ribosyltransferases in turkey erythrocytes, J. Biol. Chem., 258:4926.

INOSITOL 1,4,5-TRISPHOSPHATE AND INOSITOL 1,3,4,5-TETRAKISPHOSPHATE: TWO SECOND MESSENGERS IN A DUET OF CALCIUM REGULATION

Robin F. Irvine

AFRC Institute of Animal Physiology and Genetics Research
Babraham, Cambridge CB2 4AT, U.K.

INTRODUCTION

In the "dual second messenger" hypothesis incorporating diacylglycerol and inositol(1,4,5)trisphosphate (Berridge, 1984), it is axiomatic that Ins(1,4,5)$\underline{P}_3$ is the link between inositides and Ca^{2+} (Berridge & Irvine, 1984). However, it is now emerging that this hypothesis may be incomplete, and that there is a third second messenger derived from inositides whose presence was not even remotely suspected until recently. The fact that its very existence was unknown, and the rather enigmatic nature of its function (discussed in some detail below), have together delayed for some years its entry into the picture. But now, as the mist clears, it is emerging that Ins(1,4,5)$\underline{P}_3$ is only one half of the inositol phosphate messenger story. It is on the other half, Inositol 1,3,4,5-tetrakisphosphate, that this short review will focus.

2. DISCOVERY AND METABOLISM OF INS(1,3,4,5)$\underline{P}_4$

Ins(1,3,4,5)$\underline{P}_4$ emerged backwards as it were, in that its breakdown product, Ins(1,3,4)$\underline{P}_3$, was discovered first, at exactly the same time as Ins(1,4,5)$\underline{P}_3$'s messenger role was becoming a firm proposition (Irvine et al. 1984). The slower rate of production of Ins(1,3,4)$\underline{P}_3$ (as opposed to the 1,4,5 isomer) in stimulated cells (e.g. Irvine et al., 1985a) suggested that it might be a breakdown product, and so it proved to be with the discovery of Ins(1,3,4,5)$\underline{P}_4$ and the demonstration that the latter could act as a precursor for Ins(1,3,4)$\underline{P}_3$ (Batty et al., 1985). The structure of Ins(1,3,4,5)$\underline{P}_4$ with a 1,4,5 grouping contained within it, suggested that its route of synthesis must be by a 3-phosphorylation of either PtdIns(4,5)$\underline{P}_2$ or Ins(1,4,5)$\underline{P}_3$, and the latter proved to be so when an active and specific Ins(1,4,5)$\underline{P}_3$-3-kinase activity was discovered (Irvine et al., 1986). The significance of the 1,3,4,5 structure was not lost on others, and shortly afterwards Ins$\underline{P}_3$ kinase activity was found in a large variety of tissues (see Irvine et al. 1988 for references). Ins(1,4,5)$\underline{P}_3$-3-kinase is in fact found in almost all tissues in the animal kingdom (with a very few notable exceptions discussed below), and also probably exists in an alga, Chlamydomonas, (RFI, A.J. Letcher, D.J. Lander and A. Musgrave in preparation) and yeast (P.T. Hawkins personal communication). It is, in most tissues, Ca^{2+}-calmodulin regulated (see Irvine et al. 1988) but may also be regulated in other ways, as discussed below.

The enzymology of Ins(1,3,4,5)$\underline{P}_4$ metabolism has been discussed at length elsewhere, as have the probable reasons for the complexity of this metabolism (Irvine et al., 1988). In this chapter I wish to focus closely on what we know about Ins(1,3,4,5)$\underline{P}_4$ as a second messenger, and what the physiological significance of this might be.

3. PHYSIOLOGICAL AND PHARMACOLOGICAL EFFECTS OF INS(1,3,4,5)$\underline{P}_4$

Having found where Ins(1,3,4,5)$\underline{P}_4$ came from, it was an obvious next step to make sufficient quantities of it to test for biological activity. This approach, particularly if done by microinjection of high concentrations of inositol phosphates into cells, can lead to artefacts and misleading results, and there have certainly been some of these to add spice to the search for what is going on. The first report (Higashida and Brown, 1986) documented an effect on an ion current of undefined nature in neuroblastoma cells induced by pressure injection of Ins(1,3,4)$\underline{P}_3$ or Ins(1,3,4,5)$\underline{P}_4$. The lack of specificity (both InsP's worked) was rather worrying, and in fact inorganic phosphate will induce the same effect (D.A. Brown personal communication). A similar response, specific for Ins(1,3,4)$\underline{P}_3$, but only induced in about 50% of cells, was also found for another neuroblastoma cell line (Tertoolen et al., 1987); Ins(1,3,4,5)$\underline{P}_4$ had no effect in this assay. The ability of pharmacological concentrations of highly charged inositol phosphates to induce ionic effects is perhaps not too surprising, and at present a physiological function of Ins(1,3,4)$\underline{P}_3$ is unlikely (Irvine et al., 1988; Irvine, 1989a).

The first effect of Ins(1,3,4,5)$\underline{P}_4$ to be documented which occurred at micromolar concentrations (i.e. at levels which might be thought to exist <u>in vivo</u>), and with a clear structural specificity, was the activation of sea urchin eggs (Irvine and Moor, 1986; 1987). Elsewhere we have discussed (a) the problems associated with interpreting the effect of removing Ca^{2+} from sea urchin eggs; we originally interpreted Ins(1,3,4,5)$\underline{P}_4$'s effects as being due to Ca^{2+} entry, but intracellular mobilization could have been the major source of Ca^{2+} (see Irvine et al., 1988; Crossley et al., 1988), and (b) the difficulties encountered with seasonality (Irvine and Moor, 1987; Irvine et al., 1988). Together these problems have made sea urchin eggs unsuitable as a system for detailed study of Ins(1,3,4,5)$\underline{P}_4$'s messenger role, but the basic observation, that Ins(1,3,4,5)$\underline{P}_4$ can raise intracellular Ca^{2+} provided (and only provided) that a Ca^{2+}-mobilizing Ins$\underline{P}_3$ is present, remains the crucial starting point for our understanding of Ins(1,3,4,5)$\underline{P}_4$ as a second messenger.

A similar phenomenon, namely an increase in Ca^{2+} (apparently by Ca^{2+} influx) caused by Ins(1,3,4,5)$\underline{P}_4$, which could be modulated ("primed") by Ins(1,4,5)$\underline{P}_3$ has also been reported by Parker and Miledi (1987) using Xenopus oocytes. This observation has proved difficult for others to reproduce (e.g. Snyder et al., 1988), but there are possible reasons for this discussed below. The most clear, reproduceable, and convincing demonstration of this ability of Ins(1,3,4,5)$\underline{P}_4$ to raise Ca^{2+} (provided that an Ins$\underline{P}_3$ is present), has come from the intracellular perfusion of mouse lacrimal cells by Morris et al. (1987). In these experiments Ca^{2+} is measured by a Ca^{2+}-dependent K^+ efflux, and a remarkable synergism between Ins(1,4,5)$\underline{P}_3$ and Ins$\underline{P}_4$ (of a clarity similar to that seen by Irvine & Moor (1986)) is observed in initiating what is clearly a Ca^{2+} entry phenomenon. The exquisite specificity of this effect (for both the Ins$\underline{P}_3$ and Ins$\underline{P}_4$ parts of the "duet") has now been demonstrated, and the trivial explanation that Ins(1,3,4,5)$\underline{P}_4$ is "protecting" the Ins(1,4,5)$\underline{P}_3$ from hydrolysis has been eliminated (Changya et al., 1989). Thus these data provide, for at least one tissue, a virtually unshakeable demonstration of the ability of Ins(1,3,4,5)$\underline{P}_4$ to induce Ca^{2+} entry in an absolutely Ins(1,4,5,)$\underline{P}_3$- dependent fashion.

The data of Changya et al. (1989) also illustrate another property of Ins(1,3,4,5)$\underline{P}_4$, that it can increase the amount of Ca^{2+} <u>mobilized</u> by Ins(1,4,5)$\underline{P}_3$ i.e. it also has effects also in the absence of

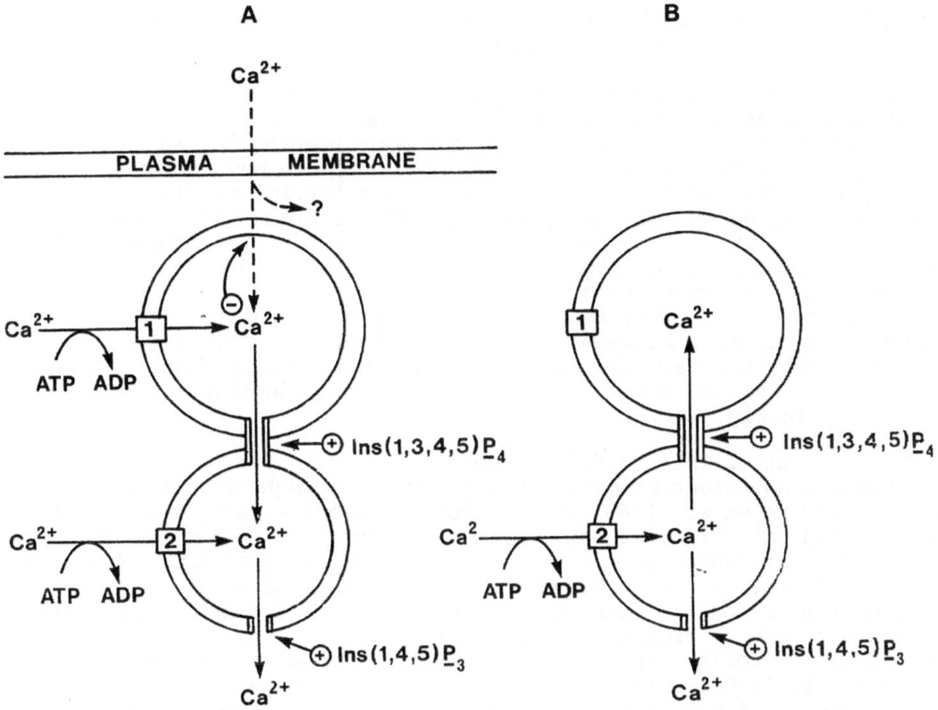

Fig. 1 Proposed control of Ca^{2+} homoeostasis by Ins P_4

This Figure depicts schematically how Ins(1,3,4,5)$\underline{P}_4$ may modulate Ca^{2+} mobilization and entry (A), and sequestration (B). In A, there are two distinct pools of Ca^{2+} with different Ca^{2+} pumps sequestering Ca^{2+} into them (see text for further details). One of these pools receives Ca^{2+} from outside the cell by a mechanism as yet not understood (Putney, 1986; Irvine, 1989a,c), and the other is the target for Ins(1,4,5)$\underline{P}_3$-induced Ca^{2+} release. Ins(1,3,4,5)$\underline{P}_4$ is suggested to link these two lengths together, thus promoting both Ca^{2+} mobilization and entry. In B, one pool is pumping Ca^{2+} either very slowly or not at all, and thus Ins(1,3,4,5)$\underline{P}_4$ can, in the absence of Ins(1,4,5)$\underline{P}_3$ (or in the presence of a sub-optimal level of Ins(1,4,5)$\underline{P}_3$) stimulate Ca^{2+} re-sequestration by increasing the capacity of the pool into which pump 2 sequesters Ca^{2+}. The conditions of B are suggested to exist in the experiments of Hill et al. (1988), and might also pertain under certain conditions in intact cells. See text for further discussion.

extracellular Ca^{2+}. This is not an all-or-nothing effect of the sort seen with Ca^{2+} entry, but it clearly shows that in addition to Ins(1,3,4,5)$\underline{P}_4$ being essential for Ins(1,4,5)$\underline{P}_3$-induced Ca^{2+} entry, it can also "help" Ins(1,4,5)$\underline{P}_3$ access more Ca^{2+} for mobilization. A very similar set of phenomena are evident in Aplysia bag cells (L. Fink, J.A. Conner, R.F. Irvine and L. Kaczmarek, unpublished observations) where Ins(1,3,4,5)$\underline{P}_4$

can "help" $InsP_3$ mobilize intracellular Ca^{2+}, but its effects are apparently more dramatic on Ca^{2+} entry.

The simplest explanation for this absolute requirement for both inositol phosphates in order that Ca^{2+} entry can occur, along with a partial requirements for Ins(1,3,4,5)P_4 for mobilization is, as far as we can see, to invoke the concept that Ins(1,3,4,5)P_4 can mediate or control Ca^{2+} entry into Ins(1,4,5)P_3 - mobilizable pools (Irvine et al. 1988). Originally we suggested this entry would be direct from outside the cell (Irvine and Moor, 1987). However, in the light of the above observations concerning intracellular mobilization, and also taking into consideration the increasing evidence in favour of a "capacitance" pool of Ca^{2+} which directly controls Ca^{2+} entry into the cell (Putney, 1986; see Irvine 1989c for recent references), it seems more likely that the entry of Ca^{2+} into Ins(1,4,5)P_3-mobilizable pools would be from other intracellular pools, and these would include Putney's "capacitance" pool (depicted very schematically in Fig. 1A).

This concept of Ins(1,3,4,5)P_4 causing Ca^{2+} transfer between pools has at present no direct evidence in its favour, though, as discussed elsewhere (Irvine et al. 1988; Irvine 1989a,b) there is some evidence for a GTP-mediated Ca^{2+} transfer between pools in some circumstances (Mullaney et al. 1987). We cannot yet rule out the possibility that Ins(1,4,5) and Ins(1,3,4,5)P_4 interact at a closely associated site on a Ca^{2+} release mechanism from one pool. Some indirect evidence for the pool-linking idea does, however, emerge from the recent data of Hill et al. (1988) obtained in permeabilized cells. In these experiments, the Ca^{2+} uptake was slow, and Hill et al. (1988) observed an apparent re-sequestration of Ca^{2+} caused by Ins(1,3,4,5)P_4 provided that (a) all the Ca^{2+} pools were not full and (b) either no $InsP_3$ or a sub-optimal dose of $InsP_3$, was present (the latter, or a straight addition of Ca^{2+}, were used to induce a new steady state Ca^{2+} level). During the development of the Ca^{2+} multiple-pool concept for Ins(1,3,4,5)P_4 action (Irvine, 1989b) it was necessary to suggest two different types of Ca^{2+} pool, one $InsP_3$ sensitive, the other not, which perhaps pump Ca^{2+} to differing degrees. Thenévod et al. (1989) have provided direct evidence for Ins(1,4,5)P_3-sensitive and -insensitive stores which pump Ca^{2+} by discrete mechanisms, and it should be noted that "calciosomes" (Volpi et al. 1988) may be one of these pools (Thenévod et al. 1989; Irvine et al., 1988; Irvine, 1989b). If, as illustrated in Fig. 1B, in the experiments of Hill et al. (1988) the pool which is accessed by Ins(1,3,4,5)P_4 is no longer pumping Ca^{2+} effectively, then the Ins(1,4,5)P_3-sensitive pool will set an entirely different steady state of Ca^{2+} depending on whether or not it has access to the other (non-pumping) pool or not. In other words, Ins(1,3,4,5)P_4 is effectively increasing the intra-pool volume, thus lowering [Ca^{2+}] inside the pools, and this will decrease Ins(1,4,5)P_3-mediated Ca^{2+} efflux (which must be a function of this [Ca^{2+}]). The result will be what was observed (Hill et al. 1988), namely, a lowering of the steady-state free Ca^{2+} in the "cytosol". Thus, while it appears at first sight that the data of Hill et al. (1988) are contradictory to all those phenomena discussed above in that Ins(1,3,4,5)P_4 seems to decrease rather than increase Ca^{2+}, in fact they can be viewed as providing independent evidence consistent with the concepts developed elsewhere of Ins(1,3,4,5)P_4's physiological function (Irvine et al. 1988; Irvine, 1989b).

Before leaving this particular topic, I should note that it could be that in vivo Ins(1,3,4,5)P_4 could also stimulate Ca^{2+} uptake if no (or suboptimal) Ins(1,4,5)P_3 were present. This will depend on the relative pumping ability of the various Ca^{2+} pools. The circumstances would be similar to Fig. 1B, i.e. if a weak-pumping, but large, Ca^{2+} pool is connected by an Ins(1,3,4,5)P_4-mediated "link", to a small, but strongly-pumping pool, the net rate of Ca^{2+} uptake will be considerably increased after the moment at which the latter pool would have been filled if there had been no link (in other words, the larger pool provides somewhere for the Ca^{2+} to go). An Ins(1,3,4,5)$\underline{P}_4$-stimulated sequestration by such a mechanism may be the explanation behind the observations of Perianin and Snyderman (1989) who noted a stimulated Ca^{2+} sequestration which was mediated through the f-Met-Leu-Phe receptor of netrophils, but which was not caused by the well-known kinase C-mediated stimulation of Ca^{2+} pumps. This concept of Ins$\underline{P}_4$-induced Ca^{2+} sequestration has ramifications in a possible generation of receptor-specific oscillations, as discussed below.

Finally, there is what is apparently an artefact which needs noting in this discussion of experimental effects of Ins(1,3,4,5)$\underline{P}_4$. This inositol phosphate does not generally mobilize Ca^{2+} on its own (e.g. Irvine et al. 1988; Changya et al. 1989). However, in some circumstances it does apparently cause Ca^{2+} mobilization (e.g. Parker and Miledi, 1987; Snyder et al., 1988; Crossley et al., 1988; Cullen et al., 1989; Fink et al., in preparation). Some of these effects might be due to Ins(1,4,5)$\underline{P}_3$ contamination of the Ins(1,3,4,5)$\underline{P}_4$, but in at least one instance this Ins(1,3,4,5)$\underline{P}_4$-induced Ca^{2+} mobilization has been shown to be almost certainly due to 3-dephosphorylation of it to form Ins(1,4,5)$\underline{P}_3$ (Cullen et al., 1989). This phenomenon depended on the method of cell permeabilization (Cullen et al. 1989). In microinjected cells it may also be condition-dependent; the effect of Ins(1,3,4,5)$\underline{P}_4$ on Ca^{2+} mobilization in Aplysia bag cells, (Fink et al., unpublished observations) is partly seasonal, and is increased by bathing the cells in Ca^{2+} free medium. Thus it needs to be stressed that, any instance where Ins(1,3,4,5)$\underline{P}_4$ mobilizes Ca^{2+} on its own should be interpreted with extreme caution.

In conclusion, now that some solid evidence for Ins(1,3,4,5)$\underline{P}_4$ being a second messenger (discussed above) exists, and now that we have a beginning of an understanding as to what it actually does, we can start to examine those instances where Ins(1,3,4,5)$\underline{P}_4$ has little effect experimentally, in a new light, and try to fit them into the overall picture.

4. WHY DOES INS(1,3,4,5)$\underline{P}_4$ NOT HAVE EFFECTS IN ALL EXPERIMENTAL SYSTEMS?

The quickest answer to this question is that in those systems Ins(1,3,4,5)$\underline{P}_4$ has no second messenger role. However, there are a number of objections to this simple answer. These must be considered before going on to discuss some of the experimental difficulties of working with Ins(1,3,4,5)$\underline{P}_4$, which may have led to some of these controversies.

Do all cells make Ins(1,3,4,5)P_4?

Making Ins(1,3,4,5)$\underline{P}_4$ uses up an ATP molecule and requires a specific and active enzyme, regulated by Ca^{2+} and possibly by other mechanisms (below), which is probably found in almost all eukaryotic cells. It is therefore pertinent to ask, why do cells go to the trouble of making it if they don't use it? There have been reports of instances where people have failed to find any detectable Ins(1,3,4,5)$\underline{P}_4$ in stimulated tissues, but many of these are open to other interpretations. It should be remembered that, as with other second messengers (e.g. cyclic AMP), pharmacological doses of agonist are often required for us (the experimenters) to see a rise in the second messenger. There are many instances where full physiological responses of tissues are seen with no detectable rise in cyclic AMP or Ins(1,4,5)$\underline{P}_3$, and only at supra-optimal agonist concentration can biochemists detect the primary signal. Ins(1,4,5)$\underline{P}_3$-3-kinase has an affinity for Ins(1,4,5)$\underline{P}_3$ of a similar order to the potency of Ins(1,4,5)$\underline{P}_3$ in Ca^{2+}-mobilization assays (0.1-0.6 μM in most tissues), and thus Ins(1,3,4,5)$\underline{P}_4$ production may reach a maximum at levels of Ins(1,4,5)$\underline{P}_3$ which are scarcely detectable, and all the "excess" Ins(1,4,5)$\underline{P}_3$ produced by extra agonist doses (which is the Ins(1,4,5)$\underline{P}_3$ we detect) may go down the phosphatase route of metabolism. Certainly there are many tissues (e.g. neutrophils, Limulus photoreceptors) where only

$Ins(1,3,4)P_3$ (the breakdown product of $Ins(1,3,4,5)P_4$) and not $Ins(1,3,4,5)P_4$ itself, can be detected, and only a few workers bother with hplc to look for $Ins(1,3,4)P_3$ which, as far as we know, is virtual proof that $Ins(1,4,5)P_3$-3-kinase exists.

There are only a handful of naturally-occurring cells for which there is really convincing evidence that there is no $Ins(1,4,5)P_3$-3-kinase (for the purpose of this discussion I exclude the mutant cell line of Horstman et al. 1988, in which many aspects of Ca^{2+} homeostasis are atypical). Cells apparently not making $Ins(1,3,4,5)P_4$ are, the slime mould Dictyostelium at the cell-aggregatory stage of its cycle (van Haastert et al. 1989), spermatozoa of ram (R.A.P. Harrison, E. Roldan, D.J. Lander and R.F. Irvine in preparation) and squid retina (e.g. Szuts et al. 1986). In the first of these, the $Ins(1,4,5)P_3$ response is transitory (Van Haastart et al. 1989) as is the Ca^{2+} signal (Abe et al. 1989), neither of which is surprising in view of the fact that the extracellular agonist (cyclic AMP) is oscillating, and a transitory response is perhaps therefore expected; most of what we understand about $Ins(1,3,4,5)P_4$'s actions (above) makes it entirely likely that in such circumstances it would not be required. Mammalian sperm does not, as far as is known, mobilize intracellular Ca^{2+} when the acrosome reaction is induced, and current evidence (see Roldan & Harrison, 1989) is that it is the diacylglycerol which is required when $PtdInsP_2$ breakdown is stimulated (probably initiated by a Ca^{2+} influx through a channel). Again, there would be no requirement for $Ins(1,3,4,5)P_4$ if $Ins(1,4,5)P_3$ is not mobilizing Ca^{2+}, so its omission is not surprising. The apparent absence of $Ins(1,3,4,5)P_4$ or its breakdown products in squid retina is surprising given the rapid production of $Ins(1,3,4)P_3$ in another invertebrate photoreceptor, Limulus (Irvine et al. 1985b), and this is discussed further below.

No doubt a few other tissues will emerge as being deficient in $Ins(1,4,5)P_3$-3-kinase, and as the exceptions to the rule these will help shed light on $Ins(1,3,4,5)P_4$'s function. But currently it is true to say that almost all animal (and maybe plant) cells probably make $Ins(1,3,4,5)P_4$; the fact that a few do not, suggests that the option not to make it exists, and therefore cells that make it, probably use it.

Experiments where $Ins(1,3,4,5)P_4$ has no apparent effect

Considering the possible mode of action of $Ins(1,3,4,5)P_4$ depicted very simply in Fig. 1A and discussed in more detail elsewhere (Irvine et al., 1988; Irvine, 1989b), it could in fact be considered a lucky break that any detectable effect is seen at all in some systems. As pointed out before (Irvine 1989b), if an indirect assay of Ca^{2+} is used (such as egg activation) then if $Ins(1,4,5)P_3$ - accessible stores contain enough Ca^{2+} to initiate the response, no $Ins(1,3,4,5)P_4$ will be required; if they do not contain enough Ca^{2+}, an absolute requirement for $Ins(1,3,4,5)P_4$ will be evident, and such a stark contrast in $Ins(1,3,4,5)P_4$-dependency caused by minor differences in Ca^{2+} distribution could well lie at the heart of the sea urchin egg story (Irvine and Moor, 1986; Crossley et al., 1988; Irvine et al. 1988).

Limitations of Ca^{2+} stores, however, will not explain apparent differences in effects on Ca^{2+} entry (e.g. Morris et al. 1987). Several groups (Llano et al. 1987; Penner et al. 1988; Ohya et al., 1988) have perfused cells with inositol phosphate solutions in a manner not unlike that of Morris et al. (1987), and, in contrast to the latter, have observed a Ca^{2+} entry induced by $Ins(1,4,5)P_3$ with no apparent dependence on $Ins(1,3,4,5)P_4$. The first of these even used the same cells as Morris et al. (albeit a different species, rat versus mouse), and these cells (in both species) produce $Ins(1,3,4,5)P_4$ on stimulation (J.W. Putney Jnr. personal communication; L. Stephens and RFI unpublished). At this stage we cannot rule out the possibility that different cells are "wired" differently; for example, in some cells Ins(1,4,5)$\underline{P}_3$ may gain access directly to the "capacitance" pool of Ca^{2+} (see Fig. 1 and legend) and this would eliminate any requirement for Ins(1,3,4,5)$\underline{P}_4$ when Ins(1,4,5)$\underline{P}_3$ induces Ca^{2+} entry. However, when one considers that this possibility could require mouse and rat lacrimal cells to be different (Llano et al. 1987; Morris et al., 1989), it seems reasonable to suggest instead that even such a striking difference in results is most likely to be due to differences in experimental procedure. More recent data on the mouse lacrimal cells (L. Changya, D.V. Gallacher, R.F. Irvine and O.H. Petersen, unpublished) have begun to throw some light on one possibility. Of all the groups cited in this context, only Morris et al. (1987) did not perfuse the cells with Ins(1,4,5)$\underline{P}_3$ when first making contact between the pipette contents and the cell (i.e. when rupturing the plasma membrane "patch" to go into "whole cell" mode). In the mouse lacrimal cell (Changya et al., unpublished), once the Ca^{2+} entry is established by the essential combination of Ins(1,4,5)$\underline{P}_3$ and Ins(1,3,4,5)$\underline{P}_4$, if the former inositol phosphate is removed and the latter left on its own, the Ca^{2+} entry reverses swiftly; however, if Ins(1,4,5)$\underline{P}_3$ is left, and only Ins(1,3,4,5)$\underline{P}_4$ is removed, the reversal of the stimulated Ca^{2+} entry is extremely slow. In vivo it seems highly unlikely that the reversal would be slow, but clearly it is under these experimental conditions. If, under the different conditions of Llano et al. (1987), Penner et al. (1988) and Ohya et al. (1989), the Ins(1,3,4,5)$\underline{P}_4$ part of the mechanism was even briefly activated, it is entirely possible that it then would not reverse. Experimentally the effects would therefore be Ins(1,3,4,5)$\underline{P}_4$-insensitive. A similar possibility was earlier suggested (Irvine, 1987) to account for Kuno and Gardner's (1987) Ins(1,4,5)$\underline{P}_3$-induced Ca^{2+} "current" in lymphocytes.

Another quirk of Ins(1,3,4,5)$\underline{P}_4$'s actions is revealed in the experiments of Hill et al. (1988). Ins(1,3,4,5)$\underline{P}_4$ only re-sets the steady state of Ca^{2+} which has been raised by Ins(2,4,5)$\underline{P}_3$ (used there as a non-metabolisable Ins$\underline{P}_3$), if it is added before the Ins(2,4,5)$\underline{P}_3$. If Ins(1,3,4,5)$\underline{P}_4$ is added after the Ins(2,4,5)$\underline{P}_3$, it has no effect. This is not due to the inability of Ins(1,3,4,5)$\underline{P}_4$ to exert its effect at raised Ca^{2+} concentrations, because it is able to re-set the Ca^{2+} steady state when the latter is induced by addition of Ca^{2+} alone. The most obvious and reasonable interpretation of this result is that if the Ins$\underline{P}_3$-accessible Ca^{2+} store is partly depleted, Ins(1,3,4,5)$\underline{P}_4$ can no longer work. Clearly that is not so in the experiments of Morris et al. (1987) where Ins(1,3,4,5)$\underline{P}_4$ can be added on top of high levels of Ins(1,4,5)$\underline{P}_3$ and can exert its effect, and so once again it seems that Ins(1,3,4,5)$\underline{P}_4$'s actions are refractory in different ways depending on the experimental system. Phenomena of the sort of those discussed here might contribute to the difference noted above betwen Parker and Miledi (1987) who found Ins(1,3,4,5)$\underline{P}_4$ could affect Ca^{2+} entry in Xenopus oocytes, and Snyder et al. (1988) who found it could not. The latter workers injected heroic concentrations of Ins(2,4,5,)$\underline{P}_3$ into the eggs, and the discussions above suggest that loading systems with high doses of an Ins$\underline{P}_3$ may, unless one is very lucky, obscure Ins(1,3,4,5)$\underline{P}_4$'s contribution.

Another probable contributory factor to all these systems presently under discussion, is the danger of producing some Ins(1,3,4,5)$\underline{P}_4$ artefactually even when using an Ins$\underline{P}_3$ such as Ins(2,4,5)$\underline{P}_3$ which is not phosphorylated by the kinase (Irvine et al. 1988; Irvine, 1989). Resting levels of Ins(1,4,5)$\underline{P}_3$ in tissues may be 0.1 μM or more (see Irvine et al., 1988, for discussion) and a Ca^{2+}-induced activation of Ins(1,4,5)$\underline{P}_3$ kinase may induce enough Ins(1,3,4,5)$\underline{P}_4$ to activate the latter's effects (which might be irreversible as discussed above). It is a significant fact that in the two cellular systems for which a synergism between Ins$\underline{P}_3$ and Ins(1,3,4,5)$\underline{P}_4$ have been most clearly documented, the mouse lacrimal gland (Morris et al. 1987; Changya et al. 1989) and Aplysia bag cells (Fink et al. unpublished), there is no necessity to use Ins(2,4,5)$\underline{P}_3$ as an

$InsP_3$ which will not form Ins(1,3,4,5)$\underline{P}_4$ - i.e. as good a synergism with Ins(1,3,4,5)$\underline{P}_4$ in raising Ca^{2+} can be shown with Ins(1,4,5)$\underline{P}_3$ as with Ins(2,4,5)$\underline{P}_3$. This means that in these systems under these conditions, Ins(1,4,5)$\underline{P}_3$-3-kinase activity is low and the absence of an artefactual generation of Ins(1,3,4,5)$\underline{P}_4$, may be the principal reason why the synergism is so easily demonstrated in these particular experiments.

In conclusion, as stated at the start of this section, these many considerations lead us to be grateful that through various combinations of circumstancees governed not least by luck, a few experimental systems do show experimentally a clear-cut and dramatic requirement for Ins(1,3,4,5)$\underline{P}_4$. It seems reasonable to suggest that the few experimental phenomena which are so clearly Ins(1,3,4,5)$\underline{P}_4$-dependent are not exceptions, except in so far as in those examples we have the right combination of experimental conditions to allow us to see easily what is actually true in almost all tissues. Those data are, I believe, telling us that in most tissues the mobilization and entry of calcium is not mediated by Ins(1,4,5)$\underline{P}_3$ alone, but rather by a duet between Ins(1,4,5)$\underline{P}_3$ and its essential partner, Ins(1,3,4,5)$\underline{P}_4$.

5. WHY TWO MESSENGERS AND NOT ONE?

Indeed, why make life so compicated? An obvious answer is as a possible safety valve (i.e. if Ins(1,4,5)$\underline{P}_3$ did regulate Ca^{2+} mobilization and entry all on its own, it might be better to introduce an extra intervening Ca^{2+}-compartment and second messenger to make the whole thing all the more controllable). A less obvious, but more more interesting answer, is to make the system deliberately complicated so that it is more subtle. If there is only one message, raised or lowered, then the cell cannot do much with it other than raise it or lower it. But for subtleties such as (a) telling one agonist from another even though both are using the same second messengers, (b) distinguishing between acute (e.g. secretion) and chronic (e.g. growth) responses using the same signals, and (c) an almost infinite range of possibilities of interacting with other second messenger systems, it may be necessary to have some complexity. The spatial and temporal aspects of Ca^{2+} signalling (Berridge et al. 1988) are almost certainly examples of this required fine-tuning and subtlety, and it is worth considering that the dual messenger hypothesis of Ca^{2+} regulation espoused here could be a principal means of generating these subtleties.

Ins(1,3,4,5)P_4 and Ca^{2+} oscillations

Of a number of possible mechanisms for the generation of Ca^{2+} oscillations (mostly discussed by Berridge et al., 1988), the Ins(1,3,4,5)$\underline{P}_4$ mechanism (Irvine, 1989a; Yale and Gallacher, 1988) has a number of attractive aspects. As Ins(1,3,4,5)$\underline{P}_4$ recruits Ca^{2+} for Ins(1,4,5)$\underline{P}_3$ to mobilize, and as Ins(1,4,5)$\underline{P}_3$-3-kinase is a Ca^{2+}-stimulated enzyme, then at a low rate of Ins(1,4,5)$\underline{P}_3$ production one can envisage a critical Ca^{2+} concentration at which Ins(1,3,4,5)$\underline{P}_4$ starts to be generated. This will raise the Ca^{2+} more, hence activating the kinase more and an autocatalytic surge of Ca^{2+} will occur. One of the major problems in an oscillation mechanism is, how is the signal terminated? The beauty of Ins(1,4,5)$\underline{P}_3$-3-kinase is that it is also a signal-terminating enzyme - if when fully activated it phosphorylates all the Ins(1,4,5)$\underline{P}_3$ then the Ca^{2+} mobilization will immediately cease, and the Ca^{2+} will begin to be pumped down again.

It is in the pumping down of Ca^{2+} to control levels that another interesting feature of the Ins(1,3,4,5)$\underline{P}_4$-generated oscillation model emerges, for, as discussed in section 4 above, this Ca^{2+} pumping might be accelerated in the presence of Ins(1,3,4,5)$\underline{P}_4$. One of the most intriguing aspects of the Ca^{2+} oscillations phenomenon is that at least in hepatocytes (see Berridge et al., 1988) individual agonists give characteristically different shapes of Ca^{2+} "spikes" differing principally by the speed at which the Ca^{2+} returns to normal. If each agonist generated a spike which terminated in the presence of different amounts of Ins(1,3,4,5)$\underline{P}_4$, then this specificity might be explained. More precisely, such a phenomenon would occur if there were some additional Ins(1,3,4,5)$\underline{P}_3$-3-kinase activity which was agonist-specific. With an agonist which caused some additional Ins(1,4,5)$\underline{P}_3$-3-kinase over and above the Ca^{2+}-dependent activity, the "crisis" of Ins(1,3,4,5)$\underline{P}_4$ generation leading to a Ca spike would occur at a lower Ins(1,4,5)$\underline{P}_3$ concentration than it would with an agonist which did not stimulate any extra Ins(1,3,4,5)$\underline{P}_4$ production; thus for the former agonist, Ins(1,3,4,5)$\underline{P}_4$ levels would actually be lower when the spike peaked, and so the "recovery" of Ca^{2+} levels would be slower (due to slower pumping of Ca^{2+}). It is worth mentioning in passing that differing levels of Ins(1,3,4,5)$\underline{P}_4$-3-phosphatase (see Cullen et al., 1988) in a cell would also lengthen a Ca^{2+} "spike" by generating some Ins(1,4,5,)$\underline{P}_3$ during the removal of Ins(1,3,4,5)$\underline{P}_4$. If Ins(1,3,4,5)$\underline{P}_4$ is involved in any way with the generation of Ca^{2+} oscillations, this might be a possible reason why squid retina do not make it (see above); this is not an oscillatory tissue, because a low dose of "agonist" (one photon) gives a discrete and acute response, and not oscillations.

Receptor-specific regulation of Ins(1,4,5)P_3-3-kinase

What evidence is there that an agonist-specific production of Ins(1,3,4,5)$\underline{P}_3$ might occur, and what might be the mechanism? There are two particularly attractive possibilites to be considered: either (i) there is a direct (via a G-protein?) regulation of a membrane-bound Ins(1,4,5)$\underline{P}_3$-3-kinase by the receptor or (ii) the protein kinase C-mediated regulation of Ins(1,4,5)$\underline{P}_3$-3-kinase suggested by Imboden and Weiss (1987) occurs with a receptor-specificity conferred on it through differing degrees, with different agonists, of diacylglycerol production. This production would have to be independent of inositol phosphates and the most likely route is by phosphatidylcholine hydrolysis.

There are a few indirect pieces of evidence that regulation of Ins(1,4,5)$\underline{P}_3$-3-kinase may occur other than by Ca^{2+}, though they do not distinguish between these two possibilities. There is, for example, the observation that Acetylcholine stimulates Ca^{2+} influx in perfused mouse lacrimal cells, whereas infused Ins(1,4,5)$\underline{P}_3$ does not unless Ins(1,3,4,5)$\underline{P}_4$ is also included (Morris et al., 1987), though these experiments may be complicated by compartmentalization problems (Changya et al. unpublished). The implication is, that only in the presence of acetylcholine is some Ins(1,3,4,5)$\underline{P}_4$ produced from Ins(1,4,5)$\underline{P}_3$. In human umbilical vein endothelial cells, thrombin and histamine give very different inositol phosphate isomer patterns despite a very similar overall rate of production (Wreggett, K.A. and R.F.I. unpublished), suggesting that some regulation of inositol phosphate production might occur post-Ins(1,4,5)$\underline{P}_3$ production. Finally, Tilly et al. (1987) observed a very weak Ins$\underline{P}_3$ increase in A431 cells in response to EGF, which was all Ins(1,3,4)$\underline{P}_3$. This could as well have been due to Ins$\underline{P}_3$ kinase stimulation by the EGF receptor as to a stimulation of PtdIns$\underline{P}_2$ phosphodiesterase. In this context, note that a possible link between a receptor-regulated Ins$\underline{P}_3$ kinase, and PtdIns$\underline{P}_3$ and the production of PtdIns(3)$\underline{P}$, is discussed elsewhere (Irvine, 1989b).

None of these examples (nor two more discussed in the next section) can be taken as strong evidence, but they are suggestive of a possible control of Ins$\underline{P}_3$ kinase separate from its regulation by Ca^{2+}. Is regulation of Ins$\underline{P}_3$-3-kinase the way that a subtlety is introduced into inositol phosphate function for the reasons discussed above? If hydrolysis of phosphatidylcholine provides cells with a means of regulating protein kinase C without inositol phosphates, might it not be that regulation of Ins(1,4,5)$\underline{P}_3$-3-kinase gives cells the chance to regulate calcium via inositol phosphates without altering protein kinase C? Somewhere, in the subtleties of cellular Ca^{2+} regulation, lies the reason(s) for two second messengers being involved rather than one, but at the moment these reasons are not clear.

6. INS(1,3,4,5)$\underline{P}_4$ AND PATHOLOGY

A dual messenger system of Ca^{2+} control will inevitably malfunction, and inappropriate Ins(1,3,4,5)$\underline{P}_4$ production due to over- or under-regulation of Ins(1,4,5)$\underline{P}_3$-3-kinase would be expected to have profound effects on cellular function. As yet there is little to go on, but some suggestions of Ins(1,3,4,5)$\underline{P}_4$ metabolism being altered as a direct or indirect result of pathological change are already appearing.

In HIV-infected cells, a change in inositol phosphate metabolism is evident, and the data suggest an increase in Ins$\underline{P}_3$-kinase in resting cells; higher Ca^{2+} levels and a lack of response to PHA are possibly the result of this (K. Nye and A. Pinching, personal communication). These effects may be entirely secondary, but a long-term change in some aspect of the control of Ins(1,4,5)$\underline{P}_3$-3-kinase is hinted at. A similar pattern of raised Ins$\underline{P}_4$ (as with the HIV-infected cells, the isomer is not known) is evident in src -transformed fibroblasts (J. Garrison, personal communication). In this set of experiments, Ins(1,4,5)$\underline{P}_3$-3-kinase that can be measured in cell homogenates is actually increased in the src transformed cells. Is an increased Ins(1,3,4,5)$\underline{P}_4$ production a consequence or cause of transformed cell growth? It is noteworthy that some phosphoinositidases C have src-like sequences in them - is this an Ins(1,4,5)$\underline{P}_3$ structure recognition sequence?

In neither of these examples has the cause-effect relationship been sorted out, and only in one has an increase in Ins(1,4,5)$\underline{P}_3$-3-kinase actually been documented, so extreme caution must be observed in interpreting these data. Nevertheless, they are intriguing, and the central role of Ins(1,3,4,5)$\underline{P}_4$ in cell function proposed here would certainly predict a pathology associated with chronic changes in its production.

7. CONCLUSIONS

The experimental demonstration of a second messenger role of Ins(1,4,5)$\underline{P}_3$ was comparatively easy, was quickly reproduced by many laboratories, and was easily fitted into a simplistic view of cellular Ca^{2+} homeostasis. For these reasons, Ins(1,4,5)$\underline{P}_3$ achieved the status of a recognized second messenger very quickly (Berridge and Irvine, 1984). Once that was so, there was really no reason to suppose any other compound was involved. Now, however, we are beginning to understand how subtle is cellular Ca^{2+} control, and more and more it makes sense to have two messengers involved, not one. Also, we are beginning to understand why it may often prove difficult (as it has done) to see experimentally the effects of Ins(1,3,4,5)$\underline{P}_4$. Ins(1,4,5)$\underline{P}_3$ causes the opening of a Ca^{2+} channel which may very well be contained in the same protein as the Ins(1,4,5)$\underline{P}_3$-binding site. Thus the actions of Ins(1,4,5)$\underline{P}_3$ are robust and easy to demonstrate. The effect of Ins(1,3,4,5)$\underline{P}_4$ is very much more complex, and is correspondingly more delicate and easily overlooked experimentally. It is my hope that in the next year or so we will begin to appreciate that the messenger role of Ins(1,3,4,5)$\underline{P}_4$ is not to the detriment of the function of Ins(1,4,5)$\underline{P}_3$, but rather is an extension and enhancement of it. If this duet in Ca^{2+} regulation (Irvine et al., 1988) is the true state of affairs in most cells, then it is a marvellous example of how subtle biology can be.

References

Abe, T., Maeda, Y. and Iijima, T., 1988, Transient increase of the intracellular Ca^{2+} concentration during chemotactic signal transduction in Dictyostelium discoideum cells, Differentiation, 39: 90.

Batty, I.R., Nahorski, S.R. and Irvine, R.F., 1985, Rapid formation of inositol 1,3,4,5-tetrakisphosphate following muscarinic receptor stimulation of rat cerebral cortical slices, Biochem. J., 232: 211.

Berridge, M.J., 1984, Inositol trisphosphate and diacylglycerol as second messengers, Biochem. J., 220: 345.

Berridge, M.J. and Irvine, R.F., 1984, Inositol trisphosphate, a novel second messenger in cellular signal transduction, Nature, 212: 849.

Berridge, M.J., Cobbold, P.H. and Cuthbertson, K.S.R., 1988, Spatial and temporal aspects of cell signalling, Phil. Trans. Roy. Soc. Lond. B, 320: 325.

Changya, L., Gallacher, D.V., Irvine, R.F., Potter, B.V.L. and Petersen, O.H., 1989, Inositol 1,3,4,5-tetrakisphosphate is essential for sustained activation of Ca^{2+}-dependent K^{+} current in single internally perfused mouse lacrimal acinar cells, J. Membrane Biol., 109: 85.

Crossley, I., Surann, K. and Whitaker, M., 1988, Activation of sea urchin eggs is independent of external calcium ions, Biochem. J, 252: 257.

Cullen, P.J., Irvine, R.F., Drøbak, B.K. and Dawson, A.P., 1989, Inositol 1,3,4,5-tetrakisphosphate causes release of Ca^{2+} from permeabilized mouse lymphoma L1210 cells by its conversion into inositol 1,4,5-trisphosphate, Biochem. J., 259: 931.

Higashida, H. and Brown, D.A., 1986, Membrane current responses to intracellular injections of inositol 1,3,4,5-tretrakisphosphate and inositol 1,3,4-trisphosphate in NG108-15 hybrid cells, FEBS Letters, 208-283.

Hill, T.D., Dean, N.M. and Boynton, A.L., 1988, Inositol 1,3,4,5-tetrakisphosphate induces Ca^{2+} sequestration in rat liver cells, Science, 242: 1176.

Horstman, D.A., Takemura, H. and Putney, J.W. Jnr., 1988, Formation and metabolism of [^{3}H]inositol phosphates in AR42J pancreatoma cells. Substance P-induced Ca^{2+} mobilization in the apparent absence of inositol 1,4,5 trisphosphate 3-kinase activity, J. Biol. Chem., 263: 15297.

Imboden, J.R. and Pattison, G., 1987, Regulation of inositol 1,4,5-trisphosphate kinase activity after stimulation of human T cell antigen receptor, J. Clin. Invest., 79: 1538.

Irvine, R.F., 1987, Inositol phosphates and calcium entry, Nature, 328: 386.

Irvine, R.F., 1989a, Functions of inositol phosphates, in: "Inositol Lipids in Cell Signalling", R.H. Michell, A.H. Drummond and C.P. Downes eds., Academic Press, London.

Irvine, R.F., 1989b, How do inositol 1,4,5-trisphosphate and inositol 1,3,4,5-tetrakisphosphate regulate intracellular Ca^{2+}? Biochem. Soc. Trans., 17: 6.

Irvine, R.F., 1989c, Inositol polyphosphates as intracellular second messengers, in: "Receptors and second messengers, implications for Drug Development", S.R. Nahorski ed., John Wiley, Chichester, U.K. (in press).

Irvine, R.F. and Moor, R.M., 1986, Micro-injection of inositol 1,3,,5-tetrakisphosphate activates sea urchin eggs by a mechanism dependent on external Ca^{2+}, Biochem. J., 240: 917.

Irvine, R.F. and Moor, R.M., 1987, Inositol (1,3,4,5) tetrakisphosphate-induced activation of sea urchin eggs requires the presence of inositol trisphosphate, Biochem. Biophys. Res. Commun., 146: 284.

Irvine, R.F., Letcher, A.J., Lander, D.J. and Downes, C.P., 1984, Inositol trisphosphates in carbachol-stimulated rat parotid glands, Biochem. J., 223: 237.

Irvine, R.F., Änggård, F.E., Letcher, A.J. and Downes, C.P., 1985a, Metabolism of inositol 1,4,5-trisphosphate and inositol 1,3,4-trisphosphate in rat parotid glands, Biochem. J., 229: 505.

Irvine, R.F., Anderson, R.E., Rubin, L.J. and Brown, J.E., 1985b, Inositol 1,3,4-trisphosphate concentration is changed by illumination of Limulus ventral photoreceptors, Biophys. J. 47: 38a.

Irvine, R.F., Letcher, A.J., Heslop, J.P. and Berridge, M.J., 1986, The inositol tris/tetrakisphosphate pathway-demonstration of inositol (1,4,5)trisphosphate-3-kinase activity in animal tissues, Nature, 320: 631.

Irvine, R.F., Moor, R.M., Pollock, W.K., Smith, P.M. and Wreggett, K.A., 1988, Inositol phosphates: prolifertion, metabolism and function, Phil. Trans. Roy. Soc. Lond B, 320: 281.

Kuno, M. and Gardner, P. Ion channels activated by inositol 1,4,5-trisphosphate in the plasma membrane of human T-lymphocytes, Nature, 326: 301.

Llano, I., Marty, A. and Tanguy, J., 1987, Dependence of intracellular effects of GTPys and inositol trisphosphate on cell membrane potential and on external Ca^{2+} ions, Pflüngers Archiv., 409: 499.

Morris, A.P., Gallacher, D.V., Irvine, R.F. and Petersen, O.H., 1987, Synergism of inositol trisphosphate and tetrakisphosphate in activating Ca^{2+}-dependent K^+ channels, Nature, 330: 653.

Mullaney, J.M., Chueh, S.-H., Ghosh, T.K. and Gill, D.L., 1989, Intracellular calcium uptake activated by GTP: evidence for a possible guanine nucleotide-induced transmembrane conveyance of intracellular calcium, J. Biol. Chem. 262: 13865.

Ohya, Y., Terada, K., Yamaguchi, K., Inoue, R., Okabe, K., Kitamura, K., Hirata, M. and Kuriyama, H., 1988, Effects of inositol phosphates on the membrane activity of smooth muscle cells of the rabbit portal vein, Pflügers Arch., 412: 382.

Parker, I. and Miledi, 1987, Injection of inositol 1,3,4,5-tetrakis-phosphate into Xenopus oocytes generates a chloride current dependent upon intracellular calcium, Proc. Roy. Soc. Lond. B, 232: 59.

Penner, R., Matthews, G. and Neher, E., 1988, Regulation of calcium influx by second messengers in rat mast cells, Nature, 334: 499.

Perianin, A. and Snyderman, R, 1989, Analysis of calcium homeostasis in activated human polymorphonuclear leukocytes. Evidence for two distinct mechanisms for lowering cytosolic calcium, J. Biol. Chem., 264: 1005.

Roldan, E.R.S. and Harrison, R.A.P., 1989, Polyphosphoinositide breakdown and subsequent exocytosis in the Ca^{2+}/ionophore-induced acrosome reaction of mammalian spermatozoa, Biochem. J., 259: 397.

Snyder, P.M., Krause, K-H. and Welsh, M.J., 1988, Inositol trisphosphate isomers but not inositol 1,3,4,5-tetrakisphosphate, induce calcium influx in Xenopus laevis oocytes, J. Biol. Chem., 263: 11048.

Szuts, E.Z., Woods, S.F., Reid, M.J. and Fein, A., 1986, Light stimulates the rapid formation of inositol trisphosphate in squid retinas, Biochem. J., 240: 929.

Tertoolen, L.G., Tilly, B.C., Irvine, R.F. and Moolenaar, W.H., 1987, Electrophysiological responses to bradykinin and microinjected polyphosphates in neuroblastoma cells. Possible role of inositol 1,3,4-trisphospahte in altering membrane potential, FEBS Letters, 214: 365.

Thévenod, F., Dehlinger-Kremer, H., Kemmer, J.P., Christian, A.L., Potter, B.V.L. and Schulz, I., 1989, Characterization of inositol 1,4,5-trisphosphate-sensitive (IsCaP) and -insensitive (IisCaP) non-mitochondrial Ca^{2+} pools in rat pancreatic acinar cells.

Tilly, B.C., van Paridon, P.A., Verlaan, I., Wirtz, K.W.A., de Laat, S.W. and Moolenaar, W.H., 1987, Inositol phosphate metabolism in bradykinin-stimulated human A431 carcinoma cells, Biochem. J., 244: 129.

Van Haastert, P.J.M., de Vries, M.J., Penning, L.C., Roovers, E., Van der Kaay, J., Erneux, C. and Van Lookern Campagne, M.M., 1989, Chemoattractant and guanine 5-[y-thio] trisphosphate induce the accumulation of inositol 1,4,5-trisphospahte in Dictyostelium cells that are labelled with [^{3}H]inositol by electroporation, Biochem. J., 258: 577.

Volpe, P., Krause, K.-H., Hashimoto, S., Zorgato, F., Pozzan, T., Meldolesi, J. and Lew, D.P., 1987, "Calciosome", a cytoplasmic organelle: the inositol 1,4,5-trisphosphate-sensitive Ca^{2+} store of non-muscle cells, Proc. Natn. Acad. Sci. USA, 85: 1091.

Yule, D.I. and Gallacher, D.V., 1988, Oscillations of cytosolic calcium in single pancreatic acinar cells stimulated by acetyl choline, FEBS Letters, 239: 358.

THE ROLE OF ENDOGENOUSLY STIMULATED PROTEIN KINASE C IN REGULATING

Susan E. Rittehouse, Warren G. King, C. Peter Downes*, and Gregory L. Kucera

Department of Biochemistry, University of Vermont College of Medicine, Burlington, VT, USA and *Smith, Kline, and French Research, Ltd., The Frythe, Herts., U.K.

Since the discovery in 1983 that Ins(1,4,5)P_3 can cause the mobilization of intracellular stores of Ca^{2+} (Streb, et al., 1983), the factors that regulate the intracellular accumulation of this inositol polyphosphate have been important subjects for study in any attempt to understand Ca^{2+}-dependent cellular activation. As is the case for any other transient compound, accumulation of Ins(1,4,5)P_3 is controlled directly by its generation and metabolism. The generation of Ins(1,4,5)P_3 occurs following the activation of cellular phosphoinositidase C (PIC), a receptor-dependent and, in most cell systems studied thus far, GTP-binding protein-dependent membrane event. Obviously, factors that perturb any of these stages will modulate the rate at which the substrate, PtdIns(4,5)P_2, is hydrolyzed by PIC. In addition, the availability of PtdIns(4,5)P_2 is also an important consideration. Once formed, Ins(1,4,5)P_3 can be acted upon by at least two enzymes, the most active and best characterized of which are 5-phosphomonoesterase (5-PME), yielding Ins(1,4)P_2 (Downes, et al., 1982; Storey, et al., 1984), and 3-kinase, yielding Ins(1,3,4,5)P_4 (Irvine, et al., 1986). Ins(1,3,4,5)P_4 is also a substrate for 5-PME, becoming Ins(1,3,4)P_3 (Batty, et al., 1985). Of the isolated myoinositol bis, tris, and tetrakisphosphates, Ins(1,4,5)P_3 has been shown to be by far the most potent facilitator of intracellular Ca^{2+} release, with Ins(1,4)P_2 and Ins(1,3,4)P_3 displaying negligible activity. The function of Ins(1,3,4,5)P_4 seems likely to be more than the removal of Ins(1,4,5)P_3, however, its role has yet to be elucidated fully.

We have been studying the regulation of inositol polyphosphate accumulation in human platelets. Platelets undergo numerous rapid metabolic and morphologic changes, many of which are dependent upon a rise in cytoplasmic Ca^{2+} concentrations, when challenged with a potent agonist such as α-thrombin. Ins(1,4,5)P_3, at μM levels, can release Ca^{2+} from sequestered stores in permeabilized platelets (Brass and Joseph, 1985) and vesicle fractions (O'Rourke, et al., 1985). Platelet InsP_3 concentrations increase rapidly and dramatically in response to thrombin, as shown in Table 1. We have measured total platelet mass of InsP_3 following the separation, by anion exchange chromatography, of InsP_3 from other inositol phosphates, and gas chromatographic analysis as described previously (Rittenhouse and Sasson, 1985; Rittenhouse, 1987). Ins(1,4,5)P_3 was quantitated using an enzyme-based assay developed in our laboratory (Tarver, et al., 1987). One notes several things after such analysis. Firstly, the non-cyclic species predominate in the first 10s. Secondly,

Table 1. Accumulation of Inositol Trisphosphates in Thrombin-Stimulated Platelets

Incubation time(s)	$Ins(1,2\ cyc\ 4,5)P_3$	pmol per 10^9 cells $Ins(1,4,5)P_3$	Total $InsP_3$
0	0 ± 1	1 ± 1	20 ± 10
10	0 ± 1	10 ± 1	220 ± 10
30	10 ± 2	7 ± 1	260 ± 20

assuming an intracellular platelet volume of <10 μl per 10^9 platelets, μM concentrations of $Ins(1,4,5)P_3$, enough to mobilize Ca^{2+}, are attained within 10s of exposure to thrombin. Finally, most of the $InsP_3$ in stimulated and unstimulated platelets is not $Ins(1,4,5)P_3$. When platelets are radiolabeled with either [^{3}H]myoinositol or [^{32}P]P_i, and the labeled inositol phosphates are resolved by HPLC, most of the labeled $InsP_3$ co-migrates with $Ins(1,3,4)P_3$ standards. Since co-migration on HPLC is not definitive of structure, we have subjected this material to exhaustive digestion with $NaIO_4/BH_4^-$ or 4-phosphomonoesterase, and concluded from analysis of the resulting products that better than 90% of it is D-$Ins(1,3,4)P_3$ (King, Downes, Prestwich, and Rittenhouse: manuscript in preparation). In view of the presence of PtdIns-3 kinase activity capable of phosphorylating PtdIns4P to $PtdIns(3,4)P_2$ in some cells (Auger, et al., 1989), we considered it important to analyze platelet $PtdInsP_2$. Although the susceptibility of $PtdIns(3,4)P_2$ to PIC has not been characterized, it was necessary to determine whether we could rule it out as a possible precursor for $Ins(1,3,4)P_3$. Therefore, [^{32}P]-labeled platelet lipids derived from control or thrombin-exposed cells were resolved into PtdIns, PtdInsP and $PtdInsP_2$ fractions by thin layer chromatography (Traynor-Kaplan, et al., 1988) and the $PtdInsP_2$ fraction was deacylated with methylamine reagent (Whitman, et al., 1988). Glycerophosphoinositol bisphosphate in the digests were resolved by HPLC on Partisphere SAX columns eluted with $(NH_4)_2HPO_4$, and the products compared with known standards as described by Auger, et al., 1989. The results are shown in Figure 1. It is seen that more than 98% of the $PtdInsP_2$ digest co-migrates with glycerophosphoinositol$(4,5)P_2$, rather than glycerophosphoinositol$(3,4)P_2$. This distribution is unaltered by exposure of platelets to thrombin. Thus, it is highly unlikely that $PtdIns(3,4)P_2$ is the precursor for the $Ins(1,3,4)P_3$ that accumulates in stimulated human platelets.

Consistent with what has been described for other cell systems, platelets contain a kinase that phosphorylates $Ins(1,4,5)P_3$. The product, digested with 5-PME, co-migrates on HPLC with $Ins(1,3,4)P_3$. We therefore consider it most probable that the kinase product is $Ins(1,3,4,5)P_4$. Saponin-permeabilized platelets incubated with $Ins(1,4,5)P_3$ form $Ins(1,4)P_2$, $Ins(1,3,4,5)P_4$, and $Ins(1,3,4)P_3$. We have investigated whether 3-kinase activity is subject to regulation by protein kinase C (PKC). Platelets contain PKC capable of phosphorylating numerous substrates in intact platelets exposed to β-phorbol diester tumor promoters or diacylglycerol, and laboratories working with other cell types have reported that PKC enhances the activity of 3-kinase (Imboden and Pattison, 1987). If platelets are preincubated for one minute with β-phorbol-12,13 dibutyrate, permeabilized with saponin, and exposed to [^{3}H]$Ins(1,4,5)P_3$ in the presence of 2,3-diphosphoglycerate (to inhibit 5-PME), higher amounts of [^{3}H]$Ins(1,3,4,5)P_4$ are formed than in Me_2SO (0.1%) controls. Kinetically, this is traceable to a two-fold increase in V_{max}, with no alteration in apparent K_m (King and Rittenhouse, 1989). Such stimulation could not be achieved by substituting α-phorbol dibutyrate for the PKC-activating compound. Taking another approach, we found that incubation

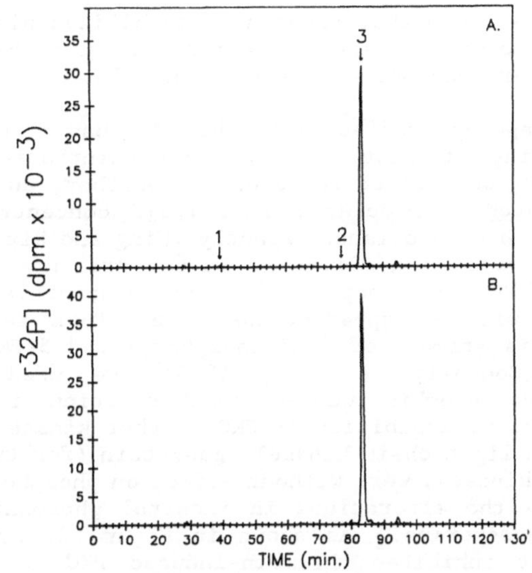

Fig. 1. HPLC analysis of glycerophosphoinositide bisphosphate from [^{32}P]-labeled human platelets. A. Buffer-exposed platelets. B. Thrombin-exposed platelets. Migration of [^{3}H] standards indicated as 1 = glycerophosphoinositol-4P, 2=glycerophosphoinositol-3,4 P_2, 3=glycerophosphoinositol-4,5 P_2.

of platelets with thrombin enhanced 3-kinase activity, whereas incubation with the PKC inhibitor staurosporine inhibited this increase. Apparently 3-kinase is subject to regulation by PKC, stimulated either endogenously or by phorbol ester. This implies that the stimulated platelet, with activated PKC, devotes more energy to removing Ins(1,4,5)P_3 in converting it to a potentially biologically active compound which, in turn, is rapidly hydrolyzed to a relatively inactive compound, Ins(1,3,4)P_3.

Other possible targets for PKC in regulating Ins(1,4,5)P_3 accumulation are evident from the above introductory paragraph. We had observed (Rittenhouse and Sasson, 1985) that exposure of platelets to phorbol myristate acetate prior to thrombin dramatically decreased the accumulation of InsP_3 in these cells. Decreases in InsP_3 caused by phorbol ester treatment were also observed when platelet activating factor, collagen (Watson and Lapetina, 1985), γ-thrombin, or thromboxane A_2 analogues (Rittenhouse, et al., 1988) were used as platelet agonists, making less likely the explanation that inhibitory effects were exerted at the receptor for the agonist. Accumulation of phosphatidic acid (PA), another, indirect, indicator of PIC activation, was also found to be inhibited by added activators of PKC (Macintyre, et al., 1985; Rittenhouse, et al., 1988). Finally, membranes derived from phorbol ester-treated platelets were found to be less active in hydrolyzing PtdIns(4,5)P_2 than were control membranes (Baldassare, et al., 1989). Thus, some stage in PIC activation, subsequent to occupancy of a stimulatory receptor, is inhibited by agonists for PKC. Since activation of PIC in platelets appears, at least in part, to be regulated by GTP-binding protein(s) (Haslam and Davidson, 1984; Brass, et al., 1986; Banga, et al., 1988; Kucera and Rittenhouse, 1988), the target could be PIC directly, or the relevant G protein(s).

An additional target for PKC, hydrolysis of Ins(1,4,5)P_3 is indicated by findings from Connolly, et al. (1986) and Molina y Vedia and Lapetina (1986). The former group found that phosphorylation of isolated 5-PME by PKC increased the hydrolysis of Ins(1,4,5)P_3 by increasing the V_{max}. The latter investigators found that saponin-permeabilized platelets that had first been exposed to phorbol ester were more active in hydrolyzing added Ins(1,4,5)P_3 than were phorbol ester-free controls.

All of the foregoing studies imply that PKC has a negative feedback function in regulating, at least, Ins(1,4,5)P_3 concentrations in stimulated platelets. However, we decided to determine whether, and at what level, endogenously activated PKC modulates Ins(1,4,5)P_3 concentrations. Some of these findings have been published recently (King and Rittenhouse, 1989). We stimulated platelets with thrombin, in the presence or absence of a potent inhibitor of PKC, staurosporine. Since phosphorylation of platelet protein migrating with an apparent MW on SDS-PAGE gels of 47KDa is observed when PKC is stimulated, and phosphorylated 5-PME also migrates in this region (Connolly, et al., 1986), we used inhibition by staurosporine of the thrombin-induced phosphorylation of this protein as a marker for successful inhibition of PKC. Other kinase inhibitors such as ML-7 (for myosin light chain kinase), genistein (for tyrosine kinase), and HA-1004 (for A kinase), were without effect on phosphorylation of P47, nor did they cause the alterations in inositol phosphate levels to be described for staurosporine. As shown in Figure 2, staurosporine, at concentrations that inhibited thrombin-induced PKC activation by 84%, potentiated thrombin-induced accumulations of Ins(1,4,5)P_3 by 2.7-fold, of Ins(1,3,4,5)P_4 by 1.9-fold, and of Ins(1,3,4)P_3 by 1.2-fold. This early (10s) potentiation did not appear, however, to be attributable to enhanced PIC activity, inasmuch as staurosporine-treated platelets exposed to thrombin for 10s exhibited the same accumulation of diacylglycerol and PA as staurosporine-free controls (Figure 3). Intracellular Ca^{2+} concentrations, monitored by fura2, were also elevated in staurosporine-exposed, thrombin-treated platelets, consistent with an elevation in Ins(1,4,5)P_3. Stimulation of inositol polyphosphate accumulation by

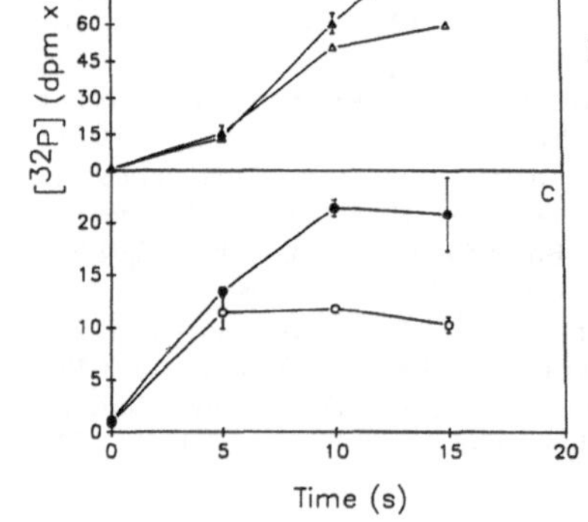

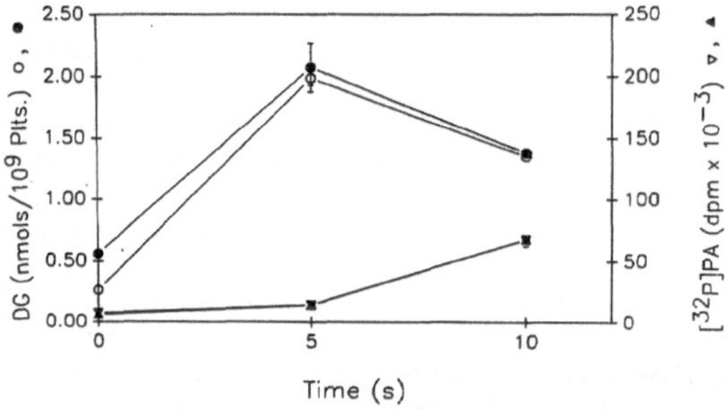

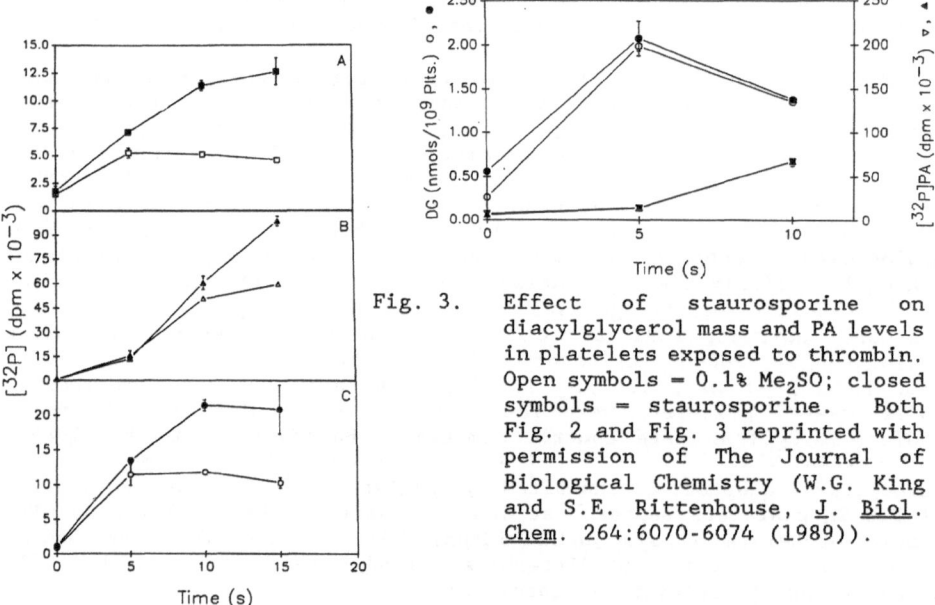

Fig. 3. Effect of staurosporine on diacylglycerol mass and PA levels in platelets exposed to thrombin. Open symbols = 0.1% Me_2SO; closed symbols = staurosporine. Both Fig. 2 and Fig. 3 reprinted with permission of The Journal of Biological Chemistry (W.G. King and S.E. Rittenhouse, J. Biol. Chem. 264:6070-6074 (1989)).

Fig. 2. Effect of staurosporine on accumulation of inositol polyphosphates in thrombin-stimulated human platelets. Open symbols = 0.1% Me_2SO vehicle; closed symbols = staurosporine. A=Ins(1,4,5)P_3; B=Ins(1,3,4)P_3; C=Ins(1,3,4,5)P_4.

staurosporine, in the absence of enhanced PA levels, is also observed when the thromboxane A_2 mimetic U46619 is substituted for thrombin as an agonist (King and Rittenhouse, unpublished observations). Finally, the thrombin-induced promotion of Ins(1,4,5)P_3 hydrolysis in saponin-permeabilized platelets is also inhibited by staurosporine.

These results, obtained during the early stages of agonist-induced platelet activation, indicate that physiologically-stimulated PKC does promote the removal of Ins(1,4,5)P_3, and at the level of 5-PME rather than at the level of PIC. The factors affecting the accumulation of Ins(1,3,4,5)P_4 are complex. On one side, 3-kinase leading to Ins(1,3,4,5)P_4 formation should be less active when PKC is inhibited. However, under these conditions there is almost 3-fold more Ins(1,4,5)P_3 substrate, more available Ca^{2+} (which stimulates platelet 3-kinase; Daniel, et al., 1988), and presumably less hydrolysis by 5-PME. The Ins(1,3,4,5)P_4 levels observed are thus the product of the action of these factors. Finally, the modestly enhanced levels of Ins(1,3,4)P_3, in the face of strongly elevated Ins(1,4,5)P_3 and Ins(1,3,4,5)P_4, again probably reflect reduced activity of 5-PME.

When we subject platelets to longer periods of incubation with thrombin, some rather different effects of staurosporine are observed, as seen in Figure 4. At 10s, accumulation of [^{3}H]InsP_3 (due primarily to Ins(1,3,4)P_3), is only slightly enhanced by staurosporine, [^{3}H]Ins(1,4)P_2 is slightly decreased, and [^{32}P]PA is unchanged. After 30s, all three parameters of PIC activity are elevated approximately two-fold. We take this as evidence in support of a feedback role for PKC in negatively modulating PIC, subsequent to its effects on 5-PME and, perhaps, 3-kinase. What is the cause of this lag? Presumably the answer lies in a protein target for PKC whose rate of phosphorylation is less rapid than that of 5-PME, or the possible accumulation of an inhibitor downstream of the phosphorylation event. One would also like to know how generally applicable this type of feedback is with respect to different platelet agonists. If one assumes its generality, any such model must account for homologous desensitization. This would entail, we would propose, the modulation by PKC of a factor localized to an occupied receptor.

In summary, endogenously-stimulated PKC in thrombin-activated platelets exerts feedback inhibitory effects on the accumulation of Ins(1,4,5)P_3 at two levels: initially at the level of Ins(1,4,5)P_3 removal, and subsequently at the level of Ins(1,4,5)P_3 generation.

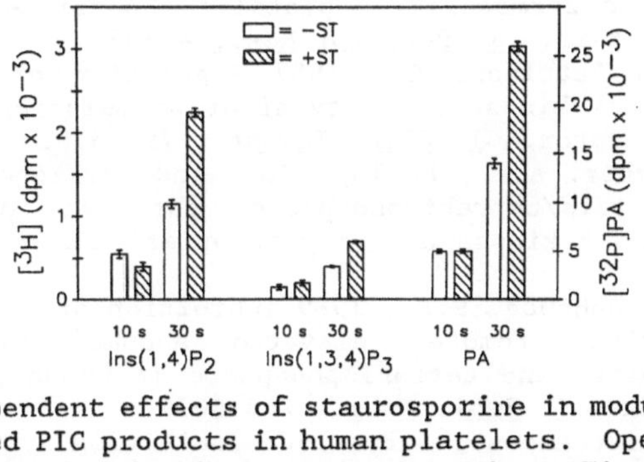

Fig. 4. Time-dependent effects of staurosporine in modulating thrombin-activated PIC products in human platelets. Open bars = 0.1% Me_2SO vehicle; hatched bars = staurosporine. Time refers to period after thrombin addition. Staurosporine was added to cells 120s prior to thrombin. Values shown are increments above thrombin-free basal levels, which were not altered by staurosporine.

REFERENCES

Auger, K.R., Serunian, L.A., Soltoff, S.P., Libby, P. and Cantley, L.C., 1989 PDGF-dependent tyrosine phosphorylation stimulates production of novel polyphosphoinositides in intact cells. Cell, **57**:167.

Baldassare, J.J., Henderson, P.A. and Fisher, G.J., 1989 Plasma membrane associated phospholipase C from human platelets: synergistic stimulation of phosphatidylinositol 4,5-bisphosphate hydrolysis by thrombin and guanosine 5'-O-(3-thiotriphosphate). Biochem., **28**:56.

Banga, H.S., Walker, R.K., Winberry, L.K. and Rittenhouse, S.E., 1988 Platelet adenylate cyclase and phospholipase C are affected differentially by ADP-ribosylation: effects on thrombin-mediated responses. Biochem.J., **252**:297.

Batty, I.R., Nahorski, S.R. and Irvine, R.F., 1985 Rapid formation of inositol 1,3,4,5-tetrakisphosphate following muscarinic receptor stimulation of rat cerebral cortical slices. Biochem. J., **232**:211.

Brass, L.F. and Joseph, S.K., 1985 A role for inositol trisphosphate in intracellular Ca^{2+} mobilization and granule secretion in platelets. J. Biol. Chem., **260**:15172.

Brass, L.F., Laposata, M., Banga, H.S. and Rittenhouse, S.E., 1986 Regulation of the phosphoinositide hydrolysis pathway in thrombin-stimulated platelets by a pertussis toxin-sensitive guanine nucleotide-binding protein. J. Biol. Chem., **261**:16838.

Connolly, T.M., Lawing, W.J., Jr., and Majerus, P.W., 1986 Protein kinase C phosphorylates human platelet inositol trisphosphate 5-phosphomonoesterase, increasing the phosphatase activity. Cell, **46**:951.

Daniel, J.L., Dangelmaier, C.A. and Smith, J.B., 1988 Formation and metabolism of inositol 1,4,5-trisphosphate in human platelets. Biochem. J., **253**:789.

Downes, C.P., Mussatt, M.C. and Michell, R.H., 1982 The inositol trisphosphate phosphomonoesterase of the human erythrocyte membrane. Biochem. J., **203**:169.

Gerrard, J.M., Townsend, D., Stoddard, S., Witkop, C.J. and White, J.G., 1977 The influence of prostaglandin G_2 on platelet ultrastructure and platelet secretion. Am. J. Pathol. **86**:99.

Hannun, Y.A., Greenberg, C.S. and Bell, R.M., 1987 Sphingosine inhibition of agonist-dependent secretion and activation of human platelets implies that protein kinase C is a necessary and common event of the signal transduction pathways. J. Biol. Chem., **262**:13620.

Haslam, R.J. and Davidson, M.L., 1984 Receptor-induced diacylglycerol formation in permeabilized platelets: possible role for a GTP-binding protein. J. Receptor Res., **4**:605.

Imboden, J.B. and Pattison, G., 1987 Regulation of inositol 1,4,5-trisphosphate kinase activity after stimulation of human T cell antigen receptor. J. Clin. Invest., **79**:1538.

Irvine, R.F., Letcher, A.J., Heslop, J.P. and Berridge, M.J., 1986 The inositol tris/tetrakisphosphate pathway: demonstration of INS(1,4,5)P_3 3-kinase activity in animal tissues. Nature(Lond), **320**:631.

King, W.G. and Rittenhouse, S.E., 1989 Inhibition of protein kinase C by staurosporine promotes elevated accumulations of inositol trisphosphates and tetrakisphosphate in human platelets exposed in thrombin. J. Biol. Chem., **264**:6070.

Kucera, G.L. and Rittenhouse, S.E., 1988 Inhibition by GDPβs of agonist-activated phospholipase C in human platelets requires cell permeabilization. Biochem. Biophys. Res. Commun., **153**:417.

Molina y Vedia, L. and Lepetina, E.G., 1986 Phorbol 12,13-dibutyrate and 1-oleoyl-2-acetyl-diacylglycerol stimulate inositol trisphosphate dephosphorylation in human platelets. J. Biol. Chem., **261**:10493.
O'Rourke, F.A., Halenda, S.P., Zavoico, G.B. and Feinstein, M.B., 1985 Inositol 1,4,5-trisphosphate releases Ca^{2+} from a Ca^{2+}-transporting membrane vesicle fraction derived from human platelets. J. Biol. Chem., **260**:956.
Rittenhouse, S.E., 1987 Measurement by capillary gas chromatography of mass changes in myo-inositol trisphosphate. in: "Methods in Enzymology, Vol 141 B, P.M. Conn and A.R. Means, eds., Academic Press, Orlando.
Rittenhouse, S.E., Banga, H.S., Sasson, J.P., King, W.G. and Tarver, A.P., 1988 Regulation of platelet phospholipase C. Phil. Trans. R. Soc. Lond. B., **320**:299.
Rittenhouse, S.E. and Sasson, J.P., 1985 Mass changes in myoinositol trisphosphate in human platelets stimulated by thrombin: inhibitory effects of phorbol ester. J. Biol. Chem., **260**:8657.
Storey, D.J., Shears, S.B., Kirk, C.J. and Michell, R.H., 1984 Stepwise enzymatic dephosphorylation of inositol 1,4,5-trisphosphate to inositol in liver. Nature, **312**:374.
Streb, H., Irvine, R.F., Berridge, M.J. and Schulz, I., 1983 Release of Ca^{2+} from a nonmitochondrial intracellular store in pancreatic acinar cells by inositol-1,4,5-trisphosphate. Nature(Lond), **306**:67.
Tarver, A.P., King, W.G. and Rittenhouse, S.E., 1987 Inositol 1,4,5-trisphosphate and inositol 1,2-cyclic 4,5-trisphosphate are minor components of total mass of inositol trisphosphate in thrombin-stimulated platelets. J. Biol. Chem., **262**:17268.
Traynor-Kaplan, A.E., Harris, A.L., Thompson, B.L., Taylor, P. and Sklar, L.A., 1988 An inositol tetrakisphosphate-containing phospholipid in activated neutrophils. Nature, **334**:353.
Watson, S.P., and Lapetina, E.G., 1985 1,2 Diacylglycerol and phorbol ester inhibit agonist-induced formation of inositol phosphates in human platelets: possible implications for negative feedback regulation of inositol phospholipid hydrolysis. Proc. Nat. Acad. Sci., U.S.A., **82**:2623.
Whitman, M., Downes, C.P., Keeler, M., Keller, T. and Cantley, L., 1988 Type I phosphatidylinositol kinase makes a novel inositol phospholipid phosphatidylinositol-3-phosphate. Nature, 332:644.

ACKNOWLEDGMENTS

The blood-drawing services of the General Clinical Research Center of The Medical Center Hospital of Vermont and the excellent secretarial assistance of Lisa Fosher are greatly appreciated. This work was supported by National Institutes of Health grant HL38622 and a NATO Collaborative Research grant.

LEUKOTRIENE D_4 INCREASES THE METABOLISM OF AN INOSITOL TETRAKISPHOSPHATE - CONTAINING PHOSPHOLIPID IN U937 CELLS

Henry M. Sarau, Maritsa N. Tzimas, James J. Foley, James D. Winkler, Matthew E. Kennedy[1] and Stanley T. Crooke[2]

Department of Pharmacology, SK&F Laboratories
P.O. Box 1539, King of Prussia, PA 19406
[1]Department of Biology Intern, Villanova University
Villanova, PA 19085
[2]Department of Pharmacology, Baylor College of Medicine
1 Baylor Plaza, Houston, TX 77030

INTRODUCTION

Substantial progress has been made over the last several years in understanding leukotriene (LT)D_4 receptors and the signal transduction process (Crooke et al., 1988; Crooke et al., 1989). LTD_4 interacts with highly selective, specific receptors for LTD_4 which are located on the plasma membrane and which have been characterized in guinea pig lung, guinea-pig heart, human lung, rat basophilic leukemia cells, human monocytic leukemia U937 cells, human mesangial cells and sheep tracheal smooth muscle cells. The receptors are coupled through guanine nucleotide binding proteins (G proteins) to a phosphatidylinositol-specific phospholipase C (PIP_2-PLC). In U937 cells, LTD_4 receptors interact with at least two types of G proteins, one being sensitive and the other insensitive to pertussis toxin. LTD_4-induced activation of PIP_2-PLC results in increased formation of inositol phosphates (IP), diacylglycerol (DAG) and transient increases in intracellular calcium ($[Ca^{2+}]_i$). Subsequently, protein kinase C is activated, phosphatidyl choline specific phospholipase A_2 is activated, enhancing the release of arachidonic acid and its metabolism to various products (Crooke et al., 1989).

LTE_4 interacts with LTD_4 receptors but functions as a partial agonist with a maximal Ca^{2+} mobilization response only 30% that of LTD_4 (Saussy et al., 1989). Exhaustive studies on the effects of LTE_4 on inositol phosphate metabolism indicated that no increase in Ins(1,4,5)P_3 could be detected at any time after treatment with varying concentrations of LTE_4. In contrast, increases in $InsP_4$ were detected at 5 sec after LTE_4 peaked by 20 sec and were maintained for >2 min. Also Ins(3,4,5)P_3 was elevated by 10 sec and remained elevated at 2 min. These results led us to look for PIP_3 in U937 cells which could be hydrolyzed by LTD_4 receptor stimulation and which would yield $InsP_4$ directly by a PIP_3-PLC. PIP_3 was recently shown to be present in activated human neutrophils (Traynor-Kaplan et al., 1988). The present study demonstrates the presence of PIP_3 in human U937 cells and shows that LTD_4 and LTE_4 stimulate the metabolism of this phosphatidylinositol trisphosphate.

METHODS

U937 Cell Culture Conditions

Human U937 cells (American Type Tissue Collection) were grown in RPMI-1640 medium supplemented with 10% (v/v) heat inactivated fetal calf serum in spinner culture at 37° C in a humidified atmosphere of 5% CO_2 and 95% air. For differentiation to "macrophage-like" cells, the U937 cells were seeded at 10^5 cells/ml in T-175 flasks in the above media with 1.3% DMSO and grown for 3 days.

Cell Labelling and Assay

Cells were washed 3 times with Krebs Ringer Hensleit (KRH) buffer, pH 7.4, without KH_2PO_4 with 0.1% bovine serum albumin (BSA). Cells were resuspended at 1-1.5 x 10^8 cells/ml and labelled with ^{32}P-orthophosphate (0.5 mCi/ml) for 1 hour at 37° C. Cells were washed three times with KRH buffer containing 0.1% BSA. Cells were resuspended in the same buffer at 4 x 10^7 cells/ml and 540 µl samples were preincubated at 37° C for 10 min. For antagonist studies the compounds were present during the preincubation. The reaction was initiated by addition of 60 µl of the agonist and terminated with 3 ml of chloroform : methanol (1:2) with butylated hydroxytoluene (0.005%).

Extraction and Separation of the Lipids

To the above assay tubes was added 2.1 ml of chloroform and 2.1 ml of 2.24 N HCl. The upper aqueous phase was washed 3 times with 1.5 ml chloroform. The lower organic phases were pooled, washed with 1.0 ml methanol/1N HCl (1:1) and dried under argon. The dried lipids were dissolved in 2x50 ul of chloroform:methanol (2:1) and spotted on heat activated oxalate - impregnated Merck silica gel 60 TLC plates. The TLC plates were developed to the top in chloroform/methanol/acetone/acetic acid/water (80/26/30/24/14, by volume) (Traynor-Kaplan et al., 1988).

Deacylation of Lipids

The phosphatidylinositol bands were identified with cold phospholipid standards run on the plate. The labeled lipids were observed by exposure (1.5 hr) to Kodak X-Omat X-ray film. Appropriate phospholipid bands were scraped into tubes and 3 ml methylamine (40%)/methanol/butanol (50/40/10) was added. Samples were sonicated and incubated at 53° C for 1 hour. Supernatant was taken to dryness under argon @ 50° C. To the residue was added 1 ml of water and 1.2 ml of n-butanol/petroleum ether/ethylformate (20:4:1). The lower aqueous phase which contains the deacylated lipids was washed with 0.75 ml of above organic mix. The pooled aqueous layers were neutralized to pH 6-7 with cyclohexylamine and stored at -20° C until analyzed (Traynor-Kaplan et al., 1988). One ml aliquots of the samples were injected onto a Beckman 10 SAX HPLC anion exchange column (4 mm x 25 cm) and the glycerol phosphoinositides separated with an ammonium formate gradient as described (Stevens et al., 1988). The total radioactivity in the peak co-migrating with Gro-PIP_3 was determined for duplicate samples and assays generally repeated 3 times.

Peroxidation of Deacylated Lipids

To identify the structure of Gro-PIP_3 the deacylated lipids were treated with $NaIO_4$. $NaIO_4$ (60 µl, 0.1 M) was added to 1.25 ml deacylated lipid samples and incubated at room temperature in the dark for 75 min. Ethyleneglycol (60 µl, 0.1%) was added, incubated for 18 min, then 800 µl of 1% dimethylhydrazine, adjusted to pH 4 with formic acid, was added and incubated for 60 min. Samples were passed through Dowex 50x8 columns (H^+form) and the elutant adjusted to pH 7 with cyclohexylamine and stored at -20° C until analyzed by HPLC (Traynor-Kaplan et al., 1988). The oxidized deacylated lipids were separated by ammonium formate gradient HPLC analysis as described above. In addition another HPLC system utilizing an ammonium phosphate gradient system was also used (Cunha-Melo et al., 1987).

Membrane isolation and assay - U937 cells were labelled for 3 days with ^{3}H-myoinositol and then washed twice with 220 mM sucrose, 15 mM acetate buffer, pH 5. Affi Gel 731 beads (Bio-Rad) were added, the cells washed with 10 mM Tris pH 7.4. The cells were broken by vortexing and the membrane bound to the beads washed 4 times and resuspended in 115 mM KCl, 20 mM NaCl, 5 mM HEPES, 1 mM $MgCl_2$, 50 nM $CaCl_2$, 10 mM glucose, pH 7.2. After incubation at 37° for 2 min, the production of ^{3}H-inositol phosphates was quantitated by HPLC as described previously (Stevens et al., 1988).

RESULTS

Presence of PIP_3 in U937 Cells

Cells were usually labelled with ^{32}P-orthophosphate to facilitate the identification of PIP_3. Silica gel TLC of the phospholipids showed many labelled phospholipid bands. One minor band was transiently increased increased by LTD_4 in a time- and concentration-dependent manner (Fig. 1). The major phosphatidylinositols were not dramatically affected by the agonists. Figure 1 shows the radioautogram from a LTD_4 concentration response study. Lane 1 was the control (no agonist) and lanes 2-8 were 1, 3.3, 10, 33, 100, 330, 1000 nM LTD_4 concentrations, respectively. Stimulation with LTD_4 was for 15 sec which coincides with the time for the maximal increase in $[Ca^{2+}]_i$ induced by LTD_4 (Crooke et al., 1988).

It was not possible to reproducibly scrape the PIP_3 band from the TLC plate without getting contamination from surrounding bands especially the minor band directly above the presumed PIP_3 (probably lyso-PIP_2). Therefore to quantitate changes in this band and to identify it as PIP_3 the phospholipid band was scraped from the TLC plate and deacylated to form the glycerol phosphoinositides. The ^{32}P glycerol phosphoinositides with added ^{3}H inositol phosphate standards were separated by HPLC and a typical profile is shown in figure 2A. The band of interest elutes between IP_3 and IP_4 as shown by Alexis E. Traynor-Kaplan et al. (1988). An aliquot of the sample run in fig. 2A was treated with 5 mM $NaIO_4$ to oxidize off the glycerol resulting in the formation of IP_4. Fig. 2B shows the HPLC profile of the periodate treated sample which shows that the Gro-PIP_3 band has been quantitatively hydrolyzed to a component which co-elutes with Ins(1,3,4,5)P_4. As further proof we ran the periodate treated sample on the ammonium phosphate gradient anion exchange HPLC system and the $NaIO_4$ treated Gro-PIP_3 peak again co-eluted with the [^{3}H]Ins (1,3,4,5)P_4 standard (data not shown). These results indicate that PIP_3 is present in the U937 cells and that LTD_4 causes a transient increase in the metabolism of this phospholipid. All pharmacological studies were quantitated after HPLC of the deacylated PIP_3 band, i.e. total radioactivity in Gro-PIP_3 peak.

Equilibrium labelling of the phospholipids was probably not achieved with the brief ^{32}P-labelling. Therefore for some studies the U937 cells were labelled with myo-[^{3}H]inositol (1 µCi/ml) for 3 days while differentiating with DMSO. The location of the phospholipids separated by TLC was identified with the Bioscan system 200 imaging scanner and the PIP_3 band scraped and deacylated. U937 cells labelled with myo-[^{3}H]inositol

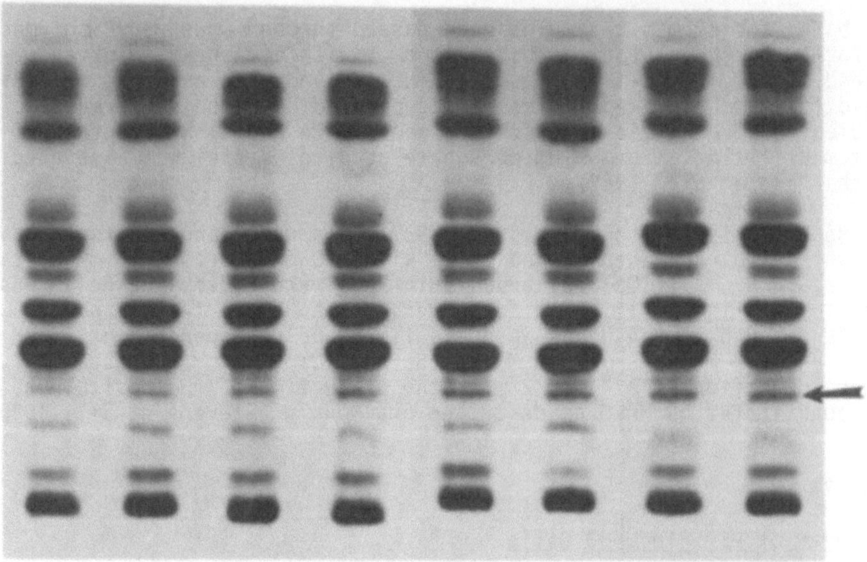

Figure 1. Autoradiograms of the TLC separations of U937 cell phospholipids labelled with ^{32}P-orthophosphate showing the LTD_4 concentration response with 15 sec stimulation. Lanes 1-8 were stimulated for 15 sec with 0, 1, 3.3, 10, 33, 100, 330 and 1000 nM LTD_4, respectively. The PIP_3 band region is indicated with the arrow. Typical chromatogram from 3 experiments with samples run in duplicate.

showed the presence of PIP_3 by the same criteria as used for the ^{32}P-labelled cells.

PIP_3 Present in U937 Cell Membranes

Additional evidence for the presence of PIP_3 in U937 cells was obtained with membranes from ^3H-myoinositol labelled cells. As shown in fig. 3, the isolated membranes produced $[^3H]InsP_1$, $[^3H]InsP_2$, $[^3H]Ins(1,4,5)P_3$ and $[^3H]Ins(1,3,4,5)P_4$. The $InsP_4$ levels amounted to 45% of the $Ins(1,4,5)P_3$ levels (fig. 3 insert). The addition of exogenous $[^3H]Ins(1,4,5)P_3$ to unlabelled U937 cell membranes did not result in any conversion of $[^3H]Ins(1,4,5)P_3$ to $[^3H]InsP_4$ (data not shown), indicating that the 3-kinase was not present in the preparation and suggesting that the $InsP_4$ produced by the ^3H-myoinositol labelled cells may be from direct hydrolysis of PIP_3 by a specific PLC.

LTD_4-Induced Formation of PIP_3

Fig. 4A, shows the quantitation of the LTD_4 concentration-response with 15 sec LTD_4 stimulation. Maximal PIP_3 formation was obtained at concentrations of 100 nM. The EC_{50} was 6 nM LTD_4, which corresponds with the EC_{50} (3 nM) of LTD_4-induced $[Ca^{2+}]_i$ mobilization in U937 cells.

The time course for both LTD_4 and LTE_4 stimulation of the formation of ^{32}P-PIP_3 is shown in fig. 4B. LTD_4 (100 nM) showed a biphasic increase in PIP_3 with peaks at 15 seconds and another at 90 sec, which remains elevated for several minutes before declining to baseline levels (data not shown). The early transient increase in PIP_3 correlated with the $[Ca^{2+}]_i$ transient induced by LTD_4 which peaks at 10-15 sec then decreases in about 60 sec to a new plateau $[Ca^{2+}]_i$ level, and remained at that plateau for several minutes. The plateau of increased $[Ca^{2+}]_i$ may relate to the persistent second increase in PIP_3 formation and the sustained increase in $InsP_4$ levels obtained after LTD_4 stimulation.

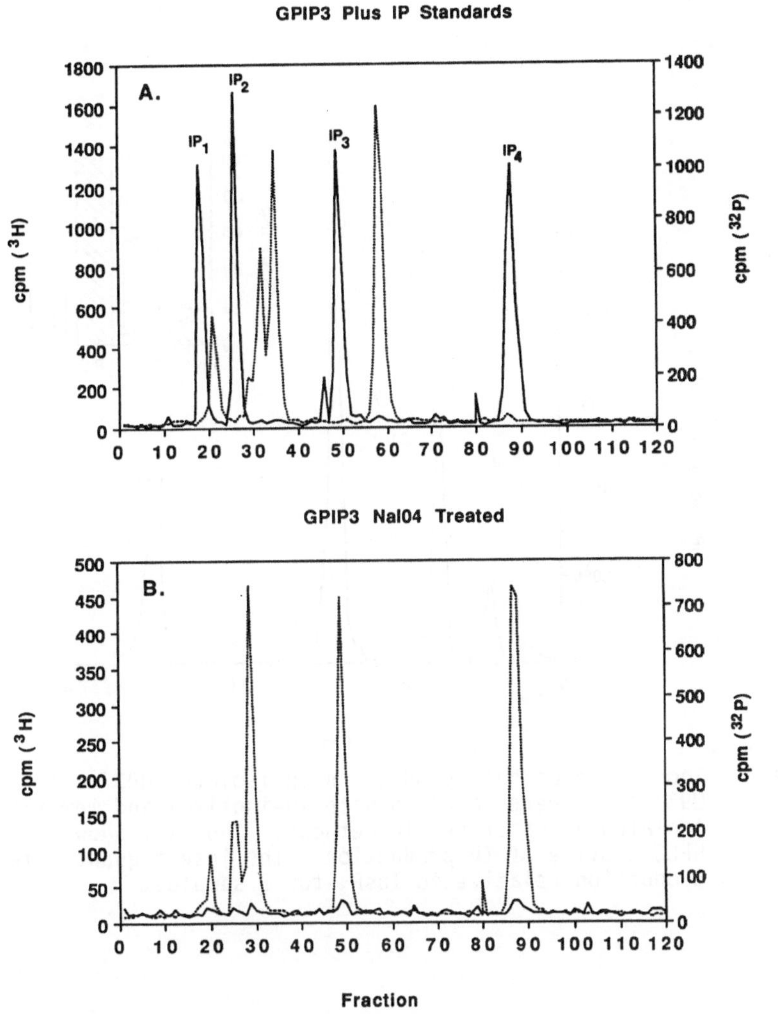

Figure 2. HPLC elution profile of the deacylated PIP_3 band from TLC plates as shown in fig. 1. A. Elution profile of deacylated ^{32}P-PIP_3 band that was spiked with $[^3H]InsP_1$, $InsP_2$, $InsP_3$ and $InsP_4$ standards. Fractions (0.4 ml) collected starting 7 min after injection and 4 ml Tru-count scintillant added to each vial for counting. B. An aliquot of the deacylated sample was treated with 5 mM $NaIO_4$. Elution profile of the $NaIO_4$ treated sample with the same HPLC systems used for fig. 2A.

LTE_4, the partial agonist, at maximally effective concentrations (10 µM) showed a rapid rise in $^{32}P-PIP_3$ formation but was without an initial transient increase (fig. 4B). The increase in $^{32}P-PIP_3$ level was sustained and then decreased slowly to basal level after 3 min (data not shown). During the time of increased PIP_3 levels, LTE_4 was shown to induce sustained elevated levels of $[Ca^{2+}]_i$ (Saussy et al., 1989). There was no early transient increase in PIP_3 metabolism after LTE_4 that would correspond with the early rapid $[Ca^{2+}]_i$ transient that occurs with LTE_4. However, there was a significant rapid rise in $^{32}P-PIP_3$ which could account for the rapid increase in $InsP_4$ obtained after LTE_4 stimulation (Saussy et al., 1989).

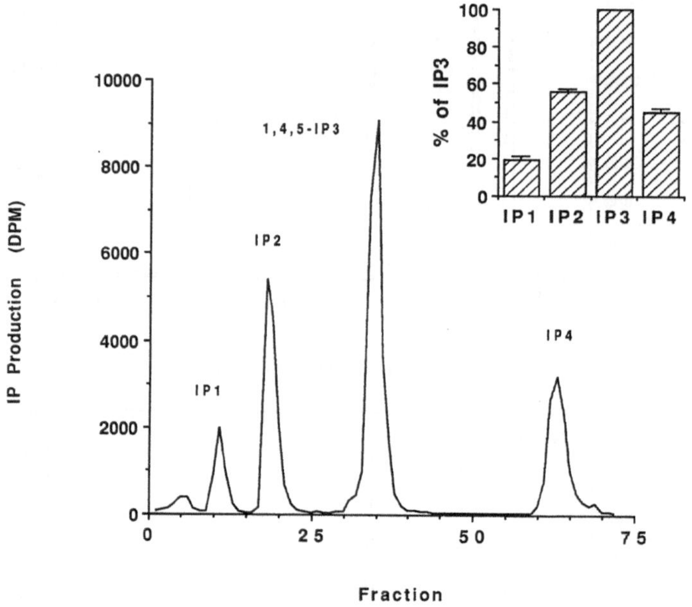

Figure 3. Inositol phosphate production in isolated U937 cell membranes. U937 cells were labelled with ^3H-inositol and membranes isolated as described in Methods. The fig. shows a typical HPLC profile of IP production. The insert quantitates the IP production relative to $InsP_3$ for 3 samples.

The effects of time and LTD_4 concentration were also evaluated in ^3H-myoinositol labelled U937 cells. There were no significant differences in the EC_{50} of LTD_4 to induce PIP_3 or in the time course for maximally effective concentrations of LTD_4 between ^3H or ^{32}P labelled cells (data not shown).

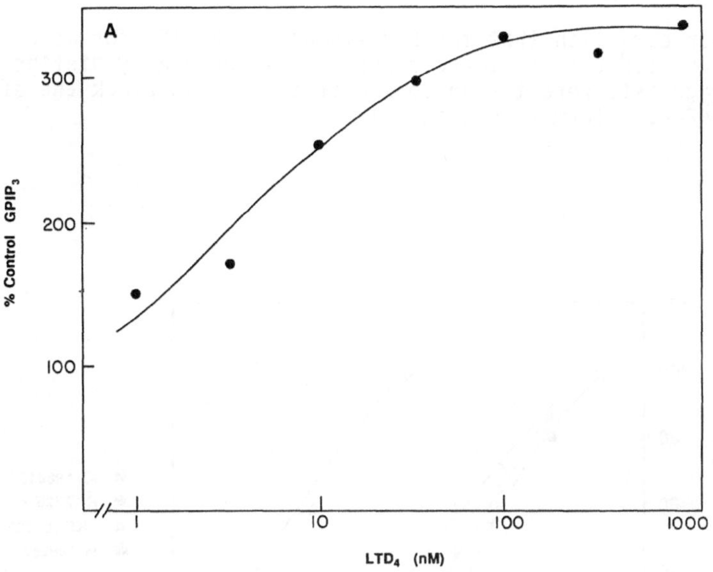

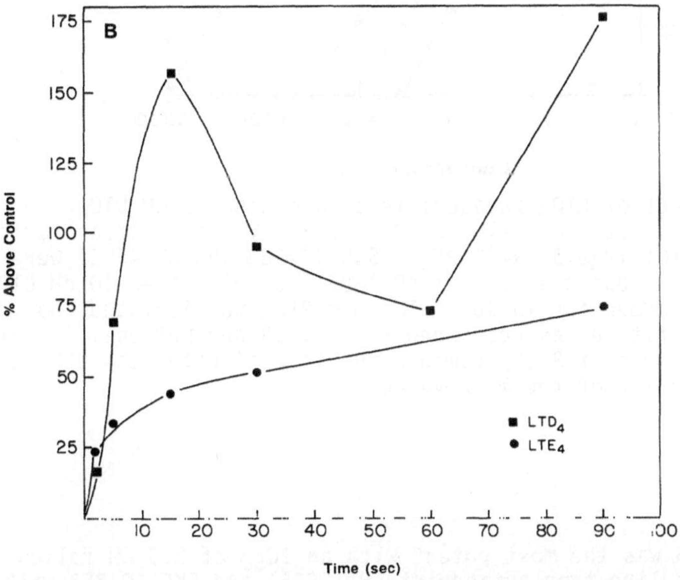

Figure 4. A. Effect of increasing concentrations of LTD_4 on the formation of ^{32}P-PIP_3. Various concentrations of LTD_4 were used to stimulate ^{32}P-PIP_3 formation for 15 sec. The PIP_3 bands on the TLC were deacylated and ammonium formate gradient HPLC system used to elute Gro-PIP_3. The total counts in the Gro-PIP_3 peak were determined and the % of the unstimulated sample calculated. B. The time course for the formation of ^{32}P-PIP_3 in U937 cells stimulated with 100 nM LTD_4 or 10 µM LTE_4 was determined as described in 4A and Methods. Points are the mean of 3 experiments run in duplicate, the SEM was less than 9% of the mean value.

Effect of LTD_4 Receptor Antagonists on LTD_4-Induced PIP_3 Formation

To further establish that the LTD_4-induced ^{32}P-PIP_3 formation was a receptor mediated effect, three classes of structurally distinct LTD_4 receptor antagonists were tested to see if they would block the effects of LTD_4 on ^{32}P-PIP_3 formation (fig. 5).

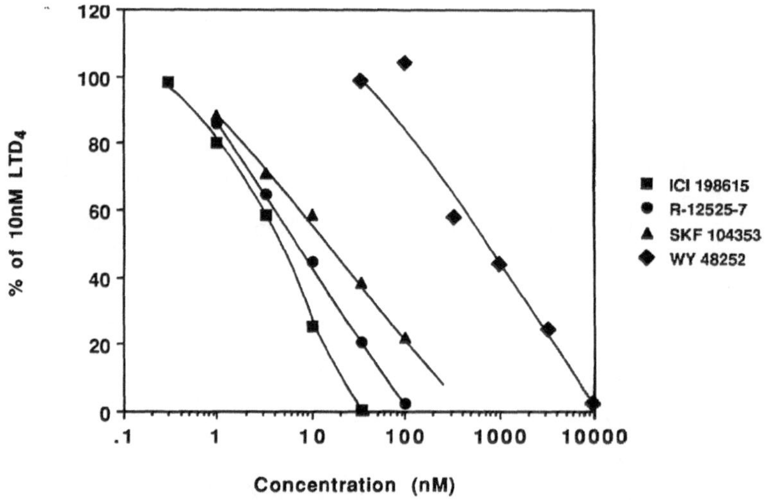

Figure 5. Effect of LTD_4 antagonists at blocking 10 nM LTD_4 stimulated formation of ^{32}P-PIP_3. Varying concentrations of ICI-198615, R-12525-7, SKF 104353 and WY 48252 were preincubated with the ^{32}P labelled cells then 10 nM LTD_4 was added for 15 sec. The Gro-PIP_3 was isolated and quantitated as described in fig. 2A and Methods. Points are the mean of 3 experiments run in duplicate, the SEM was less than 12% of the mean value.

ICI-198615 was the most potent with an IC_{50} of 3.5 nM followed by the Rorer quinoline type antagonist (R-12525) and SKF 104353 with IC_{50} values of 7.8 nM and 14.7 nM, respectively. The Wyeth quinoline type antagonist (WY-48252) and the Lilly antagonist (LY-171883, data not shown) completely blocked the LTD_4 induced effect but were weaker with IC_{50} values of 760 nM and 2,050 nM, respectively. The affinities of the antagonists in this assay were lower than those obtained with [^3H]-LTD_4 binding assays to the membranes of U937 cells, but the rank order of potency was identical. A possible reason for the lower affinity may be the presence of 0.1% BSA in the incubation media which was included to maintain cell viability. All the LTD_4 antagonists showed high protein binding affinity in the ^3H-LTD_4 receptor binding assay. Recent studies indicate that BSA is not an essential component of the PIP_3 assay system and the compounds appear more potent in the absence of BSA.

Effect of Pertussis Toxin on LTD_4 induced PIP_3 Production

Preliminary studies were run to determine the effect of pertussis toxin (10 ng/ml for 18 hr) on the LTD_4-induced increase in ^{32}P-PIP_3. The levels of ^{32}P label in PIP_3 of control U937 cells was 70% that of the pertussis toxin treated cells. When challenged with LTD_4 or LTE_4 the treated cells did not respond with increased formation of PIP_3, rather a small decrease was observed (data not shown). These results indicate that pertussis toxin blocked the LTD_4 and LTE_4 induced effects and this is being evaluated further with ^3H-myoinositol labelled U937 cells.

CONCLUSIONS

Human U937 cells contain a low level of PIP_3. LTD_4 stimulation of the U937 cells results in a time- and concentration-dependent increase in PIP_3 formation. LTE_4, the partial agonist of LTD_4 receptors, caused a rapid sustained increased level of PIP_3 which may account for the observed increase in IP_4 in the absence of increases in Ins(1,4,5)P_3 seen with this agonist. The increased PIP_3 formation was LTD_4 receptor mediated, as three structurally different LTD_4 receptor antagonists blocked the effect of LTD_4 with the proper rank order potency. This PIP_3 increase correlates with the LTD_4 induced $[Ca^{2+}]_i$ mobilization response and may be part of the mechanism of LTD_4-induced signal transduction.

REFERENCES

Crooke, S.T., Mong, S., Sarau, H.M., Winkler, J.D., and Vegesna, R.K., 1988, Mechanisms of regulation of receptors and signal transduction pathways for the peptidyl leukotrienes. Annals NY Acad. Sci. 524, 153-161.

Crooke, S.T., Mattern, M., Sarau, H.M., Winkler, J.D., Balcarek, J., Wong, A., and Bennett, C.F., 1989, The signal transduction system of the leukotriene D_4 receptor. Trends in Pharm. Sci. 10, 103-107.

Cunha-Melo, J.R., Dean, N.M., Moyer, J.D., Maeyama, K., and Beaven, M.A., 1987, The kinetics of phosphoinositide hydrolysis in rat basophilic leukemia (RBL-2H3) cells varies with the type of IgE receptor cross-linking agent used. Jour. Biol. Chem. 262, 11455-11463.

Saussy, D.L., Sarau, H.M., Foley, J.J., Mong, S., and Crooke, S.T., 1989, Mechanisms of leukotriene E_4 partial agonist activity of leukotriene D_4 receptors in differentiated U-937 cells. (Manuscript submitted).

Stevens, L., Hawkins, P.T., Carter, N., Chahwala, S.B., Morris, A.J., Whetton, A.D., and Downes, P.C., 1988, L-myo-inositol 1,4,5,6-tetrakisphosphate is present in both mammalian and avian cells.. Biochem J. 249, 271-282.

Traynor-Kaplan, A.E., Harris, A.L., Thompson, B.L., Taylor, P., and Sklar, L.A., 1988, An inositol tetrakisphosphate-containing phospholipid in activated neutrophils. Nature 334, 353-356.

IMMUNOCYTOCHEMICAL LOCALIZATION OF PROTEIN KINASE C

Susan Jaken, Susan C. Kiley, Theresa Klauck, Liqun Dong, and Susannah Hyatt

W. Alton Jones Cell Science Center
Lake Placid, NY 12946

INTRODUCTION

Protein kinase C (PKC) is a key enzyme involved in regulation of growth and differentiated function (Nishizuka, 1986). In vitro, the enzymatic activity requires both calcium and phospholipids. The neutral lipid, diacylglycerol, also interacts with PKC and decreases the K_{act} for calcium to concentrations compatible with those found in the cytoplasm. DAG is considered to be the key regulator of PKC activation state in vivo. This is due to the fact that cellular levels of DAG are tightly coupled to hormonal activation of phospholipase C. Tumor promoting phorbol esters mimic diacylglycerol with respect to binding to and activating PKC. In fact, PKC is the major cellular receptor for phorbol esters (Castagna, et al., 1982; Niedel, et al., 1983). It is for this reason that phorbol esters are so often used as specific tools for exploring the role of PKC in biological processes.

Results of cloning experiments have documented that PKC is actually a family of distinct, but closely related genes and proteins (Nishizuka, et al., 1988; Coussens, et al. 1986; Parker, et al. 1986; Knopf et al. 1986; Housey et al., 1987; Ohno, et al., 1987). Differences among the α-, β-, and γ-PKC isozymes are concentrated into five "variable" regions which are spaced between the four highly conserved "constant" regions at defined intervals (Nishizuka, et al., 1988;). The overall structure includes an N-terminal regulátory domain, which is thought to contain the phospholipid and phorbol ester binding sites, and a C-terminal catalytic domain which contains a consensus sequence for an ATP binding site. Trypsin-treatment of the purified proteins has been used to generate these two domains and demonstrate that they retain their phorbol ester binding or kinase activities, respectively. Recently, lower stringency hybridization studies have revealed three additional cross-hybridizing cDNAs known as δ, ε, and ζ (Ohno, et al., 1988; Ono et al., 1988). These differ from α, β, and γ in that they are missing the second constant region. These related cDNAs code for proteins which have Ca^{2+}-insensitive but phospholipid-dependent kinase and phorbol ester binding activities.

The reason for PKC heterogeneity is as yet unknown. Tissue distribution studies showed that α-PKC transcripts are found in all tissues, β-PKC transcripts are found in selected tissues, and γ-PKC transcripts are found only in central nervous system tissues (Nishizuka, 1988). Three PKC activities have been purified from brain cytosol (Huang, et al., 1986; Jaken and Kiley; 1987). Each PKC copurified with a phorbol ester binding activity and was shown to have distinct antigenic properties. These biochemically isolated proteins (Types 1, 2 and 3 PKC) correspond to the γ, β and α cDNAs, respectively. Biochemical characterization of Types 1, 2 and 3 PKCs revealed only subtle differences in their calcium and phospholipid requirements for binding and kinase activities (Jaken and Kiley, 1987; Huang et al., 1988; Sekiguchi et al., 1988). Differential substrate specificity among the isozymes has not yet been studied in great detail, however, in one study the EGF receptor was found to be a preferential substrate for α- vs β- and γ-PKCs (Ido, et al., 1987). The major difference between the calcium requirements for α-, β-, and γ-PKCs compared to δ, ε, and ζ PKCs suggests that regulation of the _in vitro_ activities of these two groups of PKCs may be substantially different.

We have used cell culture model systems to study the role of PKC activation in regulation of specific cellular processes and to begin to address the question of whether individual isozymes have specific functions. We have used hormone-responsive pituitary cells in which phorbol esters stimulate synthesis and secretion of prolactin to study the role of PKC in regulation of differentiated function. In order to identify potential differences in PKC that may contribute to the loss of growth regulation characteristic of transformed cells, we have compared PKC in normal and transformed fibroblasts.

Isozyme-Specific Antibodies

We have prepared several monoclonal antibodies (mAbs) to Type 3 (α) PKC (Leach et al., 1988). These mAbs were characterized with respect to their recognition of other isozymes and their interaction with either the regulatory or catalytic domains of α-PKC. Based on these criteria, three classes of mAbs were obtained:

1. Anti-catalytic domain mAbs that are specific for α-PKC.
2. Anti-catalytic domain mAbs that recognize α- and β-PKCs but not γ-PKC.
3. Anti-regulatory domain mAbs that are specific for α-PKC.

Class 1, but not Class 3 mAbs are effective inhibitors of α-PKC phosphorylation of lipocortin. In contrast, Class 3, but not Class 1 mAbs are effective inhibitors of phorbol ester binding. Interestingly, this inhibition could be competed with phosphatidylserine (PS), indicating that the epitopes recognized by these mAbs may be near the PS binding site on the regulatory domain.

Regulation of α-PKC in Hormone-Responsive Pituitary Cells

GH_4C_1 cells are a clonal line of rat pituitary cells in which the synthesis and secretion of prolactin and growth hormone can be regulated by a variety of ligands including thyrotropin-releasing hormone (TRH). TRH action is mediated by stimulation of phosphoinositide-specific phospholipase (Gershengorn, et al., 1986). Consequently, cellular levels of two PKC activators, Ca^{2+} and diacylglycerol (DAG) are increased. Phorbol esters such as phorbol dibutyrate (PDBu), bypass phospholipase C and mimic DAG by directly interacting with PKC. A major difference between TRH and PDBu cellular effects that may be relevant to activation of PKC is that TRH, but not PDBu, increases cytosolic Ca^{2+}. Nonetheless, PDBu and TRH both stimulate prolactin synthesis and secretion in these cells.

One goal of our studies has been to compare the effects of Ca^{2+} and DAG/PDBu on PKC activation. Redistribution of PKC from a soluble to a particulate fraction, as originally described by Kraft and Anderson (Kraft, et al., 1983), was used to assess the PKC activation state. Initial experiments demonstrated that GH cells are heterogeneous with respect to their PKC isozyme content. Northern analysis revealed that messages for α-, β-, and ε-PKC are present (I.B. Weinstein and L.L. Hsieh, personal communication). Furthermore, two peaks of PDBu binding and PKC activities can be separated by hydroxylapatite chromatography of GH cell cytosol. Approximately 50% of the total activity is associated with α-PKC. Our initial experiments have been directed towards describing the activation of this isozyme by Ca^{2+} and DAG/PDBu and furthermore, to identify the particulate components with which activated α-PKC associates.

When resting cultures of GH cells are lysed in chelator-containing buffers, most of the PDBu binding activity is recovered in the soluble fraction (Table 1). Treatment of cells with TRH or PDBu prior to lysis in chelator-containing buffers causes redistribution of PDBu receptors from the soluble to the particulate fraction (Table 1). That is, these PKC activators cause quantitatively similar chelator-stable association of α-PKC with the particulate fraction. The distribution between soluble and particulate fractions is Ca^{2+}-sensitive. Thus, including Ca^{2+} in the lysis buffer causes an increase in particulate and a corresponding decrease in soluble PDBu receptors. The Ca^{2+}-mediated association is, however, not chelator-stable. Thus, two pools of particulate PDBu receptors can be discerned based on the Ca^{2+}-dependence of the interaction.

Table I. Subcellular Distribution of PDBu Receptors in GH Cells

	PDBu Bound, pmol/mg* (% Control)		
Treatment	Lysate	Cytosol	Membranes
Control	1.3±0.3 (100)	5.0±1.4 (100)	0.17±0.05 (100)
TRH, 15 sec	1.3±0.4 (100)	3.3±0.9 (65)	0.50±0.11 (294)
PDBu, 10 min	1.0±0.1 (77)	2.4±0.4 (47)	0.60±0.07 (400)
Ca^{2+} Lysis	1.1±0.3 (85)	1.6±0.4 (31)	1.00±0.24 (590)

*Data shown are expressed as mean values ± SD from multiple experiments: Control (n = 11); TRH, 15 sec (n = 4); PDBu, 10 min (n = 3); and Ca^{2+} Lysis (n = 8).

Immunoblot analysis with α-PKC specific mAbs was used to compare the distribution of this isozyme with total PDBu binding activity. TRH and PDBu were equally effective in causing chelator-stable membrane association of α-PKC. Ca^{2+} was more effective than these at mediating redistribution. These results suggest that although DAG and PDBu are effective PKC activators in the absence of Ca^{2+}, Ca^{2+} is also an important determinant of the α-PKC activation state.

It is possible that α-PKC artifactually associates with particulates during cell lysis and preparation of subcellular fractions. Therefore, α-PKC distribution was also studied using immunocytochemical techniques which permit us to explore α-PKC localization with minimal disruption. Diffuse cytoplasmic fluorescence is observed in resting cells stained with anti-α-PKC mAb. No nuclear staining is observed. Treatment with TRH or PDBu decreases the area of cytoplasmic fluorescence. A corresponding increase in fluorescence intensity at cell junctions is observed indicating that α-PKC has become concentrated in these areas. These results differ from the redistribution of antigen measured by subcellular fractionation in that TRH was more effective than PDBu at causing redistribution. A major difference between the two experimental protocols is that only chelator-stable α-PKC is measured in the subcellular fractionation assays, whereas immunocytofluorescence permits analysis at ambient cytosolic Ca^{2+} levels. Thus, it is likely that the rise in cytosolic Ca^{2+} associated with TRH, but not PDBu-treatment contributes to α-PKC redistribution. This is supported by other experiments in which increased cytosolic Ca^{2+} due to K^{+} depolarization or ionomycin-treatment caused α-PKC redistribution in the immunofluorescence assay.

Differential extraction of the particulate fractions was used to identify the compartment with which α-PKC associates. Particulate fractions from TRH- or PDBu-treated cultures were first sonicated in the presence of chelators, then centrifuged. The resulting pellet was further extracted with Triton X-100 and recentrifuged. The amount of α-PKC in the supernatants and pellets was determined by immunoblots. Neither chelators nor Triton X-100 were effective at solubilizing α-PKC. Most of the antigenic activity was recovered in the chelator- and detergent-insoluble fraction.

These results suggested to us that activation of α-PKC in these cells caused association with detergent-insoluble material, commonly referred to as the cytoskeleton (CSK). We tested this hypothesis by comparing the immunofluorescence patterns in CSK preparations prepared from resting and TRH- or PDBu-stimulated cultures. Almost no fluorescence was found in CSKs from the resting cultures. In contrast, CSKs from stimulated cultures were stained very brightly.

Our results can be incorporated into a working model for α-PKC activation in GH cells. In resting cells, α-PKC is found in the cytoplasm. Activation causes α-PKC to become associated with the detergent-insoluble CSK. We propose that this redistribution is essential for interaction between α-PKC and its substrates. In support of this, we have found that activation correlates with reorganization of both microfilaments and intermediate filaments in these cells. It is likely that these changes are directly related to the increased release and synthesis of prolactin that accompanies PKC activation.

Comparison of α-PKC from Normal and Transformed Fibroblasts

Several growth factors are known to act via stimulation of phosphoinositide metabolism and PKC (Berridge et al., 1986). Consequently, controlled PKC activation is considered to be important for normal cell growth. In contrast, prolonged or uncoordinated activation of PKC may be important in transformation. In many cases, transformed cells have lost their requirement for growth factors that activate PKC. One explanation for this may be that certain transformed cells have higher basal levels of phospholipase C activity and consequently, higher levels of activated (i.e. particulate) PKC (Preiss et al., 1986; Halsey et al., 1987). An important role for PKC in transformation is also indicated by the fact that phorbol esters are tumor promoters and that they cause partial mimicry of the transformed phenotype, presumably due to increasing the activity of PKC (Weinstein et al., 1979). Furthermore, overexpression of PKC decreases cell doubling time and changes morphology (Housey et al., 1988; Persons et al., 1988).

We have compared the PKC and PDBu binding properties of normal and SV40-transformed rat embryo fibroblasts (REF-52) in order to assess the role of PKC in regulated and unregulated cell growth. Whereas the normal cells require vasopressin (VP), which stimulates phosphoinositide metabolism, for optimal cell growth, their SV40-transformed counterparts have circumvented this requirement. We, therefore, undertook a comparison of the distribution of PKC and PDBu receptors in the normal and transformed cells. Comparable levels of PDBu receptors were measured in whole cells as well as soluble and particulate fractions. Treatment with PDBu caused redistribution to the particulate fraction in both cell types. Therefore, in these cells, neither increased PKC levels or increased membrane-associated PKC activity appears to be contributing to the transformed phenotype.

Although our results do not suggest that transformation is associated with increased amounts of activated PKC, these transformed cells do show evidence for disorder within their PKC signal transduction pathway. VP stimulates phosphoinositide metabolism in the normal cells. This VP response is attenuated in the SV40-transformed cells even though they have comparable levels of VP receptors (C. Welsh, personal communication). Both VP and phorbol esters stimulate phosphatidylcholine metabolism in the normal, but not the transformed cells (Cabot et al., 1988; Welsh, et al., 1988). We, therefore, considered that inappropriate subcellular location rather than increased quantity of PKC may be an underlying factor in transformation of these cells.

Using immunocytochemical techniques, we could discern three pools of α-PKC in the normal cells; diffuse cytoplasmic, fiber-associated, and nuclear. Specificity of staining was demonstrated by absorption of antibody on a PKC affinity column. The fiber-associated staining suggested that some α-PKC may be in the CSK. Comparison of staining patterns in whole cells vs CSK preparations demonstrated that while the cytoplasmic fluorescence was removed with detergent extraction, the nuclear and fiber-associated staining remained (Fig. 1).

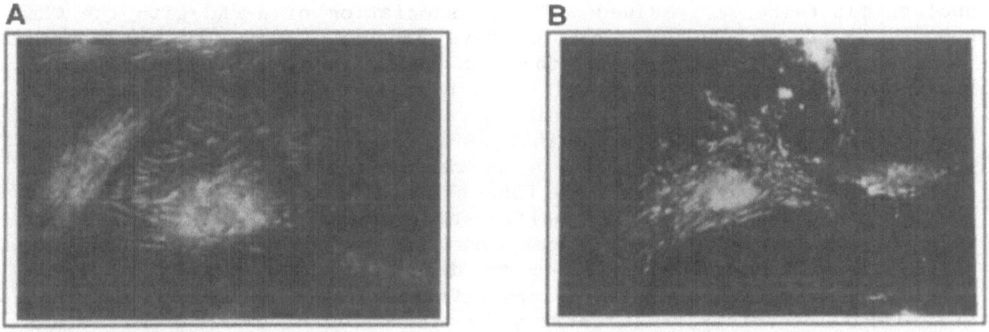

Figure 1. Immunocytochemical staining of α-PKC in whole cells and cytoskeltons of REF-52 cells. CSK were prepared by Triton X-100 extraction. Cells and CSKs were fixed in formaldehyde, permeabilized with methanol, and sequentially stained with α-PKC specific, anti-catalytic domain mAb, M4, and FITC-conjugated second antibody. A, whole cells; B, CSKs.

The α-PKC staining in CSKs was compared with that of known CSK proteins. While the pattern does not resemble that of vimentin, tubulin, or actin, it does resemble that of the focal contact protein vinculin. In fact, α-PKC staining colocalized with staining for another focal contact protein, talin (Fig. 2). The focal contact is the specialized area of the plasma membrane which mediates both the attachment of the cell to the substratum and the attachment of the microfilaments to the plasma membrane (Burridge et al., 1987; Geiger et al., 1987). This complex is composed of several structural proteins (integrin, talin, vinculin, and α-actin) and several regulatory proteins. Staining for both pp 60 src and phosphotyrosine is also concentrated in focal contacts (Burridge et al., 1987). The association of PKC with this specialized structure that mediates the interaction of the cell with its extracellular environment underscores the importance of the focal contact in signal transduction processes.

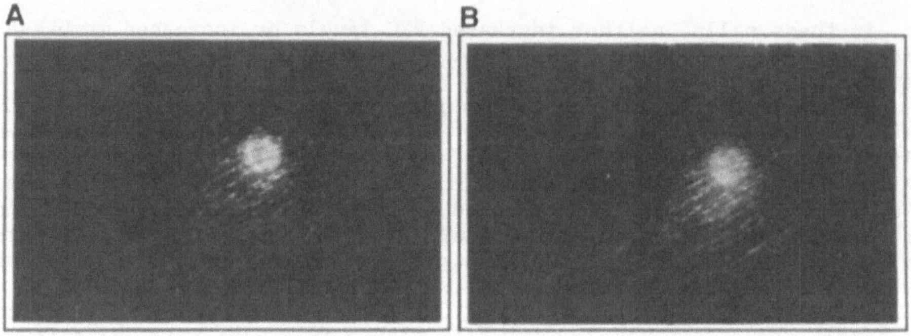

Figure 2. Colocalization of talin and α-PKC. CSKs were prepared and fixed as described in Figure 1. Preparations were incubated with both M4 and rabbit anti-talin serum (a gift from Dr. Keith Burridge, University of North Carolina). After washing, samples were incubated with both FITC- and TRITC- second antibodies. A, α-PKC; B, talin.

Immunofluorescence of α-PKC in SV40-transformed cells included the cytoplasmic and nuclear staining observed in the normal cells; however, the fiber-associated staining was missing. In CSK preparations, only nuclear fluorescence remained. Thus, association of α-PKC with the CSK appears to be a transformation-sensitive process. This does not appear to be due to a total loss of focal contacts in the transformed cells because the characteristic focal contact staining of vinculin and talin is still apparent (although diminished relative to the normal cells).

Our working hypothesis is that α-PKC is a regulatory component of the focal contact in normal cells in which it plays a role in ordered signal transduction. We propose that the association of α-PKC with the focal contact is mediated by specific interactions with target proteins. This hypothesis provides a framework for identifying relevant PKC substrates. Transformation leads to decreased ability to form focal contacts, and α-PKC is no longer concentrated in the focal contacts that are formed. Possibly, this decreased interaction of α-PKC with the CSK contributes to the disordered signal transduction characteristic of transformed cells.

SUMMARY

Results obtained with two different cell lines demonstrate that association of α-PKC with the CSK may be an important component of PKC signal transduction. Through use of an overlay assay to detect PKC binding proteins, we have developed an approach for identifying proteins that mediate α-PKC association with the CSK. Although many proteins can act as PKC substrates *in vitro*, it has been difficult to identify primary targets for PKC phosphorylation *in vivo*. The value of our model is that it allows us to focus on a specific subset of proteins as potential substrates for α-PKC. In the future, we will extend this work to address whether other PKC isozymes differ from α-PKC in their subcellular localization and association with CSK structures and proteins.

REFERENCES

1. Berridge, M.J., 1986, Intracellular signalling through inositol trisphosphate and diacylglycerol, Biol. Chem. Hoppe-Seyler. 367:447.

2. Burridge, K., Molony, L., and Kelly, T., 1987, Adhesion plaques: sites of transmembrane interaction between the extracellular matrix and the actin cytoskeleton, J. Cell. Sci. Suppl. 8:211.
3. Cabot, M., Welsh, C.J., Zhang, Z., Cao, H.-t., Chabbott, H., and Lebowitz, M., 1988, Vasopressin, phorbol diesters and serum elicit choline glycerophospholipid hydrolysis and diacylglycerol formation in nontransformed cells: transformed derivatives do not respond, Biochem. Biophys. Acta. 959:46-57.
4. Castagna, M., Takai, Y., Kaibuchi, K., Sano, K., Kikkawa, U. and Nishizuka, Y., 1982, Direct activation of calcium-activated phospholipid-dependent protein kinase by tumor-promoting phorbol esters, J. Biol. Chem. 257:7847.
5. Coussens, L., Parker, P.J., Rhee, L., Yang-Feng, T.L., Chen, E., Waterfield, M.D., U. Francke, U. and Ullrich, A., 1986, Multiple, distinct forms of bovine and human protein kinase C suggest diversity in cellular signaling pathways, Science. 233:859.
6. Geiger, B., Volk, T., Volberg, T., and Bendori, R., 1987, Molecular interactions in adherens - type contacts, J. Cell. Sci. Suppl. 8:251.
7. Gershengorn, M., 1986, Mechanism of thyrotropin releasing hormone stimulation of pituitary hormone secretion, Ann. Rev. Physiol. 48:515.
8. Halsey, D.L., Girard, P.R., Kuo, J.F., and Blackshear, P.J., 1987, Protein kinase C in fibroblasts, J. Biol. Chem. 262:2234.
9. Housey, G.M., O'Brian, C.A., Johnson, M.D., Kirschmeier, P., and Weinstein, I.B., 1987, Isolation of cDNA clones encoding protein kinase C: Evidence for a protein kinase C-related gene family, Proc. Natl. Acad. Sci. USA. 84:1065.
10. Housey, G.M., Johnson, M.D., Hsiao, W.L.W., O'Brien, C.A., Murphy, J.P., Kirschmeier, P., and Weinstein, I.B., 1988, Overproduction of protein kinase C causes disordered growth control in rat fibroblasts, Cell. 52:343.
11. Huang, K.-P., Nakabayashi, H., and Huang, F.L., 1986, Isozymic forms of rat brain Ca^{2+}-activated and phospholipid-dependent protein kinase, Proc. Natl. Acad. Sci. USA. 83:8535.
12. Huang, K.P., Huang, F.L., Nakabayashi, H., and Yoshida, Y., 1988, Biochemical characterization of rat brain protein kinase C isozymes, J. Biol. Chem. 263:14839.
13. Ido, M., Sekiguchi, K., Kikkawa, U., and Nishizuka, Y., 1987, Phosphorylation of the EGF receptor from A431 epidermoid carcinoma cells by three distinct types of protein kinase C, FEBS Lett. 219:215-218.
14. Jaken, S., and Kiley, S.C., 1987, Purification and characterization of three types of protein kinase C from rabbit brain cytosol, Proc. Natl. Acad. Sci. 84:4418.
15. Knopf, J.L., Lee, M.-H., Sultzman, L.A., Kriz, R.W., Loomis, C.R., Hewick, R.M. and Bell, R.M., 1986, Cloning and expression of multiple protein kinase C cDNAs, Cell. 46:491.
16. Kraft, A., and Anderson, W.B., 1983, Phorbol esters increase the amount of Ca^{2+}, phospholipid-dependent protein kinase associated with the plasma membrane, Nature. 301:621.
17. Leach, K.L., Powers, E.A., McGuire, J.C., Dong, L., Kiley, S.C., and Jaken, S., 1988, Monoclonal antibodies specific for Type 3 protein kinase C recognize distinct domains of protein kinase C and inhibit in vitro functional activity, J. Biol. Chem. 263:6927.
18. Niedel, J.E., Kuhn, L.J. and Vandenbark, G.R., 1983, Phorbol diester-receptor copurifies with protein kinase C, Proc. Natl. Acad. Sci. USA. 80:36.

19. Nishizuka, Y., 1986, Studies and perspectives on protein kinase C, Science. 233:305.
20. Nishizuka, Y., 1988, The molecular heterogeneity of protein kinase C and its implications for cellular regulation, Nature. 334:661.
21. Ohno, S., Kawasaki, H., Imajoh, S., Suzuki, K., Inagaki, M., Yokokura, H., Sakoh, T., and Hidaka, H., 1987, Tissue-specific expression of three distinct types of rabbit protein kinase C, Nature. 325:161.
22. Ohno, S., Akita, Y., Konno, Y., Imajoh, S., and Suziki, K., 1988, A novel phorbol ester receptor/protein kinase, nPKC, distantly related to the protein kinase C family, Cell. 53:731.
23. Ono, Y., Fujiik, T., Ogita, K., Kikkawa, U., Igarashi, K., and Nishizuka, Y., 1988, The structure, expression, and properties of additional members of the protein kinase C family, J. Biol. Chem. 263:6927.
24. Parker, P.J., Coussens, L., Totty, N., Rhee, L., Young, S., Chen, E., Stabel, S., Waterfield, M.D. and Ullrich, A., 1986, The complete primary structure of protein kinase C-the major phorbol ester receptor, Science. 233:853.
25. Persons, D.A., Wilkison, W.O., Bell, R.M., and Finn, O.J., 1988, Altered growth regulation and enhanced tumorigenicity of NIH 3T3 fibroblasts transfected with protein kinase C-1cDNA, Cell. 52:447-458.
26. Preiss, J., Loomis, C.R., Bishop, W.R., Stein, R., Niedel, J.E., and Bell, R.M., 1986, Quantitative Measurement of sn-1,2-diacylglycerols present in platelets, hepatocytes, and ras- and sis-transformed normal rat kidney cells, J. Biol. Chem. 261:8596-8600.
27. Sekiguchi, K., Tsukuda, M., Ase, K., Kikkawa, U., and Nishizuka, Y., 1988, Mode of activation and kinetic properties of three distinct forms of protein kinase C from rat brain, J. Biochem. (Tokyo) 103:759-765 (1988).
28. Weinstein, I.B., Lee, L., Fisher, P.B., Mufson, A., and Yamasaki, H., 1988, Action of phorbol esters in cell culture: mimicry of transformation, altered differentiation, and effects on cell membranes, J. of Supramol. Structure 12:195-208.
29. Welsh, C.J., Cao, H.-t., Chabbott, H., and Cabot, M.C., 1988, Vasopressin is the only component of serum-free medium that stimulates phosphatidylcholine hydrolysis and accumulation of diacylglycerol in cultured REF52 cells, Biochem. Biophys. Res. Commun. 152:565-572.

A PROTEIN KINASE C PSEUDOSUBSTRATE PEPTIDE BLOCKS G-PROTEIN AND MITOGENIC LECTIN INDUCED PHOSPHORYLATION OF THE CD3 ANTIGEN IN HUMAN T-LYMPHOCYTES

Denis R. Alexander, Jonathan D. Graves, Susan C. Lucas, J. Mark Hexham, Doreen A. Cantrell and Michael J. Crumpton

Imperial Cancer Research Fund
Lincoln's Inn Fields
London, WC2A 3PX

INTRODUCTION

Protein kinase C (PKC) activation occurs upon binding of mitogenic monoclonal antibodies (mAbs) to the antigen receptor complex (Ti-CD3) or CD2 antigen of human T-lymphocytes (Isakov et al., 1987; Truneh et al., 1985; Weiss and Imboden, 1987; Friedrich and Gullberg, 1988; Nel et al., 1987; Alexander and Cantrell, 1989). Activation of T-cell PKC is associated with changes in cell-surface expression of antigens such as CD2 (Cantrell et al., 1988), CD3 (Cantrell et al., 1985) and CD4 (Acres et al., 1986), with the phosphorylation of cytosolic 19kDa and 80kDa proteins (Friedrich and Gullberg, 1988; Friedrich et al., 1988), and with the induction of specific genes, such as the IL-2 receptor gene, which are required for a proliferative response (Isakov et al., 1987; Crabtree, 1989; Irving et al., 1989). Such data imply, though do not prove, that PKC plays a role in linking signals from Ti-CD3 and CD2 to nuclear events. In contrast, PKC is apparently not required for mediating mitogenic signals via the IL-2 receptor (Mills et al., 1988).

The pathways leading to PKC activation in normal human T-cells are poorly understood. Specific antigen, mitogenic lectins, or mAbs which bind to Ti, CD3 or CD2 lead to phosphatidylinositol (PI) phosphate turnover in T-cells, presumably by the activation of a PI-specific phospholipase C, with the consequent production of diacylglycerol (DAG) and inositol (1,4,5)-trisphosphate (IP_3) (Imboden and Stobo, 1985; Imboden et al., 1987; Pantaleo et al., 1987). Since IP_3 is known to elevate intracellular calcium by causing its release from intracellular stores (Berridge, 1987), and DAG activates PKC (Nishizuka, 1984), PI turnover appears to be an important mechanism for regulating PKC activity in T-cells. However, it is not yet known how Ti-CD3 and CD2 are coupled to changes in PI metabolism. A role for G-protein(s) has been suggested by the finding that aluminium fluoride can induce PI turnover and PKC activation in murine T-cell hybridomas (O'Shea et al., 1987), and that guanine nucleotides can regulate phospholipase C mediated PI turnover in murine thymocytes (Zilberman et al., 1987). Nevertheless, direct evidence that Ti-CD3 or CD2 mediate regulation of PI metabolism via G-protein coupling is not yet available.

Three further points underline the potential complexity of PKC regulation in T-cells. Firstly, it is now clear from work on other cells that PI metabolism represents only one of several possible sources of DAG, and that receptor mediated stimulation of PKC can occur in the absence of increased PI turnover (Whetton et al., 1988). For example, stimulation of phosphatidylcholine-specific phospholipase C and phospholipase D can also generate DAG (Pelech and Vance, 1989), so leading to PKC activation. Secondly, DAG is one of several agents that have been reported to activate PKC, and it is possible that PKC in T-cells could be regulated by the free fatty acid products which result from phospholipase A_2 action on membrane phospholipids (McPhail et al., 1984; Oishi et al., 1988). Thirdly, cDNA clones encoding 7 distinct PKC enzymes have been described so far (Nishizuka, 1988), and those sub-types which have been purified and characterised have different regulatory properties (Nishizuka, 1988; Ohno et al., 1988; Huang et al., 1988). At least two of these isozymes (PKC_α and PKC_β) are present in T-cells (Shearman et al., 1988; Yoshida et al., 1988). The PKC_β isozyme appears to be a major sub-type in human peripheral T-cells, whereas the PKC_α isozyme is relatively more abundant in the human leukaemia T-cell line Jurkat-6 (R. Marais and D. Cantrell, submitted). It is possible that the differential regulation of T-cell PKC isozymes could reflect distinct pathways of signal transduction.

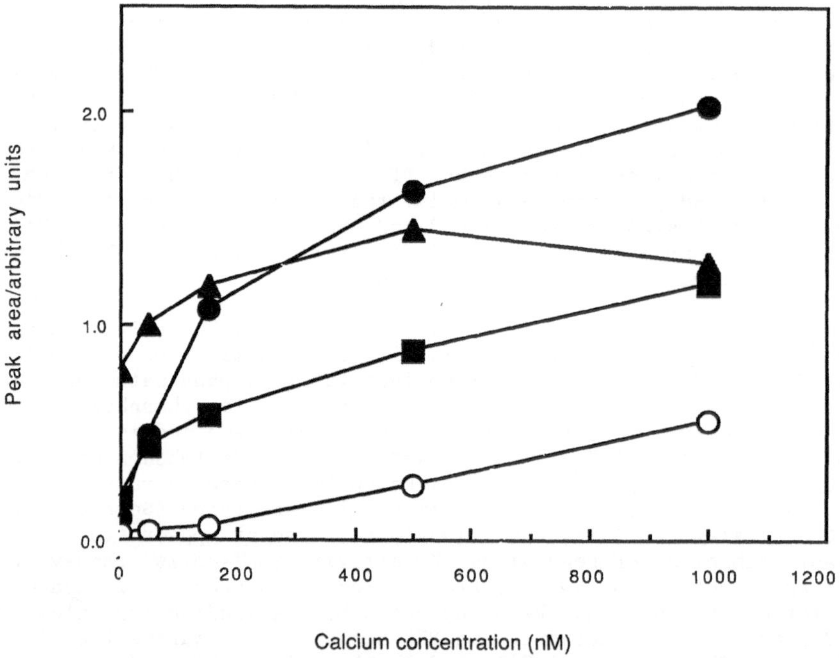

Fig. 1. **Calcium dependence of CD3 antigen γ chain phosphorylation in permeabilised T-lymphocytes.** Cells were incubated for 5 min with streptolysin-O in the presence of various concentrations of calcium, [γ-^{32}P]ATP and either 20 ng/ml Pdbu (▲), 10 μg/ml PHA (■), 100 μM GTPγS (●) or no addition (o). The autoradiograms of SDS-PAGE analyses of CD3 antigen γ chain phosphorylation were analysed densitometrically.

An immediate consequence of T-cell activation via the CD3-Ti or CD2 pathways is the phosphorylation of the CD3 antigen γ polypeptide (Cantrell et al., 1987; Breitmeyer et al., 1987). We have recently described a system using streptolysin-O permeabilised T-cells which facilitates investigation of signal transduction mechanisms, including CD3 antigen phosphorylation (Alexander et al., 1989). In this paper we show that mitogenic lectins such as phytohaemagglutinin (PHA) as well as GTPγS, phorbol esters and diacylglycerol can induce CD3 antigen phosphorylation in permeabilised cells, and that in each case phosphorylation can be effectively blocked by introducing into the cells a novel PKC pseudo-substrate peptide inhibitor. We also present evidence that the GTPγS regulated pathway of PKC activation which leads to phosphorylation of the CD3 antigen is distinct from the pathway activated by PHA.

RESULTS

Permeabilisation of T-Cells with Streptolysin-O

Streptolysin-O has previously been used to permeabilise a wide range of mammalian cells, particularly in studies on exocytosis (Gomperts, 1989). We have found that this reagent rapidly and reproducibly permeabilises human T-lymphoblasts and T-leukaemia cell lines, allowing the introduction of peptides and polypeptides into the cell interior (Alexander et al., 1989). The presence of calcium-EGTA in the permeabilisation buffer allows free calcium to be stabilised at concentrations pertaining to either resting (80-100 nM calcium) or activated (150-500 nM calcium) T-cells. Permeabilisation with streptolysin-O is followed by a loss of cytosolic markers, such as lactate dehydrogenase, from the cell interior, so that this enzyme activity is no longer detectable in the cell pellet after 15 min (Alexander et al., 1989). In the nominal absence of calcium, PKC is also lost from permeabilised cells. However, by including agents in the permeabilisation buffer which induce translocation of PKC to intracellular membranes, the enzyme is retained inside the cells, so allowing investigation of PKC mediated phosphorylation events (Alexander et al., 1989).

Phosphorylation of the CD3 Antigen in Permeabilised Cells

T-lymphoblasts (> 90% $CD8^+$) were derived from human peripheral blood as described (Cantrell et al., 1985; 1987) and maintained in culture for 7-14 days by the addition of exogenous interleukin-2 (IL-2). The IL-2 was removed from cell cultures 3 days before experiments in order to induce quiescence. Phosphorylation of the CD3 antigen γ polypeptide in streptolysin-O treated, quiescent T-lymphoblasts was carried out using [γ-^{32}P]-ATP in the permeabilisation buffer for 2 min at 37°C, before stopping the reaction by rapidly centrifuging and then lysing the cells on ice. The UCHT1 mAb on Sepharose beads was used for the immunoprecipitation of the phosphorylated Ti-CD3 complex. Phorbol-12,13-dibutyrate (Pdbu), PHA or GTPγS were found to induce γ chain phosphorylation (Fig. 1), as well as the diacylglycerol analogues 1-oleoyl-2-acetylglycerol and 1,2-diolein (data not shown). Only hydrolysis-resistant analogues of GTP induced significant phosphorylation, GTP, GDPβS, CTP and ITP being ineffective in this respect. The level of the γ chain phosphorylation induced by an optimal concentration of PHA at 500 nM calcium was only half that obtained with Pdbu or 100 μM GTPγS (Fig. 1), suggesting some differences in the signalling pathways regulated by PHA and GTPγS. Phosphorylation induced by Pdbu in the nominal absence of calcium was 50% of its value at 500 nM calcium, whereas phosphorylation induced by PHA and GTPγS was completely calcium-dependent (Fig. 1). In contrast, no Pdbu induced γ chain phosphorylation was detectable in permeabilised Jurkat-6 cells in the absence of calcium, although phosphorylation was substantial at a calcium concentration range of 50-500 nM in these cells (Alexander et al., 1989). These data are consistent with the dominant expression of the PKC_{β} isozyme, which has a substantial calcium-independent activity (Nishizuka, 1988; Ohno et al., 1988; Huang, 1988), in T-lymphoblasts, and the relatively high level of the PKC_{α} isozyme, which requires calcium for its activity, in Jurkat-6 cells (R. Marais and D. Cantrell, submitted). However, it should be stressed that both T-lymphoblasts and Jurkat-6 cells may contain other PKC isozyme(s) which could also be involved in mediating Pdbu induced CD3 antigen phosphorylation.

Inhibition of CD3 Antigen Phosphorylation by a PKC Pseudosubstrate Peptide

Since PKC is the only kinase known to be activated by Pdbu (Nishizuka, 1984), it is reasonable to assume that Pdbu induced CD3 antigen phosphorylation is a PKC mediated event. This assumption is supported by previous findings using a murine T-cell hybridoma showing that phorbol esters no longer induced CD3 antigen γ chain phosphorylation following down-regulation of PKC (Patel, 1987). However, chronic phorbol ester treatment of normal human T-cells down-regulates PKC by only 70-80% (Cantrell et al., 1988), showing that this method is not suitable for studying PKC mediated events in these cells. Since PHA or GTPγS could potentially induce CD3 antigen phosphorylation via stimulation of other kinases, we have

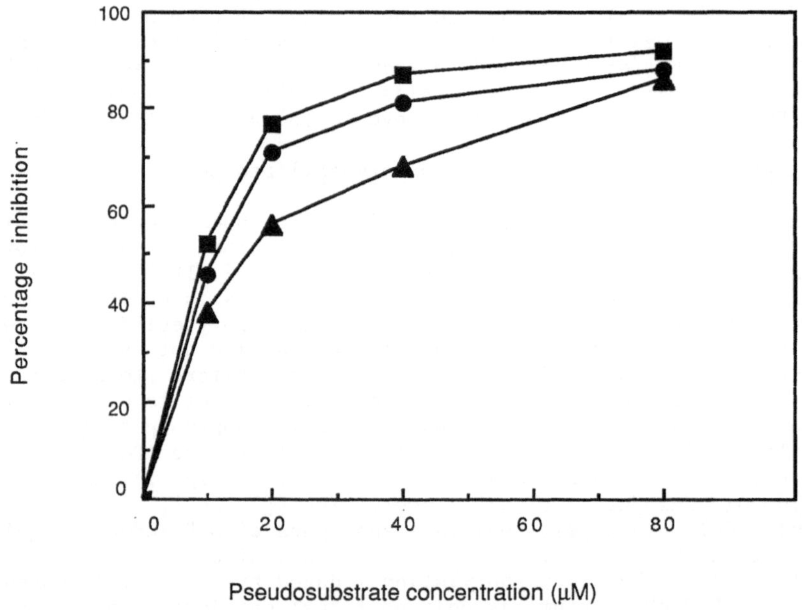

Fig. 2. **Inhibition of CD3 antigen γ chain phosphorylation by the PKC pseudosubstrate peptide in permeabilised T-lymphocytes.** Phosphorylation of the γ chain was carried out in the presence of streptolysin-0 and [^{32}P]ATP at a concentration of 400 nM calcium in the presence of 100 µM GTPγS (▲), 20 ng/ml Pdbu (●) and 10 µg/ml PHA (■). The pseudosubstrate peptide was added with the permeabilisation buffer at the concentrations shown.

established a novel technique whereby a pseudosubstrate peptide inhibitor of PKC was introduced into permeabilised T-cells. This peptide, which has the sequence Arg-Phe-Ala-Arg-Lys-Gly-Ala-Leu-Arg-Gln-Lys-Asn-Val, has previously been shown to be a potent competitive inhibitor of PKC activity at concentrations having no effect on protein kinase A or myosin light chain kinase activities (House and Kemp, 1987). Addition of 50 μM pseudosubstrate with streptolysin-O to T-lymphoblasts inhibited by > 90% phosphorylation of the CD3 antigen γ chain induced by PHA, GTPγS or Pdbu (Fig. 2), suggesting that phosphorylation was being mediated by PKC in each case. Regardless of the agent used to induce phosphorylation, the pseudosubstrate concentrations required to inhibit phosphorylation were very similar (Fig. 2), consistent with the idea that the same kinase was being activated by all three agents. Interestingly, it was also found that the relatively low level of γ chain phosphorylation induced by calcium alone (0.5-1.0 μM) in T-lymphoblasts (Alexander et al., 1989) was completely inhibited by 50 μM pseudosubstrate, as was calcium induced phosphorylation of the γ chain in Jurkat-6 cells (data not shown).

In contrast to its marked action on CD3 antigen phosphorylation, addition of 50 μM pseudosubstrate to permeabilised cells had no effect on the phosphorylation levels of 17 other polypeptides detected in whole cell lysates, showing that this peptide did not cause a generalised inhibition of kinases. Furthermore, addition of 50 μM concentrations of 4 other basic peptides, including the pseudosubstrate peptide of protein kinase A, to permeabilised T-lymphoblasts, did not affect Pdbu or 800 nM calcium induced CD3 antigen phosphorylation (data not shown). Thus, the PKC pseudosubstrate peptide appears to be specific in its inhibition of PKC, further supporting a role for activated PKC in mediating PHA and GTPγS induced CD3 antigen phosphorylation.

How do PHA and GTPγS Induce PKC Activation in Permeabilised T-Cells?

Although lectins are known to bind to a large number of T-cell surface glycoproteins (Sitkovsky et al., 1984), there is considerable evidence that the Ti-CD3 complex plays an essential role in mitogenic lectin induced T-cell activation. For example, PHA is unable to induce IL-2 secretion in T-cell mutants lacking cell-surface expression of Ti-CD3 (Weiss et al., 1984).

Since it has been proposed that signal transduction via Ti-CD3 involves G-protein coupling to phospholipase C, so inducing hydrolysis of inositol phospholipids (Weiss and Imboden, 1987), it is clearly of interest to know whether PHA and GTPγS mediate PKC activation in permeabilised T-cells via the same pathway. In order to investigate such a possibility, T-lymphoblasts were labelled with myo-[^{3}H]inositol, and inositol phosphate release from permeabilised cells was measured in the presence of PHA or GTPγS. It was found that PHA induced a 5-fold increase in inositol phosphate accumulation at 150 nM calcium, whereas the increase induced by 100 μM GTPγS was only 2-3 fold (Fig. 3(a)). There was a close correlation between the concentrations of PHA required to increase inositol phosphate levels, and the concentrations which induced CD3 antigen phosphorylation (Fig. 3(b)), consistent with the idea that PHA effects are mediated via the breakdown of inositol phospholipids, with the consequent production of DAG and activation of PKC (Nishizuka, 1984). In contrast, there was no correlation between the concentrations of GTPγS which induced increases in inositol phosphates, and the concentrations required to induce CD3 antigen phosphorylation (Fig. 4). For example, 10 μM GTPγS induced maximal inositol phosphate accumulation, yet did not cause CD3 antigen phosphorylation (Fig. 4), despite the fact that an equivalent increase in inositol phosphates induced by PHA was accompanied by concomitant CD3 antigen phosphorylation (Fig. 3).

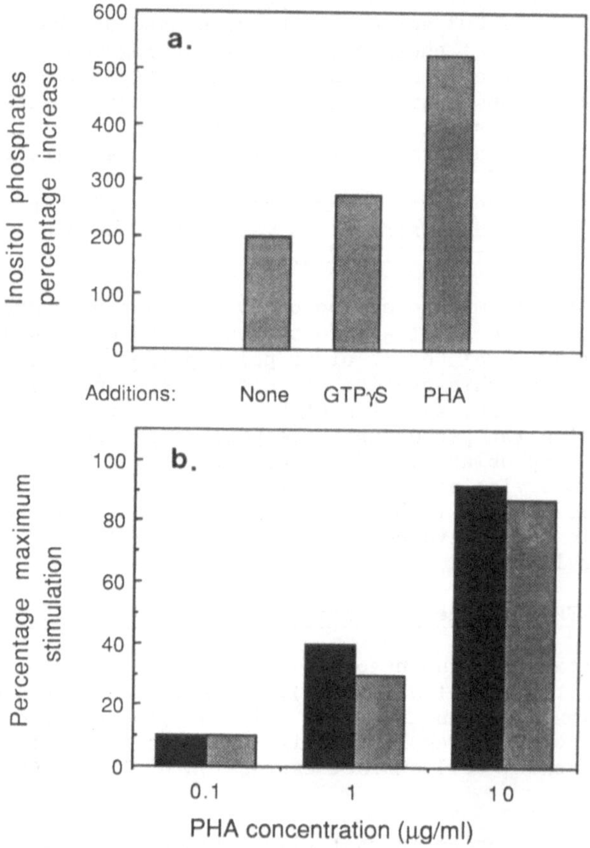

Fig. 3. **Inositol phosphate production in permeabilised T-lymphocytes.** (a) The relative increase in inositol phosphate levels induced by 100 μM GTPγS or 10 μg/ml PHA in cells permeabilised in the presence of 150 nM calcium. Total [^{3}H]inositol phosphates were extracted at 10 min and analysed by anion exchange. Data are expressed as a percentage of the inositol phosphates produced in cells permeabilised in the presence of nominal zero calcium alone. (b) A comparison between PHA induced production of inositol phosphates (▩) and CD3 antigen γ chain phosphorylation (■). Cells were incubated for 10 min before extraction of inositol phosphates and measurement of CD3 antigen phosphorylation in the presence of 500 nM calcium and the indicated amount of PHA. The data are representative of 3 separate experiments.

DISCUSSION

Our results clearly implicate PKC in mediating phosphorylation of the CD3 antigen γ polypeptide upon addition of PHA, GTPγS or Pdbu to permeabilised human T-cells. This conclusion depends largely upon the finding that CD3 antigen phosphorylation is inhibited by a PKC pseudosubstrate peptide. It is important, therefore, that further studies be continued on the effects of this peptide on other purified kinases. Nevertheless, our results so far indicate that the effects of the pseudosubstrate are confined to inhibition of PKC.

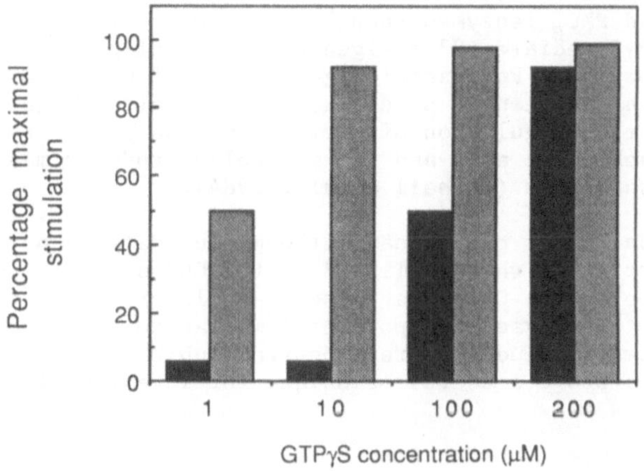

Fig. 4. **A comparison between GTPγS induced inositol phosphate production and CD3 antigen γ chain phosphorylation in permeabilised T-lymphocytes.** Details are as in the legend for Fig. 3(b), except that the indicated concentrations of GTPγS were included in the permeabilisation media. Inositol phosphates (▩); CD3 antigen γ chain phosphorylation (■).

Since PHA induced levels of inositol phosphates correlate with the extent of CD3 antigen γ chain phosphorylation, whereas no such correlation can be found for the actions of GTPγS, it appears that PHA and GTPγS induce PKC activation via pathways which may be overlapping, but which are not identical. Thus, although it is probable that both PHA and GTPγS (1-10 μM) act via a phosphoinositol specific phospholipase C, so mobilising DAG from inositol phospholipids, GTPγS at higher concentrations presumably has additional effects in activating other G-protein regulated pathways, perhaps in the process inducing DAG mobilisation from other membrane phospholipids, such as phosphatidylcholine (Whetton et al., 1988; Pelech and Vance, 1989). This additional DAG production could explain the marked γ chain phosphorylation induced by 50-100 μM GTPγS. However, it is possible that other GTPγS induced signals could have concomitant negative effects upon CD3 antigen phosphorylation, which predominate at lower concentrations of the nucleotide. This could explain our otherwise puzzling observation that 10 µM GTPγS induces a significant increase in inositol phosphate levels without inducing any detectable CD3 antigen phosphorylation. Such a finding could be due, for example, to stimulation of a G-protein regulated phosphatase, which counteracts the effects of PKC activation. In addition, GTPγS could activate a G_s regulated adenylate cyclase, so elevating cAMP, which then mediates down-regulating effects on the pathway(s) leading to CD3 antigen phosphorylation. Some support for this speculation comes from the work of Patel et al. (1987) who showed that cAMP inhibits antigen-induced PI turnover in a murine T-cell hybridoma, and also inhibits CD3 antigen γ chain phosphorylation in these cells. Furthermore, it has been reported that agents which elevate cAMP in human T-lymphocytes reduce the elevation of DAG and production of inositol phosphates induced via the CD2 pathway (Bismuth et al., 1988).

It is not yet known whether phosphorylation of the CD3 antigen is mediated via a specific isozyme of PKC, since the pseudosubstrate inhibits both the PKC_α and PKC_β isozymes (unpublished observations). It is possible that PHA and GTPγS mediate CD3 antigen phosphorylation by activating specific PKC isozymes. For example, generation of DAG's having fatty-acid side-chains of varying length or degrees of saturation might conceivably lead to differential regulation of certain PKC isozymes. Furthermore, it appears that arachidonic acid and its metabolic products may have selective effects on PKC sub-types (McPhail et al., 1984).

In conclusion, it is clear that although our data do not directly address the question of whether Ti-CD3 or the CD2 antigen are coupled to G-proteins, streptolysin-O permeabilised T-cells provide a convenient model system in which to analyse such potential mechanisms of signal transduction. The use of pseudosubstrate peptide inhibitors in permeabilised cells should also prove a useful technique for investigating other PKC mediated events.

ACKNOWLEDGEMENTS

We are indebted to Kim Richardson for typing the manuscript.

REFERENCES

Acres, R.B., Conlon, P.J., Mochizuki, D.Y., and Gallis, B., 1986, Rapid phosphorylation and modulation of the T4 antigen on cloned helper T cells induced by phorbol myristate acetate or antigen. J. Biol. Chem., 261:16210.

Alexander, D.R., and Cantrell, D.A., 1989, Kinases and phosphatases in T-cell activation. Immunol. Today, in press.

Alexander, D.R., Hexham, J.M., Lucas, S.C., Graves, J.D., Cantrell, D.A., and Crumpton, M.J., 1989, A protein kinase C pseudosubstrate peptide inhibits phosphorylation of the CD3 antigen in streptolysin-O permeabilised human T-lymphocytes. Biochem. J., in press.

Berridge, M.J., 1987, Inositol trisphosphate and diacylglycerol: two interacting second messengers. Ann. Rev. Biochem., 56:159.

Bismuth, G., Theodorou, I., Gouy, H., Le Gouvello, S., Bernard, A., and Debre, P., 1988, Cyclic AMP-mediated alteration of the CD2 activation process in human T lymphocytes. Preferential inhibition of the phosphoinositide cycle-related transduction pathway. Eur. J. Immunol., 18:1351.

Breitmeyer, J.B., Daley, J.F., Levine, H.B., and Schlossman, S.F., 1987, The T11 (CD2) molecule is functionally linked to the T3/Ti T cell receptor in the majority of T cells. J. Immunol., 139:2899.

Cantrell, D.A., Davies, A.A., and Crumpton, M.J., 1985, Activators of protein kinase C down-regulate and phosphorylate the T3/T-cell antigen receptor complex of human T lymphocytes. Proc. Natl. Acad. Sci. USA, 82:8158.
Cantrell, D.A., Davies, A.A., Londei, M., Feldman, M., and Crumpton, M.J., 1987, Association of phosphorylation of the T3 antigen with immune activation of T-lymphocytes. Nature, 325:540.
Cantrell, D.A., Verbi, W., Davies, A., Parker, P., and Crumpton, M.J., 1988, Evidence that protein kinase C differentially regulates the human T lymphocyte CD2 and CD3 surface antigens. Eur. J. Immunol., 18:1391.
Crabtree, G.R., 1989, Contingent genetic regulatory events in T lymphocyte activation. Science, 243:355.
Friedrich, B., and Gullberg, M., 1988, The role of protein kinase C in early activation vs. growth of T-lymphocytes. Eur. J. Immunol., 18:489.
Friedrich, B., Lunstrom, M., Noreus, K., Cantrell, D.A., and Gullberg, M., 1988, Activation-dependent phosphorylation of endogenous protein kinase C substrates in quiescent human T lymphocytes. Immunobiol., 176:465.
Gomperts, B.D., 1989, GTP-binding proteins and exocytotic secretion. in: "G-Proteins", L. Birnbaumer, and R. Iyengar, eds., Academic Press, New York.
House, C., and Kemp, B.E., 1987, Protein kinase C contains a pseudo-substrate prototope in its regulatory domain. Science, 238:1726.
Huang, K.P., Huang, F.L., Nakabayashi, H., and Yoshida, Y., 1988, Biochemical characterisation of rat brain protein kinase C isozymes. J. Biol. Chem., 263:14839.
Imboden, J.B., and Stobo, J.D., 1985, Transmembrane signalling by the T cell antigen. J. Exp. Med., 161:446.
Imboden, J.B., Weyand, C., and Garonzy, J., 1987, Antigen recognition by a human T cell clone leads to increases in inositol trisphosphate. J. Immunol., 138:1322.
Irving, S.G., June, C.H., Zipfel, P.F., Siebenlist, U., and Kelly, K., 1989, Mitogen-induced genes are subject to multiple pathways of regulation in the initial stages of T-cell activation. Mol. Cell. Biol., 9:1034.
Isakov, N., Mally, M.I., Scholz, W., and Altman, A., 1987, T-lymphocyte activation: the role of protein kinase C and the bifurcating inositol phospholipid signal transduction pathway. Immunol. Reviews, 95:89.
McPhail, L.C., Clayton, C.C., and Snyderman, R., 1984, A potential second messenger role for unsaturated fatty-acids: activation of Ca^{2+}-dependent protein kinase. Science, 224:622.
Mills, G.B., Girard, P., Grinstein, S., and Gelfrand, E.W., 1988, Interleukin-2 induces proliferation of T lymphocyte mutants lacking protein kinase C. Cell, 55:91.
Nel, A.E., Bouic, P., Luttanze, G.R., Stevenson, H.C., Miller, P., Dirienzo, W., Stefanini, G.F., and Galbraith, R.M., 1987, Reaction of T lymphocytes with anti-T3 induces translocation of C-kinase activity to the membrane and specific substrate phosphorylation. J. Immunol., 138:3519.
Nishizuka, Y., 1984, The role of protein kinase C in cell surface signal transduction and tumour promotion. Nature, 308:693.
Nishizuka, Y., 1988, The molecular heterogeneity of protein kinase C and its implications for cellular regulation. Nature, 334:661.
Ohno, S., Akita, Y., Konno, Y., Imajoh, S., and Suzuki, K., 1988, A novel phorbol ester receptor/protein kinase, nPKC, distantly related to the protein kinase C family. Cell, 53:731.
Oishi, K., Raynor, R.L., Charp, P.A., and Kuo, J.F., 1988, Regulation of protein kinase C by lysophospholipids. J. Biol. Chem., 263:6865.
O'Shea, J.J., Urdahl, K.B., Luong, H.T., Chused, T.M., Samelson, L.E., and Klausner, R.D., 1987, Aluminum fluoride induces phosphatidylinositol turnover, elevation of cytoplasmic free calcium, and phosphorylation of the T cell antigen receptor in murine T cells. J. Immunol., 41:3463.

Pantaleo, G., Olive, D., Poggi, A., Kozumbo, W.J., Moretta, L., and Moretta, A., 1987, Transmembrane signalling via the T11-dependent pathway of human T cell activation. Evidence for the involvement of 1,2-diacylglycerol and inositol phosphates. Eur. J. Immunol., 17:55.
Patel, M.D., Samelson, L.E., and Klausner, R.D., 1987, Multiple kinases and signal transduction. J. Biol. Chem., 262:5831.
Pelech, S.L., and Vance, D.E., 1989, Signal transduction via phosphatidylcholine cycles. Trends in Biochem., 14:28.
Shearman, M.S., Berry, N., Oda, T., Ase, K., Kikkawa, U., and Nishizuka, Y., 1988, Isolation of protein kinase C subspecies from a preparation of human T lymphocytes. FEBS Letts., 234:387.
Sitkovsky, M.V., Pasternack, M.S., Lugo, J.P., Klein, J.R., and Eisen, H.V., 1984, Isolation and partial characterization of concanavalin A receptors on cloned cytotoxic T lymphocytes. Proc. Natl. Acad. Sci. USA, 81:1519.
Truneh, A,, Albert, F., Goldstein, P., and Schmitt-Verhilst, A.M., 1985, Early steps of lymphocyte activation bypassed by synergy between calcium ionophores and phorbol ester. Nature, 313:318.
Weiss, A., Imboden, J., Shoback, D., and Stobo, J., 1984, Role of T3 surface molecules in human T-cell activation: T3-dependent activation results in an increase in cytoplasmic free calcium. Proc. Natl. Acad. Sci. USA, 81:4169.
Weiss, A., and Imboden, J.B., 1987, Cell surface molecules and early events involved in human T-lymphocyte activation. Adv. Immunol., 41:1.
Whetton, A.D., Monk, P.N., Consalvey, S.D., Huang, S.J., Dexter, T.M., and Downes, C.P., 1988, Interleukin 3 stimulates proliferation via protein kinase C activation without increasing inositol lipid turnover. Proc. Natl. Acad. Sci. USA, 85:3284.
Yoshida, Y., Huang, F.L., Nakabayashi, H., and Huang, K., 1988, Tissue distribution and developmental expression of protein kinase C isozymes. J. Biol. Chem., 263:9868.
Zilberman, Y., Howe, L.R., Hesketh, T.R., and Metcalfe, J.C., 1987, Calcium regulates inositol 1,3,4,5-tetrakisphosphate production in lysed thymocytes and in intact cells stimulated with concanavalin A. EMBO J., 6:957.

GROWTH FACTOR-STIMULATED TYROSINE

PHOSPHORYLATION OF PHOSPHOLIPASE C

Matthew Wahl[†], Nancy Olashaw[‡], and Shunzo Nishibe[†],

Jack Pledger[‡], Graham Carpenter[†§]
Departments of Biochemistry[†], Cell Biology[‡], and Medicine[§]
Vanderbilt University School of Medicine
Nashville, TN 37232

INTRODUCTION

Polypeptide growth factors are a diverse group of hormone-like molecules that elicit dramatic alterations in cellular structure, metabolism, gene expression, and proliferation (Roberts and Sporn, 1989). Receptor proteins differentially expressed on the surface of diverse cell types bind growth factors with high affinity and specificity (reviewed in Carpenter, 1987; Yarden and Ullrich, 1988). The growth factor receptors mediate the pleiotropic effects of growth factors on cellular functions; however, the molecular mechanisms of growth factor-stimulated mitogenic signaling are incompletely defined.

Identification and purification of receptors for several growth factors, including the epidermal growth factor receptor (EGF-R) and the platelet-derived growth factor receptor (PDGF-R), together with elucidation of their primary structures from complementary DNA, revealed the presence of intrinsic receptor tyrosine kinase activity (Ek et al., 1982; Tremble et al., 1986; Carpenter et al., 1978; Downward et al., 1984; Ullrich et al., 1984). Binding of the appropriate ligand to the extracellular domain of a growth factor receptor leads to stimulation of the intracellular kinase activity. Alterations of cellular function that rapidly ensue upon ligand stimulation may depend in part or in whole upon ligand-stimulated receptor many such responses are absent in cells expressing kinase-deficient growth factor receptors (Escobedo et al., 1988; Chen et al., 1987; Honegger et al., 1987a; Honegger et al., 1987b; Molenaar et al., 1988; Glenney et al., 1988). Furthermore, the proliferation of cells expressing the kinase-deficient receptors is no longer affected by the growth factor (Escobedo et al., 1988; Chen et al., 1987; Honegger et al., 1987a). Identification of physiological substrates for the tyrosine kinases may help define the biochemical mechanisms important in regulation of cell proliferation. Thus far, despite the identification of certain proteins as substrates for growth factor receptor kinases (reviewed in Carpenter, 1987), it is still unclear whether or not the phosphorylation of any of these cellular substrates on tyrosine residues plays an important role in the control of proliferation.

Two well-characterized, growth factor-sensitive cell culture systems were used in our investigation of cellular substrates for receptor tyrosine kinases that may be relevant to the mechanism of growth factor action. PDGF, a dimeric polypeptide present in serum, modulates the proliferation of many mesenchymal cell types (Pledger et al., 1977). The BALB/c 3T3 fibroblast is a non-transformed murine cell line that expresses functional PDGF-Rs on the cell surface (Bowen-Pope and Ross, 1982; Heldin et al., 1982; Huang et al., 1982). Addition of PDGF to BALB/c 3T3 cells rapidly activates the PDGF-R tyrosine kinase. Following PDGF binding, quiescent BALB/c 3T3 cells undergo alterations in a number of cellular functions, such as lipid metabolism, shape, protein synthesis, and gene transcription, collectively termed "competence" (Pledger et al., 1977). PDGF-stimulated re-entry into the cell cycle renders BALB/c 3T3 cells sensitive to the presence of "progression" factors, such as insulin-like growth factor-I or EGF, which ultimately lead to cell division (Pledger et al., 1977). EGF is a polypeptide found in a variety of biological fluids and tissues (reviewed in Carpenter and Wahl, 1989). While the physiological role of EGF is still undefined, EGF is clearly a potent modulator of the proliferation and differentiation of epithelial tissues in intact animals (Carpenter and Wahl, 1989). A-431 is a transformed cell line derived from a human epidermoid carcinoma that is particularly suited to the investigation of cellular biochemical events that ensue upon EGF binding, since it expresses a very high level of cell surface EGF-Rs (Fabricant et al., 1977). Many observations related to understanding the mechanisms of growth factor action, including EGF receptor purification and description of ligand-stimulated tyrosine kinase activity, were made initially with the A-431 cell line (reviewed in Carpenter and Wahl, 1989). By comparison of the responses elicited by PDGF in BALB/c 3T3 cells with those elicited by EGF in A-431 cells we have sought to identify common features involved in growth factor regulation of cell proliferation.

Activation of cell surface PDGF-R in BALB/c 3T3 and EGF-R in A-431 cells after ligand binding leads to rapid changes in many cellular functions; thus possible cellular substrates for the receptor tyrosine kinase include the key proteins involved in the regulation of these functions. Of considerable interest in the study of growth factor signaling mechanisms are the observations that both PDGF and EGF stimulate rapid increases in the cellular metabolism of phosphatidylinositides (PIs) (Habenicht et al., 1981; Sawyer and Cohen, 1981). A key step in the hormone-sensitive PI pathway is the hydrolysis of phosphatidylinositol 4,5-bisphosphate (PIP_2) by the enzyme phospholipase C (PLC), generating two compounds, diacylglycerol and inositol 1,4,5-trisphosphate (IP_3), which act as second messengers in hormonal action (reviewed in Abdel-Latif, 1986; Rhee et al., 1989). Diacylglycerol is a potent stimulator of protein kinase C activity and IP_3 is capable of releasing stored intracellular Ca^{2+} (Berridge, 1987; Nishizaka, 1984). Both protein kinase C activity and intracellular Ca^{2+} level are potential modulators of the mitogenic response. Thus, the sustained growth factor-stimulated production of intracellular second messengers from PIP_2 represents a mechanism by which growth factors may regulate cellular proliferation.

We have studied growth factor-stimulated PI metabolism in A-431 and BALB/c 3T3 cells in order to identify possible receptor kinase substrates. Recent data demonstrate that treatment of BALB/c 3T3 cells with PDGF or A-431 cells with EGF results in a rapid increase in tyrosine phosphorylation of the PLC-γ isozyme (Wahl et al., 1988; Wahl et al., 1989a; Wahl et al., 1989b; Meisenhelder et al., 1989). We compared the stimulation of PI metabolism with the tyrosine phosphorylation of PLC-γ after growth factor treatment of cells under several experimental conditions (Wahl et al., 1989a; Wahl et al., 1989b). The data demonstrate that growth factor-stimulated PLC-γ tyrosine phosphorylation occurs: 1) at least as rapidly as an increase in IP_3 formation 2); at low temperature (3°C) in the absence of detectable IP_3 formation; and 3) in the presence and absence of extracellular Ca^{2+}. Furthermore, PLC-γ appears to be phosphorylated on tyrosine only after treatment of cells with certain agonists known to stimulate receptor tyrosine kinase activity, rather than after treatment with any agonist of PI metabolism. We conclude from these results that increased tyrosine phosphorylation of PLC-γ is specifically associated with growth factor treatment of cells in a manner that suggests a functional relationship between PLC-γ phosphorylation and increased PI metabolism.

RESULTS AND DISCUSSION

In order to investigate the possible role of receptor tyrosine kinase activity in PDGF- and EGF-stimulated increases in PI metabolism two basic approaches were employed. In certain experiments, cells were metabolically radiolabeled with either [^{35}S]-methionine or ^{32}P-orthophosphate. After treatment in the absence or presence of growth factors, cellular proteins were extracted with a detergent-containing buffer. Metabolically radiolabeled PLC-γ protein was immunoprecipitated with specific monoclonal antibodies (Suh et al., 1988) and visualized by gel electrophoresis and autoradiography. In other experiments, cells were treated without or with growth factor and then cellular proteins were extracted with a detergent-containing buffer. The soluble extracts were then treated with antiphosphotyrosine affinity matrix (Frackelton, 1983) in order to adsorb proteins phosphorylated on tyrosine. After washing the matrix to remove non-adsorbed proteins, the specifically bound proteins were recovered from the matrix by elution with buffer containing phenylphosphate, a phosphotyrosine analog. PLC activity present in the soluble fractions was assayed using [^{3}H]-PIP_2 as the substrate, and the activity in these fractions compared.

PLC-γ Tyrosine Phosphorylation in Intact Cells

In both BALB/c 3T3 and A-431 cells treatment of ^{35}S-radiolabeled cellular proteins with monoclonal antibodies (Suh et al., 1988) prepared against three distinct PLC isozymes (PLC-ß1, PLC-γ, PLC-δ having $M_r \approx 150$ kDa, 145 kDa, 85 kDa, respectively) results in readily detectable precipitation of the ^{35}S-PLC-γ isozyme (Wahl et al., 1989a; Wahl et al., 1989b). Thus, we investigated EGF- and PDGF-stimulated phosphorylation of PLC-γ in A-431 and BALB/c 3T3 cells labeled with ^{32}P-orthophosphate. EGF and PDGF treatment increase the incorporation of radioactivity into ^{32}P-PLC-γ approx. 3.0-fold and 5.0-fold respectively (Figure 1, insets). Acid hydrolysis of the ^{32}P-radiolabeled PLC-γ and separation of phosphoamino acids by high voltage thin layer electrophoresis revealed approx. 7.0-fold increases in the PLC-γ phosphotyrosine content from growth factor-stimulated cells (Wahl et al., 1989a; Wahl et al., 1989b). These results suggest that PLC-γ is a cellular substrate for the EGF receptor in A-431 cells and the PDGF receptor in BALB/c 3T3 cells, but do not exclude the possibility that PLC-γ is a substrate for an intervening cellular tyrosine kinase. However, the EGF receptor readily phosphorylates the PLC-γ isozyme on tyrosine residues when the two purified proteins are combined in an *in vitro* kinase assay, whereas PLC-ß1 and PLC-δ are relatively poor substrates for the receptor (Nishibe et al., 1989).

Phosphoamino acid analysis of ^{32}P-PLC-γ also demonstrated that, in both A-431 and BALB/c 3T3 cells, PLC-γ is phosphorylated on serine residues, and that growth factor treatment of cells increased the phosphoserine content approx. 2-fold (Wahl et al., 1989a; Wahl et al., 1989b). Thus, PLC-γ is also a substrate for a growth factor-stimulated serine kinase.

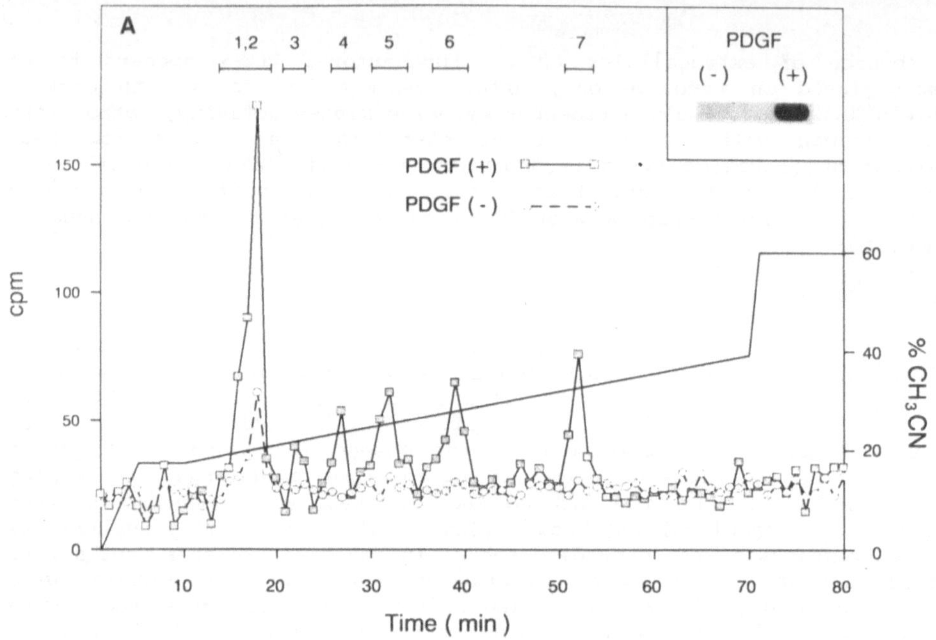

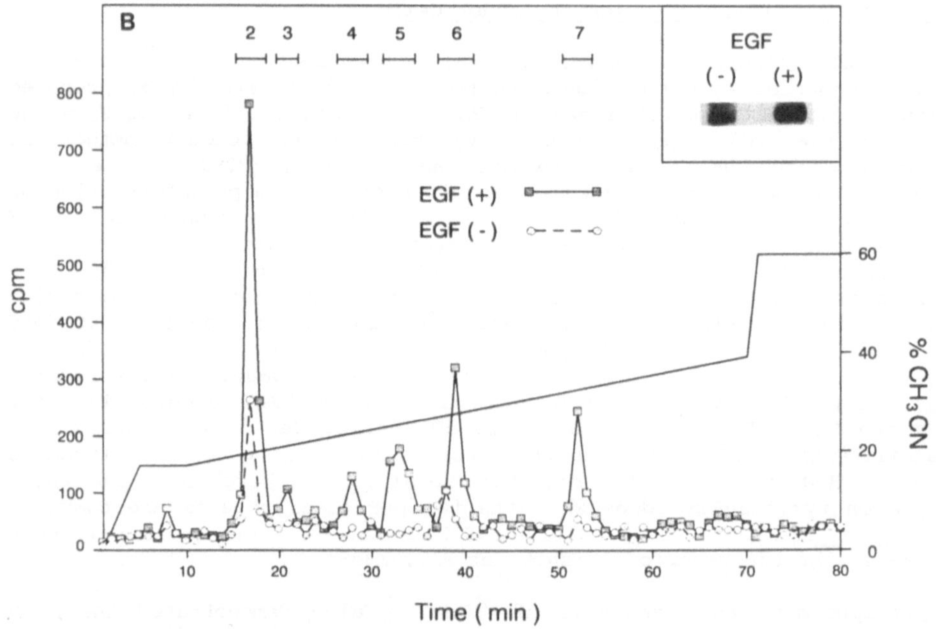

Figure 1. Comparison of phosphorylation of PLC-γ in PDGF-treated BALB/c 3T3 cells (A) with EGF-treated A-431 cells (B). The autoradiographs of ^{32}P-PLC-γ immunoprecipitated from cells treated without or with growth factor for 10 min (PDGF) or 2 min (EGF) at 37°C are shown in the insets. The gel slices were digested with trypsin and the phosphopeptides separated by HPLC using a C8 reverse phase column.

Multiple Sites of PLC-γ Phosphorylation

In order to evaluate whether PDGF and EGF stimulation resulted in the phosphorylation of PLC-γ on identical amino acid residues, the immunoprecipitated ^{32}P-radiolabeled PLC-γ protein from BALB/c 3T3 and A-431 cells was digested with excess trypsin, and the resulting peptides were separated by high performance liquid chromatography (HPLC). The PLC-γ tryptic phosphopeptides were located by measuring the ^{32}P radioactivity present in each fraction (Figure 1). Several PLC-γ phosphopeptides were identified from both EGF- and PDGF-treated cells that were not detectable in untreated cells. Thus it is likely that growth factor treatment induces phosphorylation of PLC-γ at multiple sites. Since the major tryptic phosphopeptides are common to both the human PLC-γ from EGF-treated A-431 cells and murine PLC-γ from PDGF-treated BALB/c 3T3 cells, it is likely the EGF and PDGF stimulate phosphorylation of identical sites on PLC-γ. We analyzed each of the ^{32}P-PLC=γ tryptic phosphopeptides for the presence of phosphotyrosine. A majority of the phosphopeptides (1, 4, 5, 6, 7) from growth factor-treated cells contained phosphotyrosine as judged either by high voltage thin-layer electrophoretic phosphoamino acid analysis (34) or by adsorption to phosphotyrosine antibody matrix. Certain of the ^{32}P-PLC-γ phosphopeptides (1, 2) from PDGF-treated cells were poorly resolved by HPLC separations. Phosphopeptide 2 from control and growth factor-stimulated cells did not bind to phosphotyrosine antibody matrix and contained the majority of the PLC-γ phosphoserine. It is clear from these data that growth factor treatment results in phosphorylation of multiple tyrosine residues in PLC-γ, and at least a single serine residue.

Stoichiometry of PLC-γ Tyrosine Phosphorylation

In order to estimate the percentage of PLC-γ that is phosphorylated on tyrosine after growth factor treatment of cells, detergent extracts containing ^{35}S-radiolabeled or unlabeled cellular proteins were first exposed to the phosphotyrosine immunoaffinity matrix, then the non-adsorbed and specifically eluted fractions were treated with PLC-γ antibodies. The immunoprecipitates from unlabeled cells were assayed for PLC activity, while the immunoprecipitates from radiolabeled cells were visualized by gel electrophoresis and autoradiography (data not shown). We observed that PDGF treatment of BALB/c 3T3 cells led to tyrosine phosphorylation of approx. 70% of the ^{35}S-PLC-γ protein, whereas the corresponding immunoprecipitate of unlabeled proteins contained approx. 53% of the total PLC activity precipitable with antibodies to PLC-γ. In A-431 cells, EGF induced tyrosine phosphorylation of approx. 35% of the PLC-γ, as determined by immunoprecipitation of ^{35}S-PLC-γ. Assay of the PLC activity revealed that the corresponding unlabeled immunoprecipitate contained approx. 45% of the PLC activity precipitable with PLC-γ antibodies. These data reveal that a large portion (approx. 35-70%) of the total PLC-γ pool is rapidly phosphorylated on tyrosine residues after growth factor treatment. The assay of PLC activity present in the PLC-γ immunoprecipitates indicates that no phosphorylation-associated change in PLC-γ specific activity was detected. However, the relationship between PLC activity measured in the immunoprecipitates with the PLC specific activity in growth factor-treated or control cells is unclear.

Growth Factor Regulation of PI Metabolism and PLC-γ Phosphorylation

We compared the stimulation of cellular PI metabolism and the tyrosine phosphorylation of PLC-γ under a variety of experimental conditions in order to assess the possibility of a functional relationship between the two events (Wahl et al., 1989a; Wahl et al., 1989b). To monitor cellular changes in the PI pathway, cells were labeled with [^{3}H]-

inositol. Changes in PI metabolism were then assessed by measuring [^{3}H]-inositol phosphates ([^{3}H]-IPs), which are produced by PLC-mediated hydrolysis of PIP_2 and subsequent metabolic conversions. The [^{3}H]-IPs produced in each experimental condition were quantitated after separation by anion exchange chromatography of the [^{3}H]-inositol-containing compounds present in acid extracts of cells (Wahl and Carpenter, 1988; Olashaw and Pledger, 1988). The growth factor stimulation of PLC-γ tyrosine phosphorylation was assessed by measuring the PLC activity that was extracted with phosphotyrosine antibody matrix from detergent-solubilized cell preparations.

Growth Factor Stimulation at Low Temperature

Addition of PDGF to BALB/c 3T3 cells and EGF to A-431 cells at 37°C produces increased formation of [^{3}H]-IPs and tyrosine phosphorylation of PLC-γ with time courses that are very rapid and essentially indistinguishable (Wahl et al., 1988a; Wahl et al., 1989b). Since ligand stimulation of receptor tyrosine kinase activity at 37°C is accompanied by a variety of other metabolic events, such as the rapid internalization and down-regulation of the receptor (reviewed in Carpenter, 1987), the relationship between these events, PI metabolism, and PLC-γ phosphorylation is unclear. While both PDGF and EGF are capable of stimulating receptor tyrosine kinase activity when cells are treated with growth factors at a low temperature, a variety of other receptor-associated phenomena, such as internalization and down-regulation, do not occur (Carpenter, 1987).

As demonstrated in Table 1, both EGF and PDGF were capable of stimulating tyrosine phosphorylation of PLC-γ at low temperature; however, there was no stimulation of [^{3}H]-IP_3 formation (Huang et al., 1982; Wahl et al., 1989b). The same PLC-γ amino acid residues were phosphorylated at 3°C and 37°C, as determined by tryptic digestion and HPLC resolution of ^{32}P-PLC phosphopeptides (data not shown). The rapid growth factor-induced stimulation of PLC-γ phosphorylation at 3°C suggests that PLC-γ is a readily accessible substrate for cell surface EGF and PDGF receptors. Furthermore, PLC-γ tyrosine phosphorylation clearly does not require receptor internalization or PLC-mediated hydrolysis of PIP_2, suggesting that phosphorylation is an early event in the interaction of the receptor tyrosine kinase and PI signaling pathways.

Table 1. Growth Factor Stimulation of PLC-γ Tyrosine Phosphorylation in Intact Cells at 3°C

Cell Line	Growth Factor	Fold Increase PLC Activity
BALB/c 3T3 fibroblast	PDGF	19.1
A-431 epidermoid carcinoma	EGF	14.5

Cells were treated without or with growth factor (PDGF, 100 ng/ml, 10 min; EGF, 200 ng/ml, 30 min) at 3°C, lysed in detergent buffer, and extracted with phosphotyrosine antibody matrix. The specific eluates obtained from the antibody matrix were assayed for PLC activity as described (Wahl et al., 1989a; Wahl et al., 1989b).

Influence of Extracellular Ca^{2+} in Growth Factor Stimulation

The intracellular level of free Ca^{2+} is likely to be an important regulator of cellular PLC activities, since all of the purified mammalian isozymes display Ca^{2+}-dependent hydrolysis of PIs (reviewed in Rhee et al., 1989). Growth factor treatment of cells at 37°C results in rapid increases in intracellular free Ca^{2+} levels (Hasegawa-Sasaki, 1985; Hepler et al., 1987). The initial increase in free Ca^{2+} presumably results from IP_3-mediated release of stored Ca^{2+} and occurs in the absence of extracellular Ca^{2+}. However, prolonged growth factor-stimulated increases in free Ca^{2+} levels require the presence of extracellular Ca^{2+}. The biochemical mechanisms regulating the Ca^{2+} homeostasis of hormone-treated cells are largely undefined, but represent an area of considerable research effort and interest (Berridge and Galione, 1988).

We compared the influence of extracellular Ca^{2+} in the stimulation of PLC-γ tyrosine phosphorylation and PI metabolism (Wahl et al., 1989a; Wahl et al., 1989b). Cells were treated with EGF or PDGF for a brief or prolonged period of time, in the absence or presence of extracellular Ca^{2+}. At an early time after growth factor treatment, extracellular Ca^{2+} had little effect on the increases in the cellular levels of [^{3}H]-IP_3 (Table 2)

Table 2. Growth Factor-Induced PLC-γ Tyrosine Phosphorylation and [^{3}H]-IP_3 Formation in the Absence and Presence of Extracellular Ca^{2+}

	Time (min)	Fold Increase PLC Activity	
		$-Ca^{2+}$	$+Ca^{2+}$
BALB/c 3T3 (+/-PDGF)			
†Intact Cells	5	0.28	0.35
	30	0.25	0.79
‡In Vitro	2	30.4	26.1
	30	34.9	20.1
A-431 (+/-EGF)			
†Intact Cells	1	0.30	0.29
	30	0.13	1.83
‡In Vitro	1	21.4	18.8
	30	18.8	14.7

†Cells labeled with [^{3}H]-inositol were treated without or with growth factor (PDGF, 100 ng/ml; EGF 200 ng/ml) in the absence or presence of extracellular Ca^+ (1.8 mM), lysed with 10% trichloroacetic acid, and [^{3}H]-IP_3 separated using anion exchange chromatography as described (Wahl and Carpenter, 1988; Olashaw and Pledger, 1988).

‡Cells were treated without or with growth factor (PDGF, 100 ng/ml; EGF, 200 ng/ml) in the absence or presence of extracellular Ca^{2+} (1.8 mM), lysed in detergent buffer, and extracted with phosphotyrosine antibody matrix. The specific eluates obtained from the antibody matrix were assayed for PLC activity as described (Wahl et al., 1989a; Wahl et al., 1989b).

and other [^{3}H]-IPs (data not shown). When the treatment period was extended (30 min), a marked dependence upon extracellular Ca^{2+} was observed for maximal growth factor-stimulated formation of [^{3}H]-IP_3 (Table 2) and other [^{3}H]-IPs (data not shown). Tyrosine phosphorylation of PLC-γ proceeded rapidly in the absence and presence of extracellular Ca^{2+} and remained elevated even 30 min after growth factor treatment. Thus, extracellular Ca^{2+} serves as a positive regulator of PDGF and EGF stimulation of cellular PI metabolism, but it is not required either for PLC-γ tyrosine phosphorylation or the initial formation of IP_3 that results from growth factor treatment. These results again suggest a role for increased PLC-γ phosphorylation early in the interaction of the receptor tyrosine kinase and PI signaling pathways; however, increased PLC-γ phosphorylation is not sufficient to maintain the maximal growth factor-stimulated increase in cellular IP_3 formation or Ca^{2+} mobilization.

<u>Agonist-Stimulated PI Metabolism and PLC-γ Phosphorylation</u>

Diverse ligands are capable of inducing rapid increases in cellular PI metabolism (Abdel-Latif, 1986). The mechanisms by which many different agents couple to a common signaling pathway are unknown. Many reports have implicated unidentified GTP-binding proteins as playing a functional role in receptor-PLC coupling (Cockcroft, 1987; Fain et al., 1988). Since

Table 3. Comparison of PLC-γ Tyrosine Phosphorylation and [^{3}H]-IP_3 Formation Induced by Agonists

Cell	Agonist	Fold Increase PLC Activity	
		†Intact Cells	‡In Vitro
BALB/c 3T3	PDGF (100 ng/ml)	1.90	43.9
	NaF (10 mM) + $AlCl_3$ (10μM)	1.87	0.44
A-431	EGF (200 ng/ml)	0.70	64.0
	Bradykinin (10 μM)	1.02	0.0
	ATP (20 μM)	0.72	0.0

†Cells labeled with [^{3}H]inositol were treated without or with the indicated agonist for 10 min (BALB/c 3T3) or 2 min (A-431), lysed with 10% trichloroacetic acid, and [^{3}H]-IP_3 separated using anion exchange chromatography as described (Wahl and Carpenter, 1988; Olashaw and Pledger, 1988).

‡Cells were treated without or with the indicated agonist for 10 min (BALB/c 3T3) or 2 min (A-431), lysed in detergent buffer, and extracted with phosphotyrosine antibody matrix. The specific eluates obtained from the antibody matrix were assayed for PLC activity as described (Wahl et al., 1989a; Wahl et al., 1989b).

direct tyrosine phosphorylation of a PLC isozyme represents a novel mechanism for interaction of receptor with the PI pathway, we investigated the specificity of tyrosine phosphorylation in the stimulation of PI metabolism (Wahl et al., 1989a; Wahl et al., 1989b). BALB/c 3T3 cells were treated either with PDGF or with AlF_4^- (generated by the combination of NaF and $AlCl_3$). Both of these treatments yielded similar increases in cellular formation of $[^3H]$-IP_3 (Table 3). However, while PDGF stimulated tyrosine phosphorylation of PLC-γ, no tyrosine phosphorylation was associated with the AlF_4^- treatment. This is consistent with the hypothesis that AlF_4^- modulates cellular PLC activity through direct interaction with a GTP-binding-PLC regulatory protein, without affecting cellular tyrosine kinase activity. In A-431 cells, EGF, bradykinin, and ATP were each potent stimulators of cellular $[^3H]$-IP_3 formation, yet PLC-γ tyrosine phosphorylation was associated only with EGF (Table 3). Taken together, these results suggest that tyrosine phosphorylation of PLC-γ is not merely coincident with any agonist-stimulated increases in cellular PLC activity. It is unclear whether the agents that do not induce PLC-γ tyrosine phosphorylation couple to PLC-γ by a discreet mechanism involving GTP binding proteins, or if these agents couple to distinct PLC isozymes present in these cell types. The possibility that distinct PLC isozymes are utilized cannot be ruled out, since a substantial portion of total detergent-extractable PLC activity in these cell types is not attributable to PLC-γ (data not shown).

Conclusion

The biochemical mechanisms involved in transduction of a mitogenic signal are largely undefined. We examined EGF-stimulated A-431 cells and PDGF-treated BALB/c 3T3 cells in order to identify common biochemical aspects involved in cellular responses to growth factors. Since receptor tyrosine kinase activity appears to be fundamental for cellular mitogenic responsiveness to growth factors, and since the PI signaling pathway is rapidly affected in growth factor-treated cells, we sought to identify biochemical responses common to both PDGF and EGF that may play a functional role in the interaction of these second messenger systems. Since PLC is believed to be a key enzyme in the hormone-sensitive PI pathway, we investigated whether or not PLC was a cellular substrate for the PDGF and EGF receptor kinases. Indeed, growth factor treatment of cells results in rapid tyrosine phosphorylation of the 145 kDa PLC-γ isozymes. The observations that growth factor-stimulated tyrosine phosphorylation of PLC-γ: 1) occurs rapidly at 37°C, and at 3°C in the absence of receptor internalization or PIP_2 hydrolysis; 2) does not require extracellular Ca^{2+}; and 3) is not observed with ligands that are not associated with stimulation of tyrosine kinase activity; all suggest a functional role for direct interaction of receptor and PLC-γ in the initiation of increased PIP_2 hydrolysis. Since extracellular Ca^{2+} is clearly a potent modulator of growth factor-stimulated cellular PI metabolism, and since other enzymes involved in PI metabolism such as the PI kinases are also likely sites for interaction of these two signaling pathways, the relationship of PLC-γ tyrosine phosphorylation to cellular PLC-γ activity and the mitogenic response remains uncertain.

Acknowledgements

This work was supported by National Institutes of Health research grants CA42713, CA43720, and training grants DK07563 and GM07347. We thank Susan Heaver for help in preparing this manuscript.

REFERENCES

Abdel-Latif, A. A., 1986, Calcium-mobilizing receptors, polyphosphoinositides and the generation of second messengers, Pharmacol. Rev. 38:227.

Berridge, M. J. and Galione, A., 1988, Cytosolic calcium oscillators, FASEB J. 2:3074.

Berridge, M. J., 1987, Inositol trisphosphate and diacylglycerol: Two interacting second messengers, Ann. Rev. Biochem. 56:159.

Bowen-Pope, D. F. and Ross, R., 1982, Platelet-derived growth factor. II. Specific binding to cultured cells, J. Biol. Chem. 257:5161.

Carpenter, G., 1987, Receptors for epidermal growth factor and other polypeptide mitogens, Ann. Rev. Biochem. 56:881.

Carpenter, G., King, Jr., L. and Cohen, S., 1978, Epidermal growth factor stimulates phosphorylation in membrane preparations in vitro, Nature 276:409.

Carpenter, G. and Wahl, M. I., 1989, The epidermal growth factor family of mitogens, in: "Peptide Growth Factors and their Receptors," Handbook of Experimental Pharmacology, A. B. Sporn, M. B. Roberts, ed., Springer-Verlag, Heidelberg.

Chen, W. S., Lazar, C. S., Poenie, M., Tsien, R. Y., Gill, G. N., and Rosenfeld, M. G., 1987, Requirement for intrinsic protein tyrosine kinase in the immediate and late actions of the EGF receptor, Nature 328:820.

Cockcroft, S., 1987, Polyphosphoinositide phosphodiesterase: regulation by a novel guanine nucleotide binding protein, Gp, Trends in Biol. Sci. 12:75.

Cooper, J. A., Sefton, B. M., and Hunter, T., 1983, Detection and quantification of phosphotyrosine in proteins, Meth. Enzymol. 99:387.

Downward, J., Yarden, Y., Mayes, E., Scrace, G., Totty, N., Stockwell, P., Ullrich, A., Schlessinger, J., Waterfield, M. D., 1984, Close similarity of epidermal growth factor receptor on v-erb-B oncogene protein sequences, Nature 307:521.

Ek, B., Westermark, B., Wasteson, H., and Heldin, C. H., 1982, Stimulation of tyrosine-specific phosphorylation by platelet-derived growth factor, Nature 295:419.

Escobedo, J. A., Barr, P. J., and Williams, L. T., 1988, Role of tyrosine kinase and membrane-spanning domains in signal transduction by the platelet-derived growth factor receptor, Mol. Cell. Biol. 8:5126.

Fabricant, R. N., De Larco, J. E. and Todaro, G. J., 1977, Nerve growth factor receptors on human melanoma cells in culture, Proc. Natl. Acad. Sci. USA 74:565.

Fain, J. N., Wallace, M. A., and Wojcikiewicz, K. J. H., 1988, Evidence for involvement of guanine nucleotide-binding regulatory proteins in the activation of phospholipase C by hormones, FASEB J. 2:2569.

Frackelton, Jr., A. R., 1983, Characterization of phosphotyrosyl proteins in cells transformed by Abelson murine leukemia virus: Use of a monoclonal antibody to phosphotyrosine, Cancer Cells 3:339.

Glenney, Jr., J. R., Chen, W. S., Lazar, C. S., Walton, G. M., Zokas, L. M., Rosenfeld, M. G., and Gill, G. N., 1988, Ligand-induced endocytosis to the EGF receptor is blocked by mutational inactivation and by microinjection of anti-phosphotyrosine antibodies, Cell 52:675.

Habenicht, A. J. R., Glomset, J. A., King, W. C., Nist, C., Mitchell, C. D., and Ross, R., 1981, Early changes in phosphatidyl-inositol and arachidonic acid metabolism in quiescent Swiss 3T3 cells stimulated to divide by platelet-derived factor, J. Biol. Chem. 256:12329.

Hasegawa-Sasaki, H., 1985, Early changes in inositol lipids and their metabolites induced by platelet-derived growth factor in quiescent Swiss mouse 3T3 cells, Biochem. J. 232:99.

Heldin, C. H., Westermark, B., and Wasteson, A., 1981, Specific receptors for platelet-derived growth factor on cells derived from connective tissue and glia, Proc. Natl. Acad. Sci. USA, 78:3664.

Helper, J. R., Nakahata, N., Lovenberg, T. W., DiGuiseppi, J., Herman, B., and Harden, T. K., 1987, Epidermal growth factor stimulates the rapid accumulation of inositol (1,4,5)-trisphosphate and a rise in cytosolic calcium mobilized from intracellular stores in A431 cells, J. Biol. Chem. 262:2951.

Honegger, A. M., Dull, T. J., Felder, S., Van Obberghen, E., Bellot, F., Szapary, D., Schmidt, A., Ullrich, A., and Schlessinger, J., 1987a, Point mutation at the ATP binding site of EGF receptor abolishes protein-tyrosine kinase activity and alter cellular routing, Cell 51:199.

Honegger, A. M., Szapary, D., Schmidt, A., Lyall, R., Van Obberghen, E., Dull, T. J., Ullrich, A., and Schlessinger, J., 1987b, A mutant epidermal growth factor receptor with defective protein tyrosine kinase is unable to stimulate proto-oncogene expression and DNA synthesis, Mol. Cell. Biol. 7:4568.

Huang, J. S., Huang, S. S., Kennedy, B., and Deuel, T. F., 1982, Platelet-derived growth factor: Specific binding to target cells, J. Biol. Chem. 257:8130.

Meisenhelder, J., Suh, P.-G., Rhee, S. G., and Hunter, T., 1989, Phospholipase C-γ is a substrate for the PDGF and EGF receptor protein-tyrosine kinases in vivo and in vitro, Cell, in press.

Moolenaar, W. H., Bierman, A. J., Tilly, B. C., Verlaan, I., Defize, L. H. K., Honegger, A. M., Ullrich, A., and Schlessinger, J., 1988, A point mutation at the ATP-binding site of the EGF receptor abolishes signal transduction, EMBO J. 8:707.

Nishibe, S., Wahl, M. I., Rhee, S. G., and Carpenter, G., 1989, Tyrosine phosphorylation of phospholipase C-II in vitro by the epidermal growth factor receptor, J. Biol. Chem., in press.

Nishizuka, Y., 1984, The role of protein kinase C in cell surface signal transduction and tumor promotion, Nature 308:693.

Olashaw, N. E., and Pledger, W. J., 1988, Epidermal growth factor stimulates formation of inositol phosphates in BALB/c 3T3 cells pretreated with cholera toxin and isobutylmethylxanthine, J. Biol. Chem. 263:1111.

Pledger, W. J., Stiles, C. D., Antoniades, H. N., and Scher, C. D., 1977, Induction of DNA synthesis in BALB/c 3T3 cells by serum components: Re-evaluation of the commitment process, Proc. Natl. Acad. Sci. USA 74:4481.

Rhee, S. G., Suh, P.-G., Ryu, S.-H., and Lee, S. Y., 1989, Studies of inositol phospholipid-specific phospholipase C, Science, 244:546.

Roberts, A. B., and Sporn, M. B., ed., 1989, Peptide growth factors and their receptors, in: Handbook of Experimental Pharmacology, Springer-Verlag, Heidelberg.

Sawyer, S. T., and Cohen, S., 1981, Enhancement of calcium uptake and phosphatidylinositol turnover by epidermal growth factor in A-431 cells, Biochemistry 22:6280.

Suh, P. G., Ryu, S. H., Choi, W. C., Lee, K. Y., and Rhee, S. G., 1988, Monoclonal antibodies to three phospholipase C isozymes from bovine brain, J. Biol Chem. 263:14497.

Tremble, P. M., Chen, E. Y., Ando, M. E., Harkins, R. N., Francke, U., Fried, V. A., Ullrich, A., and Williams, L. T., 1986, Structure of the receptor for platelet-derived growth factor helps define a family of closely related growth factor receptors, Nature 323:226.

Ullrich, A., Coussens, L., Hayflick, J. S., Dull, T. J., Gray, A., Tam, A. W., Lee, J., Yarden, Y., Libermann, T. A., Schlessinger, J., Downward, J., Mayes, E. L. V., Whittle, N., Waterfield, M. D., and Seegurg, P. H., 1984, Human epidermal growth factor receptor cDNA sequence and aberrant expression of the amplified gene in A431 epidermoid carcinoma cells, Nature 309:418.

Wahl, M., and Carpenter, G., 1988, Regulation of epidermal growth factor-stimulated formation of inositol phosphates in A-431 cells by calcium and protein kinase C, J. Biol. Chem. 263:7581.

Wahl, M. I., Daniel, T. O., and Carpenter, G., 1988, Antiphosphotyrosine recovery of phospholipase C activity after EGF treatment of A-431 cells, Science 241:968.

Wahl, M. I., Nishibe, S., Pann-Ghill, S., Rhee, S. G., and Carpenter, G., 1989a, Epidermal growth factor stimulates tyrosine phosphorylation of phospholipase C-II independently of receptor internalization and extracellular calcium, Proc. Natl. Acad. Sci. USA 86:1568.

Wahl, M. I., Olashaw, N. E., Nishibe, S., Rhee, S. G., Pledger, W. J., and Carpenter, G., 1989b, Platelet-derived growth factor induces rapid and sustained tyrosine phosphorylation of phospholipase C-γ in BALB/c 3T3 cells, Molec. Cell. Biol., in press.

Yarden, Y., and Ullrich, A., 1988, Growth factor receptor tyrosine kinases, Ann. Rev. Biochem. 57:443.

EICOSANOID FORMATION AND REGULATION OF PHOSPHOLIPASE A_2

Eduardo G. Lapetina

Division Of Cell Biology
Burroughs Wellcome Co.
Research Triangle Park, N.C. 27709

ARACHIDONIC ACID IS LIBERATED FROM ARACHIDONOYL - PHOSPHOLIPIDS BY PHOSPHOLIPASE A_2

The oxygenated derivatives of arachidonic acid that are biologically active are defined as eicosanoids (Needleman et al., 1986). Among the eicosanoids are prostaglandins, including prostacyclin, thromboxanes, leukotrienes, and various hydroxy acids (Fig. 1). The eicosanoid precursor, arachidonic acid, is esterified in the 2-position of several phospholipids, and it must be hydrolyzed before the eicosanoids can be synthesized (Fig. 1). The liberated arachidonic acid can be enzymatically oxygenated by a membrane-bound cyclooxygenase, a microsomal cytochrome p450 or a cytosolic lipoxygenase with the formation of unstable intermediate products (Fig.1). These intermediate products include endoperoxides for prostaglandin production and epoxides and hydroperoxides for leukotriene and hydroxy acid formation (Fig 1). The type of arachidonate oxygenation is characteristic of the enzymes that each cell contains.

The eicosanoid precursors are not found as free substances in most cells, and liberation of arachidonic acid from the sn-2-position of phospholipids by phospholipase A_2 must precede metabolic oxygenation (Lapetina, 1982, 1984). The generation of free eicosanoids from phospholipids (Fig. 1) is thus a rate-limiting process and occurs as a consequence of the stimulation of cell-surface receptors. Phosphatidylcholine, phosphatidylethanolamine, phosphatidylinositol, phosphatidic acid, plasmenylethanol-amine, and 1-O-alkyl-2-arachidonoyl-sn-glycero-3-phosphocholine are all known to release arachidonic acid in different cells. Phospholipase A_2 has been detected in almost any cell in which its presence has been investigated. This activity requires an alkaline pH (7.8-9.5) and is stimulated by Ca^{2+}-ions. Glucocorticoids are able to inhibit phospholipase A_2 in different cells (Flower, 1981). It is thought that this action is due to the glucocorticoid-induced synthesis of lipocortin (Flower, 1981). However, it seems that lipocortin produces depletion of the phospholipid substrate in *in vitro* studies. Therefore the basis for a physiological role for lipocortins *in vivo* is questionable.

CHARACTERIZATION OF EICOSANOIDS

The basic technique for characterization of eicosanoids is the incubation of the specific cell type with radiolabeled arachidonic acid. These incubations are carried out in the presence and absence of specific cell agonists. The products that are formed are then identified by thin-layer or high-performance liquid chromatography. The structure is determined by gas chromatography and mass spectrometry (GC-MS) and UV spectrometry (Samuelsson, 1983). Radioimmunoassays for various eicosanoids have been developed for the detection of minute amounts of eicosanoids in biological samples. Bioassays (Salmon, 1983) that use the characteristic response of tissues to specific eicosanoids are still widely used.

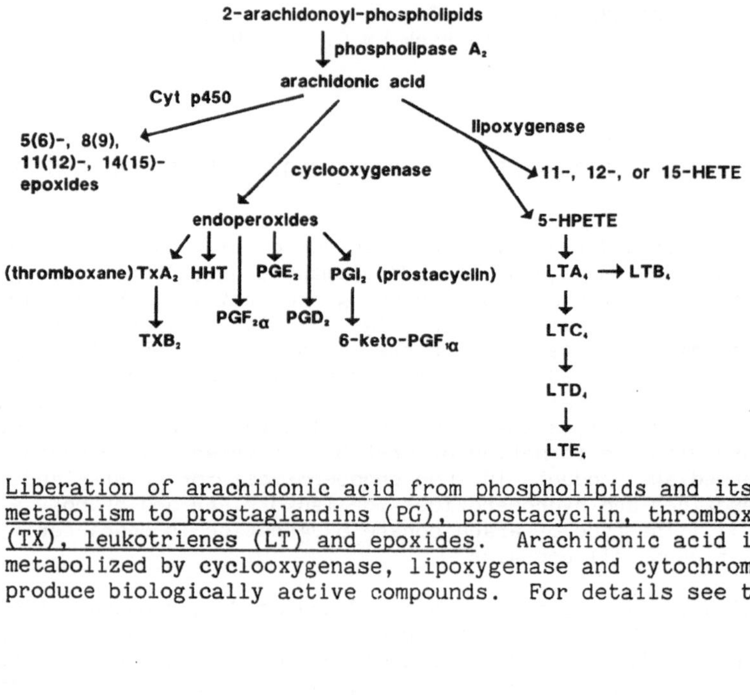

Fig. 1. Liberation of arachidonic acid from phospholipids and its metabolism to prostaglandins (PG), prostacyclin, thromboxanes (TX), leukotrienes (LT) and epoxides. Arachidonic acid is metabolized by cyclooxygenase, lipoxygenase and cytochrome p450 to produce biologically active compounds. For details see text.

THE FORMATION OF ENDOPEROXIDES, THROMBOXANES, AND PROSTAGLANDINS BY CYCLOOXYGENASE

Cyclooxygenase (Fig. 1) is a heme-containing membrane-bound enzyme that requires endoperoxides and molecular oxygen to oxygenate arachidonic acid (Needleman, 1986). Other fatty acids such as eicosapentaenoic acid are not as efficiently oxygenated as arachidonic acid by cyclooxygenase. This activity leads to the formation of endoperoxides after the insertion of two oxygens in the arachidonate molecule. The first oxygen is inserted at the C-11 position of arachidonate to form the 11-peroxy derivative (11-HPETE). The second oxygen is added at C-15 to yield 15-hydroperoxy-9,11-endoperoxide with a cyclopentane ring (PGG_2). This is reduced to the 15-hydro derivative (PGH_2) as a consequence of the peroxidase activity that is present in cyclooxygenase.

In blood platelets the main product of the endoperoxides is thromboxane A_2 (TXA_2) (Samuelsson, 1981), which induces vascular smooth muscle contraction and platelet aggregation. The main product of endoperoxides in vascular endothelial cells is prostacyclin (PGI_2) (Gryglewski, 1976), which is spontaneously hydrolyzed to 6-keto-$PGF_{1\alpha}$. Prostacyclin is a vasodilator and inhibits platelet aggregation. Prostacyclin stimulates adenylate cyclase, which increases cellular cyclic AMP levels.

Endoperoxides are also converted to other prostaglandins such as PGE_2, PGD_2, and $PGF_{2\alpha}$ or to the 12-hydroxy-heptadeca-trienoic acid (HHT) as indicated in Fig 1. Acetylation of cyclooxygenase by aspirin produces total inhibition of cyclooxygenase.

LIPOXYGENASE PRODUCES HYDROXY ACIDS AND LEUKOTRIENES

Lipoxygenases are cytosolic enzymes that catalyze the incorporation of one molecule of oxygen to yield hydroperoxy derivatives of arachidonic acid. The hydroperoxy group could be in carbon 5, 11, 12, or 15 of arachidonic acid (Fig. 1). The reduction of the hydroperoxy derivatives to the hydroxy analogs is due to a peroxidase activity associated to the lipoxygenase activity. The platelet 12-lipoxygenase was the first lipoxygenase described in mammalian tissues.

The 5-lipoxygenase leads to formation of the leukotrienes (Lewis and Austen, 1984) that are potent, biologically active compounds. Leukotrienes are mediators of allergic, inflammatory, and other pathological events. They are synthesized in neutrophils, eosinophils, monocytes, mast cells, macrophages, and other tissues. The 5-hydroperoxyeicosatetraenoic acid (5-HPETE) is dehydrated to an unstable epoxytriene, LTA_4, which produces LTC_4 by addition of the thiol group of glutathione across the epoxide ring by a glutathione transferase activity. LTC_4 is further metabolized to LTD_4 by removal of glutamate by γ-glutamyl transpeptidase, which, in turn, yields LTE_4 by a loss of glycine by a dipeptidase. LTA_4 can also be enzymatically hydrated to LTB_4, which is a potent chemotactic agent (Fig. 1). The 5-lipoxygenase is stimulated by calcium, and agents like the calcium ionophore A23187 greatly stimulate this activity.

ARACHIDONATE METABOLITES PRODUCED BY CYTOCHROME P450

Microsomal cytochrome p450 (Fig. 1) produces the epoxidation of the double bonds of the arachidonic acid to yield the 5(6) oxido-, 8(9) oxido-, 11(12) oxido-, and 14(15) oxido- eicosatrienoic acid (Needleman, 1986). Microsomal epoxide hydrolase can convert the epoxides to vic-diols. These diols can be further oxidize by cytochrome p450 in C-20 (ω oxidation) or C-19 (ω-1 oxidation) to produce trihydroxy fatty acids.

REGULATION OF PHOSPHOLIPASE A_2

It is not well understood how receptor binding is able to activate phospholipase A_2. It has been shown that GTP-binding proteins are involved in the activation of phospholipase C and, also, phospholipase A_2 (Cockroft and Gomperts, 1985; Lapetina, 1986). These findings are based on the ability of GTP analogues to stimulate phospholipase C and phospholipase A_2 in permeabilized cells and membrane preparations or by pertussis toxin to inhibit their activities. Pertussis toxin ADP-ribosylates and inactivates G_i, which is the G-protein that inhibits adenylate cyclase and comprises three polypeptide subunits known as α, β, and γ. Therefore, a role for G_i or $G_i\beta\gamma$ has been suggested in those cases where pertussis toxin inhibits

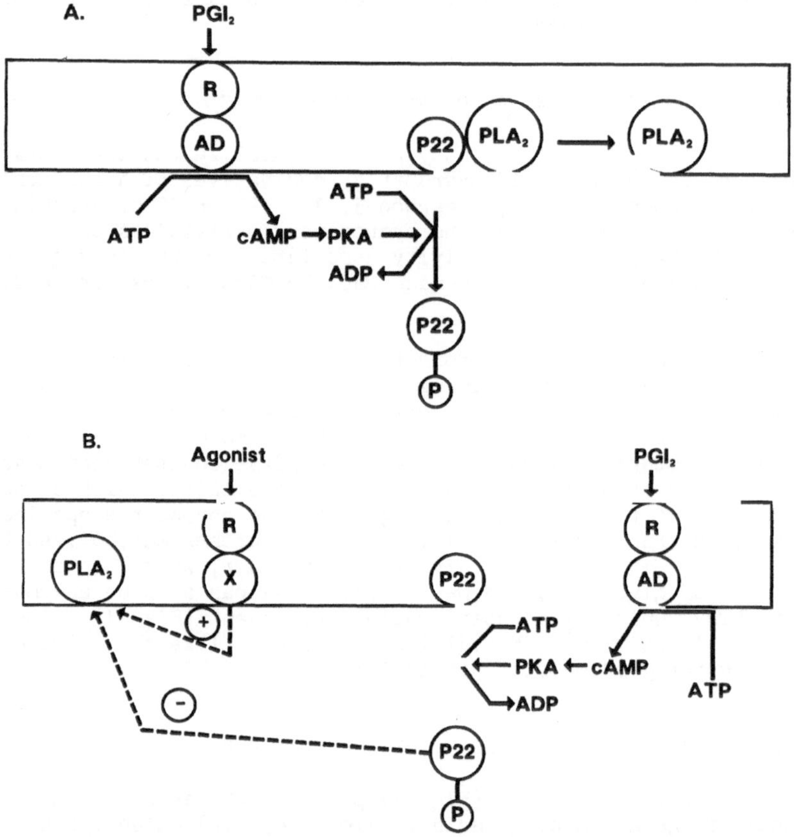

Fig. 2. A ras-related protein (p22) might influence the activity of phospholipase A_2. Cyclic AMP or prostacyclin (PGI_2) produce the A-kinase (PKA) phosphorylation of a ras protein (p22) which can then be translocated from the membrane. This might leave membrane-bound phospholipase A_2 in an inhibited state (A). Alternatively, the translocated and phosphorylated p22 protein might inhibit membrane-bound phospholipase A_2. For details see text.

agonist-induced stimulation of phospholipases. However, in those cells where pertussis toxin is ineffective at inhibiting phospholipases, another G-protein is believed to be involved. The identity of the specific G-protein that activates phospholipases has not yet been determined (Crouch and Lapetina, 1988a). However, our recent information suggests that a G-protein that is affected by phosphorylation is related to the regulation of phospholipase A_2 (Lapetina et al., 1989). The increase of cellular levels of cyclic AMP in platelets causes the inhibition of phospholipase A_2 (Crouch and Lapetina, 1988b). We have found that cyclic AMP-dependent protein kinase phosphorylates a G-protein that is immunologically recognized by *ras* p21-antibodies (Lapetina et al., 1989). This 22 kDa (p22) *ras*-related protein is present in the membrane, and it is translocated to the cytosolic fraction after phosphorylation. It is possible that the phosphorylation and translocation of the 22 kDa (p22) *ras*-related protein regulate phospholipase A_2. This proposed interrelationship is described in Figure 2. In platelets, cyclic AMP levels can be increased by prostacyclin (PGI_2) or prostacyclin analogs such as iloprost. The interaction of prostacyclin with its platelet membrane receptor causes the activation of adenylate cyclase and the increase of cyclic AMP levels. This increase of cyclic AMP correlates with the inhibition of phospholipase A_2 and the phosphorylation of the 22 kDa (p22) *ras*-related protein. The ways that the 22 kDa (p22) *ras*-related protein might be regulating phospholipase A_2 are shown in Figure 2. The unphosphorylated 22 kDa (p22) *ras*-related protein might normally have a stimulatory effect on the phospholipase A_2. When the 22 kDa (p22) *ras*-related protein is phosphorylated and translocated, phospholipase A_2 remains in the membrane in an inhibited state (Figure 2A). Alternatively, the phosphorylated 22 kDa (p22) *ras*-related protein might have an inhibitory action on the membrane-bound phospholipases as shown in Figure 2B.

The actual interrelationship between phospholipase A_2, G_i protein and the *ras*-related protein needs further investigation.

REFERENCES

Cockcroft, S., and Gomperts, B. D., 1985, Role of guanine nucleotide binding protein in the activation of polyphosphoinositide phosphodiesterase, Nature, 314:534.

Crouch, M. F., and Lapetina, E. G., 1988a, A role of Gi in control of thrombin receptor-phospholipase C in coupling human platelets, J. Biol. Chem., 263:3363.

Crouch, M. F., and Lapetina, E. G., 1988b, No direct correlation between Ca_{2+} mobilization and dissociation of Gi during platelet phospholipase A_2 activation, Biochem. Biophys. Res. Comm., 153:21.

Flower, R., 1981, Glucocorticoids, phospholipase A_2 and inflammation, Trends Pharmacol. Sci., 2:186.

Gryglewski, R. J., Bunting, S., Moncada, S., Flower, R. J., and Vane, J. R., 1976, Arterial walls are protected against deposition of platelet thrombi by a substance (prostaglandin X) which they make from prostaglandin endoperoxides, Prostaglandins, 12:685.

Lapetina, E. G., 1982, Regulation of arachidonic acid production: Role of phospholipases C and A_2, Trends Pharmacol Sci., 3:115.

Lapetina, E. G., 1984, Phospholipases, in: "Annual Reports Medicinal Chemistry," 19:213.

Lapetina, E. G., 1986, Inositide-dependent and -independent mechanisms in platelet activation, in: "Phosphoinositides and Receptor Mechanisms," J. W. Putney, Jr., ed., p. 271, Alan R. Liss, Inc., New York.

Lapetina, E. G., Lacal, J. C., Reep, B. R., and Molina y Vedia, L., 1989, A ras-related protein is phosphorylated and translocated by agonists that increase cyclic AMP levels in human platelets, Proc. Natl. Acad. Sci., 86:3131.

Lewis, R. A., and Austen, K. F., 1984, The biologically active leukotrienes: Biosynthesis, metabolism, receptors, functions, and pharmacology, J. Clin. Invest., 73:889.
Needleman, P., Tunk, J., Jakschik, B. A., Morrison, A. R., and Lefkowith, J. B., 1986, Arachidonic acid metabolism, Ann. Rev. Biochem., 55:69.
Salmon, J. A., 1983, Measurement of eicosanoids by bioassay and radioimmunoassay, Br. Med. Bull., 39:227.
Samuelsson, B., 1981, Prostaglandins, thromboxanes, and leukotrienes: Formation and biological roles, in: "The Harvey Lectures," Series 75, Academic Press, New York, NY, 1:40.
Samuelsson, B., 1983, Leukotrienes: Mediators of immediate hypersensitivity reactions and inflammation, Science, 220:568.

INHIBITION OF PHOSPHOLIPASE A_2 ACTIVATION IN P388D_1 MACROPHAGE-LIKE CELLS

Keith B. Glaser, Mark D. Lister and Edward A. Dennis

Department of Chemistry
University of California at San Diego
La Jolla, California 92093 U.S.A.

Introduction

The biosynthesis of eicosanoids is limited by the availability of free arachidonic acid as substrate for the cyclooxygenase and lipoxygenase enzymes. The level of free arachidonic acid in a resting cells is maintained very low (Kunze and Vogt, 1971) and upon stimulation, many cells are capable of releasing arachidonic acid from membrane phospholipids where it is found predominantly esterified in the *sn*-2 position of the phospholipids. A likely phospholipase candidate for the release of arachidonic acid would be a membrane-associated phospholipase A_2 which could directly cleave arachidonic acid from the *sn*-2 position of membrane phospholipids (Dennis, 1987). However, recently it has become apparent that there are other possible pathways for the release of arachidonic acid in the whole cell. Activation of either phospholipase A_1 or phospholipase C coupled with secondary enzymes, lysophospholipase or diacyl glycerol lipase, respectively, would also result in the release of arachidonic acid from membrane phospholipids (Dennis, 1987). Due to the complexity of the possible mechanisms for the release of arachidonic acid, the elucidation of the relevant enzyme(s) involved in the release mechanism depends upon (a) the isolation and characterization of the relevant enzymes, (b) evaluation of specific inhibitors of the relevant enzymes *in vitro*, (c) correlation of the effects of these inhibitors on the enzymes *in vitro* with their effects on eicosanoid biosynthesis in the whole cell and (d) demonstration that these inhibitors have little or no effect on the activities of the other enzymes which may be involved in arachidonic acid release (Dennis, 1987).

The P388D_1 macrophage-like cell line has served as a model to study the enzymes possibly involved in arachidonic acid release (Ulevitch *et al.*, 1988). The P388D_1 cell line has both the enzymatic and cell surface markers of a normal peritoneal macrophage and functional aspects of a peritoneal macrophage. (Koren *et al.*, 1975). Most importantly, is the production of arachidonic acid metabolites in response to various stimuli (Nitta and Suzuki, 1982; Aussel and Fehlman, 1986; Lister *et al.*, 1989). From the P388D_1 cell, the various phospholipases have been characterized and several purified to homogeneity (Ulevitch *et al.*, 1988; Zhang and Dennis, 1988).

A likely candidate for the relevant enzyme in arachidonic acid release is the membrane associated, Ca^{2+}-dependent phospholipase A_2 (Ross *et al.*, 1985). This enzyme has been purified to near homogeneity (Ulevitch *et al.*, 1988) and has been characterized kinetically (Lister *et al.*, 1988). In this manuscript, we report the evaluation of various potential phospholipase A_2 inhibitors *in vitro* on the partially purified membrane-associated phospholipase A_2 and the correlation of these effects with inhibitor effects on prostaglandin production in the intact P388D_1 cell (Lister *et al.*, 1989).

Results and Discussion

Prostaglandin Production in $P388D_1$ *cells*

The $P388D_1$ macrophage-like cells respond to various stimuli, e.g. calcium ionophore A23187, melittin and platelet-activating factor (PAF), to produce PGD_2 as the major cyclooxygenase metabolite (Table 1). PGE_2 is produced to a lesser extent than PGD_2 and even lesser quantities of PGI_2 (measured as 6-keto-$PGF_1\alpha$) and TXA_2 (measured as TXB_2) are produced upon cellular activation. The predominance of PGD_2 as the major cyclooxygenase product (Nitta and Suzuki, 1982; Lister *et al.*, 1989) as compared to PGE_2 or PGI_2, the major product of resident peritoneal macrophages, may reflect the transformed phenotype of the cell line or the clonal expansion of a subpopulation (McGuire *et al.*, 1985). The predominance of PGD_2 and PGE_2 in the arachidonic acid metabolic profile suggest the resemblence of these cells to resident peritoneal macrophages rather than the elicited type of macrophage where TxB_2 is conserved as the major cyclooxygenase metabolite (Tripp *et al.*, 1985).

TABLE 1

PROSTANOID PRODUCTION BY $P388D_1$ MACROPHAGE-LIKE CELLS[a]

Stimulus	Eicosanoid (ng mg protein^{-1})[b]			
	PGE_2	PGD_2	6-keto-$PGF_{1\alpha}$	TxB_2
NONE	3.2	35.7	1.1	N.D.[c]
A23187 (0.5 μM)	15.7	138.2	2.3	0.71
MELITTIN (0.5 μg ml^{-1})	31.5	311.0	1.5	0.62
PAF[d] (1 μM)	10.9	55.7	2.1	0.42

[a] Adapted with permission from Lister *et al.* 1989. *J. Biol. Chem. 264*, 8520-8528.
[b] $P388D_1$ cells were stimulated for 4 hr at 37 °C and the released arachidonate metabolites were measured by RIA.
[c] N.D. - below detection limit of RIA.
[d] PAF contained 1.0 μM cytocholasin B to prevent chemokinesis; cytocholasin B alone had no effect on prostanoid production.

Effects of p-bromophenenacyl bromide

Evaluation of inhibitors began with the antimalarial agent quinacrine and *p*-bromophenacyl bromide (BPB). Quinacrine had no significant effect on the $P388D_1$ phospholipase A_2 activity at concentrations as high as 0.8 mM (Fig. 1). This is consistent with the lack of inhibition of phospholipase A_2 by quinacrine in most *in vitro* phospholipase A_2 assays. *p*-Bromophenacyl bromide is an irreversible inhibitor of extracellular phospholipase A_2 which covalently modifies the active site histidine residue. BPB either in preincubation or dose response studies produced no greter than 50% inhibition of phospholipase A_2 activity at a concentration of 500 μM, the solubility limit of this compound in aqueous systems (Fig. 1). BPB did not significantly affect phospholipase A_2 activity at concentrations of 100 μM or less. Because the substrate concentration in this *in vitro* assay was 100 μM, and only 50% inhibition of phospholipase A_2 activity was observed at 500 μM, a 5:1 ratio of inhibitor to substrate, it is difficult to dissociate any effect of BPB on the P388D$_1$ phospholipase A$_2$ from the disruptive effects of BPB on the substrate, which would also reduce phospholipase A$_2$ activity.

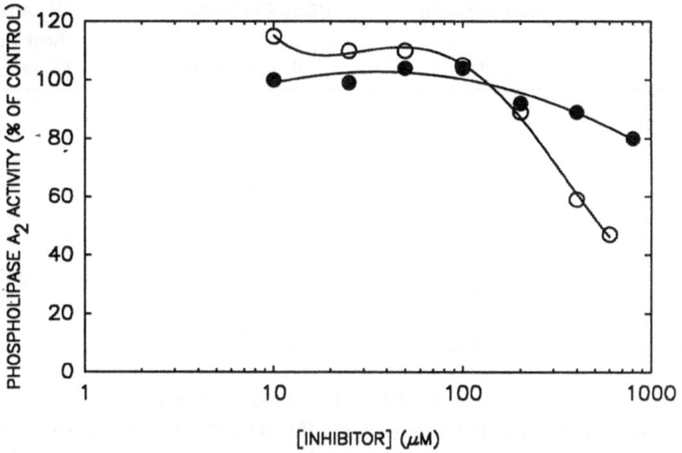

Figure 1. Dose response curve for quinacrine (●) and *p*-bromophenacyl bromide (BPB, ○), on phospholipase A$_2$ activity toward vesicles of dipalmitoyl phosphatidylcholine. Adapted with permission from Lister *et al.* 1989, *J. Biol Chem. 264*, 8520-8528.

The effect of BPB on PGE$_2$ production and arachidonic acid release from [^{3}H] arachidonic acid prelabeled P388D$_1$ cells is shown in Table 2. BPB at concentrations of 1.0 μM or less had no significant effect on PGE$_2$ production in response to calcium ionophore A23187 stimulation. At 10 μM, BPB apparently reduced PGE$_2$ production to 74% of control. However, there was also a similar reduction in the metabolism of exogenously added arachidonic acid (2 μM, data not shown) which is suggestive of effects of BPB on the cyclooxygenase enzyme. There was also no dose dependent reduction in the release of [^{3}H] arachidonic acid from prelabeled P388D$_1$ cells between 1 and 10 μM BPB, implying no effect of BPB on the arachidonic acid release mechanism in P388D$_1$ cells.

BPB is a potent inactivator of extracellular type phospholipase A$_2$ enzymes, (Roberts *et al.*, 1977) however, the lack of inhibition of the intracellular P388D$_1$ phospholipase A$_2$ marks a contrast between the effects of this inhibitor on the intracellular enzyme and the more commonly studied extracellular enzyme. The effects of BPB on different phospholipase A$_2$ sources has been variable (DeWinter *et al.*, 1982; Taniquchi *et al.*, 1988). Intracellular phospholipase A$_2$ appears to be less sensitive to inhibition by BPB as demonstrated herein and in recent studies on the intracellular phospholipase A$_2$ from RAW 264.7 cells (Leslie *et al.*, 1988). Therefore, caution should be observed when translating inhibitor effects on extracellular phospholipase A$_2$ with its effects on intracellular phosoholipase A$_2$ and arachidonic acid metabolism in the intact cell.

Effects of Manoalide and Manoalogue

Manoalide (de Silva and Scheuer, 1980) is a marine natural product which has been shown to have potent antiinflammatory activity (Jacobs *et al.*, 1985). Manoalide is also an irreversible inhibitor

TABLE 2

EFFECT OF INHIBITORS ON PGE_2 PRODUCTION AND $[^3H]$-ARACHIDONIC ACID RELEASE IN $P388D_1$ CELLS[a]

Compound	Concentration (μM)	PGE_2 Production[b] (% of Control)	$[^3H]$-Arachidonic Acid Release[c] (% of Control)
Indomethacin	0.1	0	n.d.[e]
BW755c	1.0	59	n.d.
	10.0	0	91
BPB[d]	0.5	92	n.d.
	1.0	86	89
	10.0	74	86

[a] Adapted with permission from Lister *et al.* 1989, *J. Biol. Chem.* *264*, 8520-8528.
[b] PGE_2 production measured by RIA in response to A23187 (0.5 μM) stimulation under conditions in which minimal toxicity was observed. When this was below the level of detection for the RIA, 0% is indicated.
[c] $[^3H]$-Arachidonic acid release measured in the presence of 0.1% BSA (essentially fatty acid free) from 1 x 10^6 $P3888D_1$ cells prelabeled for 18 hr with $[^3H]$-arachidonic acid and stimulated for 4 hr with A23187 (0.5 μM).
[d] Concentrations of BPB greater than 1 μM were toxic to $P388D_1$ cells.
[e] n.d. - not determined.

of extracellular phospholipase A_2 (de Freitas *et al.*, 1984; Lombardo and Dennis, 1985; Glaser and Jacobs, 1986) and has been shown to inhibit eicosanoid production and arachidonic acid release in murine resident peritoneal macrophages (Mayer *et al.*, 1988). Manoalogue is a synthetic analog which irreversibly inhibits cobra venom phospholipase A_2 and retains similar properties of the parent compound manoalide (Reynolds *et al.*, 1988). Both of these compounds were evaluated on the purified $P388D_1$ phospholipase A_2 and in the intact $P388D_1$ cell (Lister *et al.*, 1989). Manoalide and manoalogue dose-dependently inhibit the activity of the $P388D_1$ phospholipase A_2. The IC_{50} for manoalide and manoalogue are 16 and 26 μM, respectively, and therefore can be considered potent inhibitors of this enzyme. In contrast to the inhibition of extracellular phospholipase A_2 by these compounds, the inhibition of the intracellular $P388D_1$ phospholipase A_2 appears to be reversible (Lister *et al.*, 1989).

In the intact $P388D_1$ cells, manoalide and manoalogue dose dependently inhibit the production of PGE_2 in response to calcium ionophore A23187 stimulation (Fig. 2). There was an apparent IC_{50} for manoalide of 0.6 μM and an apparent IC_{50} of 1 μM for manoalogue, with 70% inhibition of PGE_2 production a dose of 2 μM. Higher doses of manoalide and manoalogue resulted in loss of cell viability as measured by either trypan blue exclusion or release of lactate dehydrogenase. $P388D_1$ cells which were prelabeled with $[^3H]$ arachidonic acid, release $[^3H]$ arachidonic acid when stimulated with calcium imophore A23187 (approx. 5% of the total label incorporated in released over a 4 hr stimulation period). Manoalide and manoalogue inhibited the release of $[^3H]$ arachidonic acid and this inhibition appears to correlate with the observed inhibition of PGE_2 production (Fig. 2).

Conclusion

The P388D$_1$ macrophage-like cell line has been used as a macrophage model (Koren *et al.*, 1975) for the study of monokine production (Mizel *et al.*, 1978) and macrophage function (Koren *et al.*, 1975). This cell line has the capacity to produce arachidonic acid metabolites to various stimuli (Nitta and Suzuki, 1982; Aussel and Fehlman, 1986; Lister *et al.*, 1989) and produces PGD$_2$ as the major cyclooxygenase product. The predominance of PGE$_2$ over TXA$_2$ production by these cells suggest a similarity to the resident type of peritoneal macrophage rather than the elicited type. These observations make this cell line compatable to the study of the relevant enzyme(s) for arachidonic acid release. A likely phospholipase candidate has been purified and therefore inhibitors can be evaluated both on the phospholipase A$_2$ *in vitro* and in the intact P388D$_1$ cell. The P388D$_1$ macrophage-like cell phospholipase A$_2$ has been previously characterized (Ulevitch *et al.*, 1988; Lister *et al.*, 1988) and is one of the first intracellular phospholipase A$_2$ to be purified and kinetically analyzed in detail (Lister *et al.*, 1988)

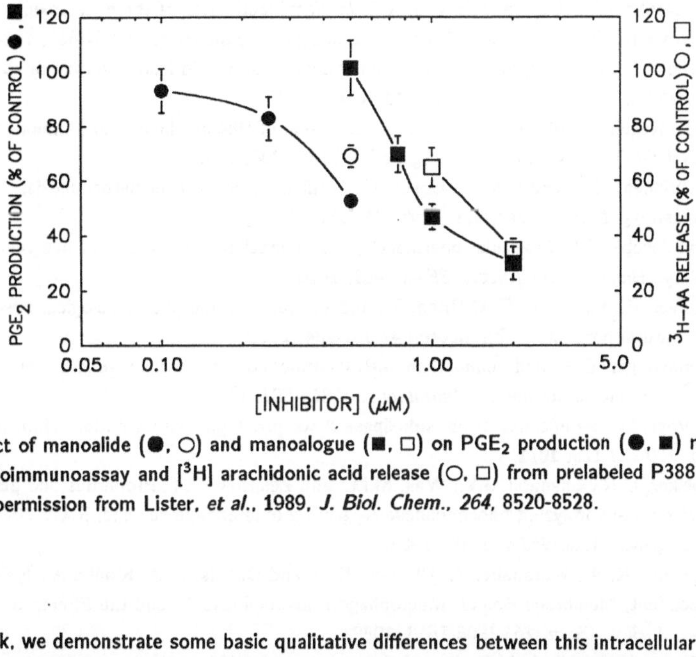

Figure 2. Effect of manoalide (●, ○) and manoalogue (■, □) on PGE$_2$ production (●, ■) measured by specific radioimmunoassay and [^{3}H] arachidonic acid release (○, □) from prelabeled P388D$_1$ cells. Adapted with permission from Lister, *et al.*, 1989, *J. Biol. Chem.*, *264*, 8520-8528.

In this work, we demonstrate some basic qualitative differences between this intracellular P388D$_1$ phospholipase A$_2$ and the more commonly studied extracellular forms of phospholipase A$_2$. The P388D$_1$ phospholipase A$_2$ is insensitive to inhibition by BPB and prostaglandin production in the intact P388D$_1$ cell is also insensitive to inhibition by BPB. This facet marks a contrast between intra- and extracellular forms of phospholipase A$_2$. This insensitivity to inhibition by BPB is also evident in another macrophage intracellular phospholipase A$_2$ from RAW 264.7 cells (Leslie *et al.*, 1988). Secondly, extracellular phospholipase A$_2$ are irreversibly inhibited by both manoalide and manoalogue whereas the P388D$_1$ phospholipase A$_2$ appears to be reversibly inhibited by these agents. These differences between the P388D$_1$ phospholipase A$_2$ and extracellular phospholipase A$_2$ are qualitatively different enough to indicate caution in the extrapolation of either inhibitor or "activator"effects from studies with extracellular phospholipase A$_2$ directly to the intracellular form of phospholipase A$_2$. The insensitivity of both the phospholipase A$_2$ and PGE$_2$ production in the intact cell to inhibition by BPB coupled with the apparent correlation of the inhibition of [^{3}H] arachidonic acid release and PGE_2 production in P388D_1 cells by manoalide and manoalogue suggests an involvement of the calcium dependent membrane-associated phospholipase A_2 in the arachidonic acid release mechanism in P388D_1 cells. The final determination of the relevant enzyme(s) for arachidonic acid release awaits further evaluation of these inhibitors on the other phospholipases from the P388D_1 cell (Ross *et al.*, 1985) which may be implicated in arachidonic acid release. These studies demonstrate an ongoing effort to correlate inhibition studies on phospholipase A_2 *in vitro* and PGE_2 production from the same cell which is necessary to elucidate the relevant mechanism for arachidonic acid release.

Acknowledgements

Financial support for the studies reported herein was provided by NIH grants GM-20,501 and AI-15,136 and the Lilly Research Laboratories.

REFERENCES

Aussel, C., and Fehlman, M. α-Fetoprotein stimulates leukotriene synthesis in P388D_1 macrophages.. *Cell. Immunol. 101* 415-425, 1986.

de Freitas, J.C., Blankemeier, L.A., and Jacobs, R.S. *In vitro* inactivation of the neurotoxic action of β-bungarotoxin by the marine natural product, manoalide.. *Experientia 40* 864-865, 1984.

de Silva, E.D., and Scheuer, P.J. Manoalide, an antibiotic sesterterpenoid from the marine sponge *Luffariella variabilis*.. *Tetrahedron Lett. 21* 1611-1614, 1980.

Dennis, E. A. The Regulation of Eicosanoid Production: Role of Phospholipases and Inhibitors. *BIO/TECHNOLOGY* (Nature Publishing Co.) 5 1294-1300, 1987.

DeWinter, J.M., Vianen, G.J., and van den Bosch, H. Purification of rat liver mitochondrial phospholipase A_2. *Biochim. Biophys. Acta. 712* 332-341, 1982.

Glaser, K.B., and Jacobs, R.S. Molecular pharmacology of manoalide: inactivation of bee venom phospholipase A_2.. *Biochem. Pharmacol. 35* 449-453, 1986.

Jacobs, R.S., Culver, P., Langdon, T., O'Brien, T., and White, S. Some pharmacological observation on marine natural products.. *Tetrahedron 41* 981-984, 1985.

Koren, H.S., Handwerger, B.S., and Wunderlich, J.R. Identification of macrophage-like characteristics in a cultured murine tumor line.. *J. Immunol. 114* 894-897, 1975.

Kunze, H., and Vogt, W. Significance of phospholipase A for prostaglandin formation. *Ann. N. Y. Acad. Sci. 180* 123-125, 1971.

Leslie, C.C., Voelker, D.R., Channon, J.Y., Wall, M.M., and Zelarney, P.T. Properties and purification of an arachidonyl-hydrolyzing phospholipase A_2 from a macrophage cell line, RAW 264.7.. *Biochim. Biophys. Acta. 963* 476-492, 1988.

Lister, M. D., Deems, R. A., Watanabe, Y., Ulevitch, R. J, and Dennis, E. A. Kinetic Analysis of the Ca^{2+} Dependent, Membrane-Bound, Macrophage Phospholipase A_2 and the Effects of Arachidonic Acid. *J. Biol. Chem. 263* 7506-7513, 1988.

Lister, M. D., Glaser, K. B., Ulevitch, R. J., and Dennis, E. A. Inhibition Studies on the Membrane-Associated Phospholipase A_2 *in vitro* and Prostaglandin E_2 Production *in vivo* of the Macrophage-Like P388D_1 Cells: Effects of Manoalide, 7,7-Dimethyl-5,8-eicosadienoic Acid, and *p*-Bromophenacyl Bromide, *J. Biol. Chem. 264*, 8520-8528. 1989.

Lombardo, D., and Dennis, E. A. Cobra Venom Phospholipase A_2 Inhibition by Manoalide: A Novel Type of Phospholipase Inhibitor. *J. Biol. Chem. 260* 7234-7240, 1985.

Mayer, A. M. S., Glaser, K. B., and Jacobs, R. S. Regulation of eicosanoid biosynthesis *in vitro* and *in vivo* by the marine natural product manoalide: a potent inactivator of venom phospholipases. *J. Pharm. Exp. Ther. 244* 871-874, 1988.

McGuire, J.C., Richard, K.A., Sun, F.F., and Tracey, D.E. Production of prostaglandin D_2 by murine macrophage cell lines.. *Prostaglandins 30* 949-967, 1985.

Mizel, S.B., Oppenheim, J.J., and Rosenstreich, D.L. Characterization of lymphocyte-activating factor (LAF) produced by a macrophage cell line, P388D_1. *J. Immunol.* *120* 1504-1508, 1978.

Nitta, T., and Suzuki, T. Receptor-mediated prostaglandin synthesis by a murine macrophage cell line (P388D_1).. *J. Immunol.* *120* 2527-2532, 1982.

Reynolds, L. J., Morgan, B. P., Hite, G. A., Mihelich, E. D., and Dennis, E. A. Phospholipase A_2 Inhibition and Modification by Manoalogue. *J. Am. Chem. Soc.* *110* 5172-5177, 1988.

Roberts, M. F., Deems, R. A., Mincey, T. C., and Dennis, E. A. Chemical Modification of the Histidine Residue in Phospholipase A_2 (*Naja naja naja*): A Case of Half-Site Reactivity. *J. Biol. Chem.* *252* 2405-2411, 1977.

Ross, M. I., Deems, R. A., Jesaitis, A. J., Dennis, E. A., and Ulevitch, R. J. Phospholipase Activities of the P388D_1 Macrophage-Like Cell Line. *Arch. Biochem. Biophys.* *238* 247-258, 1985.

Taniquchi, K., Urakami, M., and Takanaka, K. Effects of various drugs on superoxide generation, arachidonic acid release and phospholipase A_2 in polymorphonuclear leukocytes. *J. Pharmacol.* *46* 275-284, 1988.

Tripp, C.S., Leahy, K.M., and Needleman, P. Thromboxane synthase is preferentially conserved in activated mouse peritoneal macrophages.. *J. Clin. Invest.* *76* 898-901, 1985.

Ulevitch, R. J., Sano, M, Watanabe, Y., Lister, M. D., Deems, R. A., and Dennis, E. A. Solubilization and Characterization of a Membrane-Bound Phospholipase A_2 from the P388D_1 Macrophage-Like Cell Line. *J. Biol. Chem.* *263* 3079-3085, 1988.

Zhang, Y., and Dennis, E. A. Purification and Characterization of Lysophospholipase from P388D_1 Macrophage-Like Cell Line. *J. Biol. Chem.* *263* 9965-9972, 1988.

THE ROLE OF GTP-BINDING PROTEINS IN THE REGULATION OF PHOSPHOLIPASE A_2 AND C IN BOVINE RETINAL ROD OUTER SEGMENTS

Carole L. Jelsema and Alice D. Ma*

Laboratory of Cell Biology, National Institute of Mental Health, Bethesda, MD 20892; * Department of Internal Medicine University of Pennsylvania, Philadelphia, PA 29103

INTRODUCTION

Signal-transducing GTP-binding proteins (G proteins) are membrane-bound coupling proteins that link the activation of cell surface receptors to changes in specific enzyme activities intracellularly, resulting in altered cellular behavior. These G proteins are heterotrimers which, upon binding of ligand-activated receptors, exchange the GDP bound to the α subunit for GTP and dissociate into the component GTP-bound α and βγ subunits. These dissociated G protein subunits then interact with and modulate the activity of specific effector systems to which they are linked. Effector systems known to be coupled by G proteins include adenylate cyclase (Gilman, 1987); cAMP and cGMP phosphodiesterase (Elks et al., 1983; Fung et al., 1981); the Ca^{++}, $Na^+$and K^+ ion channels (see review, Dunlap et al., 1987); phospholipase C (Cockcroft and Gomperts, 1985; Ohta et al., 1985) and phospholipase A_2 (Jelsema, 1987; Fuse and Tai, 1987; Axelrod et al., 1988).

While the G proteins responsible for the stimulation (G_s) and inhibition (G_i) of adenylate cyclase and the stimulation of retinal cGMP phosphodiesterase (transducin or G_t) have been identified, the identities of the G proteins regulating the ion channels and the phospholipases remain controversial. This controversy may be due, in part, to modulation of these systems by multiple G proteins, either directly or indirectly (Axelrod et al., 1988). The identity of the specific G proteins involved in phospholipase regulation has, however, proven particularly elusive since the isolated, purified enzymes are fully active. Addition of isolated G proteins or their subunits to the purified enzymes in vitro, therefore, has no effect on enzyme activity even though guanine nucleotides and bacterial toxins alter phospholipase activities in vivo, indicating a role for G proteins in their regulation. Thus, the G protein regulation of the phospholipases appears to be indirect, acting either at the level of non-G protein regulatory components that maintain the phospholipases in an inhibitory state or via other G protein-mediated second messengers that are the actual phospholipase modulators. Phospholipase A_2 (PLA_2), for example, responds to G protein-dependent second messengers such as Ca^{++} (Loeb and Gross, 1986) and cAMP (Samuelsson et al., 1978) and phospholipase C (PLase C) also is sensitive to cAMP (Yada et al., 1989). Until it is known whether the G protein regulation of the phospholipases occurs secondary to the action of other G protein-activated second messengers and/or via interaction with negative modulators, attempts at identifying the relevant G proteins by in vitro reconstitution experiments using purified receptors, G proteins and phospholipases will continue to be futile.

Marginal success at demonstrating G protein interaction with phospholipases has been reported in add-back experiments using membranes in which the action of the native G proteins regulating the phospholipases can be inhibited either by toxin pretreatment of the membranes or by partially depleting the membranes of the G protein of interest. A role for G_i in the activation of PLase C has been suggested by their ability to stimulate the enzyme when added to pertussis toxin-treated rat brain membranes (Ueda et al., 1989) and a link between stimulation of PLA_2 and activation of G_t has been demonstrated in similar experimants adding G_t back to G_t-depleted rod outer segments (ROS) of bovine retina (Jelsema,1987; Jelsema and Axelrod, 1987a, 1987b). In this system, either exogenous G_t added under dissociating conditions or $G_{t\beta\gamma}$ subunits added alone caused stimulation of PLA_2 activity. In contrast, the α subunit inhibited the βγ-induced effect, presumably by facilitating reassociation of the inactive heterotrimer. Since $G_{o\beta\gamma}$ also stimulated PLA_2 activity (Jelsema et al., 1989) and the γ subunit appears unique to G_t while the β subunit is one of two β subunits shared by the signal-transducing G proteins (Roof et al., 1985), a role for the 36 kDa β subunit in PLA_2 stimulation is suggested. This would have implications for all G proteins in either the stimulation of PLA_2 or its inhibition by other G_α subunits. Thus, a multiplicity of G proteins may have the capacity to modulate PLA_2 activity secondarily, if not directly. In fact, there is evidence of G protein-dependent inhibition of PLA_2 in bovine photoreceptors (Jelsema, 1987), although whether it relates to the presence of a specific G protein inhibitory to PLA_2 or to secondary effects elicited by other G protein-mediated second messengers is not known. To further investigate the role of other G protein subunits in PLA_2 activity, the effect of isolated subunits of specific G proteins (G_i, G_o, G_s and G_t) on the PLA_2 activity of dark-adapted, G_t-poor ROS was examined under conditions that either promoted or inhibited subunit dissociation. Our studies indicate that specific G protein α subunits appear to modulate PLA_2 directly while others exert their effects strictly by interaction with G protein βγ subunits. More importantly, the stimulatory effects of βγ on PLA_2 activity were found to occur independent of their interaction with G protein α subunits. This eliminates the possibility that the stimulatory effect of the βγ subunits occurs by blocking the action of an inhibitory G_α and indicates a more direct role for these subunits in regulation of PLA_2.

PLAse C was also found to be activated by light or guanosine-5'-(3-O)-thio-triphosphate (GTPγS) in bovine photoreceptors (Jelsema and Ma, 1989). Since the end-products of PLA_2 activity can lead to activation of PLase C and vice versa (Irvine et al., 1979; Mobley and Tai, 1985), the question arose as to the independence of the two enzyme activities and the uniqueness of their G protein regulation. To investigate this possibility, the effects of guanine nucleotides (GTPγS and GDPβS), bacterial toxins (pertussis and cholera toxin) and the PLase C inhibitor, neomycin (Schacht, 1976), were assessed for their effects on both basal and light-activated PLase C relative to their effects on PLA_2 activity in bovine ROS. In addition, the effects of G_t subunits on the PLase C activity of G_t-depleted, dark-adapted ROS was analysed to determine whether G_t similarly functions in the light activation of PLase C. Our results indicate that in bovine photoreceptors, both phospholipases are activated by light and under dual regulation by both inhibitory and stimulatory G proteins, with G_t functioning in the light activation of both enzymes. Despite these similarities, however, the two phospholipases appear to be activated by different mechanisms and their G protein regulation is fundamentally different.

METHODS

Materials. L-α-1-palmitoyl-2-arachidonyl-[1-^{14}C]-phosphatidylcholine (54.5 mCi//mmol), [5,6,8,9,11,12,14,15-^{3}H(N)] arachidonic acid (91.2 mCi/mmol), and L-α-phosphatidyl-[2-^{3}H(N)]-inositol (10 ci/mmol) were purchased from New England Nuclear. L-*myo*-[U-^{14}C] inositol 1-phospate (52.5 mCi/mmol) was purchased from Amersham. Cholera toxin and pertussis toxin were obtained from List Biological Laboratories (Campbell, CA). GTPγS and GDPβSwere from Boehringer Mannheim. CL-4B sepharose and DEAE-sephacel were obtained from Pharmacia. The G_s, G_i, G_o subunits and the G protein, arf, were kindly provided by Dr. R. Kahn.

Preparation of rod outer segments and isolation of transducin. Dark-adapted rod outer segments (ROS) of bovine retina were isolated and purified using retina dissected under dim red light. G_t-depleted ROS were then preparedby extensive washing (six times each) with first isotonic, then hypotonic buffers (Jelsema, 1989). This procedure routinely extracted 75 to 80% of the membrane G_t, based on the loss of GTPγS-binding from the ROS membranes (Jelsema, 1987). In contrast, similar washing of purified ROS prepared from light-activated membranes did not remove G_t unless GTP was added to the membranes (Kuhn, 1980).

G_t was, therefore, isolated from purified, light-activated ROS membranes that had been extensively washed then extracted with GTP (Sternweis and Gilman, 1982) and the GTP was separated from G_t by chromatography on DEAE-sephacel (Baehr et al., 1979). G_t accounted for 95% of the protein extracted by GTP, as determined using 12% polyacrylamide gels that had been silver stained and, based on the absence of cross-reactivity with available antibodies to other G proteins, 100 % of the GTP-binding protein extracted.

Isolation of transducin subunits. G_t subunits were separated chromatographically as previously described (Jelsema and Axelrod, 1987). Fractions containing the isolated α and βγ subunits, once identified by gel electrophoresis, were pooled, concentrated by filtration (Amicon YM10), then dialyzed overnight as previously described (Jelsema and Axelrod, 1987). The dialysis step removed the cholate and/or NaCl as well as residual GTP, Al^{3+}, and F^- ions which interfered with either the PLA_2 assay or with subunit reassociation experiments. The G_t subunits were stored at 4°C and used within 3 days.

Phospholipase A_2 assay. In vitro PLA_2 assays were performed on isolated ROS using 1-palmitoyl-2-[^{14}C-arachidonyl]-phosphatidylcholine as the substrate. A micellar suspension of radiolabeled substrate (10^7 dpm/ml) was prepared by evaporating the solvent under N_2, adding 0.12 M tris HCl, pH 8.8, dispersng the lipids by a 5 min sonication (Bronson sonicating water bath) and a 15 sec sonication (Kontes microtip probe) immediately prior to the assay (Jelsema et al., 1989). Reactions were initiated by addition of radiolabeled micellar substrate (20 µl) to dark-adapted ROS membranes (20 µg) in an 80 µl reaction mixture containing a final concentration of 4 mM glutathione, 0.6 mM NaCl, 40 mM $MgCl_2$, 5 mM $CaCl_2$, 30 mM tris HCl, pH 8.8. Incubations were at 37°C under dim red light or white light (300 watts) in the presence or absence of 100 µM GTPγS or GDPβS and/or 50 ng of G protein subunits. Reactions were stopped with 100 µl 1N formic acid at zero time, to determine the amount of non-enzymatic hydrolysis, and at 10 min, during the linear phase of the reaction (Jelsema, 1987). The [^{14}C]-arachidonic acid released was separated from radiolabeled substrate as previously described (Jelsema et al., 1989) and counted with 10 ml Biofluor (New England Nuclear) in a LKB scintillation counter programmed to correct for quenching, counting efficiency, and the spill of [^{14}C]-dpm into the [^{3}H]-channel. To correct for loss of [^{14}C]-dpms during the lipid extraction step, [^{3}H]-arachidonic acid (2000 dpm/50 µl) was added prior to the extraction step and the percent recovery determined. Following subtraction of the zero-time control and correction for recovery, PLA_2 activity was expressed as nmol [^{14}C]-arachidonic acid released/min/mg protein.

Phospholipase C assay. In vitro assays of PLase C were performed on isolated, dark-adapted ROS using L-3-phosphatidylinositol [2-^{3}H-inositol] as the substrate. Unlabeled phosphoinositides and phosphatidylserine (0.08 mg/ml) in a 2:1 ratio were added to the radiolabeled substrate (10^7 dpm/ml), and a micellar suspension prespared as described above. Reactions were initiated by addition of 20 µl of the sonicated, radiolabeled substrate to dark-adapted ROS membranes (20 µg, unless otherwise specified) in a reaction mixture containing a final concentration of 4 mM glutathione, 0.6 mM NaCl, 40 mM $MgCl_2$, 5 mM $CaCl_2$, 30 mM tris HCl, pH 7.5, in a final volume of 80 µl. Incubations were at 37°C under either dim red light or white light (300 watts). Reactions were stopped at zero time, to determine the amount of non-enzymatic hydrolysis, and at a time during the linear phase of the reaction (10 min) by addition of 1.5 ml methanol:chloroform (2:1, v/v). Samples were extracted by a modified Bligh and Dyer procedure (Jelsema and Ma, 1989). Samples were then vortexed and centrifuged (1000 x g, 10 min) to separate the phases. Aliquots (1 ml) of the water-soluble upper phase were analyzed for the release of [^{3}H]-dpms following evaporation of the methanol and addition of 10 mls Biofluor (New England Nuclear). The radioactivity of the samples was measured as described above. To correct for the loss of the released [^{3}H]-inositol 1-phosphate during lipid extraction, [^{14}C]-myo-inositol-1-phosphate (2000 dpm/50 µl) was added to each sample prior to the final extraction step, and the percent [^{14}C]-dpm recovered was used to correct for the loss of [^{3}H]-dpm. Following subtraction of the nonspecific hydrolysis observed in the zero-time control, PLase C activity was expressed as pmol [^{3}H]-inositol phosphate released/min/mg protein.

Toxin treatment. ROS membranes were treated with activated pertussis (10 ng/ml) or cholera toxin (50 ng/ml) in the absence of light for 2 hrs at 30°C, as previously described (Jelsema and Ma, 1989). The toxins were present at a 1:4 dilution in the subsequent assays.

RESULTS

Phospholipases A_2 and C are both activated by light and dually regulated by stimulatory and inhibitory G proteins. In photoreceptor membranes of bovine retina, both the PLA_2 and PLase C activities were stimulated severalfold upon exposure of dark-adapted membranes to light in in vitro assays using exogenous [^{14}C]-arachidonylphosphatidylcholine and phosphatidyl-[^{3}H]-inositol, respectively, as the substrates (Fig. 1A,B; Jelsema, 1987; Jelsema and Ma, 1989). Basal PLA_2 activity was significantly higher than basal PLase C activity, releasing 29.4 nmol [^{14}C]-arachidonate/min/mg protein relative to 0.7 pmol [^{3}H]-inositol phosphate released/min/mg protein in dark-adapted ROS. The light-induced increase in PLA_2 activity was also greater than that observed with PLase C, reaching levels 4 to 5-fold that of basal (i.e., dark-adapted) activity while PLase C activity increased a maximum of 2 to 3-fold.

To determine whether GTP-binding proteins functioned in the regulation of the retinal phospholipases, the effect of GTPγS was examined. GTPγS, which induces dissociation of G proteins, stabilizing them in their active, dissociated state (Gilman, 1987), markedly increased the PLA_2 and PLase C activities of dark-adapted ROS, mimicking the effect of light (Fig. 1; Jelsema, 1987; Jelsema and Ma, 1989). These results suggest G proteins may function in the light activation of both phospholipases. Addition of GTPγS to light-activated ROS, however, inhibited the light stimulation of the enzymes (Fig. 1; Jelsema and Ma, 1989), suggesting the existence of (an)other G protein(s) able to inhibit the light-stimulated activities. Such dual regulation of an enzyme by both a stimulatory and an inhibitory G protein is reminiscent of the G protein regulation of adenylate cyclase, where activation of G_i leads to inhibition of G_s-stimulated adenylate cyclase without affecting basal activity (Gilman, 1987).

Effects of guanine nucleotides, bacterial toxins and neomycin indicate the two phospholipases are independent and exhibit fundamental differences in their regulation by G proteins. The similarities in the effects of light and/or GTPγS on the two phospholipases raised questions as to their independence since, as previously indicated, either of the two phospholipases can be activated by prior action of the other. To address this question, modulators of the two enzymes were examined for their effects on both phospholipases. The PLase C inhibitor, neomycin (Schacht, 1976), inhibited both basal and activated PLase C (Fig. 1B), but had no effect on PLA_2 activity (Fig. 1A). Thus, in this tissue, stimulation of PLA_2 did not occur by prior activation of PLase C. Demonstrating that increased PLase C activity could occur without prior action of phospholipase A_2 proved more problematic since none of the PLA_2 inhibitors proved unique to this phospholipase.

The effects of GDPβS and the bacterial toxins, however, clearly indicated fundamental differences in the G protein regulation of the two phospholipases, supporting the independence of the two enzymes. While both PLA_2 and PLase C were similarly stimulated upon exposure to light and/or GTPγS, addition of GDPβS had markedly different effects on the two phospholipases. GDPβS, which competes with endogenous GTP to stabilize the inactive, associated state of G proteins (Cassel and Selinger, 1977), inhibited the

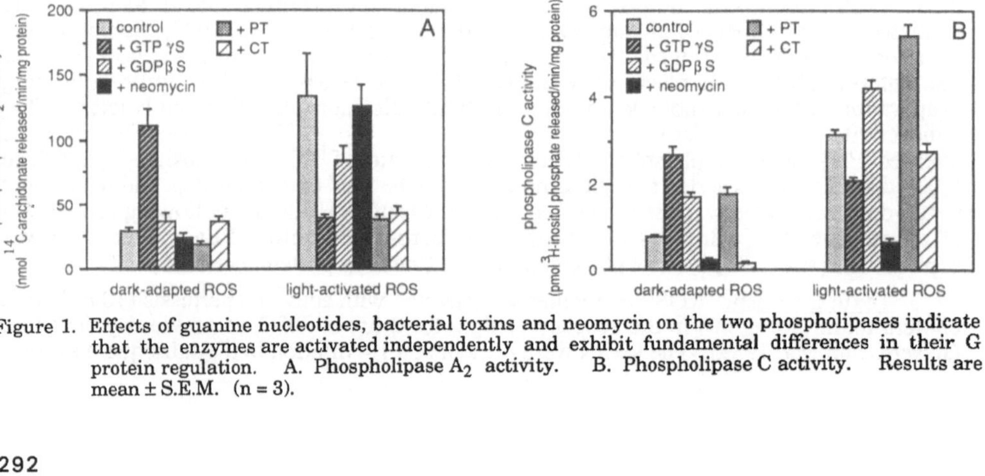

Figure 1. Effects of guanine nucleotides, bacterial toxins and neomycin on the two phospholipases indicate that the enzymes are activated independently and exhibit fundamental differences in their G protein regulation. A. Phospholipase A_2 activity. B. Phospholipase C activity. Results are mean ± S.E.M. (n = 3).

light-activated but not the basal state of PLA_2 (Fig. 1A), consistent with a role for a stimulatory G protein in PLA_2 activation. In contrast, GDPβS stimulated both basal and light-activated PLase C (Fig. 1B), suggesting this enzyme may be under tonic inhibitory control by another G protein, with stimulation occurring by removal of the inhibitory control.

Transducin (G_t), the major G protein of retina, binds photolysed rhodopsin (Bourne and Stryer, 1986), making it a likely candidate for the G protein that functions in the light activation of PLA_2 and PLase C. This possibility was examined by assessing the effect of bacterial toxins on the phospholipase activities since G_t serves as a substrate for both pertussis toxin (PT) and cholera toxin (CT). PT-induced ADP-ribosylation stabilizes G_t in its associated, inactive state (Van Dop et al., 1984) while CT ADP-ribosylates G_t only in the presence of light or guanosine-5'-(β,γ-imino)triphosphate, a nonhydrolyzable GTP analog (Abood et al., 1982), suggesting CT acts on dissocated, activated G_t.

Treatment of dark-adapted ROS with PT inhibited both basal and light-induced PLA_2 (Fig. 1A; Jelsema, 1987), consistent with a role for G_t in the light activation of PLA_2. The inhibition of basal PLA_2 activity is suggested to result from the fact that dark-adapted ROS are in a partially-activated state due to the presence of red light (Jelsema, 1987). In contrast, PT stimulated both basal and light-activated PLase C, similar to the effect of GDPβS (Fig. 1B; Jelsema, 1989). This supports the concept that PLase C is under tonic inhibitory control by a G protein and indicates the inhibitory G protein is PT-sensitive, but does not exclude a role for G_t in PLase C activation. CT, rather than stimulating the phospholipases as expected were the toxin acting to stabilize G_t in its activated state, proved inhibitory to both phospholipases. CT inhibited light-stimulated PLA_2 (Fig. 1A) while inhibiting only basal PLase C activity (Fig. 1B). These CT effects on the two enzymes are inconsistent with a CT effect on G_t and neither support nor refute a role for G_t in phospholipase activation.

Transducin functions as the stimulatory G protein in the light activation of both phospholipases. To more directly address the potential role of G_t in the light activation of the phospholipases, G_t-poor ROS were analyzed for phospholipase activities in the presence or absence of light and/or GTPγS. Using [^{35}S]-GTPγS-binding as a measure of the G_t content of the membranes, removal of 80% of the membrane G_t was paralleled by a loss of both the light- and GTPγS-induced activation of each phospholipase (Table 1; Jelsema, 1987; Jelsema and Ma, 1989). Furthermore, removal of G_t unmasked the G protein observed to inhibit light-activated PLA_2 (Fig. 1A) such that addition of GTPγS to G_t-poor ROS led to inhibition rather than activation of PLA_2 (Table 1). The loss of light- and GTPγS-induced PLA_2 activity in G_t-poor ROS appeared to correlate with the loss of G_t since there was no loss of PLA_2 activity per se. This was demonstrated by the fact that stimulation of ROS membranes with melittin, a PLA_2 activator (Habermann, 1972), led to comparable activity in both G_t-rich and G_t-poor ROS (Table 1). In addition, exogenous G_t restored the light- and GTPγS-induced activation of PLA_2 in dark-adapted, G_t-poor ROS (Fig. 2A), further supporting a role for G_t in the light activation of PLA_2.

Table 1. Phospholipase A_2 and phospholipase C activities of transducin-rich and transducin-poor ROS in the presence or absence of light and/or the phospholipase activators, melittin and deoxycholate.

	phospholipase A_2 activity (nmol ^{14}C-arachidonate released/min/mg protein) ± melittin		phospholipase C activity (pmol ^{3}H-inositol phosphate released/min/mg protein) ± deoxycholate	
	-	+	-	+
transducin-rich ROS:				
dark:	29.4 ± 2.6	275.2 ± 22.1	0.78 ± 0.03	5.23 ± 0.65
+ light:	133.6 ± 24.2	280.4 ± 18.1	3.13 ± 0.12	5.65 ± 0.82
transducin-poor ROS:				
dark:	46.1 ± 2.7	237.5 ± 24.7	0.43 ± 0.07	1.58 ± 0.11
+ light:	63.8 ± 3.5	262.7 ± 14.8	0.47 ± 0.05	1.62 ± 0.14

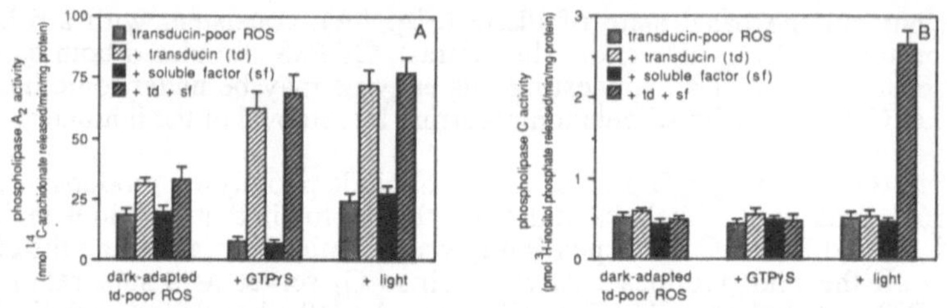

Figure 2. Transducin alone restores the light- and GTPγS-induced increase in phospholipase A_2 activity in transducin-depleted ROS but restoration of light-activated phospholipase C requires an added factor. A. Phospholipase A_2 activity. B. Phospholipase C activity. Results are mean ± S.E.M. (n = 3).

While removal of G_t also led to loss of light- and GTPγS-stimulated PLase C activity (Table 1), further examination revealed that deoxycholate-stimulated PLase C activity itself appeared to be lost upon repeated washing of the dark-adapted membranes (Table 1; Jelsema and Ma, 1989). The loss of light- and GTPγS-induced phospholipase C activity, therefore, cannot be directly attributed to the removal of G_t, although it may be a factor. Furthermore, addition of exogenous G_t alone was unable to restore the activation of phospholipase C (Fig. 2B). However, the light activation, but not the GTPγS-induced stimulation, of PLase C could be restored by addition of exogenous transducin along with a soluble factor present in concentrated supernatant from the isotonic washes, although the soluble factor alone had no effect on PLase C activity (Fig. 2B). The added requirement for a soluble factor and the inability to restore the GTPγS-induced increase in PLase C activity even upon addition of both G_t and concentrated supernatant, again indicate a fundamental difference in the regulation of the two phospholipases.

Transducin α and βγ subunits have contrasting roles in the modulation of PLA_2 and PLase C activity. The isolated, purified subunits of G_t were added to G_t-poor ROS membranes in the presence or absence of light to determine the role of the individual subunits in the activation of the two phospholipases. As previously reported (Jelsema, 1987) and illustrated here (Fig. 3A), the βγ subunits were found to stimulate PLA_2. The α subunit, while slightly stimulatory when added alone, was able to inhibit the βγ-induced increase in phospholipase A_2 when added under nondissociating conditions (Fig. 3A). These results suggest that the inhibitory effect of the α subunit on the βγ-induced increase in PLA_2 activity occurs as a result of facilitating subunit reassociation.

In contrast, PLase C activity was stimulated by addition of $G_{t\alpha}$ while the βγ subunits were inhibitory when added in the absence of light, regardless of whether GTPγS was present (Fig. 3B). In the presence of light, however, the βγ subunits stimulated the enzyme and, when added together with $G_{t\alpha}$, the stimulatory effects of the α and βγ subunits were additive (Fig. 3B). These results suggest that the βγ subunits have an effect on PLase C activity independent of the action of the α subunit. In view of the PT- and GDPβS-induced stimulation of basal and activated PLase C (Figs. 3A, 4A) and the GTPγS-mediated inhibition of the light-activated enzyme (Fig. 3A), there appears to be not only tonic inhibition of PLase C by a G protein, but also inhibition of the light-activated enzyme upon activation of a G protein. It is unclear, however, whether inhibition is mediated by the α or

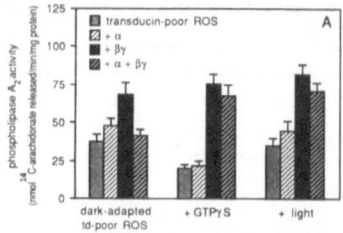

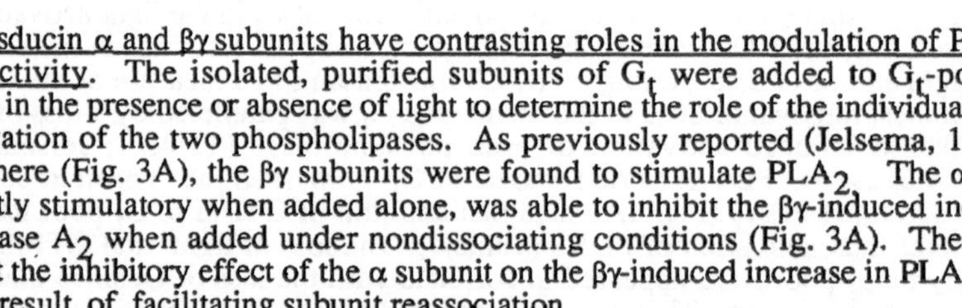

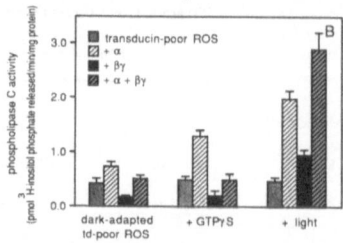

Figure 3. Transducin mediates the activation of phospholipases A_2 and C, but by action of different subunits. A. The βγ subunits stimulate phospholipase A_2 activity while the α subunit inhibits this βγ effect. B. The α subunit stimulates phospholipase C activity while the βγ subunits either inhibit or stimulate the enzyme activity depending on the presence or absence of light. Results are mean ± S.E.M. (n = 3).

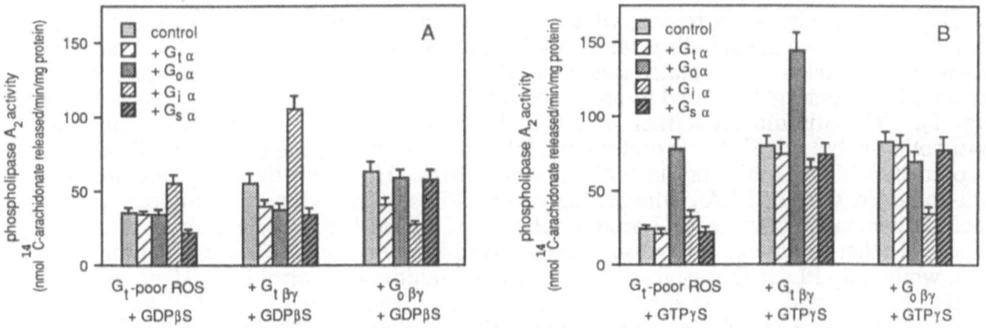

Figure 4. Effect of G protein subunits on the phospholipase A_2 activity of dark-adapted, transducin-poor ROS in the presence of (A) GDPβ S or (B) GTPγS. Results are mean ± S.E.M. (n = 3).

βγ subunits of this inhibitory G protein. The ability of the βγ subunits of G_t to inhibit PLase C activity even in the presence of GTPγS and their ability to prevent the α-induced increase in PLase C activity in the absence of light (Fig. 3B), together with the tonic inhibition of the enzyme suggest that inhibition of PLase C may be a function of the βγ subunits which are present in membranes in excess of the G protein α subunits.

G protein βγ subunits stimulate PLA_2 while the effects of the α subunits vary depending on their state of association or dissociation. The βγ subunits of both G_t and G_o stimulated PLA_2 when added to G_t-poor ROS. Since these G proteins share a common β but have different γ subunits (Roof et al., 1985), it is suggested that the 36 kDa β subunit is responsible for stimulation of PLA_2, thereby implicating most G proteins in regulation of PLA_2. $G_{t\alpha}$ (T_α), when added under nondissociating conditions, inhibited the βγ-induced increase in PLA_2 (Figs. 3A, 4A). The inhibitory effect of $G_{t\alpha}$ is, therefore, suggested to occur by facilitating subunit reassociation, which terminates the signal. To examine the potential for modulation of PLA_2 by other G_α subunits, their effect on PLA_2 was examined in G_t-poor ROS under dissociating or non-dissociating conditions, i.e., with either GTPγS or GDPβS added. When $G_{t\alpha}$ was added alone to G_t-poor ROS with GDPβS present, it had little effect on PLA_2, but inhibited PLA_2 stimulation by either $G_{t\beta\gamma}$ ($T_{\beta\gamma}$) or $G_{o\beta\gamma}$ (Fig. 4). In contrast, $G_{s\alpha}$ was slightly inhibitory when added alone to G_t-poor ROS in the presence of GDPβS and inhibited $G_{t\beta\gamma}$-, but not $G_{o\beta\gamma}$-stimulated PLA_2 activity (Fig. 4). In the presence of GDPβS, $G_{o\alpha}$ also inhibited PLA_2 activity induced by $G_{t\beta\gamma}$, but not $G_{o\beta\gamma}$ (Fig. 4). However, $G_{o\alpha}$ was found to stimulate PLA_2 in the presence of GTPγS (Fig. 4B), and this stimulation was additive with the stimulatory effect of $G_{t\beta\gamma}$, but not $G_{o\beta\gamma}$ (Fig. 4). $G_{i\alpha}$ also appeared able to both inhibit and stimulate PLA_2. $G_{i\alpha}$ stimulated PLA_2 when added in the presence of GDPβS, suggesting it may inhibit in its dissociated state, and the stimulation by GDPβS-treated $G_{i\alpha}$ could be blocked by addition of $G_{t\beta\gamma}$, but not $G_{o\beta\gamma}$ (Fig. 4).

DISCUSSION

In bovine ROS, PLA_2 and PLase C are both light-activated enzymes that are under dual regulation by G proteins, with G_t functioning to couple the light activation of both phospholipases. These similarities and the fact that by-products of one phospholipase can lead to activation of the other (Loeb and Gross, 1986; Irvine et al., 1979) initially raised concerns about the independence of the two enzyme activities. However, our results indicate that despite these similarities, the two phospholipases are activated independently of one another and exhibit fundamental differences in their G protein regulation. Not only do the phospholipases respond differently to the bacterial toxins, which modulate the activities of their G protein substrates, but the α and βγ subunits of G_t have markedly different effects on the two phospholipases, indicating that even though they share a common G protein, the mechanism of action for G_t is distinct for the two enzymes.

Neomycin, a potent and specific inhibitor of PLase C (Schacht, 1976), inhibited both basal and light-stimulated PLase C activity (Fig. 1B) but had no effect on PLA_2 activities (Fig. 1A). Activation of PLA_2, therefore, did not require prior activation of PLase C. The independence of the two phospholipase activities was further indicated by their responses to GDPβS and the bacterial toxins, which are a reflection of differences in the G protein regulation of the two phospholipases. Both GDPβS and PT, which function to stabilize G proteins in their inactive, associated state (Van Dop et al., 1984; Cassel and Selinger, 1977), increased basal and light-induced PLase C activity but inhibited the light-activation of PLA_2 (Fig. 1). The stimulatory effect of PT and GDPβS on PLase C suggests that this phospholipase, but not PLA_2, is under tonic inhibitory control by a PT-sensitive G protein. In contrast, CT was inhibitory to both phospholipases. However, the toxin inhibited the light-activated state of PLA_2 whereas only basal PLase C activity was inhibited by CT. Since CT generally activates its G protein substrate (Gilman, 1987), the inhibitory effects of CT suggest that G protein activation is involved in inhibition of the two phospholipases. Thus, while only PLase C appears to be under tonic inhibitory control by a PT-sensitive G protein, both phospholipases appear to be susceptible to inhibition by (a) CT-sensitive G protein(s). Whether or not the phospholipases share a common CT-sensitive inhibitory G protein, it is clear there are basic differences in the mode of action on the two enzymes since CT inhibits the stimulated state of the one phospholipase and the basal state of the other.

G_t appears to function in the light activation of both phospholipases but employs different mechanisms of action. A role for G_t in the light activation of PLA_2 in bovine retinal ROS was indicated by: 1) the ability of GTPγS to mimic and GDPβS to prevent the light activation of PLA_2 (Fig. 1A; Jelsema, 1987), demonstrating a role for a light-sensitive G protein in PLA_2 activation; 2) the inhibitory effect of PT on the light activation of G_t (Fig. 1A); Jelsema, 1987), consistent with the PT-induced ADP-ribosylation of $G_{t\alpha}$ and the subsequent stabilization of G_t in its inactive, associated state (Van Dop et al., 1984); 3) the loss of light- and GTPγS-induced PLA_2 activity upon removal of G_t from dark-adapted ROS without concomitant loss of melittin-stimulated PLA_2 activity (Table 1; Jelsema, 1987); and 4) the ability of exogenous G_t to restore the light- and GTPγS-induced increase in PLA_2 in dark-adapted, G_t-poor ROS (Fig. 2A; Jelsema, 1987; Jelsema and Axelrod, 1987). In contrast, the only support for a role for G_t in the light activation of PLase C is the ability of exogenous G_t to restore the light activation of PLase C in G_t-depleted ROS and this also required the presence of a soluble factor (Fig.2B; Jelsema and Ma, 1989). While the soluble fraction had some PLase C activity associated with it, addition of the soluble fraction in the absence of G_t had no effect on the PLase C activity of the ROS (Fig. 2B). Furthermore, even in the presence of the soluble fraction, G_t was unable to restore the GTPγS-induced increase in PLase C (Fig. 2B). The requirement for an additional factor and the ability to restore only the light-induced and not the GTPγS-induced increase in PLase C activity of G_t-poor ROS adds support to the concept that the two enzymes, while requiring light and G_t for activation, are regulated by different mechanisms. The differing roles for G_t in the activation of the two phospholipases is, however, best illustrated by the contrasting roles of the α and βγ subunits of G_t in the modulation of the two phospholipase activities (Fig. 3).

In the modulation of PLA_2 by G_t, the βγ subunits appear to function in the activation of the enzyme, based on their stimulation of PLA_2 in dark-adapted, G_t-poor ROS (Fig. 3A; Jelsema and Axelrod, 1987). The α subunit, in contrast, blocked the βγ-induced increase in PLA_2 (Fig. 3A; Jelsema, 1987). This inhibition appears to occur by facilitating reassociation of the inactive heterotrimer, since the inhibitory effect of the α subunit on the βγ-induced increase in PLA_2 activity was lost under conditions that cause subunit dissociation (Fig. 3A; Jelsema and Axelrod, 1987). The βγ-induced stimulation of PLA_2, however, was observed in the presence of GTPγS, which prevents subunit reassociation. The stimulatory effect of the βγ subunits on PLA_2 activity is, therefore, independent of its interaction with other G protein α subunits, suggesting that the βγ subunits, not the α subunit of another G protein, are the primary effectors in the light activation of PLA_2 in the ROS. The observed increase in basal activity of dark-adapted, G_t-poor ROS (Table I) may, in fact, reflect the preferential retention of $G_{t\beta\gamma}$ during the washing procedures employed to remove G_t (Sternweis, 1985).

In the modulation of PLAse C activity by G_t, the βγ subunits primarily have an inhibitory role while the α subunit consistently stimulates PLase C activity in the presence or absence of light and/or GTPγS (Fig. 3B). The effect of the βγ subunits, although inhibitory in this case, was again observed to occur in the presence of GTPγS, indicating that the βγ effect on PLase C is also independent of its interaction with other G protein α subunits. In the presence of light, however, the βγ subunits had a stimulatory rather than an inhibitory effect and their stimulation of PLase C was additive with the effect of the α subunit. The additive nature of their stimulation of PLase C indicates that the α and βγ subunits increase PLase C activity by different mechanisms. Since light, but not GTPγS, is able to convert the βγ inhibition of PLase C to a stimulation, light appears to have an additional role that is unrelated to its effect on G protein dissociation. One possible explanation is the light-induced association of inhibitory βγ subunits with another protein, such as the 33 kDa protein described by Lee et al., (1987) which appears to bind βγ in the light when βγ is dissociated from $G_{t\alpha}$. This would account for the stimulatory effect of the βγ subunits in the presence of light and not GTPγS, and would also involve a mechanism distinct from the effect of $G_{t\alpha}$. A different hypothesis, however, is required to explain how the presence of a soluble fraction allows G_t to restore the light activation of PLase C but not its activation by GTPγS. The light activation of PLase C occurs only in the presence of a soluble fraction known to contain PLase C activity but whose addition alone had no effect on the membrane-associated PLase C activity. One hypothesis to explain the differing effects of light and GTPγS on PLase C activity is that light not only dissociates the G protein responsible for PLase C activation, but also induces a shift in location of PLase C from the cytosol to the membrane, thus allowing for its activation by the dissociated G protein. A requirement for light to bring the enzyme in contact with its activated G protein would explain why addition of G_t in the presence of GTPγS was insufficient for PLase C activation.

In terms of the dual regulation of PLA_2 activity by G proteins, inhibition of light-activated PLA_2 appears to occur by activation of another CT-sensitive G protein, as indicated by the ability of either GTPγS or CT to inhibit light-stimulated but not basal enzyme activity (Figs. 1A). Although G_t is a substrate for CT (Bourne and Stryer, 1986), the inhibitory action of CT is independent of its effects on G_t since CT treatment of exogenous G_t had no inhibitory effect on its ability to restore the light activation of PLA_2 (Jelsema et al., 1989) despite its ability to inhibit light-activated PLA_2 in G_t-rich ROS (Fig.1A). The mechanism for this G protein-mediated inhibition of light-activated PLA_2 is suggested to occur by interaction of the dissociated α subunit of this CT-sensitive, "inhibitory" G protein (possibly G_s) with the βγ subunits of G_t, resulting in inhibition of the βγ-activated PLA_2. In support of a role for G_s in regulation of PLA_2, $G_{s\alpha}$ was able to inhibit the stimulation of PLA_2 induced by $G_{t\beta\gamma}$ (but not $G_{o\beta\gamma}$) in G_t-poor ROS (Fig. 4A). Similar to the inhibition by $G_{t\alpha}$, inhibition by $G_{s\alpha}$ occurred only in the presence of GDPβS (Fig. 4), suggesting its capacity for blocking stimulated PLA_2 activity also occurs via its interaction with $G_{t\beta\gamma}$.

The presence of a G protein that more directly inhibits PLA_2 was observed upon removal of G_t from dark-adapted ROS. In G_t-poor ROS, GTPγS was found to inhibit rather than stimulate PLA_2 (Fig. 1A; Table 1). Since this inhibition occurs upon addition of GTPγS, which prevents subunit reassociation (Gilman, 1987), this G protein-mediated inhibition occurs independent of the βγ-induced stimulation of PLA_2, suggesting action of an inhibitory G_α. A potential candidate for this inhibitory effect is $G_{i\alpha}$. PLA_2 activity of G_t-poor ROS was stimulated by addition of $G_{i\alpha}$ in the presence of GDPβS (Fig. 4A), which stabilizes the inactive state of G proteins. This suggests that dissociated $G_{i\alpha}$ may function in an inhibitory capacity. Moreover, the stimulation of PLA_2 activity by $G_{i\alpha}$ when added in the presence of GDPβS was additive with that observed upon addition of $G_{t\beta\gamma}$ but not $G_{o\beta\gamma}$ (Fig. 4), further indicating that the $G_{i\alpha}$ effect occurs independent of an interaction with $G_{t\beta\gamma}$.

The ability of $G_{o\alpha}$ to inhibit βγ-induced stimulation of PLA_2 in the presence of GDPβS, similar to the effects of $G_{s\alpha}$ and $G_{t\alpha}$ (Fig. 4A), suggests that $G_{o\alpha}$ may also function as an inhibitor of activated PLA_2 by facilitating removal of the stimulatory βγ subunits. In contrast to effects of other G_α subunits, however, $G_{o\alpha}$ was found to stimulate PLA_2 activity upon dissociation, <u>i.e.</u>, in the presence of GTPγS. Since the stimulation observed with GTPγS-treated $G_{o\alpha}$ was additive with the stimulation by $G_{t\beta\gamma}$ (but not $G_{o\beta\gamma}$), $G_{o\alpha}$ is suggested to directly stimulate PLA_2 (Fig.4B). These results indicate that multiple G proteins are able to modulate PLA_2 activity, which may account for differences in the toxin sensitivity of PLA_2 in various systems (Jelsema, 1987; Burch et al., 1988). Some of the inhibitory G_α effects on PLA_2 were observed only in the presence of $G_{t\beta\gamma}$ or required the presence of GDPβS, suggesting PLA_2 inhibition is a result of their interaction with the stimulatory β subunit. Other G_α effects, such as PLA_2 stimulation by GDPβS-inactivated $G_{i\alpha}$ and GTPγS- activated $G_{o\alpha}$, were additive with the stimulatory effects of $G_{t\beta\gamma}$ but not

$G_{o\beta\gamma}$, suggesting more direct roles for $G_{i\alpha}$ and $G_{o\alpha}$ in the inhibition and stimulation of PLA_2, respectively. The different effects of G protein α subunits when interacting with $G_{t\beta\gamma}$ and $G_{o\beta\gamma}$ further suggest that either there are fundamental differences between the 35 and 36 kDa forms of β or the accompanying γ subunit may markedly alter G protein interactions.

The observation that the βγ–induced stimulation of PLA_2 occurs in the presence of GTPγS (Fig. 3B; Fig. 4B), which prevents subunit reassociation, precludes the βγ subunits acting by removal of an inhibitory G protein α subunit. The βγ subunits also do not appear to directly activate the enzyme, based on in vitro assays using soluble pancreatic PLA_2 (Jelsema et al., 1989) or partially-purified PLA_2 (unpublished results). While absence of a βγ effect on soluble or partially-purified membrane-associated PLA_2 does not definitively eliminate their possible interaction with a membrane-bound form of the enzyme, these results fail to support direct action of the βγ subunits on PLA_2. Othe. pctential mechanisms of action include interaction of the βγ subunits with non-G protein PLA_2 modulators. βγ subunits have been reported to interact with calmodulin (Katada et al., 1987), a potential activator of PLA_2. However, the Ca^{++}-mediated regulation of PLA_2 is independent of calmodulin (Withnall et al., 1984). Furthermore, βγ interaction with calmodulin inhibits rather than stimulating calmodulin-dependent enzymes, as indicated by the βγ-induced inhibition of calmodulin- activated adenylate cyclase (Katada et al., 1987).

Another possible mechanism of action for the βγ-induced stimulation of PLA_2 was suggested by a report that calpactin, a PLA_2 inhibitor thought to act by binding phospholipid substrate (Davidson et al., 1987), was present in bovine ROS (Huang et al., 1986). In an in vitro system using soluble pancreatic PLA_2 and calpactin isolated from human placenta, calpactin inhibited PLA_2 activity 40% and the βγ subunits of either G_t or G_o were able to block the calpactin-induced inhibition of PLA_2 (Jelsema et al., 1989). These results suggest that stimulation of PLA_2 by the βγ subunits may occur by blocking the action of a non-G protein PLA_2 inhibitor. Actin, a cytoskeletal protein that binds calpactin (Glenney, 1986), also prevented the calpactin-mediated inhibition of PLA_2 supporting the cancept that interaction with calpactin can prevent its inhibition of PLA_2. While experiments are underway to determine whether the βγ subunits do in fact similarly interact with calpactin, their reported interaction with calmodulin, another Ca^{++}-binding protein (Katada et al., 1987), makes this an attractive hypothesis although the physiological relevance of the calpactin-induced inhibition of PLA_2 has been challenged on the basis of its interaction with the phospholipid substrate rather than the enzyme (Davidson et al., 1987).

In this report, we demonstrate that in bovine ROS, both PLA_2 and PLase C are susceptible to regulation by multiple G proteins, either directly or indirectly, via subunit interactions. While G_t appears to function in the light activation of both phospholipases, the βγ subunits stimulate PLA_2 activity while inhibiting PLase C. Both of these βγ effects occur in the presence of GTPγS, indicating the βγ effects on both PLA_2 and PLase C are independent of their interaction with other G_α subunits. In the case of PLA_2 stimulation, the βγ effect is suggested to involve inactivation of a non-G protein PLA_2 inhibitor, rather than direct activation of the enzyme. Multiple G proteins ($G_{t\alpha}$, $G_{s\alpha}$, $G_{o\alpha}$), by interaction with the stimulatory βγ subunits, appear able to inhibit activated PLA_2. The potential for more direct roles for $G_{i\alpha}$ and $G_{o\alpha}$ in PLA_2 inhibition and stimulation, respectively, was suggested by the additive effects of GDPβS-inactivated $G_{i\alpha}$ and GTPγS-activated $G_{o\alpha}$ on PLA_2 stimulation induced by $G_{t\beta\gamma}$

In the regulation of PLase C, the tonic inhibitory control exerted by a PT-sensitive G protein is suggested to occur by action of the βγ subunits of G_t, which are present in the membranes in excess of G_α subunits (Sternweis, 1986). PT and GDPβS, which stabilize the heterotrimeric forms of G proteins, would stimulate PLase C activity by facilitating subunit reassociation and thereby removing the dissociated, inhibitory βγ subunits. The CT-induced inhibition of basal PLase C activity would then result from the added release of βγ subunits upon activation of G_s by CT. CT would not be expected to have an inhibitory effect on PLase C in the light because the βγ subunits are no longer inhibitory in the light (Fig. 3B). The capacity for light to reverse the inhibitory effect of the βγ subunits is suggested to result from the light-stimulated interaction of the βγ subunits with a 33 kDa protein, which may increase PLase C activity by blocking the inhibition by βγ. The stimulation by $G_{t\alpha}$, however, occurs in the presence of GTPγS and is additive with the stimulatory effect of the βγ subunits in the presence of light (Fig. 3B). Thus, $G_{t\alpha}$ stimulation does not appear to involve interaction with the inhibitory βγ subunits. The light activation of the PLase C in bovine ROS may, therefore, result from both a stimulation on the part of $G_{t\alpha}$ and a light-induced removal of βγ-induced tonic inhibition. Thus, while G_t is employed in the light activation of both PLA_2 and PLase C, it is by fundamentally different mechanisms of action.

REFERENCES

Abood, M. E., Hurley, J. B., Pappone, M.-C., Bourne, H. R., and Stryer, L., 1982, Functional homology between signal-transducing proteins. Cholera toxin inactivates the GTPase activity of transducin, J. Biol. Chem., 257: 10540-10543.

Axelrod, J., Burch, R. M., and Jelsema, C. L., 1988, Receptor-mediated activation of phospholipase A_2 via GTP-binding proteins: arachidonic acid and its metabolites as second messengers, Trends Neurosci.,11: 117-123.

Baehr, W., Devlin, M. J., and Applebury, M. L., 1979, Isolation and characterization of cGMP phosphodiesterase from bovine rod outer segments, J Biol. Chem., 254: 11669-11677.

Bourne, H., and Stryer, L., 1986, G proteins: A family of signal transducers, Ann. Rev. Cell Biol., 2: 397-420.

Burch, R. M., Jelsema, C. L., and Axelrod, J., 1988, Cholera toxin and pertussis toxin stimulate prostaglandin E_2 synthesis in a murine macrophage cell line, J. Pharmacol. Exp. Therapeut., 244: 765-773.

Cassel, D., and Selinger, Z., 1977, Mechanism of adenylate cyclase activation by cholera toxin: inhibition of GTP hydrolysis at the regulatory site, Proc. Natl. Acad. Sci. U. S. A., 74: 4185-4189.

Cockcroft, S., and Gomperts, B. D., 1985, Role of guanine nucleotide binding protein in the activation of polyphosphoinositide phosphodiesterase, Nature, 314: 534-536.

Davidson, F.F., Dennis, E.A., Powell, M., and Glenney, J.R., 1987, Inhibition of phospholipase A_2 by "lipocortins" and calpaactins, J. Biol. Chem., 262: 1698-1705.

Dunlap, K., Holz, G. G., and Rane, S. G., 1987, G proteins as regulators of ion channel function, Trends Neurosci., 10: 241-244.

Elks, M. L., Watkins, P. A., Manganiello, V.C., Moss, J., Hewlett, E., and Vaughan, M., 1983, Selective regulation by pertussis toxin of insulin-induced activation of particulate cAMP phosphodiesterase activity in 3T3-L1 adipocytes, Biochem. Biophys. Res. Commun., 116: 593-598.

Fung, B. K.-K., Hurley, J. B., and Stryer, L., 1981, Flow of information in the light-triggered cyclic nucleotide cascade of vision, Proc. Natl. Acad. Sci. U. S. A., 78: 152-156.

Fuse, I., and Tai, H.-H., 1987, Stimulation of arachidonate release and inositol-1,4,5-triphosphate fromation are mediated by distinct G-proteins in human platelets, Biochem. Biophys. Res. Commun., 146: 659-665.

Gilman, A. G., 1987, G proteins: Transducers of receptor- generated signals, Ann. Rev. Biochem., 56: 615-649.

Habermann, E., 1972, Bee and wasp venoms, Science, 177: 314-322.

Huang, K.-S., Wallner, B. P., Mattaliano, R. J., Tizard, R., Burne, C., Frey, A., Hession, C., McGray, P., Sinclair, K. L., Chow, E. P., Browning, J. L., Ramachandran, K. L., Tang, J., Smart, J. E., and Pepinsky, R. B., 1986, Two human 35 kd inhibitors of phospholipase A_2 are related to substrates of $pp60^{v\text{-}src}$ and of the epidermal growth factor receptor/kinase, Cell, 46: 191-199.

Irvine, R. F., Letcher, A. J., and Dawson, R. M. C., 1979, Fatty acid stimulation of membrane phosphatidylinositol hydrolysis by brain phosphatidylinositol phosphodiesterase, Biochem. J., 179: 97-100.

Jelsema, C. L., 1987, Light activation of phospholipase A_2 in rod outer segments of bovine retina and its modulation by GTP-binding proteins, J. Biol. Chem., 262: 163-168.

Jelsema, C. L., and Axelrod, J., 1987, Stimulation of phospholipase A_2 activity in bovine rod outer segment by the βγ subunits of transducin and its inhibition by the α subunit, Proc. Natl. Acad. Sci. U. S. A., 84: 3623-3627.

Jelsema, C. L., and Axelrod, J., 1987, Light activation of phospholipases in rod outer segments, in: "Sensory Transduction, Discussions in Neurosciences," Vol. 4: 79-84, A. J. Hudspeth, P. R. MacLeish, F. L. Margolis and T. N. Wiesel, eds., Foundation for the Study of the Nervous System, Geneva, Switzerland.

Jelsema, C. L., Burch, R. M., Jaken, S., Ma, A. D., and Axelrod, J., 1989, Modulation of phospholipase A_2 activity in rod outer segments of bovine retina by G protein subunits, guanine nucleotides, protein kinases and calpactin, Neurol. and Neurobiol., 49: 25-45.

Jelsema, C. L., and Ma, A. D., 1989, Regulation of phospholipase A_2 and phospholipase C in rod outer segments of bovine retina involves a common GTP-binding protein but different mechansims of action, in: "Arachidonic Acid Metabolism in the Nervous System: Physiological and Pathological Significance," Vol. 559: 158-177, A. I. Barkai, and N. G. Bazan, eds., New York Acad. of Sciences, New York.

Katada, T., Kusakabe, K., Oinuma, M., and Ui, M., 1987, A novel mechanism for the inhibition of adenylate cyclase via inhibitory GTP-binding proteins. Calmodulin-dependent inhibition of the cyclase catalyst by the βγ subunits of GTP-binding proteins, J. Biol. Chem., 262: 11897-11900.

Kuhn, H., 1980, Light- and GTP-regulated interaction of GTPase and other proteins with bovine photoreceptor membranes, Nature, 283: 587-589.

Lee, R. H., Lieberman, B. S., and Lolley, R. N., 1987, A novel complex from bovine visual cells of a 33,000-dalton phosphoprotein with beta- and gamma-transducin: purification, and subunit structure, Biochemistry, 26: 3983-3890.

Loeb, L. A., and Gross, R. W., 1986, Identification and purification of sheep platelet phospholipase A_2 isoforms. Activation by physiologic concentrations of calcium ion, J. Biol. Chem., 261: 10467-10470.

Mobley, A., and Tai, H.-H., 1985, Synergistic stimulation of thromboxane biosynthesis by calcium ionophore and phorbol ester or thrombin in human platelets, Biochem. Biophys. Res. Commun., 130: 717-723.

Ohta, H., Okajima, F., and Ui, M., 1985, Inhibition by islet-activating protein of a chemotactic peptide-induced early breakdown of inositol phospholipids and Ca^{2+} mobilization in guinea pig neutrophils, J. Biol. Chem., 260: 15771-15780.

Roof, D. J., Applebury, M. L., and Sternweis, P. C., 1985, Relationships between the family of GTP-binding proteins isolated from bovine central nervous system. J. Biol. Chem., 260: 16242- 16249.

Samuelsson, B., Goldyne, M., Granstrom, E., Hamberg, M., and Hammerstrom, S., 1978, Prostaglandins and thromboxanes, Ann. Rev. Biochem., 47: 997-1029.

Schact, J., 1976, Biochemistry of neomycin ototoxicity, J. Acoust. Soc. Am., 59: 940-944.

Sternweis, P.C., 1986, The purified alpha subunits of G_o and G_i from bovine brain require beta gamma for association with phospholipid vesicles, J. Biol. Chem., 261: 631-637.

Sternweis, P. C., and Gilman, A. G., 1982, Aluminum: a requirement for activation of the regulatory component of adenylate cyclase by fluoride, Proc. Natl. Acad. Sci. U. S. A., 79: 4888-4891.

Sternweis, P. C., and Robishaw, J. D., 1984, Isolation of two proteins with high affinity for guanine nucleotides from membranes of bovine brain, J. Biol. Chem., 259: 13806-13813.

Ueda, H., Yoshihara, Y., Misawa, H., Fukushima, N., Katada, T., Ui, M., Takagi ,H., Satoh, M., 1989, The kyotorphin (tyrosine-arginine) receptor and a selective reconstitution with purified G_i, measured with GTPase and phospholipase C assays, J. Biol. Chem., 264: 3732-3741.

Withnall, M.T., Brown, T.J., and Diocee, B.K., 1984, Calcium regulation of phospholipase A_2 is independent of calmodulin, Biochem. Biophys. Res. Commun., 121: 507-513.

Van Dop, C., Yamanaka, G., Steinberg, F., Sekura, R. D., Manclark, C. R., Stryer, L., and Bourne, H. R., 1984, ADP-ribosylation of transducin by pertussis toxin blocks the light-stimulated hydrolysis of GTP and cGMP in retinal photreceptors, J. Biol. Chem., 259: 23-26.

Yada, Y., Nagao, S., Okano, Y., and Nozawa, Y., 1989, Inhibition by cyclic AMP of guanine nucleotide-specific phospholipase C in human platelets, F. E. B. S . Lett., 242: 368-372.

STIMULUS-COUPLED ACTIVATION OF PHOSPHOLIPASE D

M. Motasim Billah, John C. Anthes, Robert W. Egan,
Theodore J. Mullmann and Marvin I. Siegel

Department of Allergy and Inflammation
Schering-Plough Corporation
Bloomfield, New Jersey 07003 USA

SUMMARY

Phospholipase D (PLD) acting on phosphatidylcholine (PC) exhibits two distinct enzymatic activities. In addition to the hydrolytic activity that generates phoshatidic acid (PA), PLD also catalyzes a transphosphatidylation reaction whereby the phosphatidyl moiety of PC is transferred to primary alcohols to produce phosphatidylalcohols. Using novel methodologies to label the endogenous PC of human granulocytes, we have obtained conclusive evidence for PLD-catalyzed hydrolysis and transphosphatidylation in stimulated granulocytes. Dimethylsulfoxide-differentiated HL-60 granulocytes or human peripheral granulocytes have been double-labeled in 1-0-alkyl-PC(alkyl-PC) by incubating cells simultaneously with [^{3}H]alkyl-lysoPC and alkyl-lyso[^{32}P]PC. Upon stimulation with the chemotactic peptide, fMet-Leu-Phe (fMLP), phorbol diesters such as phorbol-12-myristate-13-acetate (PMA), synthetic 1,2-diacylglycerols such as 1-oleoyl 2-acetyl-sn-glycerol and the Ca^{2+} ionophore, A23187, these double-labeled cells produce PA containing both ^{3}H and ^{32}P. In the presence of ethanol, stimulated cells also generate double-labeled phosphatidylethanol (PEt). Both PA and PEt are formed within seconds after stimulation. Because the cellular ATP contains no [^{32}P]ATP, [^{32}P]PA and [^{32}P]PEt must be derived from [^{32}P]PC by PLD action. The ^{3}H/^{32}P ratios of double-labeled PA and PEt formed during stimulation are very similar to that of double-labeled PC, demonstrating that PA and PEt are formed predominantly by PLD action on PC. The protein kinase C inhibitor K252a causes only partial inhibition of PLD activation by various agents including fMLP and PMA, suggesting that mechanisms other than protein kinase C activation are predominant. Post-nuclear homogenates from granulocytes selectively hydrolyze exogenous double-labeled PC by PLD to produce double-labeled PA and PEt. This hydrolysis requires the combined presence of Ca^{2+} and nonhydrolyzable GTP analogs, suggesting PLD regulation by Ca^{2+} and GTP-binding proteins. Rapid activation of PLD by diverse agents coupled with the fact that PA, and its dephosphorylated product, diglyceride exhibit diverse biologic functions underscore the importance of PLD in signal transduction.

INTRODUCTION

Many cells produce phosphatidic acid (PA) upon stimulation of specific receptors. This PA has been thought to be formed by the combined activities of phospholipase C (PLC) and diglyceride (DG) kinase (Michell, 1975). However, phospholipase D (PLD) can act directly on phospholipids to hydrolyze the terminal phosphodiester bond and thereby generate PA and free bases directly. Nevertheless, the involvement of PLD in receptor-linked formation of PA has been overlooked primarily because of the prevailing notion of PLD being inconsequential in mammalian cells.

Initial studies on PLD have utilized preparations from plant tissues. Subsequent work has indicated that PLD, like other phospholipases, is distributed throughout the phylogenetic scale (reviewed by Heller, 1978; Kanfer, 1980). PLD activities have been partially purified from rat brain (Taki and Kanfer, 1979) and human eosinophils (Kater et al, 1976), and have been detected in other mammalian cells (Chalifour and Kanfer, 1980). Mammalian PLD is membrane-bound and prefers phoshatidylcholine (PC) as substrate (Kanfer, 1980; Bocckino et al., 1987a; Anthes et al., 1989), although, under certain conditions, phosphatidylethanolamine and phosphatidylinositol may also be degraded (Taki and Kanfer, 1979; Witter and Kanfer, 1985; Balsinde et al., 1988). Recent studies demonstrate that endogenous PC undergoes rapid turnover upon specific stimulation (reviewed by Exton, 1988; Pelech and Vance, 1989) and that this turnover is, at least in part, mediated by an activatable PLD (see below). This chapter will focus on the regulation of mammalian PLD with particular emphasis on granulocyte activation.

Identification of PLD in Stimulated Cells

Cellular PC can be readily labeled with radiolabeled choline (Cho), fatty acids or glycerol. PC turnover can then be followed by measuring Cho, phosphocholine, PA and DG. As illustrated in Fig. 1, these products could be formed as a result of PC hydrolysis by either PLC or PLD. Thus, product analysis and kinetic comparisons of these products are largely inconclusive regarding the actual identity of the phosphodiesterase activity(ies) involved.

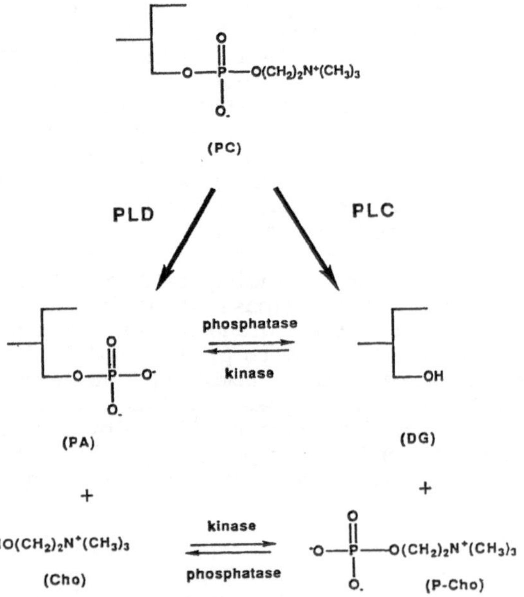

Fig. 1. Interconversion of products derived from phosphatidylcholine (PC) hydrolysis by phospholipase D (PLD) and phospholipase C (PLC)

In order to conclusively identify PLD activities in intact cells, two approaches have recently been developed (Bocckino et al, 1987b; Pai et al, 1988a,b; Liscovitch, 1989). One approach is based on the unique ability of PLD to transfer the phosphatidyl moiety of the phospholipid substrate to primary alcohols to produce phosphatidylalcohols (Dawson, 1967; Yang et al, 1967; Kobayashi and Kanfer, 1987). This transphosphatidylation reaction is believed to involve a phosphatidyl-PLD intermediate (Heller, 1978) and can be exploited to study PLD using phospholipids labeled anywhere in the phosphatidyl moiety (Pai et al, 1988a; Liscovitch, 1989).

The second approach involves labeling of endogenous PC with ^{32}P (Pai et al, 1988b). Because formation of [^{32}P]PA from [^{32}P]PC must occur by PLD action, this approach provides the conclusive proof for PLD activity. A crucial consideration in this approach is that labeling must be done in the absence of [^{32}P]ATP, because in the presence of [^{32}P]ATP, [^{32}P]PA could also be formed by the PLC/DG kinase pathway. Therefore, it is critical that cellular phospholipids not be labeled with exogenous [^{32}P]orthophosphate, because this labeling occurs through ATP.

Alkyl-PC (1-0-alkyl-PC) accounts for up to 50% of the total Cho-containing phosphoglycerides in certain cells such as neutrophils (Mueller et al, 1982). This phospholipid is conveniently labeled by incubating cells with radiolabeled alkyl-lysoPC (1-0-alkyl-2-lyso-PC), which readily enters the cell and becomes acylated into alkyl-PC (Mueller et al, 1983; Albert and Snyder, 1983; Billah et al, 1986). The stability of the ether bond against hydrolysis by cellular lysophospholipases ensures retention of the label in alkyl-lysoPC. To label cellular alkyl-PC with ^{32}P, alkyl-lyso[^{32}P]PC has been synthesized in a series of enzymatic and chemical steps (Pai et al, 1988b). Incubation of neutrophils with this synthetic alkyl-lysoPC labels endogenous alkyl-PC, but not ATP, with ^{32}P (Pai et al, 1988b). Over 90% of lipid-associated ^{32}P is recovered in alkyl-PC and the remainder in alkyl-lysoPC.

Receptor-Linked Activation of PLD

Upon stimulation with the chemotactic peptide, formyl-Met-Leu-Phe (fMLP), human neutrophils produce PA in large quantities (1 nmol/10^7 cells) (Cockcroft, 1984). The fatty acid composition of this PA differs from that of phosphatidylinositol (Cockroft and Allan, 1984) and is similar to that of PC (Walsh et al, 1983). These observations suggest that PC, and not phosphatidylinositol, is the source of the PA formed by fMLP- stimulated neutrophils. Furthermore, neutrophils in which cellular ATP has been labeled with ^{32}P by incubating the cells with [^{32}P]orthophosphate produce [^{32}P]PA upon fMLP stimulation. However, the specific activity of this [^{32}P]PA is much lower than that of [^{32}P]ATP (Cockcroft, 1984), indicating that PA is formed by mechanisms different from that involving phosphoinositide-specific PLC and DG kinase.

In addition, neutrophils labeled in alkyl-PC by preincubation of cells with 1-0-[^3H]alkyl-lysoPC produce 1-0-[^3H]alkyl-PA([^3H]alkyl-PA) and 1-0-[^3H]alkyl-DG ([^3H]alkyl-DG) in response to fMLP (Pai et al, 1988a; Agwu et al, 1989). [^3H]Alkyl-PA formation precedes [^3H]alkyl-DG generation. Furthermore, in the presence of ethanol, fMLP-stimulated HL-60 granulocytes produce 1-0-[^3H]alkyl-phosphatidylethanol (alkyl-PEt) in close parallel with [^3H]alkyl-PA (Pai et al, 1988a). These observations strongly suggest that in granulocytes, alkyl-PC is degraded by PLD.

Conclusive proof for activation of granulocyte PLD by fMLP has been obtained using cells containing alkyl-[^{32}P]PC (Pai et al, 1988b).

Activation of PLD by Phorbol Esters

Tumor promoting phorbol esters, such as 4β-phorbol-12-myristate-13-acetate (PMA), block receptor-mediated activation of phosphoinositide-specific PLC (Smith et al, 1987). However, in several cultured cells including fibroblasts (Mufson et al, 1981; Grove and Schimmel, 1982; Besterman et al, 1986; Muir and Murray, 1987; Takuwa et al, 1987; Kiss and Anderson, 1989), Madine-Darby kidney cells (Daniel et al, 1986), NG108-15 neuroblastoma cells (Liscovitch et al, 1987) and REF52 cells (Cabot et al, 1988a,b), PMA induces rapid generation of Cho, phosphocholine, PA and DG, suggesting PC hydrolysis. In addition, several different cells including lymphocytes, HL-60 cells (Pai et al, 1987; Tottenborn and Mueller, 1988) and NG108-15 neuroblastoma cells (Liscovitch, 1989) labeled with ^{3}H-labeled fatty acids produce [^{3}H]PEt upon stimulation with phorbol esters in the presence of ethanol, indicating PLD activation by phorbol esters. Conclusive proof that phorbol esters activate PLD has accrued from experiments in which HL-60 granulocytes have been labeled with ^{32}P by incubating the cells with alkyl-lyso[^{32}P]PC (Billah et al, 1989). These ^{32}P-labeled cells generate alkyl-[^{32}P]PA when treated with biologically active phorbol esters, but not with inactive analogs (Fig. 3). Furthermore, in the presence of ethanol, bioactive phorbol esters induce alkyl-[^{32}P]PEt generation. Thus, in granulocytes, phorbol ester-stimulated PLD can also catalyze the transphosphatidylation reaction.

Relative Contribution of PLD to PA and PEt Formation

Mammalian cells contain a PC-specific PLC (Wolf and Gross, 1985), and receptor-mediated activation of this enzyme has been suggested (Besterman et al, 1986; Muir and Murray, 1987; Irving and Exton, 1987; Exton, 1988; Welsh et al, 1988; Grillone et al, 1988). Therefore, it is important to determine the relative contribution of PLD and PLC to PA and PEt formation. This has been achieved in granulocytes by simultaneous labeling of endogenous alkyl-PC with ^{32}P and ^{3}H (alkyl chain). While alkyl-[^{32}P]PA is formed exclusively via PLD, [^{3}H]alkyl-PA could be formed by both PLD and the combined activities of PLC and DG kinase. The same $^{3}H/^{32}P$ ratio in the alkyl-PA and alkyl-PEt formed as in the starting alkyl-PC would indicate the exclusive catalysis by PLD, whereas a contribution from a PLC pathway would increase the $^{3}H/^{32}P$ ratios of alkyl-PA and alkyl-PEt above that of alkyl-PC. Based on these ratio determinations, it is concluded that in granulocytes stimulated with fMLP and phorbol esters, PLD is the primary route to alkyl-PA formation and that alkyl-PEt is formed exclusively by PLD (Pai et al, 1988b; Billah et al, 1989). This approach coupled with mass measurements should provide critical information regarding the identity of the specific phosphodiesterases as well as their relative contributions to product formation.

Regulation of PLD Activation

The mechanisms by which receptors are coupled to PLD activation in intact cells are not clear. Ca^{2+} influx appears to be an important component, because extracellular Ca^{2+} is essential for fMLP stimulation of PLD in granulocytes (Pai et al, 1988b) and also because the Ca^{2+} ionophore, A23187, activates this PLD (Billah et al, 1989). The observation that phorbol diesters and synthetic DG are potent activators of granulocyte PLD suggests a role for endogenously produced DG in PLD activation. Furthermore, A23187 acts synergistically with phorbol esters and DG to promote this PLD activity (Billah et al, 1989). A role for GTP-binding proteins in PLD activation is also suggested by the observation that pertussis toxin inhibits PLD activation in fMLP-stimulated granulocytes (Pai et al, 1988b) and that nonhydrolyzable GTP analogs activate PLD activity in cell lysates from HL-60 cells

Stimulation of these ^{32}P-labeled cells with fMLP induces alkyl-[^{32}P]PA production (Fig. 2). Since these cells contain no cellular [^{32}P]ATP, alkyl-[^{32}P]PA must be derived from alkyl-[^{32}P]PC by PLD action. In the presence of 0.5% ethanol, alkyl- [^{32}P]PEt is also formed. Alkyl-[^{32}P]PEt formation parallels alkyl-[^{32}P]PA generation and is accompanied by reduced accumulation of alkyl-[^{32}P]PA. These observations clearly demonstrate that fMLP receptor activation leads to rapid stimulation of PLD and that this PLD is capable of catalyzing the transphosphatidylation reaction. Clearly, in human granulocytes, PEt formation is a valid measure of PLD activity.

Although the granulocyte system remains best characterized with respect to PLD activation, evidence suggests receptor-linked PLD activation in other cells as well. PEt is formed by thrombin-stimulated platelets (Rubin, 1988) and by rat hepatocytes stimulated with vasopressin, angiotension II or epinephrine (Bocckino et al, 1987b). In addition, rat hepatocytes generate PA and DG in response to a variety of hormones including vasopressin, angiotension II and epinephrine, and PA formation precedes DG generation. On the basis of fatty acid composition , both PA and DG resemble PC (Bocckino et al, 1987a). Furthermore, REF52 rat embryo cells stimulated with vasopressin or serum generate Cho, PA and DG (Cabot et al, 1988a; Welsh et al, 1988). Again, the fact that PA and Cho appear before DG indicates PLD action on PC. Bradykinin stimulation of cultured endothelial cells also induces rapid release of Cho paralleling PA formation, due presumably to PLD activation (Martin and Michaelis, 1988).

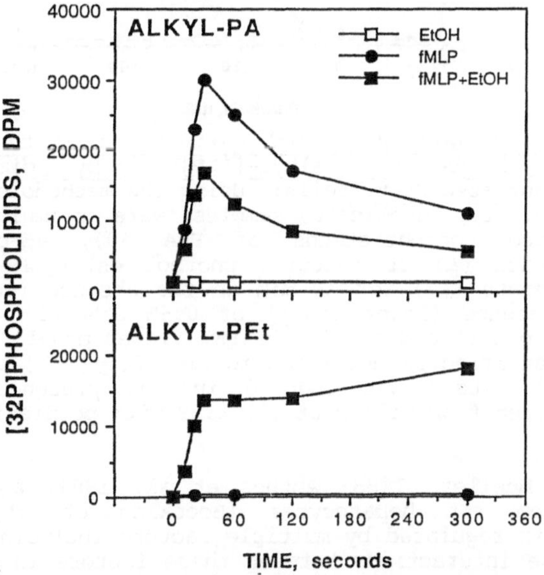

Fig. 2. Parallel formation of alkyl-[^{32}P]PA and alkyl-[^{32}P]PEt by fMLP-stimulated neutrophils: Neutrophils isolated from freshly drawn human blood were prelabeled in alkyl-PC by incubating cells with synthetic alkyl-lyso[^{32}P]PC as described previously (Pai et al, 1988b). Duplicate samples (0.5 ml) containing 5×10^6 ^{32}P-labeled cells, 5 μM cytochalasin B and 1 mM $CaCl_2$ were preincubated for 5 min before stimulating with 0.1 μM fMLP for additional times in the absence or presence of 0.5% ethanol (EtOH). Lipids were extracted and separated by thin-layer chromatography, and the radioactivity quantified by liquid scintillation counting. For further details, see Pai et al 1988b.

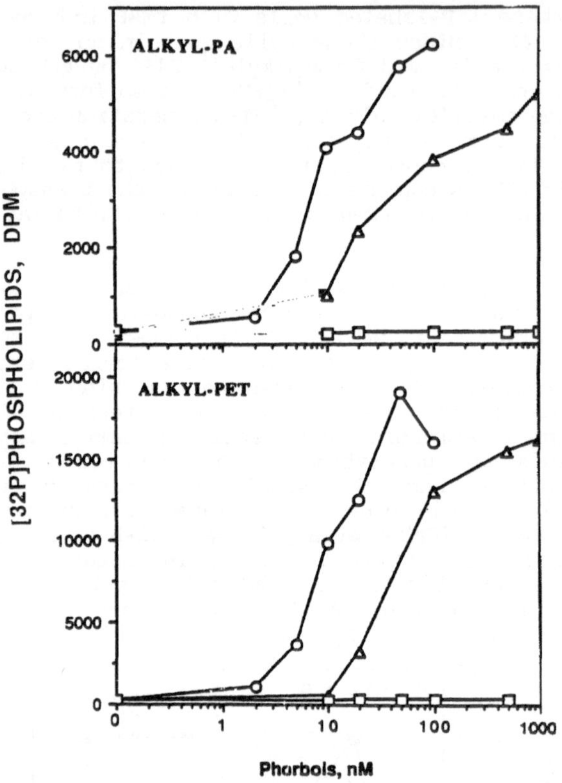

Fig. 3. Effects of various phorbol derivatives on the formation of alkyl-[^{32}P]-PA and alkyl-[^{32}P]PEt by dimethylsulfoxide differentiated HL-60 cells: Using the methodologies described in Fig. 2., duplicate samples were stimulated with the indicated concentrations of PMA (0), 4β-phorbol,-12,13-dibutyrate (Δ) or inactive phorbol ester 4α-phorbol-12,13-didecanoate ([]) for 10 min in the absence (upper panel) or the presence (lower panel) of 0.5% ethanol. Alkyl-[^{32}P]PA (upper panel) and alkyl-[^{32}P]PEt (lower panel) were separated and quantified as described in Fig. 2. [^{32}P]PA production by phorbol esters was reduced in the presence of ethanol. Reproduced from Billah et al, 1989 with permission.

(Tottenborn and Mueller, 1988; Anthes et al, 1989) and in membrane preparations from rat hepatocytes (Bocckino et al, 1987a, b). Apparently, PLD is regulated by multiple factors including Ca^{2+}, DG and GTP. The precise interactions between these factors in activating PLD remain to be established.

Phorbol esters and synthetic DGs activate PLD with a potency order similar to that for their activation of protein kinase C (Billah et al, 1989; Liscovitch, 1989). In addition, fMLP activates PLD under experimental conditions (extracellular Ca^{2+}, cytochalasin B) (Pai et al, 1988b) that also favor protein kinase C translocation and activation (Nishihira et al, 1986). PMA stimulates PEt formation by lysates from HL-60 cells but only in the presence of ATP (Tottenborn and Mueller, 1988). However, the protein kinase inhibitor, K252a, causes only partial inhibition of PLD activation by fMLP or PMA under conditions where PMA-induced protein phosphorylation is completely blocked (Billah et al, 1989). Thus, although protein kinases might be involved in PLD activation, mechanisms different from those involving protein kinases appear to predominate. Phorbol esters and DG may, therefore, interact directly with PLD in much the same way as they do with protein kinase C. On the other hand, in neuroblastoma cells, PMA-stimulated PEt formation is blocked by protein kinase C inhibitors (Liscovitch, 1989). Apparently, cell-specific mechanisms for PLD activation exist. As has been documented for phospholipases A_2 and C, PLD might also exist in multiple molecular forms with distinct regulatory controls.

Illustrated in Fig. 4 is a proposal for receptor-mediated activation of PLD in granulocytes. According to this proposal, initial activation of the phosphoinositide-specific PLC mobilizes Ca^{2+} and DG (Berridge, 1984), which then act in concert with a G protein(s) to activate PLD.

Functional Significance of PLD

Recent findings that cells contain a receptor-linked PLD and that this PLD is rapidly stimulated upon receptor stimulation indicate a critical role for PLD in cell activation. PA exhibits ionophoretic properties in liposome preparations (Serhan et al, 1981) and in intact cells (Putney et al, 1980). In the presence of low Ca^{2+} concentrations, PA exhibits fusogenic properties (Sundler and Papahadjopoulos, 1981), raising the possibility that PA may be involved in granule secretion. There are specific cell surface receptors for PA (Murayama and Ui, 1987) and PA can induce cell growth and DNA synthesis (Moolenaar et al, 1986). In addition, PLD-derived PA can be further degraded by either a specific phospholipase A_2 (Billah et al, 1981) to generate arachidonate or a

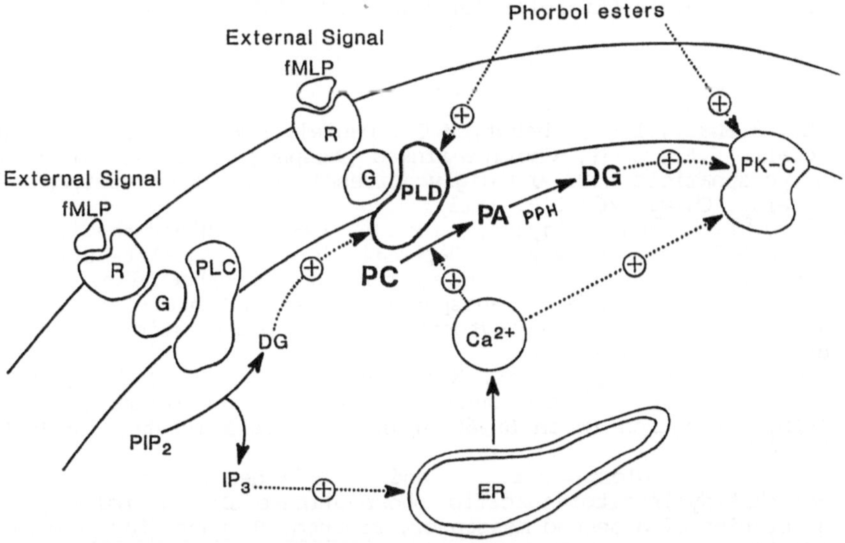

Fig. 4. Proposed regulation of receptor-linked PLD activation: Agonists (e.g. fMLP) interact with specific receptors (R) to activate phosphoinosi- tide-specific phospholipase C via a GTP-binding regulatory (G) protein. As a consequence, phosphatidylinositol-4,5-bisphosphate (PIP_2) is degraded to produce inositol-1,4,5-trisphosphate (IP_3) and DG. IP_3 releases Ca^{2+} from endoplasmic reticulum (ER). Ca^{2+} and DG then act synergistically to activate PLD by a mechanism involving a G protein. The resultant PA can be dephos- phorylated by PA phosphohydro- lase (PPH) to produce DG. This DG can then activate protein kinase C (PK-C). Phorbol esters may interact directly with PLD leading to activation.

phosphohydrolase (Brindley, 1984) to produce DG. The latter reaction sequence might contribute significantly to stimulus-induced DG generation in various cells including neutrophils (Pai et al, 1988a; Agwu et al, 1989), endothelial cells (Martin, 1988) and REF52 cells (Welsh et al, 1988). This large amount of DG can then activate protein kinase C (Nishizuka, 1984). Recent studies further indicate that neuronal cells might utilize PLD-derived Cho for acetylcholine synthesis (Hattori et al, 1987).

CONCLUSIONS

Cellular PC is rapidly hydrolyzed by specific phosphodiesterases in response to a variety of agents including hormones, growth factors and phorbol esters. Using two novel methodological approaches, it has been unequivocally demonstrated that in chemoattractant-stimulated human granulocytes, the phosphodiesteratic cleavage of PC is catalyzed by PLD. Granulocyte PLD is also activated by tumor-promoting phorbol esters, synthetic DGs and ionophore A23187. PLD activation occurs via both protein kinase C-dependent and independent mechanisms, and involves multiple factors including Ca^{2+}, DG and G proteins. Evidence suggests PC hydrolysis by PLD in various other cells including hepatocytes, neuroblastoma cells, endothelial cells and embryo REF52 cells. PLD is, therefore, emerging as a widespread component of signal transduction.

ACKNOWLEDGEMENT

We thank Virginia Citarella for fine editorial assistance.

REFERENCES

Agwu, D.E., McPhail, L.C., Chabot, M.C., Daniel, L.W., Wykle, R.L., and McCall, C.E., 1989, Choline-linked phosphoglycerides. A source of phosphatidic acid and diglycerides in stimulated neutrophils, J. Biol. Chem., 264:1405-1413.

Albert, D.H., and Snyder, F., 1983, Biosynthesis of 1-alkyl-2-acetyl-sn-glycero-3-phosphocholine (platelet-activating factor) from 1-alkyl-2-acyl-sn-glycero-3-phosphocholine by rat alveolar macrophages. Phospholipase A2 and acetyltransferase activities during phagocytosis and ionophore stimulation, J. Biol. Chem., 258:97-102.

Anthes, J.C., Eckel, S., Siegel, M.I., Egan, R.W., and Billah, M.M., 1989, Role of Ca^{2+} and guanine nucleotides on phospholipase D (PLD) activation in an HL-60 granulocyte cell lysate, FASEB J., 3:A1087.

Balsinde, J., Diez, E., and Mollinedo, F., 1988, Phosphatidylinositol-specific phospholipase D: a pathway for generation of a second messenger, Biochem. Biophys. Res. Commun., 154:502-508.

Berridge, M.J., 1984, Inositol trisphosphate and diacylglycerol as second messengers, Biochem. J., 220;345-360.

Besterman, J.M., Duronio, V. and Cuatrecasas, P., 1986, Rapid formation of diacylglycerol from phosphatidylcholine: a pathway for generation of a second messenger, Proc. Natl. Acad. Sci., USA, 83:6785-6789.

Billah. M.M., Eckel, S., Myers, R.F., and Siegel, M.I., 1986, Metabolism of platelet-activating factor (1-0-alkyl-2-acetyl-sn-glycero-3-phosphocholine) by human promyelocytic leukemic HL60 cells. Stimulated expression of phospholipase A2 and acetyltransferase requires differentiation, J. Biol. Chem., 261:5824-5831.

Billah, M.M., Lapetina, E.G., and Cuatrecasas, P., 1981, Phospholipase A2 activity specific for phosphatidic acid. A possible mechanism for the production of arachidonic acid in platelets, J. Biol. Chem., 256:5399-5403.

Billah, M.M., Pai, J.-K., Mullmann, T.J., Egan, R.W., and Siegel, M.I., 1989, Regulation of phospholipase D in HL-60 granulocytes. Activation by phorbol esters, diglyceride, and calcium ionophore via protein kinase C-independent mechanisms, J. Biol. Chem., 9069-9076.

Bocckino, S.B., Blackmore, P.F., Wilson, P.B., and Exton, J.H., 1987a, Phosphatidate accumulation in hormone-treated hepatocytes via a phospholipase D mechanism, J. Biol. Chem., 262:15309-15315.

Bocckino, S.B., Wilson, P.B., and Exton, J.H., 1987b, Calcium-mobilizing hormones elicit phosphatidylethanol accumulation via phospholipase D activation, FEBS Lett., 225:201-204.

Brindley, D.N., 1984, Intracellular translocation of phosphatidate phosphohydrolase and its possible role in the control of glycerolipid synthesis, Prog. Lipid Res., 23:115-133.

Cabot, M.C., Welsh, C.J., Czhang, Z.-C., Cao, H.-T., Chabbott, H., and Lebowitz, M., 1988a, Vasopressin, phorbol diesters and serum elicit choline glycerophospholipid hydrolysis and diacylglycerol formation in nontransformed cells: transformed derivatives do not respond, Biochim. Biophys. Acta, 959:46-57.

Cabot, M.C., Welsh, C.J., Cao, H., and Chabbott, H., 1988b, The phosphatidylcholine pathway of diacylglycerol formation stimulated by phorbol diesters occurs via phospholipase D activation, FEBS Lett., 233:153-157.

Cockcroft, S., 1984, Ca^{2+}-dependent conversion of phosphatidylinositol to phosphatidate in neutrophils stimulated with fMet-Leu-Phe or ionophore A23187, Biochim. Biophys. Acta., 795:37-46.

Cockroft, S. and Allan, D., 1984, The fatty acid composition of phosphatidylinositol, phosphatidate and 1,2-diacylglycerol in stimulated human neutrophils, Biochem. J., 222:557-559.

Chalifour, R.J., and Kanfer, J.N., 1980, Microsomal phospholipase D of rat brain and lung tissues, Biochem. Biophys. Res. Commun., 96:742-747.

Chalifour, R.J., and Kanfer, J.N., 1980, Phosphatidylglycerol formation via transphosphatidylation by rat brain extracts, Can. J. Biochem., 58:1189-1196.

Chalifour, R.J, and Kanfer, J.N., 1982, Fatty acid activation and temperature perturbation of rat brain microsomal phospholiapse D, J. Neurochem., 39:299-305.

Daniel, L.W., Waite, M., and Wykle, R.L., 1986, A novel mechanism of diglyceride formation. 12-0-tetradecanoylphorbol-13-acetate stimulates the cyclic breakdown and resynthesis of phosphatidylcholine, J. Biol. Chem., 261:9128-9132.

Dawson, R.M.C., 1967, The formation of phosphatidylglycerol and other phospholipids by the transferase activity of phospholipase D, Biochem. J., 102:205-210.

Exton, J.H., 1988, Mechanisms of action of calcium-mobilizing agonists: some variations on young theme, FASEB J., 2:2670-2676.

Grillone, L.R., Clark, M.A., Godfrey, R.W., Stassen, F. and Crooke, S.T., 1988, Vasopressin induces V1 receptors to activate phosphatidylinositol- and phosphatidylcholine-specific phospholipase C and stimulates the release of arachidonic acid by at least two pathways in the smooth muscle cell line, A-10, J. Biol. Chem., 263:2658-2663.

Grove, R.I., and Schimmel, S.D., 1982, Effects of 12-0-tetradecanoylphorbol 13-acetate on glycerolipid metabolism in cultured myoblasts, Biochim. Biophys. Acta, 711:272-280.

Hattori, H., Kanfer, J.N., and Massarelli, R., 1987, Stimulation of phospholipase D activity and indication of acetylcholine synthesis by oleate in rat brain synaptosomal preparations, Neurochem. Res., 12:687-692.

Heller, M., 1978, Phospholipase D, Adv. Lipid Res., 16:267-326.
Irving, H.R., and Exton, J.H., 1987, Phosphatidylcholine breakdown in rat liver plasma membranes. Roles of quanine nucleotides and P2-purinergic agonists, J. Biol. Chem., 262:3440-3443.
Kanfer, J.N., 1983, The base exchange enzyme and phospholipase D of mammalian tissue, Can. J. Biochem., 58:1370-1380.
Kater, L.A., Goetzl, E.J., and Austen, K.F., 1976, Isolation of human eosinophil phospholipase D, J. Clin. Invest., 57:1173-1180.
Kiss, Z., and Anderson, W.B., 1989, Phorbol ester stimulates the hydrolysis of phosphatidylethanolamine in leukemic HL-60, NIH 3T3, and baby hamster kidney cells, J. Biol. Chem., 264:1483-1487.
Kobayashi, M., and Kanfer, J.N., 1987, Phosphatidylethanol formation via transphosphatidylation by rat brain synaptosomal phospholipase D, J. Neurochem., 48:1597-1603.
Liscovitch, M., 1989, Phosphatidylethanol biosynthesis in ethanol-exposed NG108-15 neuroblastoma X glioma hybrid cells. Evidence for activation of a phospholipase D phosphatidyl transferase activity by protein kinase C, J. Biol. Chem., 264:1450-1456.
Liscovitch, M., Blusztajn, J.K., Freese, A., and Wurtman, R.J., 1987, Stimulation of choline release from NG108-15 cells by 12-0-tetradecanoylphorbol 13-acetate, Biochem. J., 241:81-86.
Martin, T.W., 1988, Formation of diacylglycerol by a phospholipase D-phosphatidate phosphatase pathway specific for phosphatidylcholine in endothelial cells, Biochim. Biophys. Acta, 962:282-296.
Martin, T.W., annd Michaelis, K.C., 1988, Bradykinin stimulates phosphodiesteratic cleavage of phosphatidylcholine in cultured endothelial cells, Biochem. Biophys. Res. Commun., 157:1271-1279.
Moolenaar, W.H., Kruijer, W., Tilly, B.C., Verlaan, I., Bierman, A.J., and de Latt, S.W., 1986, Growth factor-like action of phosphatidic acid, Nature, 323:171-173.
Mueller, H.W., O'Flaherty, J.T., and Wykle, R.L., 1982, Ether lipid content and fatty acid distribution in rabbit polymorphonuclear neutrophil phospholipids, Lipids, 17:72-77.
Mueller, H.W., O'Flaherty, J.T., and Wykle, R.L., 1983, Biosynthesis of platelet activating factor in rabbit polymorphonuclear neutrophils, J. Biol. Chem., 258:6213-6218.
Mufson, R.A., Okin, E., and Weinstein, I. B., 1981, Phorbol esters stimulate the rapid release of choline from prelabelled cells, Carcinogenesis, 2:1095-1102.
Muir, J.G., and Murray, A.W., 1987, Bombesin and phorbol ester stimulate phosphatidylcholine hydrolysis by phospholipase C: evidence for a role of protein kinase C, J. Cell. Physiol., 130:382-391.
Murayama, T., and Ui, M., 1987, Phosphatidic acid may stimulate membrane receptors mediating adenylate cyclase inhibition and phospholipid breakdown in 3T3 fribroblasts, J. Biol. Chem., 262:5522-5529.
Nishihira, J., McPhail, L.C., and O'Flaherty, J.T., 1986, Stimulus-dependent mobilization of protein kinase C, Biochem. Biophys. Res. Commun., 134:587-594.
Nishizuka, Y., 1986, Studies and perspectives of protein kinase C, Science, 233:305-312.
Pai, J.-K., Liebl, E.C., Tettenborn, C.S., Ikegwuonu, F.I., and Mueller, G.C., 1987, 12-0-tetradecanoylphorbol-13-acetate activates the synthesis of phosphatidylethanol in animal cells exposed to ethanol, Carcinogenesis, 8:173-178.
Pai, J.-K., Siegel, M.I., Egan, R.W., and Billah, M.M., 1988a, Phospholipase D catalyzes phospholipid metabolism in chemotactic peptide-stimulated HL-60 granulocytes, Biochem. Biophys. Res. Commun., 150:355-364.
Pai, J.-K., Siegel, M.I., Egan, R.W., and Billah, M.M., 1988b, Activation of phospholipase D by chemotactic peptide in HL-60 granulocytes, J. Biol. Chem., 263:12472-12477.

Pelech, S.L., and Vance, D.E., 1989, Signal transduction via phosphatidylcholine cycles, Trends Biochem. Sci., 14:28-30.

Putney, J.W., Jr., Weiss, S.J., van de Walle, C.M., and Haddas, R.A., 1980, Is phosphatidic acid a calcium ionophore under neurohumoral control?, Nature, 284:345-347.

Rubin, R., 1988, Phosphatidylethanol formation in human platelets: evidence for thrombin-induced activation of phospholipase D, Biochem. Biophys. Res. Commun., 156:1090-1096.

Serhan, C., Anderson, P., Goodman, E., Dunham, P., and Weissman, G., 1981, Phosphatidate and oxidized fatty acids are calcium ionophores. Studies employing arsenazo III in liposomes, J. Biol. Chem., 256:2736-2741.

Slivka, S., Meier, K.E., and Insel, P.A., 1988, Alpha 1-adrenergic receptors promote phosphatidylcholine hydrolysis in MDCK-D1 cells. A mechanism for rapid activation of protein kinase C, J. Biol. Chem., 263:12242-12246.

Sundler, R., and Papahadjopoulos, D., 1981, Control of membrane fusion by phospholid head groups, Biochim. Biophys. Acta., 649:743-750.

Smith, C.D., Uhling, R.J., and Snyderman, R., 1987, Nucleotide regulatory protein-mediated activation of phospholipase C in human polymorphonuclear leukocytes is disrupted by phorbol esters, J. Biol. Chem., 262:6121-6127.

Taki, T., and Kanfer, J.N., 1979, Partial purification and properties of a rat brain phospholipase, J. Biol. Chem., 254:9761-9765.

Takuwa, N., Takuwa, Y., and Rasmussen, H., 1987, A tumour promoter, 12-0-tetradecanoylphorbol 13-acetate, increases cellular 1,2-diacylglycerol content through a mechanism other than phosphoinositide hydrolysis in Swiss-mouse 3T3 fibroblasts, Biochem. J., 243:647-653.

Tottenborn, C.S., and Mueller, G.C., 1987, Phorbol esters activate the pathway for phosphatidylethanol synthesis in differentiating HL-60 cells, Biochim. Biophys. Acta, 931:242-250.

Tottenborn, C.S., and Mueller, G.C., 1988, 12-0-tetradecanoylphorbol-13-acetate activates phosphatidylethanol and phosphatidylglycerol synthesis by phospholipase D in cell lysates, Biochem. Biophys. Res. Commun., 135:249-255.

Walsh, C.E., DeChatelet, L., Chilton, F.H., Wykle, R.L., and Waite, B.M., 1983, Mechanism of arachidonic acid release in human polymorphonuclear leukocytes, Biochem. Biophys. Acta., 750:32-40.

Welsh, C.J., Cao, H-L, Chabbott, H., and Cabot, M.C., 1988, Vasopressin is the only component of serum-free medium that stimulates phosphatidylcholine hydrolysis and accumulation of diacylglycerol in cultured REF52 cells, Biochem. Biophys. Res. Commun., 152:565-572.

Witter, B., annd Kanfer, J.N., 1985, Hydrolysis of endogenous phospholipids by rat brain microsomes, J. Neurochem., 44:155-162.

Yang, S., Freer, S., and Benson, A., 1967, Transphosphatidylation by phospholipase D, J. Biol. Chem., 242:477-484.

THE ORCHESTRATION OF STIMULATORY SIGNALS FOR SECRETION IN RAT BASOPHILIC LEUKEMIA (RBL-2H3) CELLS

Michael A. Beaven, Russell Ludowyke, Heloisa M.S. Gonzaga[1], Dolores Collado-Escobar, Michihiro Hide[2], and Hydar Ali

Laboratory of Chemical Pharmacology, National Heart Lung, and Blood Inst., NIH, Bethesda, MD 20892

INTRODUCTION

Cells that express high affinity receptors for immunoglobulin E (IgE) on their surface, when stimulated with antigen, release substances that cause a wide range of antigen-mediated disorders, some of which are merely annoying but afflict a large proportion of the population and others of which are disabling and sometimes fatal. Obviously an understanding of the cascade of events that lead to release of inflammatory mediators from these cells would be useful from the viewpoint of treating these disorders. These cells, however, have become in themselves interesting experimental models for the study of stimulus-secretion coupling mechanisms.

The high affinity IgE-receptors are the exclusive domain of the blood basophil and the tissue mast cell of which there are two major subtypes, the so called mucosal mast cell and the connective tissue or serosal mast cell (Stevens et al., 1986). The two subtypes differ in the nature of the proteases and proteoglycans that are present in the secretory granules of these cells and in their responsiveness to chemical degranulating agents such as compound 48/80. The serosal mast cell, of which the rat peritoneal mast cell is the best studied, is fully responsive to compound 48/80 whereas the mucosal mast cell and blood basophil are not. All three types of cell, however, can be stimulated to secrete by multivalent binding of the appropriate antigen to the antigen-specific IgE that normally occupies the IgE-receptors. Each cell contains several hundred thousand receptors for IgE and the cross-linking of a small proportion of receptor bound IgE by antigen is sufficient to induce secretion (Maeyama et al.,1986).

The actual trigger for degranulation is the aggregation of two or more receptors for IgE (Metzger et al., 1984). Although the cloning of the genes for the subunits of these receptors (Blank et al., 1989) and their expression in cells that do not normally bear these subunits (Miller et al., 1989) have been accomplished recently, the precise

---

[1]Recipient of a CAPES Fellowship from the Ministry of Education, Brazil.
[2]Recipient of a Scholarship from the Rotary Found. of Rotary Intl.

molecular details of how the aggregated receptor communicates with GTP-regulated and effector systems within the plasma membrane are still unknown. We do, however, have a reasonably complete picture, at least in general terms, of the major stimulatory events that are triggered by the aggregation of the receptors.

Secretion from these cells occurs via a Ca^{2+}-dependent process of exocytosis to release a rich array of constituents which include primarily histamine, serotonin, [^{3}H]serotonin (if cells have been previously incubated with the labeled amine), β-N-acetylglycosaminidase, proteases and proteoglycans - anyone of which can be used as a quantitative marker for secretion (Serafin et al., 1987). Alternatively, with single cells that have been patched in the whole cell configuration, the infusion of intracellular mediators or external application of extracellular stimulants leads to well defined, stepwise changes in the electrical capacitance of the membrane - each step corresponding to the fusion of the perigranule membrane of one granule with the plasma membrane (Fernandez et al., 1984).

The mast cell utilizes the same general mechanisms as do other cells that secrete by a Ca^{2+}-dependent exocytotic processes. These include; the recruitment of GTP-binding proteins (G-proteins), phospholipase C, other phospholipases, protein kinase C, and ion channels, and as, a consequence the mobilization of intracellular (via inositol 1,4,5-trisphosphate) and extracellular (via ion channels) calcium (Beaven and Cunha-Melo, 1988). The latter event appears to be crucial for the amplification and maintenance of both stimulatory and secretory responses as will be discussed later.

Much of our understanding of the interplay between these stimulatory processes has come from studies with the rat basophilic leukemia (RBL-2H3) cells. This cell line, which is closely related to the rat mucosal mast cell (Seldin et al., 1985), contains receptors for IgE and can be stimulated with antigen. The inherent advantage of these cells is that they have no IgE attached to the receptors. Accordingly they can be primed by incubating them with monoclonal IgE of defined antigen-specificity or mixtures of IgE to achieve whatever occupancy of receptors (with IgE) that is desired.

RBL-2H3 cells do not respond to compound 48/80, which normally evokes in rat peritoneal mast cells, a pertussis toxin-sensitive hydrolysis of inositol phospholipids (Nakamura and Ui, 1985), generation of diacylglycerol (Kennerly et al., 1979; Kennerly, 1987), activation of protein kinase C (Katakami et al., 1984), activation of cation and anion channels (Penner et al.,1988) and degranulation. Evidence has been presented that stimulation with compound 48/80 utilizes at least two G-proteins, Gp (for the activation of phospholipase C) and Ge (for the activation of exocytosis). Ge remains undefined but it is thought to activate a process distal to the Ca^{2+}-dependent step(s) (Gomperts et al., 1986).

EVENTS TRIGGERED BY AGGREGATION OF IgE-RECEPTORS

Hydrolysis of Membrane Inositol and Other Phospholipids

Most studies of the early events that are induced by antigen, as opposed to compound 48/80, have been conducted with the RBL-2H3 cell. Because the majority of these antigen-induced events have been observed in rat peritoneal mast cells and other mast cell - derived lines, "mast cell" will be used as a generic term for IgE-receptor bearing cells for the remainder of this article except for a few specific instances.

In general, the hydrolysis of the inositol phospholipids (particularly phosphatidylinositol 4,5-bisphosphate) by the catalytic unit, phospholipase C is thought to generate two chemical messengers, inositol 1,4,5-trisphosphate for the release of intra cellular Ca^{2+} and diacylglycerol which activates protein kinase C (Berridge, 1987; Nishizuha, 1986). This appears to be the case in antigen-stimulated mast cells. The hydrolysis of the inositol phospholipids correlates with the rise in $[Ca^{2+}]_i$ and secretion of histamine (Beaven et al., 1984a and b). Furthermore, manipulations that increase or decrease the rate of hydrolysis result in analogous changes in the intensity of the calcium signal and secretory response (Maeyama et al., 1986 and 1988; WoldeMussie et al., 1986; Lo et al., 1987).

The phenomenon in RBL-2H3 cells is not entirely analogous to the phosphoinositide response in many other types of cells in that the hydrolysis is initially dependent on the presence of high concentrations (~1 mM) of external Ca^{2+}. Once hydrolysis is underway, however, the process no longer requires external Ca^{2+} (Maeyama et al., 1988). Therefore a Ca^{2+}-dependent step, is still required in the early stages of stimulation of intact cells. The hydrolysis of inositol phospholipids is tightly coupled to the aggregation of IgE-receptors as determined by the following criteria: 1) the rapidity and extent of hydrolysis is dependent on the number of receptors aggregated and the efficacy of the cross-linking agent in aggregating receptors. 2) the response immediately stops upon displacement of antigen with monovalent hapten and 3) the hydrolysis can be stimulated by antigen in washed, permeabilized cells in the virtual absence of Ca^{2+} (Ali et al., 1989). The phosphoinositide response of permeabilized cells to antigen is amplified by subsequently raising $[Ca^{2+}]_i$ to 0.1 μM.

The antigen-induced hydrolysis of inositol phospholipids has the characteristics of being mediated by an unidentified G-protein (Gp) (Gomperts et al., 1986). It is suppressed by guanosine 5'-O-(thiodiphosphate) (GDPβS) in a concentration dependent-fashion (Ali et al., submitted). The hydrolysis and secretion can also be induced in permeabilized cells by guanosine 5'O-(3-thiotriphosphate) (GTPγS) and other nonhydrolyzable analogs of GTP (Ali et al., 1989). However, at least in the RBL-2H3 cell, antigen-induced hydrolysis is resistant to the actions of either pertussis or cholera toxin (unpublished data).

The entire phosphoinositide cycle has been mapped in the RBL-2H3 cell (Beaven and Cunha-Melo, 1988). The salient features are that all three inositol phospholipids appear to be hydrolyzed during antigen stimulation (Ali et al., 1989), that inositol 1,3,4,5-tetrakisphosphate is converted back to inositol 1,4,5-trisphosphate by a membrane 3'-phosphomonoesterase (Cunha-Melo et al., 1988) and that, as in other cells, multiple pathways exist for the dephosphorylation of the inositol polyphosphates (Cunha-Melo et al., 1987). We have speculated that the conversion of inositol 1,3,4,5-tetrakisphosphate back to inositol 1,4,5-trisphosphate might be related to the reliance of these cells on external sources of Ca^{2+} but for reasons discussed later Ca^{2+}-influx is dependent on additional processes. It is, however, clear that inositol 1,4,5-trisphosphate is produced in more than sufficient amounts to release Ca^{2+} from intracellular stores of bound Ca^{2+} (Ali et al., 1989 and unpublished data) and does so in a highly cooperative manner (Meyer et al., 1988) that might account for the generation of repetitive spikes of $[Ca^{2+}]_i$ that have been observed in RBL-2H3 cells (Millard et al, 1988) and mast cells (Neher and Almers, 1986).

The other major product of hydrolysis of inositol phospholipids, diacylglycerol, is not derived exclusively from this source in mast cells (Kennerly, 1987) or other types of cells (Exton, 1988). A phosphatidyl choline-specific phospholipase C appears to be activated by hormones in many types of cells. Our studies indicate quite clearly phosphosphatidyl-inositol is but a minor source of diacylglycerol in antigen-stimulated RBL-2H3 cells. Also reactions that are characteristic of a phospholipase D are activated within seconds of addition of antigen (unpublished data). The function of these reactions, apart from the production of diacylglyerol, are uncertain, but they remain candidates for additional, unidentified stimulatory signals for secretion (Beaven et al., 1987).

Activation of Protein Kinase C

Diacylglycerol is a known activator of protein kinase C. This enzyme, which is ubiquitously distributed in body tissues, is present in both mast cells and the cultured analogs of mast cells. Of the various subspecies of the enzyme that have been identified, two types (II and III) are present in RBL-2H3 cells both of which translocate to the membrane upon exposure to an exogenous activator of the enzyme, phorbol myristate acetate (PMA). The type II isozyme is, however, rapidly down regulated by proteolysis, whereas the type III isozyme is down-regulated slowly (Huang et al., 1989) and is primarily responsible for the synergistic and inhibitory effects of PMA in these cells (Cunha-Melo et al., submitted). Also these two subtypes are recruited in a much more selective manner by antigen than by PMA (unpublished studies).

The original studies with rat peritoneal mast cells by Nishizuka and colleagues provided the first demonstration that stimulatory signals can be transduced via protein kinase C. The synergy between PMA and Ca^{2+} ionophore in stimulating secretion was noted (Katakami et al.,1984), as had been demonstrated previously in human basophils (Schleimer et al., 1982), but it was also shown that the same spectrum of proteins were phosphorylated whether the cells were stimulated via IgE-receptors or with PMA. The phosphorylation of myosin light and heavy chains at specific sites that are diagnostic of a protein kinase C-catalyzed phosphorylation also provides unambiguous evidence that protein kinase C is activated in antigen stimulated RBL-2H3 cells. Furthermore, the extent of this phosphorylation correlates with the secretory response to antigen (Ludowyke et al., 1989).

Studies with PMA and Ca^{2+}-ionophores indicate that the translocation of protein kinase C and the elevation of $[Ca^{2+}]_i$ individually provide weak stimulatory signals for secretion. Both agents together can elicit maximal secretory responses in mast cells (Beaven et al., 1987). While it is quite clear that activation of protein kinase C by PMA can provide a synergistic signal for secretion and can down-regulate antigen mediated signals (Cunha-Melo et al., submitted), the role of protein kinase C in actually transducing antigen-mediated signals is less clear. Carefully conducted experiments indicate that only a small proportion (<5%) of the total protein kinase C in RBL-2H3 cells is translocated to the membrane fraction following stimulation with antigen (White and Metzger, 1988). Also potent inhibitors of the enzyme are surprisingly weak inhibitors of antigen-induced secretion. One of them, staurosporine, at nanomolar concentrations effectively blocks secretion induced by the combination of PMA and A23187, but it suppresses antigen-induced secretion only at high concentrations (20 to 50 nM) and does so by inhibiting directly phospholipase C. Moreover a reduced but still substantial secretory response can be evoked in RBL-2H3 cells that are depleted of protein kinase C by prolonged exposure to PMA (Cunha-Melo et al., submitted). These and other data have led us to propose that additional cryptic signals are generated by antigen stimulation (Beaven et al., 1987).

Activation of Ion Channels and Depolarization of the plasma membrane

Many electrically nonexcitable cells, including the mast cell, exhibit initially a transient increase in $[Ca^{2+}]_i$ which does not require external Ca^{2+} and is usually diagnostic of release of intracellular Ca^{2+} by inositol 1,4,5-trisphosphate. This initial increase is sustained by influx of external Ca^{2+}. Both physiological and electrophysiological studies suggest that the influx is mediated not by voltage-dependent calcium channels but by nonspecific channels that are activated by second messengers (Penner et al., 1988). The nonspecific channels in mast cells are not activated by internal application of inositol phosphates or by elevated $[Ca^{2+}]_i$, but because they are activated by GTPγS, they may be gated by G-proteins. These channels operate only when the membrane is in a hyperpolarized state. Although some depolarization is observed shortly after stimulation of mast cells, repolarization does occur subsequently through the development of a prominent outward rectifying chloride current (Penner et al., 1988).

Other studies indicate that influx of $^{45}Ca^{2+}$ is blocked when mast cells are depolarized by high $[K^+]_o$. Because influx of $^{45}Ca^{2+}$, but not the release of intracellularly bound $^{45}Ca^{2+}$, is suppressed under these conditions it would appear that Ca^{2+}-influx is driven by the electrochemical gradient across the plasma membrane (Mohr and Fewtrell, (1987). There is also evidence that this $^{45}Ca^{2+}$-influx pathway is regulated in cholera toxin-treated cells, by a mechanism that does not involve cyclic AMP. Both the influx of $^{45}Ca^{2+}$ and the increase in $[Ca^{2+}]_i$ are markedly enhanced in cholera toxin-treated cells(Narasimhan et al.,1988). In operational terms, influx of $^{45}Ca^{2+}$ and release of intracellular Ca^{2+} can be separated into two distinct events.

In addition to the fact that the ion channels are permeable to Na^+ ions in the absence, but not in the presence, of external Ca^{2+}, Na^+ ions play an important role in exocytosis in their own right. A Na^+-dependent alkalinization of the cytosol and the inhibition of secretion by amiloride and other blockers of Na^+-transport in antigen-stimulated RBL-2H3 cells provide an indirect indication of Na^+-uptake in these cells. The relative potencies of the different blockers in suppressing secretion suggest that a Na^+/Ca^{2+} antiporter was the vehicle for Na^+-uptake (Stump et al., 1987). However, the alkalinization indicates that a Na^+/H^+ antiporter might be activated as well. The alkalinization is of slow onset but it reaches a maximum once the Ca^{2+} signal begins to decay (Ali et al., submitted). The significance of this reciprocal increase in pHi and decay in $[Ca^{2+}]_i$ will be discussed later.

RELATIVE IMPORTANCE OF INTERNAL AND EXTERNAL SOURCES OF CALCIUM IN GENERATING A SIGNAL FOR SECRETION

There is now little reason to doubt that Ca^{2+} ions are mobilized from both internal stores (via inositol 1,4,5-trisphosphate) and external sources (via ion channels) in stimulated mast cells. There is also little doubt that influx of Ca^{2+} ions is necessary for a sustained secretory response. Some secretion is observed in rat peritoneal mast cells in the absence of external Ca^{2+} (Pearce, 1982), but secretion from RBL-2H3 cells is totally dependent on external Ca^{2+} (Maeyama et al., 1988). The subtle question is whether the activation of the phosphoinositide cascade provides all the necessary signals for activation of the Ca^{2+}-influx pathway (the sequential model) or that activation of the influx-pathway is an independent event (the parallel model). The current evidence points to a model in which the influx of Ca^{2+} ions serves to replenish intracellular stores of Ca^{2+} and, at the same time, amplify and sustain both stimulatory and secretory responses to antigen.

The parallel model is supported by the following observations. The enhancement of Ca^{2+}-influx in cholera toxin-treated cells that we alluded to earlier is not associated with increased rates of hydrolysis of inositol phospholipids (Narasimhan et al., 1988). In addition, antigen-stimulated hydrolysis of the inositol phospholipids is not impaired in cells depolarized with high $[K^+]_o$, but influx of Ca^{2+} and secretion are totally blocked (unpublished data). However, the best indication of the importance of the Ca^{2+}-influx pathway in amplifying stimulatory and secretory responses to antigen has come from studies with adenosine and its analogs.

Agonists of adenosine A_2-receptors have little secretagogue activity in RBL-2H3 cells. They do produce, however, a transient release of inositol phosphates and a transient but substantial increase in $[Ca^{2+}]_i$ but no sustained influx of $^{45}Ca^{2+}$. We have no indication that cyclic AMP is involved in these effects. Of these compounds, 5' (N-ethylcarbo-xamido)-adenosine (NECA) is the most potent. Because all the effects of NECA are blocked by cholera and pertussis toxin even in washed permeabilized cells, the apparent activation of phospholipase C by NECA and that by antigen appears to involve different G-proteins. Moreover, the activation of phospholipase C and the mobilization of Ca^{2+} by NECA are insufficient to produce a secretory response without a sustained influx of Ca^{2+}. Together with NECA, the whole cascade of events in response to low concentrations of antigen are markedly amplified (manuscript in preparation).

PUTTING THE PIECES TOGETHER: RECONSTITUTION OF THE SECRETORY RESPONSE IN PERMEABILIZED RBL-2H3 CELLS

We have successfully permeabilized RBL-2H3 cells with streptolysin O (Howell and Gomperts, 1987) with full retention of the phosphoinositide (Ali et al., 1989) and secretory (Ali et al., submitted) responses to antigen. Washing the permeabilized cells, however, leads to loss of protein kinase C (Cunha-Melo et al., submitted) and the secretory response. More extensive washing eventually leads to loss of the phosphoinositide response as well (Ali et al., submitted).

The primary stimulatory events can be reproduced in unwashed permeabilized cells by buffering $[Ca^{2+}]$ and pH at elevated levels. In addition, protein kinase C can be retained by washing the permeabilized cells in the presence of PMA (Cunha-Melo et al., submitted). The effects of these maneuvers indicate that the synergy between the rise in $[Ca^{2+}]_i$ and activation of protein kinase C could provide the necessary and initial combination of signals for secretion in RBL-2H3 cells. The subsequent rise in pH_i may reinforce these signals by sensitizing the cell to the actions of Ca^{2+}. Once the pH is elevated (7.4), for example, antigen-induced secretion in permeabilized RBL-2H3 cells is sustained at near maximal rates at much lower levels of $[Ca^{2+}]$ than is required at low values of pH (7.0)(Ali et al., submitted).

The phosphorylation of proteins by Ca^{2+}-dependent kinases and protein kinase C are generally thought to be the transducing signals for mechanical responses in cells. This area of research requires detailed investigation but there are some interesting leads. The site-specific phosphorylation of the light and heavy chains of myosin in intact RBL-2H3 cells has been mentioned. The same phosphorylation reactions can be observed in permeabilized cells that retain their ability to secrete (Ludowyke et al., 1989). In chromaffin cells, the requirement for Ca^{2+} can be obviated entirely by prior treatment of cells with the nonhydrolyzable ATP analog, adenosine-O-(3-thiotriphosphate) - a possible indication that the thiophosphorylation of a protein(s) latches the system into a more receptive state for other obligatory signals (Vu and Wagner, 1989).

In summary, the accumulation of diacylglycerol, hydrolysis of inositol phospholipids, a rise in $[Ca^{2+}]_i$ which is initiated by inositol 1,4,5-trisphosphate and sustained through activation of ion channels, the activation of protein kinase C and the phosphorylation of proteins are universal responses to IgE-dependent stimuli. Rapid stimulation of the metabolism of other phospholipids and changes in pH_i are also likely but these events require additional scrutiny. The challenges for future research are to determine the molecular details of the interaction of the IgE-receptor-aggregates with G-proteins and, in turn, the G-proteins with the effector systems.

ACKNOWLEDGEMENT

The studies with protein kinase C were conducted with Drs. F. Huang and K.-P Huang (NICHD) and those on the phosphorylation of myosin were carried out with Drs. Robert Adelstein and I. Peleg (NHLBI).

REFERENCES

Ali, H., Cunha-Melo, J., and Beaven, M., 1989, Receptor-mediated release of inositol 1,4,5-trisphosphate and inositol 1,4-bisphosphate in rat basophilic leukemia RBL-2H3 cells permeabilized with streptolysin O. Biochim. Biophys. Acta., 1010:88.

Ali, H., Collado-Escobar, D. M., and Beaven, M. A., 1989, The rise in concentration of free Ca^{2+} and pH provide sequential,synergistic signals for secretion in antigen-stimulated rat basophilic leukemia (RBL-2H3) cells. J. Immunol., vol. submitted.

Beaven, M. A., Rogers, J., Moore, J. P., Hesketh, T. R., Smith, G. A., and Metcalfe, J. C., 1984, The mechanism of the calcium signal and correlation with histamine release in 2H3 cells. J. Biol. Chem., 259:7129.

Beaven, M. A., Moore, J. P., Smith, G. A., Hesketh, T. R., and Metcalfe, J. C., 1984, The calcium signal and phosphatidylinositol breakdown in 2H3 cells. J. Biol. Chem., 259:7137.

Beaven, M. A., Guthrie, D. F., Moore, J. P., Smith, G. A., Hesketh, T. R., and Metcalfe, J. C., 1987, Synergistic signals in the mechanism of antigen-induced exocytosis in 2H3 cells: Evidence for an unidentified signal required for histamine release. J. Cell Biol., 105:1129.

Beaven, M. A., and Cunha-Melo, J. R., 1988, Membrane phosphoinositide-activated signals in mast cells and basophils. Progr. Allergy., 42:123.

Berridge, M. J., 1987, Inositol trisphosphate and diacylglycerol: Two interacting second messengers. Annu. Rev. Biochem., 56:159.

Blank, U., Ra, C., Miller, L., White, K., Metzger, H., and Kinet, J.-P., 1989, Complete structure and expression in transfected cells of high affinity IgE receptor. Nature., 337:187.
Cunha-Melo, J. R., Dean, N. M., Moyer, J. D., Maeyama, K., and Beaven, M. A., 1987, The kinetics of phosphoinositide hydrolysis in rat basophilic leukemia (RBL-2H3) cells varies with the type of IgE receptor cross-linking agent used. J. Biol. Chem., 262:11455.
Cunha-Melo, J. R., Dean, N. M., Ali, H., and Beaven, M. A., 1988, Formation of inositol 1,4,5-trisphosphate and inositol 1,3,4-trisphosphate from inositol 1,3,4,5-tetrakisphosphate and their pathways of degradation in RBL-2H3 cells. J. Biol. Chem., 263:14245.
Cunha-Melo, J. R., Gonzaga, H. M. S., Ali, H., Huang, F. L., Huang, K.-P., and Beaven, M. A., 1989, Studies of protein kinase C in the rat basophilic leukemia(RBL-2H3) cell reveal that antigen-induced signals are not mimicked by the actions of phorbol myristate acetate and Ca^{2+} ionophore. J. Immunol., submitted.
Exton, J. H., 1988, Mechanisms of action of calcium-mobilizing agonists: Some variations on a young theme. FASEB J., 2:2670.
Fernandez, J. M., Neher, E., and Gomperts, B. D., 1984, Capacitance measurements reveal stepwise fusion events in degranulating mast cells. Nature., 312:453.
Gomperts, B. D., Barrowman, M. M., and Cockcroft, S., 1986, Dual role for guanine nucleotides in stimulus-secretion coupling. Fed. Proc.,45:2156.
Howell, T. W., and Gomperts, B. D., 1987, Rat mast cells permeabilised with streptolysin O secrete histamine in response to Ca^{2+} at concentrations buffered in the micromolar range. Biochim. Biophys. Acta, 927:177.
Huang, F. L., Yoshida, Y., Cunha-Melo, J. R., Beaven, M. A., and Huang, K.-P., 1989, Differential down-regulation of protein kinase C isozymes. J. Biol. Chem., 264:4238.
Katakami, Y., Kaibuchi, K., Sawamura, M., Takai, Y., and Nishizuka, Y., 1984, Synergistic action of protein kinase C and calcium for histamine release from rat peritoneal mast cells. Biochem. Biophys. Res. Commun., 121:573.
Kennerly, D. A., Sullivan, T. J., Sylwester, P., and Parker, C. W., 1979, Diacylglycerol metabolism in mast cells: A potential role in membrane fusion and arachidonic acid release. J. Exp. Med., 150:1039.
Kennerly, D. A., 1987, Diacylglycerol metabolism in mast cells. Analysis of lipid metabolic pathways using molecular species analysis of intermediates. J. Biol. Chem., 262:16305.
Lo, T. N., Saul, W., and Beaven, M. A., 1987, The actions of Ca^{2+} ionophores on rat basophilic (2H3) cells are dependent on cellular ATP and hydrolysis of inositol phospholipids. A comparison with antigen stimulation. J. Biol. Chem., 262:4141.
Ludowyke, R. I., Peleg, I., Beaven, M. A., and Adelstein, R. S., 1989, Antigen-induced secretion of histamine and the phosphorylation of myosin by protein kinase C in rat basophilic leukemia cells. J.Biol. Chem., in press.
Maeyama, K., Hohman, R. J., Metzger, H., and Beaven, M. A., 1986, Quantitative relationships between aggregation of IgE receptors, generation of intracellular signals, and histamine secretion in rat basophilic leukemia (2H3) cells. Enhanced responses with heavy water. J. Biol. Chem., 261:2583.
Maeyama, K., Hohman, R. J., Ali, H., Cunha-Melo, J. R., and Beaven, M. A., 1988, Assessment of IgE-receptor function through measurement of hydrolysis of membrane inositol phospholipids.

New insights on the phenomena of biphasic antigen concentration-response curves and desensitization. J. Immunol., 140:3919.
Metzger, H., Alcaraz, G., Hohman, R., Kinet, J. P., Pribluda, V., and Quarto, R., 1986, The receptor with high affinity for immunoglobulin E. Annu. Rev. Immunol., 4:419.
Meyer, T., Holowka, D., and Stryer, L., 1988, Highly cooperative opening of calcium channels by inositol 1,4,5-trisphosphate. Science, 240:653.
Millard, P. J., Gross, D., Webb, W. W., and Fewtrell, C., 1988, Imaging asynchronous changes in intracellular Ca^{2+} in individual stimulated tumor mast cells. Proc. Natl. Acad. Sci. USA., 85:1854.
Miller, L., Blank, U., Metzger, H., and Kinet, J.-P., 1989, Expression of high-affinity binding of human immunoglobulin E by transfected cells. Science., 244:334.
Mohr, F. C., and Fewtrell, C., 1987, Depolarization of rat basophilic leukemia cells inhibits calcium uptake and exocytosis. J. Biol., 104:783.
Nakamura, T., and Ui, M., 1985, Simultaneous inhibitions of inositol phospholipid breakdown, arachidonic acid release, and histamine secretion in mast cells by islet-activating protein, pertussis toxin. A possible involvement of the toxin-specific substrate in the Ca^{2+}-mobilizing receptor-mediated biosignaling system. J. Biol. Chem., 260:3584.
Narasimhan, V., Holowka, D., Fewtrell, C., and Baird, B., 1988, Cholera toxin increases the rate of antigen-stimulated calcium influx in rat basophilic leukemia cells. J. Biol. Chem., 263:19626.
Neher, E., and Almers, W., 1986, Fast calcium transients in rat peritoneal mast cells are not sufficient to trigger exocytosis. EMBRO J., 5(1):51.
Nishizuka, Y., 1986, Studies and perspectives of protein kinase C. Science, 233:305.
Pearce, F. L., 1982, Calcium and histamine secretion from mast cells. Prog. Med. Chem., 19:59.
Penner, R., Matthews, G., and Neher, E., 1988, Regulation of calcium influx by second messengers in rat mast cells. Nature., 334:499.
Schleimer, R. P., Gillespie, E., Daiuta, R., and Lichtenstein, L. M., 1982, Release of histamine from human leukocytes stimulated with the tumor-promoting phorbol diesters. II. Interaction with other stimuli. J. Immunol., 128:136.
Seldin, D. C., Adelman, S., Austen, K. F., Stevens, R. L., Hein, A., Caulfield, J. P., and Woodbury, R. G., 1985, Homology of the rat basophilic leukemia cell and the rat mucosal mast cell. Proc. Natl. Acad. Sci. USA, 82:3871.
Serafin, W. E., and Austen, K. F., 1987, Mediators of immediate hypersensitivity reactions. N. Engl. J. Med., 317:30.
Stevens, R. L., Katz, H. R., Seldin, D. C., and Austen, K. F., 1986, Biochemical characteristics distinguished subclasses of mammalian mast cells, in: "Mast Cell Differentiation and Heterogeneity," A. D. Befus, J. Bienenstock and J. A. Denburg eds., Raven Press, New York, 183.
Stump, R. F., Oliver, J. M., Cragoe, E. J., and Deanin, G. G., 1987, The control of mediator release from RBL-2H3 cells: Roles for Ca^{2+}, Na^{+}, and protein kinase C. J. Immunol., 139:881.
Vu, N. D., and Wagner, P. D., 1989, Thiophosphorlation causes Ca^{2+}-independent secretion of norepinephrine from permeabilized PC12 cells. J. Cell Biol., 107:338 (abstract).
White, K. N., and Metzger, H., 1988, Translocation of protein kinase C in rat basophilic leukemic cells induced by phorbol ester or by aggregation of IgE receptors. J. Immunol., 141:942.

WoldeMussie, E., Maeyama, K., and Beaven, M. A., 1986, Loss of secretory response of rat basophilic leukemia (2H3) cells at 40°C is associated with reversible suppression of inositol phospholipid breakdown and calcium signals. J. Immunol., 137:1674.

SYNCHRONIZED Ca^{2+} TRANSIENTS INDUCED BY GLUCAGON IN FURA2 LOADED HEPATOCYTES

Jan B. Hoek, Kathleen E. Coll, Thomas A. Rooney and Andrew P. Thomas

Department of Pathology and Cell Biology
Thomas Jefferson University
Philadelphia, Pa. 19107

INTRODUCTION

A wide range of agonists exert their effects on liver cells by the receptor-mediated activation of phospholipase C, leading to the formation of diacylglycerol (DAG) and inositol-1,4,5-trisphosphate (Ins-1,4,5-P_3). The latter compound acts as a second messenger to release Ca^{2+} from intracellular storage sites, resulting in an elevation of cytosolic free Ca^{2+} levels (for reviews, see Williamson et al,1985; Putney, 1989). Secondary adjustments in cellular Ca^{2+} homeostasis include the Ca^{2+} release from the cell by activation of the ATP-driven plasma membrane Ca^{2+} pump, the activation of Ca^{2+} influx into the cell required for maintaining the total cellular Ca^{2+} content and the reuptake of Ca^{2+} by intracellular organelles; other inositol phosphate derivatives may contribute to the control of some of these processes (Putney et al, 1989). A variety of agonists operate through this pathway in liver, including vasopressin, α_1-adrenergic agonists and angiotensin II. Glucagon also causes an increase in cytosolic free Ca^{2+} levels, associated with a small increase in Ins-1,4,5-P_3 levels, although it is not clear whether it activates phospholipase C as a secondary consequence of the formation of cAMP and the activation of A-kinase (Staddon and Hansford, 1989) or by interacting with a separate glucagon receptor (GR_1-receptor) which is specifically coupled to the activation of phospholipase C (Wakelam et al, 1986).

AGONIST-INDUCED SINGLE CELL CA^{2+} OSCILLATIONS

Much of the information regarding the control of receptor-mediated changes in cellular Ca^{2+} homeostasis is based on studies using suspensions of isolated cells as an experimental system. However, some years ago, the work of Woods et al (1986,1987) using single hepatocytes loaded with aequorin demonstrated that individual cells responded to agonist-induced stimulation with oscillatory changes in cytosolic Ca^{2+} concentration rather than with the sustained response which is apparent in cell suspensions. These findings have recently been confirmed and extended with a variety of other cells loaded with aequorin or with Ca^{2+}-sensitive fluorescent indicators and using a range of agonists coupled to phospholipase C activation (see Berridge and Gallione (1988) for a review). We have used the fluorescent Ca^{2+} indicator fura2 to study the control of agonist-induced Ca^{2+} oscillations in single hepatocytes in culture. In earlier studies from our laboratory (Rooney et al, 1989) the effects of phenylephrine and vasopressin were reported. Here, we will summarize some of the characteristics of the oscillatory responses to these agonists and contrast those with the Ca^{2+} responses observed after stimulation of the cells with glucagon.

CALCIUM MEASUREMENTS BY DIGITAL FLUORESCENCE IMAGING MICROSCOPY

The methodology used in digital fluorescence imaging microscopy and in loading hepatocytes with fura2 and calibrating the Ca^{2+} concentration have been described in recent publications (Prentki et al, 1988; Rooney et al, 1989). Briefly, rat hepatocytes were isolated by collagenase perfusion and kept on ice until use. Two hours prior to each individual incubation, the cells were plated on polylysine-coated glass coverslips in Williams E medium with dexamethasone (4 μg/ml), glutamine (2 mM) and 2% bovine serume albumin (BSA). The cells were then loaded with fura2/AM (3.2 μM) in the presence of 0.03% Pluronic F-127 for 20 min at 37°C, washed and studied under a Zeiss IM35 inverted microscope (16x objective) in a thermostatted open incubation chamber. The incubation medium was a modified Krebs-Ringer medium containing 10 mM HEPES, 5 mM $NaHCO_3$ and 0.25 % BSA. Cells were stimulated by completely replacing the incubation medium in the chamber with a hormone-containing medium, over a period of 1-2 sec. Fluorescence images were collected with a liquid nitrogen cooled CCD camera (Photometrics), digitized at 12 bit resolution and stored and analyzed with Heurikon HK68/M10 computer. Excitation wavelengths were selected by appropriate filters at 340 nm and 380 nm (10 nm bandwidth) for an emission wavelength of 460-600 nm. The time course of fluorescence changes in a particular cell could be followed by taking a series of successive 340 and 380 nm images of a microscope field containing 20-40 cells, with an exposure time per image of 300 msec. The mean fluorescence intensity for an individual cell was then calculated for each image of a time series within a user-selected region of the field covering most of that cell. The mean cellular Ca^{2+} concentration was obtained from the 340/380 ratio as detailed by Prentki et al (1988) and Rooney et al (1989). In many experiments, the time resolution was improved by measuring only the 380 nm fluorescence during the run, taking a 340 nm image only at the beginning and end of a run for purposes of calibration. As discussed elsewhere (Rooney et al, 1989), the Ca^{2+} concentrations can be calculated relative to the initial free Ca^{2+} level provided a small correction for photobleaching and dye leak over the time of the experiment is taken into account. In some experiments, corrections for background fluorescence were made on the basis of residual fluorescence levels after the addition of inonomycin (10 μM) and $MnCl_2$(2 mM). In other experiments, the initial Ca^{2+} concentration was assumed to be at 200 μM, which is close to the mean cytosolic Ca^{2+} level found in isolated hepatocytes.

CHARACTERISTICS OF AGONIST-INDUCED Ca^{2+} OSCILLATIONS IN FURA2-LOADED SINGLE HEPATOCYTES

In earlier studies by our group (Rooney et al, 1989) and others (Woods et al, 1986, 1987), the actions of phenylephrine and vasopressin on cytosolic free Ca^{2+} levels in single hepatocytes were found to be characterized by a set of distinctive features, which are summarized below and which are compared and contrasted to the Ca^{2+} mobilization induced by glucagon.

1. Dose-dependent frequency modulation. In isolated hepatocytes in culture, phenylephrine induced a series of Ca^{2+} transients which were characterized by a dose-dependent increase in frequency, but little or no variation in the amplitude of the Ca^{2+} response. Sequential transients in an individual cell all exhibited the same rate of rise in Ca^{2+} (55 $\pm$ 10 nM/s, measured at half peak height) and achieved the same peak Ca^{2+} level (670 $\pm$ 50 nM in a typical experiment, independent of the concentration of agonist). There was a latent period before the first Ca^{2+} transient, the

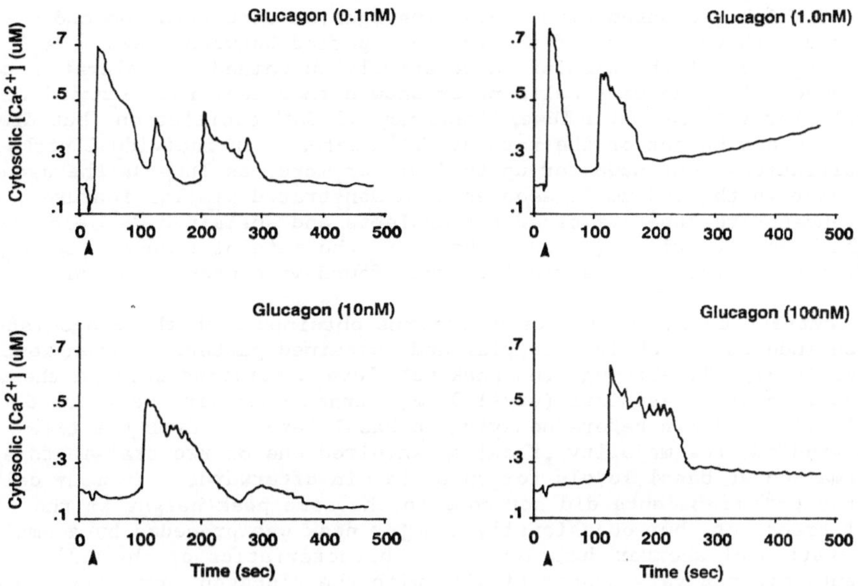

Fig. 1 Ca^{2+} transients induced by different concentrations of glucagon.

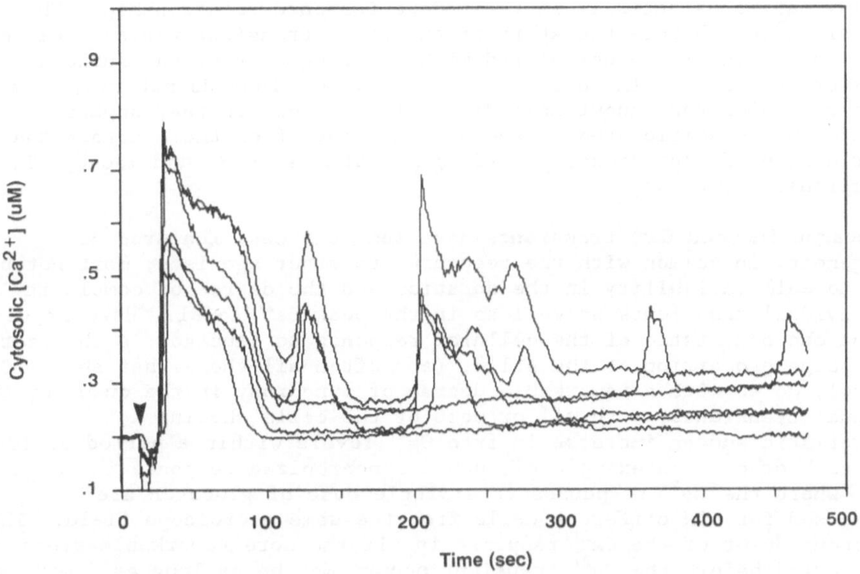

Fig. 2 Synchronous Ca^{2+} transients in response to glucagon.

length of which decreased with increasing agonist concentration and correlated with the length of the recovery period between peaks. At saturating doses of the agonist, a relatively sustained Ca^{2+} elevation was often found, although many cells never showed that feature. Removal of extracellular Ca^{2+} caused a lower frequency of Ca^{2+} oscillation, but did not affect the peak height or the rate of Ca^{2+} increase in individual peaks. The oscillations continued for up to 8 min or more, as long as the agonist was present in the medium. Vasopressin demonstrated similar features, but had a tendency to generate broader transients and sustained Ca^{2+} elevations at higher concentrations (>1 nM). However, the rate of rise of $[Ca^{2+}]_{cyt}$ in individual peaks was identical to that found with phenylephrine.

In contrast to the oscillation patterns obtained with these agonists, glucagon induced a much less regular and sustained pattern of transients, as shown in Fig. 1. Although the peak Ca^{2+} levels attained were in the same range as with other agonists (0.4-1.0 μM), transients were very broad, often lasting 2-3 min before decaying to basal levels. Over the time course studied, the majority of cells exhibited one or two transients and then remained at basal levels for up to 15 min afterwards. In many cells, the later Ca^{2+} transients did not come to the same peak height as the initial transient, but occasionally a major peak was preceded by a smaller Ca^{2+} elevation of shorter duration. The characteristics of the Ca^{2+} transients did not vary systematically with the glucagon concentration over a range of 0.01-100 nM, whereas the cAMP level was increased almost 10-fold over this concentration range. At low doses of glucagon, the predominant dose-dependent variable appeared to be the number of cells responding to glucagon, but even at 0.1 nM often 80-90 % of the cells showed a Ca^{2+} response. The hormone concentration also had little systematic effect on the length of the latent period or on the shape of the Ca^{2+} transients exhibited by the majority of the cells. A sustained elevation of cytosolic Ca^{2+} was scarce at any glucagon concentration in the range of 0.01-100 nM.

2. Temporal and spatial heterogeneity. Individual cells in a single microscope field stimulated by phenylephrine or vasopressin display a remarkable heterogeneity in the Ca^{2+} oscillations. The cells are not only out of phase with each other, but also differ in the frequency, and often the amplitude and the complexity of the Ca^{2+} oscillation, and individual cells may vary substantially in their dose response relationship. The period of latency before the start of the first transient varies from cell to cell and is inversely correlated with the frequency of the subsequent oscillatory response. In addition, the Ca^{2+} elevations do not rise in a concerted fashion throughout an individual cell, but instead appear to originate in a specific area in the cell and move from there across the cell at a rate of approximately 10-20 μm/s (Thomas, A.P. and Rooney, T, in preparation).

Glucagon-induced Ca^{2+} transients have some of these features of heterogeneity in common with the responses to other agonists, most notably a cell-to-cell variability in the duration and the degree of complexity of the individual transients as well as in the peak Ca^{2+} level. However, a dominant characteristic of the cellular response to glucagon is the fact that a large proportion of the cells, (and often all those that show a Ca^{2+} response), do so with a remarkable degree of synchrony in the onset of the individual transients, with all or most of the cells showing a characteristic sudden increase in free Ca^{2+} levels within a period of 10-20 sec of each other. An example of such a synchronized response is shown in Fig. 2, where the Ca^{2+} responses to a single dose of glucagon are superimposed for six different cells from the same microscope field. This synchronous onset of the Ca^{2+} response is all the more remarkable since the latent period before the Ca^{2+} transient occurs may be as long as 2 min, with subsequently over 80 % of the cells responding together. In preparations showing multiple transients, the later peaks often were not exhibited by

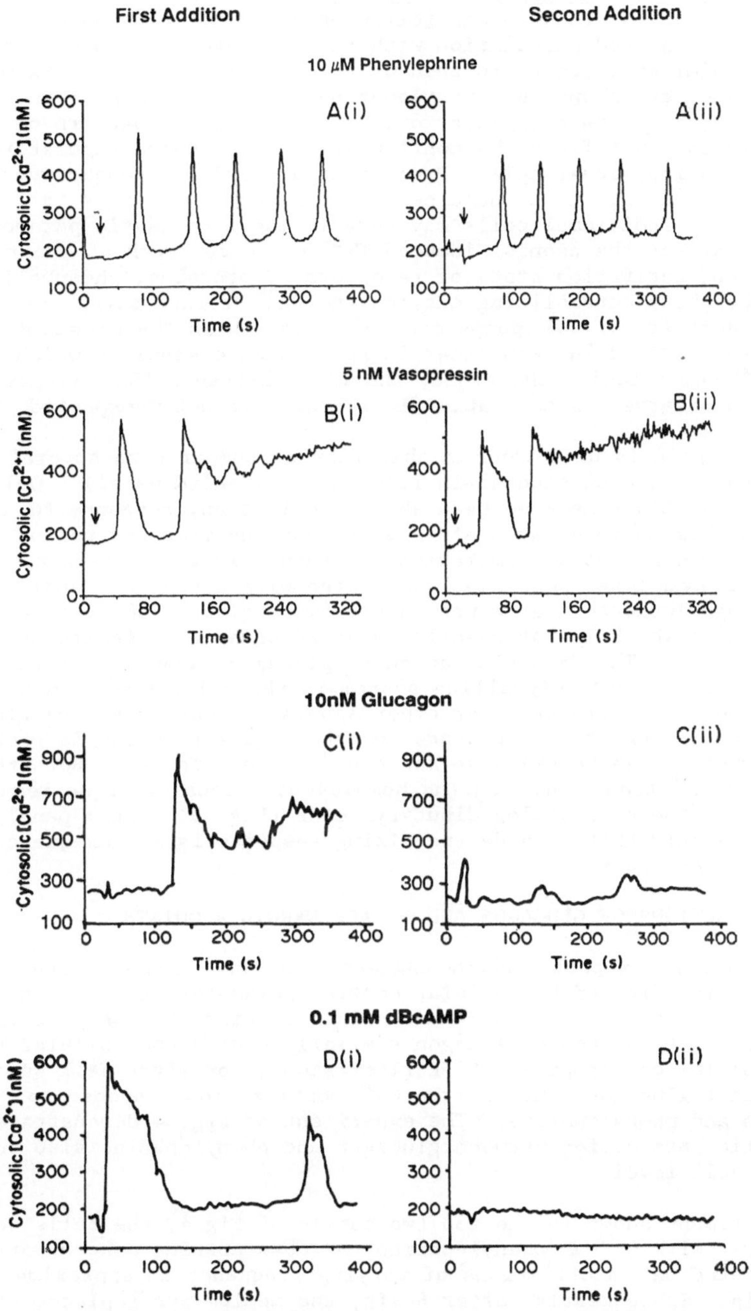

Fig. 3. Cellular Ca^{2+} transients in response to repeated stimulation with the same agonist.

all the cells, but for those that do, the transients tend to be grouped together (see Fig.2).

3. Cell-specific Ca^{2+} fingerprint patterns. When cells are allowed to recover for a period of 30-40 min following treatment with phenylephrine or vasopressin, a renewed stimulation with the same agonist can elicit essentially similar patterns in each of the individual cells. Examples of this phenomenon are shown for phenylephrine and vasopressin in the top two panels of Fig. 3. There appears to be a cell-specific pattern of Ca^{2+} responses which may reflect the configuration of feedback regulatory features of an individual cell. Prentki et al (1988) have observed this phenomenon in cultured insulinoma cells and referred to it as 'calcium-fingerprints'. Individual cells may vary in many stochastic parameters which could affect the agonist-induced Ca^{2+} response, such as their complement and activation state of receptors, G-proteins, phospholipase C, inositol phosphate metabolizing enzymes and Ca^{2+} storage sites, or in the number and activity of Ca^{2+} pumps and Ca^{2+} channels in the plasma membrane, the endoplasmic reticulum and other intracellular organelles which control overall Ca^{2+} homeostasis; therefore, the fact that such 'Ca^{2+} fingerprint' patterns are observed as cell-specific features is not unexpected.

Glucagon again is different in this respect compared to several other agonists that have been studied in liver (Fig.3, third panel). Only in a few cell preparations have we been able to elicit any response to a second addition of this hormone, even after a washout and recovery period of up to one hour. Where a second stimulation did result in a Ca^{2+} response, it was usually to a much lower peak level and often with a single transient. These findings indicate that glucagon treatment gives rise to a persistent desensitization which is not readily reversed under the conditions used in our experiments. The desensitization to glucagon cannot be attributed to a deficiency of the Ca^{2+} signalling system in the cell, since these cells still respond to phenylephrine or other agonists. Moreover, the glucagon-induced potentiation of the response to other agonists which is probably mediated by cAMP (see below) also appears to be unaffected. Both the synchronized Ca^{2+} transients and the homologous desensitization response can be mimicked by the cAMP analog dibutyryl cAMP (Fig. 3, bottom panel). Thus, it is likely that this desensitizing response is a consequence of the activation of A-kinase.

SYNERGISTIC ACTIONS OF GLUCAGON AND α_1-ADRENERGIC AGONISTS

The experiments reported above suggest that the glucagon-induced Ca^{2+} mobilization is affected by cellular control parameters that are not operative in the same way with α_1-adrenergic agonists or vasopressin. The question arises then: would glucagon similarly modify the cellular response to phenylephrine or vasopressin? Earlier studies on liver cell suspensions had shown that glucagon enhances the Ca^{2+} mobilization response to vasopressin and phenylephrine. The experiment of Fig. 4 demonstrates that a synergistic interaction between glucagon and phenylephrine also occurs at the single cell level.

In the experiment shown in the top two panels of Fig 4, the cells were treated first with 0.2 uM phenylephrine, a subsaturating concentration which generated Ca^{2+} oscillations of varying frequency in approximately 50% of the cells. Subsequently, after 6 min, the medium was replaced with one containing a combination of phenylephrine and glucagon (1 nM). After another 6 min, the medium was replaced again with one containing only phenylephrine. Two representative examples are shown of cells that differed substantially in their sensitivity to phenylephrine. In both these cells (as in essentially all other cells in the sample), the addition of glucagon did not increase the peak free Ca^{2+} level, but increased the frequency of the response and in both cells its effect became apparent only after about 90 sec. In the first cell, the phenylephrine-induced

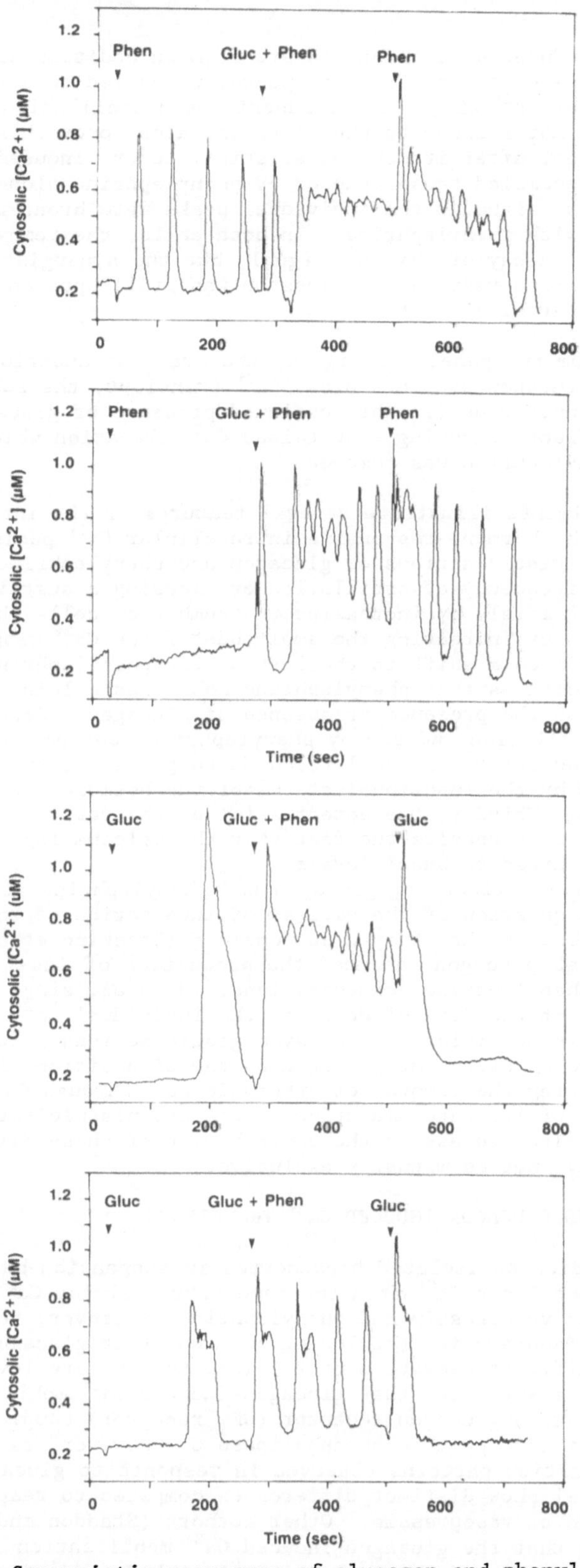

Fig. 4 Synergistic responses of glucagon and phenylephrine.

oscillation continued undisturbed (apart from an addition artefact) prior to the appearance of the glucagon response, which led to a sustained elevation of free Ca^{2+} with a small superimposed oscillation. In the second cell, which did not respond to phenylephrine alone over this period, glucagon addition, after its characteristic latency, induced an oscillatory response that resembled those induced by phenylephrine alone in more responsive cells, although the individual peaks were broader than was normally found with phenylephrine. In both cells, the removal of glucagon caused a gradual decay of the Ca^{2+} signal, but the synergistic enhancement of the phenylephrine response by glucagon was still apparent, despite the washout of the latter agonist.

In the bottom two panels of Fig. 4, the order of additions was reversed. In this case, glucagon caused a broad Ca^{2+} transient; the subsequent addition of phenylphrine (in the continued presence of glucagon) now showed an immediate effect, inducing a sustained Ca^{2+} elevation which was reversed as soon as phenylephrine was removed.

These experiments illustrate several features of the interaction of glucagon with the hormone-responsive intracellular Ca^{2+} pools. In the first place, the synergistic actions of glucagon and phenylephrine operate by increasing the frequency of oscillation or imposing a sustained elevated response and not merely by increasing the number of cells that respond to phenylephrine or by increasing the amplitude of the Ca^{2+} response. Thus, the synergism mimics a shift to the left of the phenylephrine dose-response curve. This indicates that phenylephrine mobilizes calcium from the same storage sites, in the presence or absence of glucagon. Secondly, an ongoing Ca^{2+} oscillation induced by phenylephrine does not appear to affect the latent period before the cell responds to glucagon; thus, the latency is not affected by the functional status of the hormone-sensitive Ca^{2+} response system. Thirdly, the potentiation of the response to phenylephrine occurs despite the fact that the calcium increase caused by glucagon has returned to basal levels.

The synergistic effects of glucagon and phenylephrine at the single cell level raise the question of the site(s) of interaction of these agonists. Earlier experiments on hepatocyte suspensions (Bocckino et al, 1985) had demonstrated that glucagon enhanced the production of Ins-1,4,5-P_3 induced by submaximal phenylephrine concentrations. However, since glucagon also appeared to affect the rate of decay of the individual Ca^{2+} spikes (see Fig 4, bottom), it is possible that the synergistic actions illustrated in Fig. 4 are related to an effect of glucagon on the disposition of Ca^{2+} released, e.g., by inhibiting the removal of cytosolic Ca^{2+} through Ca^{2+} pumps in the plasma membrane or its reaccumulation in the endoplasmic reticulum. Further studies are required to assess the contribution of these different sites of action which may not be mutually exclusive.

MECHANISM OF THE GLUCAGON-INDUCED Ca^{2+} TRANSIENTS

Several studies on isolated hepatocytes in suspension have indicated that glucagon mobilizes Ca^{2+} from the same intracellular Ca^{2+} pool as other agonists such as vasopressin and phenylephrine; moreover, there is evidence from different laboratories (including our own) that glucagon induces a small, but significant elevation of the inositol phosate levels. Wakelam et al (1986) have suggested that glucagon-induced Ca^{2+} mobilization occurs by way of a separate glucagon receptor (GR_1 receptor) coupled to the polyphosphoinositide-specific phospholipase C. However, as shown above, the Ca^{2+} mobilization patterns observed in response to glucagon at the single cell level show distinct differences compared to responses induced by phenylephrine or vasopressin. Other authors (Staddon and Hansford, 1989) suggested that the glucagon-induced Ca^{2+} mobilization is secondary to the synthesis of cAMP and the activation of cAMP-dependent protein kinase and this interpretation is supported by the finding that forskolin or cAMP analogs also induce a calcium elevation. However, in our experiments, the Ca^{2+} elevation was near-maximal in response to glucagon concentrations of 10^{-10} M, where the increase in cAMP formation was hardly detectable. Although at this point it cannot be excluded that a small elevation of cAMP (as measured in the total cell extract) is sufficient to generate a saturating Ca^{2+} elevation (Sistare et al, 1985), the data shown above suggest that either of these models may give an incomplete picture of the glucagon-induced Ca^{2+} response.

The dose-response relationship for glucagon-induced Ca^{2+} mobilization suggests that the primary effect of increasing hormone concentrations in the range of 10^{-11}-10^{-10}M is an increase in the number of cells responding; in a substantial number of experiments, we found no systematic evidence in the *responsive* cells for an increased frequency of Ca^{2+} transients or a decreased latency period over the concentration range of 10^{-11} to 10^{-7}M glucagon (where cAMP levels increased five to ten-fold over control levels). These parameters were strongly dose-dependent with phenylephrine and vasopressin. Moreover, the sustained Ca^{2+} elevation that is typical of saturating concentrations of other agonists was found only rarely with glucagon, even at the highest concentrations used (10^{-7}M).

The lack of correlation between cAMP formation and calcium elevation could be interpreted to mean that cAMP does not play a role in the intracellular Ca^{2+} release. On the other hand, the patterns of Ca^{2+} elevation induced by added cAMP analogs in hepatocytes are similar to those caused by glucagon, with a few broad transients rather than a sustained oscillation; these patterns are distinctly different from those induced by phenylephrine or vasopressin.

A possible explanation to account for these observations is that the glucagon-induced Ca^{2+} increase is subject to a set of contrary controls. It is conceivable that glucagon can cause Ca^{2+} mobilization both directly through receptor-mediated activation of phospholipase C and indirectly through the formation of cAMP and activation of A-kinase. At higher glucagon levels (>10^{-10}M) cAMP may decrease the rate of formation of inositol phosphates, as proposed by Wakelam et al. (1986); at the same time, it could be more effective in mobilizing Ca^{2+} from intracellular stores, or enhance the Ca^{2+} elevation by interfering with Ca^{2+} transport systems in plasma membrane and endoplasmic reticulum, simlar to its synergistic effect on phenylephrine-induced Ca^{2+} mobilization.

Even with a model of complex stimulatory and inhibitory inputs at the level of phospholipase C it is difficult to account for the remarkable synchrony that appeared to dominate the Ca^{2+} mobilization responses at all hormone concentrations. Direct cell-to-cell contact was not required for synchronizing the responses in these preparations of freshly plated liver cells, where individual cells had not yet had the time to establish contacts and develop gap junctions. In a microscope field containing 20-40 cells, those located on opposite sides of the field were as likely to respond together as those in close juxtaposition. Thus, the nature of this synchronized response is intrinsically different from the gap-junctional transmittance of Ins-1,4,5-P_3 and Ca^{2+} recently reported in hepatocytes interconnected by gap junctions (Saez et al, 1989). It is conceivable that the liver cells respond to glucagon by producing a soluble factor which can migrate to neighboring cells and stimulate phospholipase C to trigger the mobilization of Ca^{2+} in a concentration-dependent manner. However, so far we have no indication of either the existence or the nature of such a factor. A further dissection of the unexpected complications of the glucagon-induced signalling events will be required to approach this problem.

ACKNOWLEDGEMENTS

This work was supported by United States Public Health Service grants DK38461, AA 07186 and AA 07215.

REFERENCES

Berridge, M.J. and Galione, A., 1988, Cytosolic calcium oscillators. FASEB J., 2:3074-3082.

Bocckino, S.B., Blackmore, P.F. and Exton, J.H., 1985, Stimulation of 1,2-diacylglycerol accumulation in hepatocytes by vasopressin, epninephrine and angiotensin II. J. Biol. Chem. 260:14201-14207.

Graf, P., vom Dahl, S. and Sies, H., 1987, Sustained oscillations in extracellular calcium concentrations upon hormonal stimulation of perfused rat liver. Biochem.J., 241:933-936.

Mauger, J.P. and Claret, M., 1986, Mobilization of intracellular calcium by glucagon and cyclic AMP analogues in isolated rat hepatocytes. FEBS Lett. 195:106-110.

Prentki, M., Glennon, M.C., Thomas, A.P., Morris, R.L., Matschinsky, F.M. and Corkey, B.E., 1988, Cell-specific patterns of oscillating free Ca^{2+} in carbamylcholine-stimulated insulinoma cells. J.Biol.Chem. 263:11044-11047.

Putney, J.W., Takemura, H., Hughes, A.R., Horstman, D.A. and Thastrup, O., 1989, How do inositol phosphates regulate calcium sgnaling? FASEB J., 3:1899-1905.

Rooney, T.A., Sass, E.J. and Thomas, A.P., 1989, Characterization of cytosolic calcium oscillations induced by phenylephrine and vasopressin in single fura 2-loaded hepatocytes. J.Biol.Chem., in press.

Saez, J.C., Connor, J.A., Spray, D.C. and Bennet, M.V.L., 1989. Hepatocyte gap junctions are permeable to the second messnger, inositol 1,4,5-trisphosphate, and to calcium ions. Proc. Natl. Acad. Sci., U.S.A., 86:2708-2712.

Staddon, J.M. and Hansford, R.G., 1989, Evidence indicating that the glucagon-induced increase in cytoplasmic free Ca^{2+} concentration in hepatocytes is mediated by an increase in cyclic AMP concentration. Eur.J.Biochem. 179:47-52.

Sistare, F.D., Picking, R.A. and Haynes, R.C., 1985, Sensitivity of the response of cytosolic calcium in quin-2-loaded rat hepatocytes to glucagon, adenine nucleosides and adenine nucleotides. J.Biol.Chem., 260:12744-12747.

Wakelam, M.J.O., Murphy, G.J., Hruby, V.J. and Houslay, M.D., 1986, Acivation of two signal transduction systems in hepatocytes by glucagon. Nature, 323:68-71

Williamson, J.R., Cooper, R.H., Joseph, S.K. and Thomas, A.P., 1985, Inositol trisphosphate and diacylglycerol as intracellular second messengers in liver. Am.J.Physiol. 248:C203-C216.

Woods, N.M., Cuthbertson, K.S.R. and Cobbold, P.H., 1986, Repatitive transient rises in cytoplasmic free calcium in hormone-stimulated hepatocytes. Nature, 319:600-602.

Woods, N.M., Cuthbertson, K.S.R. and Cobbold, P.H., 1987, Agonist-induced oscillations in cytoplasmic free calcium concentration in single rat hepatocytes. Cell Calcium, 8:79-100

CALCIUM MOBILIZATION MECHANISMS IN BOVINE CHROMAFFIN CELLS BY VASOACTIVE PEPTIDES: ANGIOTENSIN, BRADYKININ AND ENDOTHELIN

Kathy Rasmussen and Morton P. Printz

Department of Pharmacology M-036
University of California San Diego
La Jolla, CA. 92093 USA

ABSTRACT

Exocytosis of catecholamines from cultured bovine adrenal medullary cells is influenced by both cholinergic ligands and peptides. To define mechanisms of intracellular calcium ($Ca^{2+}{}_i$) mobilization by angiotensins (ANG) II & III, bradykinin (BKY) and endothelin, studies using intracellular calcium ($Ca^{2+}{}_i$) probes were performed on cells in suspension. Complex dose-dependent $Ca^{2+}{}_i$ kinetics were observed consisting of two components; a rapid spike and an extended plateau. The order of potency for both components of the response is BKY > ANG III ≥ ANG II. At low doses of ANG and BKY, the spike is absent but a slow and sustained plateau remains. Removal of extracellular calcium decreases the BKY response to < 50% with loss of the spike, while the effect on the ANG response is minimal. The $Ca^{2+}{}_i$ response to ANG and BKY was not altered in cells treated with pertussis or cholera toxin although phorbol ester (PMA) reduced the response 50% after overnight treatment. The response to endothelin (up to 1 μM) was minimal, but a synergistic response was seen in the presence of KCl (55 mM). Nicotine and low doses of the peptides show additivity in the plateau phase. We conclude that physiological doses of peptides (esp. ANG & BKY) may markedly influence adrenal medullary responses to nicotinic cholinergic stimulation.

INTRODUCTION

The adrenal medulla is an organ intimately involved in the stress response (Kvetnansky, et al., 1978). It receives primary neural innervation via sympathetic preganglionic cholinergic neurons acting on nicotinic receptors (Burgoyne, 1984; Cena, et al., 1983). Sympathetic activation leads to the release of adrenal catecholamines and other substances, including chromogranins and peptides such as enkephalin (Phillips, et al., 1987). The multiplicity of substances released may have endocrine, autocrine or paracrine functions; however, it is generally held that they either contribute to the stress response or exert feedback modulation on the adrenal medullary response.

Catecholamine secretion in chromaffin cells is also stimulated by many peptide hormones. Angiotensin and bradykinin have been shown to stimulate release of catecholamines from the adrenal gland (Feldberg and Lewis, 1964) and in cultured chromaffin cells (Noble, et al., 1988). Peach (1971) reported that in vivo administration of angiotensins I and II stimulated release of adrenal catecholamines in a dose-dependent manner. More recently, the presence of specific receptors for angiotensin in the intact adrenal medulla and in isolated membranes have been demonstrated (Healy, et al., 1985).

Cultures of isolated adrenal medullary chromaffin cells have become a widely used model system to study the influence of a variety of endogenous and exogenous agents on secretory activity (Strittmatter, 1988; Burgoyne, 1984). Boyd (1987) identified specific angiotensin II receptors on cultured bovine chromaffin cells and found that approximately 50% of the cultured cells express a high affinity angiotensin receptor. In addition, she demonstrated a dose-dependency between receptor binding, receptor occupancy and fractional secretory response suggesting that secretion was proportional to receptor occupancy. Her studies also identified an angiotensin-mediated transient and dose-dependent increase in intracellular calcium [Ca_i^{2+}] (Boyd, 1987). Zimlichman et al., (1987) also reported that angiotensin II elicits a dose-dependent elevation of Ca_i^{2+}.

The relationship between the angiotensin receptor, other neuropeptide receptors and the classical cholinergic/nicotinic secretory pathway is far from defined. In addition, second messenger pathways important in angiotensin receptor-mediated cellular responses have not yet been fully elucidated. The present studies were undertaken to extend Boyd's initial findings on the effect of angiotensin II on [Ca_i^{2+}] and mechanisms regulating calcium-dependent exocytosis.

METHODS

Cell isolation: Chromaffin cells were isolated from bovine adrenal glands obtained from a local slaughterhouse. Cells were isolated by a modification of the procedures of Greenberg and Zinder (1982). Glands were retrogradely perfused and incubated at 37° for 30 minutes with calcium-free Locke's calcium-free Locke's buffer containing 0.3% collagenase (Boehringer Mannheim, Indianapolis IN) and 0.0015% DNAse (Sigma Chemical Company, St. Louis, MO) . This procedure was repeated after which the cortex and medulla were separated and the medullae finely minced and incubated in fresh collagenase solution in a spinner flask for 45-60 minutes at 37°. Dissociated cells were filtered, washed with buffer containing 1% BSA (Pentex Fraction 5, ICN, Lisle, IL) and gently pelleted. Following resuspension in BSA-calcium-free Locke's buffer the cells were filtered sequentially through 210 μM and 105 μM nylon mesh. Chromaffin cells were enriched by Percoll (Pharmacia LKB, Uppsala, Sweden) gradient centrifugation at 28,000 rpm (20,000 x g) for 17 minutes and collected from the band near the middle of the tube which we have shown contains highly enriched, viable chromaffin cells (Boyd, 1987). Cells were resuspended in MEM media containing 1% non essential amino acids, 1% L-glutamine, 10% fetal calf serum and antibiotics (100 U/ml penicillin, 100 μg/ml streptomycin and 0.25% fungizone). For calcium signalling studies, cells were maintained in suspension in a spinner flask overnight at 37°, 5% CO_2/95% air.

Calcium measurement: Cells were resuspended at a concentration of 2 x 10^6 viable cells/ml in Locke's buffer containing 5 mg/ml BSA and 2.2 mM $CaCl_2$. Cell viability (as determined by trypan blue exclusion) was 90-95%. The cells were incubated for 20-30 minutes at 23° C with 1 μM Fura-2 AM (Molecular Probes, Eugene OR). For each analysis, a two ml aliquot of cells was diluted to 5 ml with Lockes buffer with 2.2 mM $CaCl_2$, centrifuged for 5 min (low speed) and resuspended to 2 x 10^6 cells/ml in the same buffer. Calcium measurements were conducted using a SPEX fluorimeter with excitation wavelengths of 340 and 380 nm (slit = 1.25 mm) and emission at 505 nm (slit = 1.25 mm). Cells were stirred gently to inhibit settling of the cells and minimizing bleaching of the dye. The fluorescence spectra was recorded and cation calculation performed using SPEX DM3000CM software in the method described by Grynkiewicz et al. (1985). The K_D used for Fura 2 was 224 nM. Fluorescence measurements were obtained at 22-25° and a stable baseline was obtained before addition of the drug. When measurements were made under calcium-free conditions, the buffer contained 0 added calcium and 1 mM EGTA.

Materials: Angiotensin and bradykinin peptides were obtained from BaChem (Torrance, CA). Endothelin was kindly provided by T. Watanabe (Peptide Institute, Osaka, Japan). Pertussis and cholera toxins were from List Biologicals (Campbell, CA). Phorbol myristate (PMA) and other reagents from either Calbiochem (La Jolla, CA) or Sigma (St. Louis. MO).

RESULTS

Angiotensin II: The time course of the intracellular calcium response to angiotensin II is dose-dependent (Fig. 1 top pattern) with different response patterns between high (1 μM) and low dose (1 nM). Relating the maximum level of Ca_i^{2+} to dose yields an EC_{50} approximately 10^{-7} M. No elevation above basal levels of Ca_i^{2+} is evident at 100 pM while at 1 and 10 (not shown) nM, the Ca_i^{2+} response develops slowly and exhibits an extended time course (Fig. 1). At angiotensin II concentrations greater than 10 nM, the response pattern is similar with a rapid spike whose rate of development increases with increasing dose. Concomitantly, with

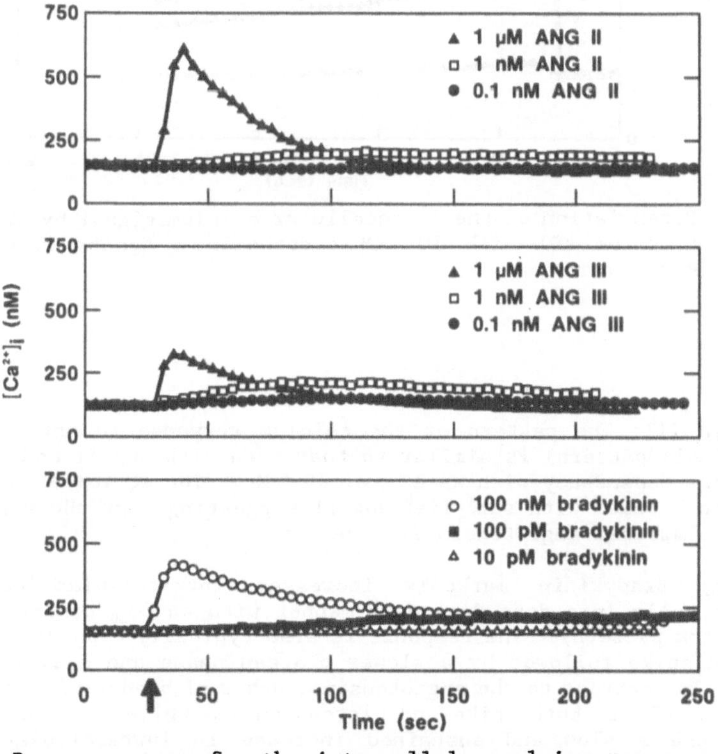

Figure 1. Dose-responses for the intracellular calcium response elicited by angiotensin II, angiotensin III, and bradykinin. Peptides were added at 30 sec (denoted by arrow).

increasing dose faster reversal kinetics are evident. Kinetic analysis of the reversal phase (from peak level) at 1 μM fits a single exponential decay with $t_{1/2}$ approx. 30 sec. A rapid return to baseline suggests possible activation of mechanisms which enhance removal of intracellular calcium.

In the absence of extracellular calcium, the dose-dependent spike is still evident, although there is slight attenuation of the response (to be reported in detail elsewhere). Sar^1-ile^8-angiotensin II, which exhibits partial agonist/antagonist properties, completely blocks the response to ang II suggesting that the intracellular calcium response is mediated through the angiotensin II receptor.

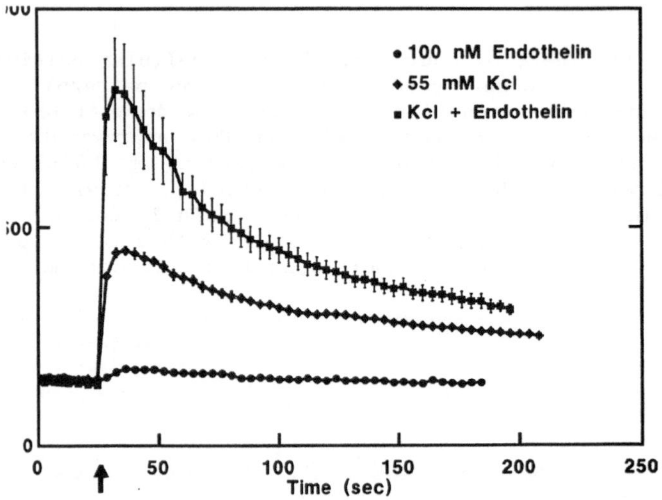

Figure 2. Potentiation of the intracellular calcium signal by simultaneous addition of 55 mM KCl with 100 nM endothelin. Agents were added as indicated by the arrow.

Angiotensin III: The pattern of the calcium response to angiotensin III (Fig. 1 middle pattern) is similar to that seen with ang II and exhibits a similar dose-dependency with an EC_{50} of 10^{-7} M . The action of angiotensin III is also blocked with sar^1-ile^8-ang II suggesting that the peptide acts through a classical angiotensin II receptor.

Bradykinin: Bradykinin markedly increases intracellular calcium in chromaffin cells in a dose-dependent manner with an EC_{50} of approx. 10^{-8} M (Fig 1 bottom pattern). The response is also typically biphasic consisting of a rapid spike followed by a slower plateau phase and a slow return to baseline. In contrast to the angiotensins, doses of bradykinin greater than 1 nM already elicit both spike and plateau phases while lower doses (100 - 1 pM) cause a slow and sustained increase in intracellular calcium. Reversal kinetics at 100 nM bradykinin fits to a single exponential decay with $t_{1/2}$ approx. 60 sec.

Bradykinin appears to be acting through a B2 receptor type. The B2 antagonist, β-[2-thienyl-ala[5,8] D-phe[7]]-bradykinin blocks the Ca_i^{2+} response to bradykinin. No response is evident with des-arg[9]-bradykinin and the B1 antagonist, des-arg[9], leu[8]-bradykinin does not interfere with the calcium response.

In contrast to angiotensins II and III, the Ca_i^{2+} response to bradykinin is strongly dependent on extracellular calcium. In the absence of extracellular calcium, the spike phase is lost and the response is attenuated by greater than 50%.

<u>Endothelin:</u> The ability of endothelin to mobilize Ca_i^{2+} is very ineffectual even up to doses of 1 μM (Fig. 2). A strikingly different result is evident if the cells receive, at the same time, a depolarizing dose (55 mM) of KCl. Although 55 mM KCl exhibits a strong calcium mobilization, totally dependent on extracellular calcium, a marked synergism is evident with 100 nM endothelin (Fig. 2).

<u>Additivity of angiotensin and bradykinin:</u> In contrast to results reported for NG115-401L or transfected 401L-C3 cells by Jackson et al. (1988), sequential addition of angiotensin and bradykinin fail to exhibit sequence-dependent heterologous desensitization. In fact, low doses of the peptides show additivity in terms of the calcium response. With combined addition of high doses (1 μM) of the peptides, no additivity is evident due probably to maximization of the calcium response. However, even at these doses, no heterologous desensitization is evident (data to be reported elsewhere).

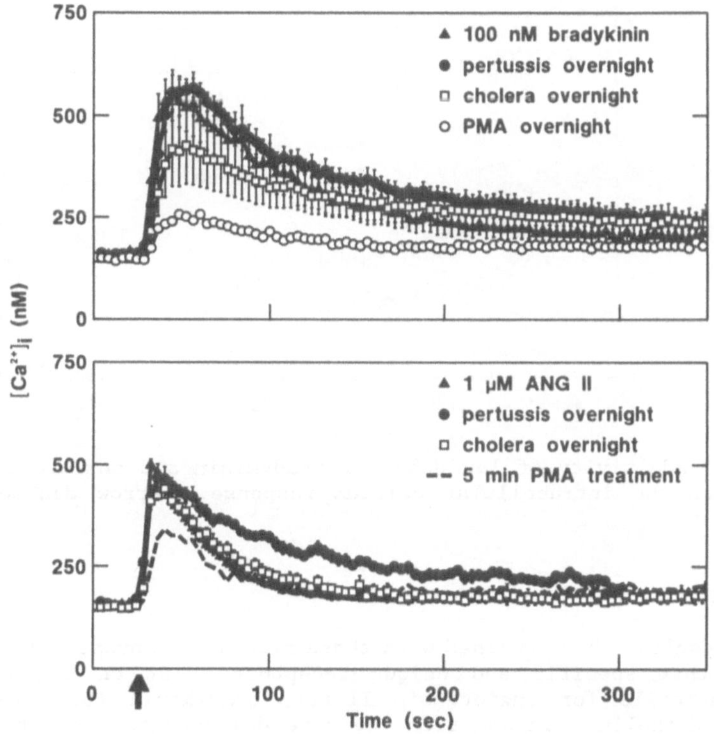

Figure 3. Effect of treatment with pertussis toxin, cholera toxin and PMA on the calcium signals elicited by bradykinin and angiotensin II. Arrow denotes time of addition.

Effects of toxin treatment on calcium response: Pertussis toxin treatment (125 ng/ml overnight) had no significant effect on the calcium response to 1-100 nM bradykinin nor were any effects evident on the angiotensin II mediated calcium signal (Fig. 3). In addition, cholera toxin treatment (100 ng/ml overnight) did not significantly attenuate the intracellular calcium response to either bradykinin or angiotensin II. In contrast, short exposure (at least 5 min) to phorbol ester (10 μM) attenuated the response to both angiotensin II and bradykinin while overnight treatment decreased the calcium response to both peptides by approx. 60-70% (Fig. 3).

Additivity of nicotine and angiotensin II/bradykinin: Nicotine addition (at doses > 1 μM) exhibits a sharp calcium spike with a rapid reversal (Fig. 4). The response to 5 μM nicotine and a low dose of either bradykinin (0.1 nM) or angiotensin II (1 nM) exhibits a response which is at least completely additive (Fig. 4).

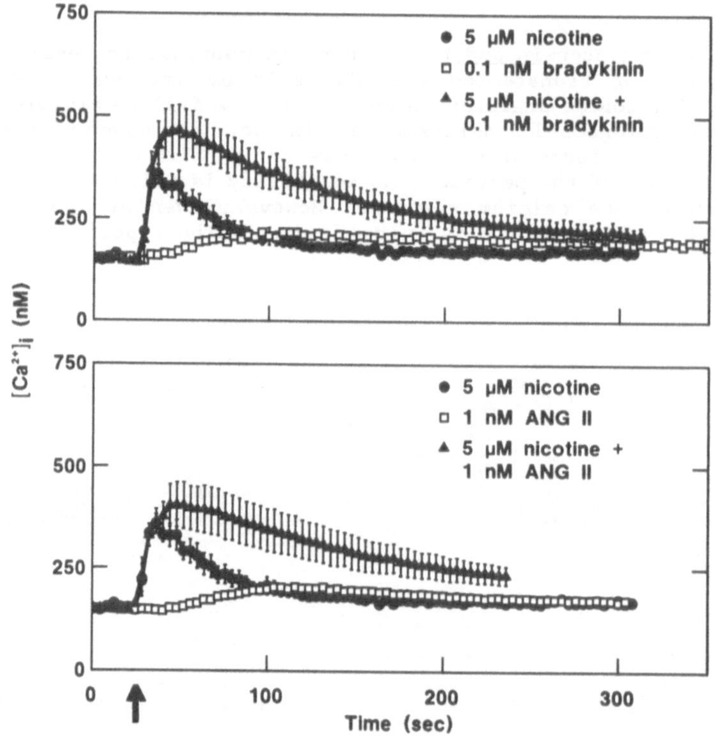

Figure 4. Additivity of low doses of bradykinin and angiotensin II with nicotine on the intracellular calcium response. Arrow denotes time of addition.

DISCUSSION

Our results, when combined with those from other investigators, clearly indicate that specific and unique receptors exist on cultured bovine chromaffin cells for angiotensin II/III, bradykinin (B2 receptor) and likely, endothelin. In addition, we have demonstrated that heterologous desensitization proposed for angiotensin II and bradykinin (Jackson et al., 1988), is not evident for the bovine chromaffin cell.

Chromaffin cells from the adrenal medulla have been studied extensively as a neuroendocrine model for stimulus-secretion coupling. In addition, because of its focal point in the circulation and its role as a source of neurohumoral agents, such as epinephrine, the adrenal medulla has long been implicated in the acute and chronic response to stress. (Kvetnansky, et al., 1978). The adrenal medulla, derived embryonically from the neural crest (Weston, 1970), receives cholinergic innervation from the splanchnic nerve of the sympathetic nervous system. Acetylcholine released from the splanchnic nerve interacts with both nicotinic and muscarinic receptors on the chromaffin cell to elicit exocytosis of catecholamines, proteins and peptides. Nicotinic receptor stimulation leads to an increase in the sodium conductance and a depolarization of the cell (Burgoyne, 1984; Cena, et al., 1983). Douglas and coworkers (see review by Rink and Knight, 1988) first demonstrated the importance of an influx of extracellular calcium in the exocytotic process of the adrenal gland. It has been shown that a rise in intracellular calcium is essential for catecholamine secretion in the chromaffin cell (Fisher, et al., 1981).

It has been shown that angiotensin and bradykinin stimulate in chromaffin cells the release of catecholamines (Boyd, 1987; Zimlichman, et al., 1987; Noble et al., 1988) and the turnover of inositol phosphates (Plevin and Boarder, 1988). The data presented here complement and extend earlier studies on the influence of angiotensin and bradykinin on Ca_i^{2+}. A important distinction between angiotensin and bradykinin is that in the absence of extracellular calcium, the increase in $[Ca_i^{2+}]$ is only slightly attenuated for the angiotensin II response but greater than 50% for the bradykinin response. This demonstrates that mobilization of Ca_i^{2+} from intracellular stores occurs upon interaction of the peptides with their receptors. However, in addition, our results indicate that complex calcium kinetics occurs in the chromaffin cell following peptide receptor activation. The results presented suggest that with high (super-physiological) doses of either peptide, activation of calcium reversal mechanisms appears to occur. At low (near physiological) doses of the peptides, the Ca_i^{2+} response is greatly extended in time and may well potentiate normal cholinergic mechanisms. The significant potentiation evident when endothelin is combined with a cell depolarizing stimulus suggests this phenomena may be extended to other modulatory peptides.

Evidence from membrane binding studies performed in our laboratory suggests the involvement of a G-protein with the angiotensin receptor. However, treatment of the whole cells with pertussis or cholera toxin did not significantly alter the $[Ca_i^{2+}]$ signal elicited by either angiotensin II or bradykinin. Yet, down regulation of protein kinase C with PMA did significantly attenuate the $[Ca_i^{2+}]$ response to both angiotensin II and bradykinin. This raises questions as to the site of action of the phorbol. Further studies are currently being performed to elucidate the role of protein kinase C in these cells.

In conclusion, we have demonstrated that angiotensin, bradykinin and endothelin, all vasoactive and potentially stress peptides, modulate and influence adrenal medullary chromaffin cells. Their role in both normal adrenal medullary physiology and in cardiovascular pathophysiology involving the adrenal medulla may involve their ability to influence and modulate intracellular calcium kinetics.

ACKNOWLEDGEMENTS: We wish to acknowledge the assistance and/or advice of Margaret Dentlinger, Jacquelyn Nguyen and Dr. Victoria Boyd. Supported by National Institutes of Health grants HL 35018 and NS 22954.

REFERENCES

Boyd, V.L., 1987, Identification and characterization of an angiotensin II receptor on cultured bovine adrenal chromaffin cells, Ph.D. Dissertation, University of California, San Diego.

Burgoyne, R.D., 1984, Mechanisms of secretion from adrenal chromaffin cells, Biochim Biophys Acta, 779:201-216.

Burgoyne, R.D., Morgan, A., and O'Sullivan, A.J., 1988, A major role for protein kinase C in calcium-activated exocytosis in permeabilized adrenal chromaffin cells, FEB 238:151-155.

Cardenes, A.M., Montiel C., Esteban C., Borges R., and Garcia A.G., 1988, Secretion from adrenaline- and noradrenaline-storing adrenomedullary cells is regulated by a common dihydropyridine-sensitive calcium channel, Brain Research 456:364-366.

Cena, V., Nicolas, G.P., Sanchez-Garcia, P., Kirpekkar, S.M. and Garcia, A.G., 1983, Pharmacological dissection of receptor-associated and voltage-sensitive ionic channels involved in catecholamine release, Neuroscience 10:1455-1462.

Feldberg, W. and Lewis, G.P., 1964, The action of peptides on the adrenal medulla. Release of adrenaline by bradykinin and angiotensin, J. Physiol. 171:98-108.

Fisher, S.K., Holz R.W., and Agranoff, B.W., 1981, Muscarinic receptors in chromaffin cell cultures mediate enhanced phospholipid labelling but not catecholamine secretion, J. Neurochem. 37:491-497.

Fonteriz, R.I., Gandia, L., Lopez, M.G., Artalejo, C.R., and Garcia, A.G., 1987, Dihydropyridine chirality at the chromaffin cell calcium channel, Brain Research 408:359-362.

Geisow, M.J., and Burgoyne, R.D., 1987, An Integrated Approach to Secretion. Phosphorylation and Calcium Dependent Binding of Proteins Associated with Chromaffin Granules, Annals New York Academy of Sciences 493:563-576.

Go, M., Numura, H., Tatsuro, K., Koumoto, J., Kikkawa, U., Saito, N., Tanaka, C., and Nishizuka, Y., 1987, Inositol Phospholipid Turnover and Protein Phosphorylation in Secretory Responses, Annals of New York Academy of Sciences 493:552-562.

Grynkiewicz, G., Poenie, M., and Tsien, R.Y., 1985, A new generation of calcium indicators with greatly improved fluorescence properties, J. Biol Chem. 260:3440-3450.

Healy, D.P., Maciejewski, A.R., and Printz, M.P., 1985, Autoradiographic localization of 125-I angiotensin ii binding sites in the rat adrenal gland, Endocrinology 116:1221-23.

Jackson, T.R., Blair, L.A.C., Marshall, J., Goedert M., and Hanley, M.R., 1988, The mas oncogene encodes an angiotensin receptor, Nature 335:437-440.

Jackson, T.R., Hallam, T.J., Downes, C.P., and Hanley, M.R., 1987, Receptor coupled events in bradykinin action: rapid production of inositol phosphates and regulation of cytosolic free Ca^{2+} in a neural cell line, EMBO J. 6:49-54.

Kvetnansky, R., Weise, V.K., Thoa, N.B., and Kopin, I.J., 1978, Effects of chronic guanethidine treatment and adrenal medullectomy on plasma levels of catecholamines and corticosterone in forcibly immobilized rats, J. Pharm and Exper. Therap. 209:287-291.

Noble, E.P., Bommer, M., Liebisch, D., and Herz, A., 1988, H_1-histaminergic activation of catecholamine release by chromaffin cells, Biochemical Pharmacology 37:221-8.

Peach, M.J., Cline, W.H. and Watts, D.T., 1966, Release of adrenal catecholamines by angiotensin II, Circ. Res. 19:571-75.

Peach, M.J., 1971, Adrenal medullary stimulation induced by angiotensin I, angiotensin II and analogues, Circ. Res. 28 Supp 2:107-117.

Penner, R., and Neher, E., 1988, The role of calcium in stimulus secretion coupling in excitable and non-excitable cells, J. Exp. Biology 139:329-345.

Phillips, J.H. and Pryde, J.G., 1987, The chromaffin granule: a model system for the study of hormones and neurotransmitters, Annals New York Academy of Sciences 493:26-42.

Plevin, R., and Boarder, M.R., 1988, Stimulation of formation of inositol phosphates in primary cultures of bovine adrenal chromaffin cells by angiotensin II, histamine, bradykinin, and carbachol, J. Neurochem. 51:634-41.

Putney, J.W., 1988, The role of phosphoinositide metabolism in signal transduction in secretory cells, J Exp Biology 139::135-150.

Regoli, D., and Barabe, J., 1980, Pharmacology of bradykinin and related kinins, Pharm. Reviews 32:1-40.

Rink, T.J., and Knight, D.E., 1988, Stimulus-Secretion Coupling: A perspective highlighting the contributions of Peter Baker, J Exp Biology 139:1-30.

Strittmatter, W.J., 1988, Molecular Mechanisms of Exocytosis: The Adrenal Chromaffin cell as a model system, Cellular and Molecular Neurobiology 8:19-25.

Waschek, J.A., Pruss, R.M., Siegel, R.E., Eiden, L.E., Bader, M.F., and Aunis, D., 1987, Regulation of Enkephalin, VIP, and Chromogranin Biosynthesis in Actively Secreting Chromaffin Cells, Annals of New York Academy of Sciences. 493:308-323

Weston, J.A., 1970, The migration and differentiation of neural crest cells, Advances in Morphogenesis. 8:41-114.

Yokohama, H., Negishi, M., Sugama, K., Hayashi, H., Ito, S., and Hayaishi, O., 1988. Inhibition of prostaglandin E2-induced phosphoinositide metabolism by phorbol ester in bovine adrenal chromaffin cells. Biochem J 255:957-962.

Zimlichman, R., Goldstein, D.S., Zimlichman, S., Stull, R., and Keiser, H.R., 1987, Angiotensin II increases cytosolic calcium and stimulates catecholamine release in cultured bovine adrenomedullary cells, Cell Calcium 8:315-325.

INFLUENCE OF GDP(B)S ON AGONIST-INDUCED CALCIUM MOBILIZATION AND PLATELET FUNCTION

Gundu H. R. Rao, J.S. Cox, V.G. Mahadevappa and James G. White

Department of Laboratory Medicine and Pathology, UMHC Box 198

University of Minnesota, Minneapolis, MN 55455

INTRODUCTION

Recent studies suggest that agonist-induced receptor mediated signal transduction occurs via GTP-binding proteins (G-proteins). These special plasma membrane associated proteins are composed of three subunits, alpha, beta and gamma (Neer et al, 1988). The GTP bound, activated alpha subunit appears to be the primary effector of the system (Litosch, 1987). This subunit contains the binding site for GTP, exhibits GTPase activity and possesses the site for the action of bacterial toxins. Agonist-induced platelet activation seems to be mediated by at least two pertussis toxin sensitive G-proteins (Brass et al, 1988).

Much of what we know of platelet G-proteins comes from data obtained by the use of agents that inhibit or stimulate G-protein mediated events in permeabilized cells (Haslam and Davidson, 1984; Brass et al, 1986). However, recent investigations have raised questions concerning the specificity of GDP(B)S as a G-protein antagonist (Krishnamurthi et al, 1988; Authi et al, 1988; Rao et al, 1988). Results of our studies suggest that GDP(B)S does influence calcium flux, but the effect is nonspecific and not limited to "G" proteins on the inside of the platelet surface membrane.

MATERIALS AND METHODS

Arachidonic acid as sodium salt was obtained from NuChek Prep, Elysian, MN. Guanosine 5'-[B-thio] diphosphate (GDP(B)S), D-myo-inositol-1, 4,5-triphosphate (IP_3), and D-myo-inositol-1,4,5,6-tetrakisphosphate (IP_4), were purchased from Boehringer Mannheim Biochemicals, Indianapolis, IN. Platelet activating factor (PAF) was from Calbiochem, San Diego CA. Fluorescent probes, Quin-2 AM and Fura-2 AM were purchased from Molecular Probes, Eugene, OR. Diltiazem was a gift from Marion Laboratories, Kansas City, MO. Injectable adrenalin and bovine thrombin were from Parke-Davis Company, Detroit, MI. Collagen was purchased from Chronolog Corporation, Havertown, PA. U46619 was a gift from the Upjohn Company, Kalamazoo, MI. Unless otherwise mentioned, all other chemicals were from Sigma Chemical Company, St. Louis, MO.

Blood for these studies was obtained from healthy adult volunteers who had not taken any medication for at least two weeks prior to venesection. Platelets were separated and aggregation studies carried out as described from our laboratory (Rao and White, 1985). Platelet samples were also fixed for studies by the electron microscope (White, 1974).

For measurement of cytosolic free calcium, washed platelets suspended in appropriate buffers were used (Rink et al, 1982; Grynkiewicz et al, 1985). Ionized calcium was monitored using either Fura 2 AM or Quin 2 AM as calcium specific fluorophores (Rao et al, 1985; Rao et al, 1986; Rao, 1988; Hallam and Rink, 1985).

Inhibitors were added to the platelet suspension 10 minutes prior to the challenge by threshold concentrations of agonists. Permeabilized platelets were obtained using saponin as the detergent (Authi et al, 1988). All experiments were repeated a minimum of three times with blood obtained from different donors. Tracings presented as figures for aggregation experiments are typical representations of actual studies.

RESULTS

Influence of GDP(B)S on agonist-induced aggregation of human platelets

Platelets obtained from normal donors aggregated irreversibly when stirred with agonists such as epinephrine (5 uM) adenosine diphosphate (ADP, 3 uM), thrombin (0.2 u/ml) and platelet activating factor (2 uM). GDP(B)S addition to PRP exerted a dose-dependent inhibitory effect on thrombin-induced platelet aggregation. It was also effective in blocking PAF (0.5 uM)-induced aggregation of platelets (Fig. 1).

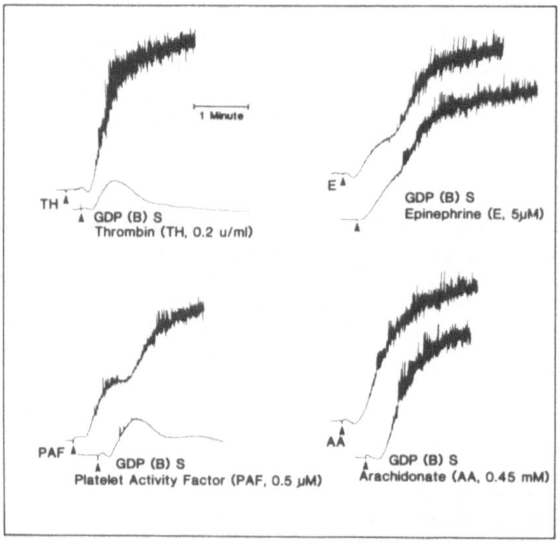

FIGURE 1. Normal control platelets aggregated irreversibly to the action of epinephrine, arachidonate, thrombin and platelet activating factor (PAF). GDP(B)S effectively blocked thrombin and PAF-induced platelet aggregation but had no such inhibitory effect on the action of epinephrine and arachidonate. However, in a concentration at which it exerted the maximum inhibitory effect (1.5 mM), epinephrine and arachidonate could still cause irreversible aggregation.

Influence of GDP(B)S on thrombin-induced cytosolic calcium mobilization in human platelets

Thrombin (0.2 u/ml) induced a significant elevation of cytosolic calcium in Fura 2 loaded intact platelets. GDP(B)S exerted a dose

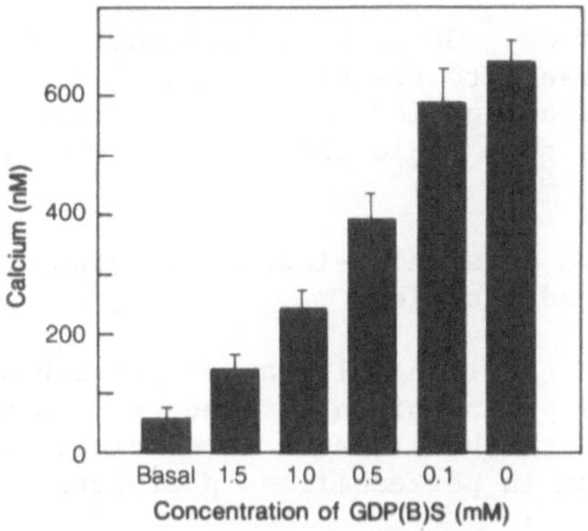

FIGURE 2. Thrombin-induced significant elevation of cytosolic free calcium in Fura-2 loaded intact human platelets. GDP(B)S exerted a dose dependent inhibitory effect on thrombin-stimulated calcium mobilization.

TABLE 1. Influence of adenine and guanine nucleotides on thrombin-induced cytosolic calcium mobilization in Fura-2 loaded human platelets

	Calcium (nM)*
Basal calcium	117 ± 16
Thrombin (0.2 u/ml)	664 ± 26
GDP(B)S; 1.5 mM + thrombin	119 ± 12
GDP; 1.5 mM + thrombin	615 ± 24
GDP; 1.5 mM + thrombin	846 ± 28
ATP; 1.5 mM + thrombin	188 ± 14

*Mean and the standard error (n = 3)

Thrombin-induced significant elevation of cytosolic free calcium in Fura-2 loaded intact human platelets. GDP(B)S and adenosine triphosphate (ATP) effectively blocked thrombin-stimulated calcium mobilization in human platelets.

dependent inhibition of thrombin-induced calcium mobilization (Fig. 2). It was more effective in inhibiting thrombin-induced calcium mobilization in the presence of extracellular calcium (1 mM) than in the absence of calcium in the suspending medium. GDP and GTP (1.5 mM) did not exert an inhibitory effect on thrombin-induced calcium mobilization. ATP, however, exerted a significant inhibitory effect (Table 1).

Influence of GDP(B)S on thrombin-induced calcium mobilization from the external medium

In view of the fact that the inhibitory effect of GDP(B)S was more pronounced in the presence of calcium, the possibility of thrombin-induced calcium entry from the medium was explored. To follow the ability of thrombin to induce calcium uptake into platelets during activation, we used Quin 2 AM loaded cells and monitored fluorescence quenching by the uptake of manganese ions entering from the medium (Hallam and Rink, 1985). Addition of manganese immediately after thrombin induced a rapid drop in the

Quin 2/calcium fluorescence. GDP(B)S significantly prevented thrombin-induced entry of manganese into platelets. Similar studies with lanthanum (1 mM), a potent calcium antagonist, also demonstrated that the thrombin-induced activation leads to a rapid entry of cations from the outside of the platelets.

Influence of GDP(B)S on agonist-induced calcium mobilization in saponin permeabilized platelets

Addition of saponin (10-15 ug/ml) caused permeabilization of washed human platelets. Over a period of three minutes, saponin by itself caused a three-fold increase in the basal level of calcium. IP_3 induced significant elevation of calcium in permeabilized platelets (Fig. 3).

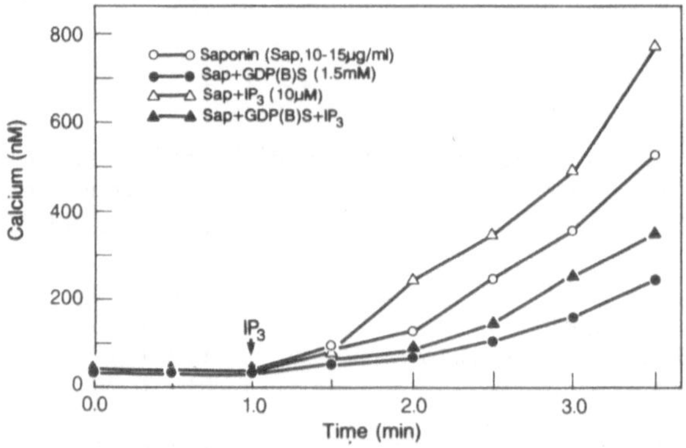

FIGURE 3. Addition of saponin permeabilized platelets for the entry of cations and low molecular weight compounds. Inositol 1,4,5-triphosphate (IP_3) caused significant elevation of cytosolic calcium in permeabilized Fura-2 loaded platelets. GDP(B)S effectively blocked IP_3-induced calcium mobilization as well as entry of calcium due to saponin action.

Other agonists such as PAF, U46619 and thrombin also significantly increased cytosolic calcium in Fura 2 loaded permeabilized platelets. GDP(B)S at a 1 mM concentration effectively blocked calcium mobilization by all agonists and also by saponin (Table 2).

TABLE 2. Influence of GDP(B)S on agonist-induced calcium mobilization in saponin permeabilized platelets.

	Calcium (nM)*				
	Basal	Saponin (10-15 ug/ml)	Thrombin (0.2 u/ ml)	U46619 (2 ug/ ml)	PAF (2 um)
Untreated	36 ± 2.4	137 ± 6.8	304 ± 12.6	236 ± 8.6	174 ± 8.6
GDP(B)S (1 mM)	32 ± 2.6	33 ± 2.8	94 ± 4.6	70 ± 4.2	55 ± 3.8

*Mean and the standard error (n = 3)

Addition of saponin to washed Fura-2 loaded human platelets caused permeabilization of plasma membrane to allow cations and low molecular weight compounds. All the agonists tested caused significant elevation of cytosolic calcium in saponin permeabilized human platelets. GDP(B)S effectively blocked agonist-induced calcium mobilization as well as saponin mediated calcium entry.

Influence of calcium antagonists on agonist-induced platelet aggregation

Since GDP(B)S effectively prevented thrombin-induced calcium mobilization from the external source, known calcium antagonists such as lanthanum, verapamil and diltiazen were tested to see if calcium entry was essential for agonist-induced platelet function. Verapamil, a putative calcium channel blocker, effectively blocked epinephrine and PAF-induced platelet aggregation. At the concentration tested (100 uM), it did not block the effect of thrombin or arachidonate (Fig. 4).

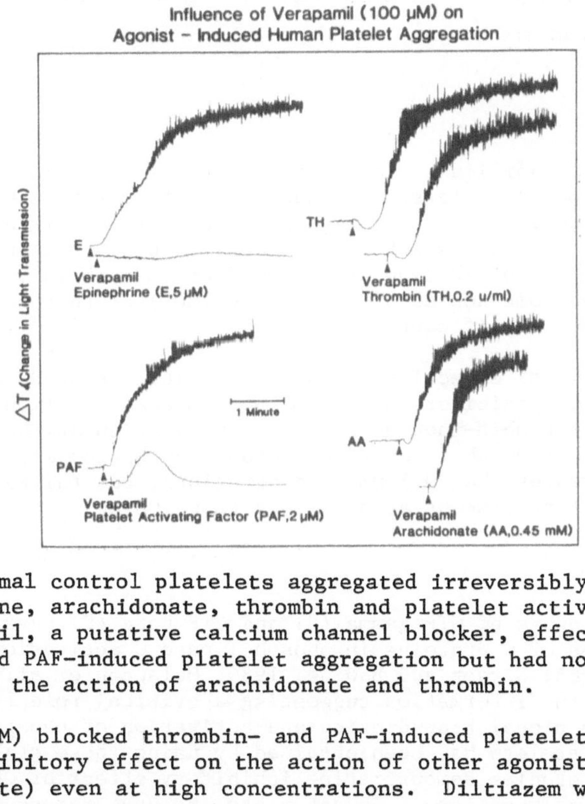

FIGURE 4. Normal control platelets aggregated irreversibly when stirred with epinephrine, arachidonate, thrombin and platelet activating factor (PAF). Verapamil, a putative calcium channel blocker, effectively blocked epinephrine and PAF-induced platelet aggregation but had no such inhibitory effect on the action of arachidonate and thrombin.

Lanthanum (1 mM) blocked thrombin- and PAF-induced platelet aggregation, but had no inhibitory effect on the action of other agonists (epinephrine and arachidonate) even at high concentrations. Diltiazem was a poor inhibitor of agonist-induced platelet aggregation (100 uM). However, at greater than 200 uM concentration, it effectively blocked the action of all agonists tested.

Influence of calcium antagonists on thrombin-induced calcium mobilization in human platelets

Thrombin (0.2 u/ml) caused significant elevation of cytosolic free calcium in Fura-2 loaded platelets (Table 3). Thrombin induced significant elevation of cytosolic calcium in Fura-2 loaded intact human platelets. Lanthanum effectively blocked thrombin-stimulated calcium elevation in human platelets at the highest concentration tested (1 mM). It was not very effective in preventing thrombin-stimulated calcium mobilization at 0.5 mM (data not shown). Verapamil and diltiazem exerted some degree of inhibition (20%) at this concentration but the effect was not significant.

TABLE 3. Influence of calcium antagonists on thrombin-induced (0.2 u/ml) calcium mobilization in human platelets

Time (mins.)	Untreated	Lanthanum (1 mM)	Verapamil (0.5 mM)	Diltiazem (0.5 mM)
0.0	43 ± 2.4	52 ± 2.6	63 ± 3.0	70 ± 3.2
0.5	312 ± 12.4	64 ± 3.2	314 ± 12.4	334 ± 10.4
1.0	297 ± 10.8	65 ± 2.8	252 ± 12.0	283 ± 8.8
2.0	235 ± 10.4	64 ± 3.0	268 ± 11.8	282 ± 8.6

* Mean and the standard error (n = 3)

Thrombin-induced significant elevation of cytosolic calcium in Fura-2 loaded intact human platelets. Lanthanum effectively blocked thrombin-stimulated calcium elevation in human platelets at the highest concentration tested (1 mM). It was not very effective in preventing thrombin stimulated calcium mobilization at 0.5 mM (data not shown). Verapamil and diltiazem exerted some degree of inhibition (20%) at this concentration but the effect was not significant.

Lanthanum (1 mM) effectively blocked thrombin-induced calcium mobilization. It had no inhibitory effect at 500 uM concentrations. Verapamil did not inhibit thrombin-induced calcium mobilization at the highest concentration tested (500 uM). Diltiazem caused an approximately 20 percent inhibition at greater than 0.5 mM concentrations, but failed to exert an inhibitory effect at concentrations lower than 200 uM.

DISCUSSION

Stable analogues of GTP gamma (S) and GTP beta (S) have been used for probing the roles of G-proteins in phospholipase C activation (Cockfort and Gomperts, 1985; Stryer and Bowine, 1986; Melancon et al, 1987; Gilman, 1987). Much of the information suggesting a critical role for G-proteins in transmembrane signal transduction and activation of phospholipase in permeabilized platelets has been obtained by using these stable analogues of GTP. Recent studies demonstrating inhibitory effect of GDP(B)s on agonist-induced aggregation of intact platelets have raised questions regarding the usefulness of this compound as a specific G-protein antagonist (Krishnamurthi et al, 1988; Authi et al, 1988).

GDP(B)S added to platelets suspended in plasma prevented the action of thrombin in a dose-dependent manner. The effect could be reversed by increasing the concentration of thrombin (Authi et al, 1988). Earlier studies have demonstrated an inhibitory effect of GDP(B)S on aggregation and secretion induced by thrombin, collagen, 1,2-dioctanoyl-SN-glycerol (DIC-8) and U46619 in washed platelet suspensions (Krishnamurthi et al, 1988). In this study, GDP(B)S inhibited the action of thrombin and platelet activating factor (PAF) but had no effect on the action of epinephrine or arachidonate.

Since secretion of granule contents is presumed to be dependent upon cytosolic calcium elevation, agonist-induced calcium mobilization was followed in Fura-2 loaded platelets. Thrombin induced a significant elevation of cytosolic calcium. GDP(B)S blocked thrombin-induced calcium mobilization in intact platelets. GDP and GTP had no effect on this process, whereas ATP was a potent inhibitor. An inhibitory effect of ATP on platelet activation by agonists has been reported in earlier studies (Mills, 1973; MacFarlane and Mills, 1975; Krishnamurthi et al, 1988). GDP(B)S was more effective in preventing thrombin-induced calcium flux in the presence of 1 mM calcium in the medium than its absence (Authi et al, 1988).

In view of the observation that GDP(B)S was more effective in preventing calcium mobilization in the presence of external calcium, the possibility that GDP(B)S might have a direct influence on thrombin-induced calcium entry was explored using manganese as the chase ion (Hallam and Rink, 1985). Addition of manganese (1 mM) to thrombin exposed Quin-2 loaded platelets rapidly lowered the Quin-2/calcium fluorescence, suggesting the entry of Mn^{2+} from the medium and competition with calcium from Quin-2/calcium complex. In this experimental condition, GDP(B)S significantly blocked thrombin-induced entry of Mn^{2+}. Results of these studies and of others suggest that one of the primary sites of action for GDP(B)S may be to prevent development of agonist-induced cation channels (Authi, et al, 1988).

To see if stimulation of cation channels was essential for agonist-induced activation of platelets, several calcium antagonists were tested. Lanthanum, verapamil, nifedipine and diltiazem have been used to inhibit the voltage-activated, inward displacement of calcium ions (Nayler, 1983). In this study, lanthanum, verapamil and diltiazem exerted inhibitory effects on thrombin-induced calcium mobilization at the highest concentration tested (1 mM). At this concentration, verapamil and diltiazem induced significant damage to the platelets. Lanthanum, although effective against thrombin and PAF, did not inhibit platelet aggregation induced by epinephrine and arachidonate. If thrombin and PAF require the stimulation of cation channels, then the lack of inhibitory effect by GDP(B)S and lanthanum on the action of other agonists suggests that they do not require this mechanism for inducing platelet activation.

Verapamil significantly inhibited the action of epinephrine and PAF at 100 uM concentration, but had no inhibitory effect on other agonists tested. The specific inhibitory action of verapamil may be related to its affinity for the binding sites of these agonists (Nayler et al, 1982). Diltiazem was an effective inhibitor of aggregation and calcium mobilization at greater than 0.5 mM, but was poor/reversible inhibitor at lower concentrations. This drug is known to inhibit influx of calcium from extracellular fluid as well as prevent release of calcium from intracellular stores (Britt, 1985). In this study, diltiazem was not very effective in preventing calcium mobilization at pharmacological doses. Indeed, none of the agents tested in this study exerted any inhibitory effect on thrombin-induced calcium mobilization at pharmacological doses, although at megadoses ($>$ 1 mM) all of them were potent inhibitors and the mechanism by which they exert their inhibitory effect at such high concentrations is not known.

Calcium-sensitive voltage activated channels have been reported in smooth muscle cells. In these cells, part of the calcium needed for contraction is derived from the extracellular fluid (Nayler et al, 1983). Calcium antagonists like verapamil, nifedipine and diltiazem have been shown to prevent calcium-mediated function in other cell systems, but little is known about their ability to prevent agonist-induced calcium mobilization in platelets (Weiss, 1985). Since none of the compounds exerted any inhibition on thrombin-induced calcium mobilization at pharmacological doses, their influence on platelet function at higher concentrations may be nonspecific, similar to the action of GDP(B)S and lanthanum.

In conclusion, results of these studies have demonstrated that GDP(B)S at high concentrations prevents thrombin and PAF-induced aggregation and calcium mobilization. It effectively blocks the entry of calcium from the extracellular medium. However, at the inhibitory concentration, it does not prevent epinephrine and arachidonate-induced platelet aggregation, suggesting that these agonists may not depend on the same mechanism to initiate activation. Furthermore, lanthanum, a known calcium antagonist, mimicked GDP(B)S in its action, prevented thrombin and PAF-induced calcium mobilization and aggregation without affecting the action of epinephrine and arachidonate. Other calcium antagonists tested (verapamil and diltiazem) were poor inhibitors of calcium mobilization at pharamcological concentrations (100-200 uM). GDP(B)S does influence calcium flux, but the effect is nonspecific and not limited to 'G' proteins on the interior of the platelet surface membrane. Receptor-mediated transmembrane signal transduction may be an essential mechanism for cell activation, but the conclusions drawn through use of antagonists like GDP(B)S need further evaluation.

Acknowledgments

The authors thank Ms. Julie Lee for her technical help and Susan Schwarze for her help with the manuscript. Supported by USPHS HL-11880, CA21737, GM-22167, HL30217, a grant from the March of Dimes, 1-886, and the Minnesota Medical Foundation (HDRF 38-85, HRF47-87, HRF57-88).

REFERENCES

Authi, K.S., Rao, G.H.R., Evenden, B.J., Crawford, N., 1988, Action of guanosine 5'-[beta-thio] diphosphate on thrombin-induced activation and Ca^{2+} mobilization in saponin-permeabilized and intact human platelets, Biochem. J. 255: 885-893.

Brass, L.F., Woolkalis, M.J., Manning, D.R., 1988, Interactions in platelets between G proteins and the agonists that stimulate phospholipase C and inhibit adenylyl cyclase, J. Biol. Chem. 263: 5348-5355.

Britt, B.A., 1985, Diltiazem, Can. Anesth. Soc. J. 32: 30-44.

Church, J., Zoster, J.T., 1980, Calcium antagonistic drugs. Mechanism of action, Can. J. Physiol. Pharmacol. 58: 254-264.

Cockford, S., Gomperts, B.D., 1985, Role of guanine nucleotide binding protein in the activation of polyphosphoisitide phosphodieterase, Nature 314: 534-536.

Gilman, A.G., 1987, G-proteins: transducers of receptor-generated signals. Ann. Rev. Biochem. 56: 615-649.

Grynkiewicz, G., Poenie, M., Tsien, R.Y., 1985, A new generation of Ca^{2+} indicators with greatly improved fluorescence properties, J. Biol. Chem. 260: 3440-3450.

Hallam, T.J., Rink, T.J., Agonists stimulate divalent cation channels in the plasma membrane of human platelets, FEBS Lett. 186: 175-179.

Haslam, R.J., Davidson, M.M.L., 1984b, Receptor induced diacyglycerol formation in permeabilized platelets; possible role for a GTP-binding protein, J. Recept. Res. 9: 605-629.

Henry, P.D., 1980, Comparative pharmacology of calcium antagonists: nifedipine, verapamil and diltiazem, Am. J. Cardiol. 46: 1047-1059.

Hrbolich, J.K., Culty, M., Haslam, R.J., 1987, Activation of phospholipase C associated with isolated rabbit platelet membranes by guanosine 5'-(gamma-thio) triphosphate and by thrombin in the presence of GTP, Biochem. J. 243: 457-465.

Krishnamurthi, S., Patel, Y., Kakkar, V.V., 1988, Inhibition of agonist-induced platelet aggregation, Ca^{2+} mobilization and granule secretion by guanosine 5'-(beta-thio) diphosphate and GDP in intact platelets, Biochem. J. 250: 209-214.

Litosch, I., 1987, Regulatory GTP-binding proteins: Emerging concepts on their role in cell function, Life Sci. 41: 251-258.

MacFarlane, D.E., Mills, D.C.B., 1975, The effects of ATP on platelets: Evidence against the role of released ADP in primary aggregation, Blood 46: 309-320.

Melancon, P., Glick, B.S., Malhotra, V., Weidman, P.F., Serafini, T., Gleason, M.L., Orci, L., Rothman, J.E., 1987, Involvement of GTP-binding "G" proteins in transport through the Golgi stack, Cell 51: 1053-1062.

Mills, D.C.B., 1973, Changes in the adenylate energy charge in human blood platelets induced by adenosine diphosphate, Nature 243: 220-222.

Nayler, W.G., 1983, Calcium antagonists, Med. J. Aust. 2: 506-513.

Neer, E.J., Clapham, D.E., 1988, Roles of G protein subunits in transmembrane signalling, Nature 333: 129-134.

Rao, G.H.R., White, J.G., 1985, Disaggregation and reaggregation of "irreversibly" aggregated platelets: A method for more complete evaluation of antiplatelet drugs. Agents and Actions 16: 425-434.

Rao, G.H.R., Peller, J.D., White, J.G., 1985, Measurement of ionized calcium in blood platelets with a new generation of calcium indicator, Biochem. Biophys. Res. Comm. 132: 652-657.

Rao, G.H.R., Peller, J.D., Semba, C.P., White, J.G., 1986, Influence of the calcium sensitive fluorophore Quin 2 on platelet function, Blood 67: 354-361.

Rao, G.H.R., 1988, Measurement of ionized calcium in normal human blood platelets, Anal. Biochem. 169: 400-404.

Rao, G.H.R., Authi, K.S., Crawford, N., White, J.G., 1988, Variable mechanisms of Ca^{2+} mobilization in human platelets. FASEB J. 2(4): 598A.

Rink, T.J., Smith, S.W., Tsien, R.Y., 1982, Cytoplasmic free Ca^{2+} in human platelets: Ca^{2+}-independent activation of shape change and secretion, FEBS Lett. 148: 21-26.

Stryer, L., Bowine, H.R., 1986, G proteins: A family of signal transducers, Ann. Rev. Cell Biol. 2: 391-419.

White, J.G., 1974, Electron microscopic studies of platelet secretion, Prog. Haemostas. Thromb. 2: 49-98.

REGULATION OF THE PRODUCTION OF LIPOXYGENASF PRODUCTS AND THE ROLE OF EICOSANOIDS IN SIGNAL TRANSDUCTION

Anthony Ford-Hutchinson

Department of Pharmacology
Merck Frosst Centre for Therapeutic Research
P.O. Box 1005
Pointe Claire-Dorval, Québec, Canada, H9R 4P8

INTRODUCTION

The leukotrienes are products of the metabolism of arachidonic acid through the 5-lipoxygenase enzyme pathway (Samuelsson, 1983; Ford-Hutchinson, 1985). This enzyme catalyses, first, the insertion of oxygen at carbon 5 to produce the intermediate, 5-hydroperoxy-6,8,11,14 eicosatetraenoic acid (5-HPETE) and, secondly, a dehydrase step which converts 5-HPETE to the unstable epoxide intermediate, 5,6-oxido-7,9,11,14-eicosatetraenoic acid (leukotriene A_4). Leukotriene A_4 may then be converted either to 5S, 12R dihydroxy-6,8,10,14(Z,E,E,Z) eicosatetraenoic acid (leukotriene B_4) by a specific enzyme, leukotriene A_4 hydrolase (Radmark et al., 1984) or to 5-S-hydroxy-6R,S-glutathionyl-7,9,11,14-(E,E,Z,Z)-eicosatetraenoic acid (leukotriene C_4) by another specific enzyme, leukotriene C_4 synthase (Bach et al., 1984). Subsequent metabolism of leukotriene C_4 by membrane-bound γ-glutamyl transferase results in the formation of leukotriene D_4 through cleavage of L-glutamic acid (Orning and Hammarstrom, 1980). Leukotriene D_4 can then be converted to leukotriene E_4 with loss of L-glycine by specific membrane-bound dipeptidases (Anderson et al., 1982).

LEUKOTRIENE RECEPTORS

Leukotriene B_4 has potent effects on leukocytes and for example stimulates the chemotaxis, chemokinesis and aggregation of polymorphonuclear leukocytes (Ford-Hutchinson et al., 1980; Ford-Hutchinson, 1985). These effects are mediated through high affinity, structurally specific receptor sites (Goldman and Goetzl, 1982). LTB_4 receptors have been shown to be coupled to pertussis toxin sensitive G proteins with resulting activation of PI turnover, elevation of intracellular calcium, activation of cellular contractile mechanisms and activation of arachidonic acid metabolism (Goldman et al., 1985; Winkler et al., 1988). Leukotriene D_4 also has high affinity, structurally specific G protein-coupled receptors which are associated with a variety of biological activities including smooth contraction (Pong and DeHaven, 1983). There is no evidence for specific leukotriene E_4 receptor and this autocoid appears to be a partial agonist for the leukotriene D_4 receptors. Leukotriene C_4

binding sites have been described in a variety of tissues but is not clear whether these are actual functional receptors or whether they simply reflect binding of leukotriene C_4 to proteins such as glutathione-S-transferase (Sun et al., 1986). A complicated series of secondary transduction processes has been proposed for the leukotriene D_4 receptor which can interact with more than one G protein, only one of which is sensitive to pertussis toxin. Ca^{2+} mobilization is proposed to occur through at least two processes including G protein-activated phosphoinositide-specific phospholipase C and a receptor operated Ca^{2+} channel. A number of second messengers arise from these interactions including products of the PI cycle and diacylglycerol which in turn results in activation of protein kinase C and topoisomerase I. This can lead to the stimulation of the release and further metabolism of arachidonic acid through both the cyclooxygenase and 5-lipoxygenase pathways (Crooke et al., 1989).

5-Lipoxygenase

As described above the enzyme 5-lipoxygenase catalyses a two step process to produce 5-HPETE and then leukotriene A_4. Considerable progress has been in the purification of rat, porcine and human 5-lipoxygenase (Rouzer et al., 1986; Ueda et al., 1986; Hogaboom et al., 1986) and the human enzyme has been cloned (Dixon et al., 1988) and expressed in a human osteosarcoma cell line (Rouzer et al., 1988), the expressed enzyme having all the characteristics of the original purified enzyme from human leukocytes. These studies show that human 5-lipoxygenase is a soluble protein of 78,000 daltons molecular weight with a moderately hydrophobic central portion although the protein is overall hydrophilic (average hydrophobicity, -0.28). The sequence contains neither membrane-spanning domains nor signal sequences consistent with lack of membrane association during the unactivated state. However, there is strong homology with the interfacal binding domains of human lipoprotein lipase and rat hepatic lipase (Dixon et al., 1988) indicating the possibility of membrane association. The sequence contains no strong homology with any ATP binding sites but two regions exhibit weak homology to the consensus Ca^{2+} binding sites from lipocortin and calmodulin respectively. There are significant homologies between human 5-lipoxygenase and other lipoxygenases as shown in Table 1.

Table 1. Homologies between human 5-lipoxygenase and other lipoxygenase enzymes (from R. Dixon)

	% homology	Molecular Weight	Number of Residues
Human 5-lipoxygenase		78 kD	673
Rat 5-lipoxygenase	93	78 kD	670
Human 15-lipoxygenase	39 (61)	75 kD	663
Soybean LOX-1	22 (42)	94 kD	838

Activation of 5-Lipoxygenase

The activity of 5-lipoxygenase is highly dependent on Ca^{2+} and is stimulated by both ATP and phosphatidylcholine micelles. In addition the human enzyme has been shown to be stimulated by a number of undefined protein fractions and a membrane preparation (Rouzer and Samuelsson, 1985; Rouzer et al., 1985). These types of results suggest a regulatory process for activation of 5-lipoxygenase which is consistent with the fact that leukotriene synthesis only occurs in cells following exposure to defined stimuli which all result in a considerable elevation in intracellular Ca^{2+}. A number of observations have suggested a potential role for membranes in 5-lipoxygenase activation. First, the substrate arachidonic acid is derived from membrane phospholipids through the action of a specific phospholipase A_2. Secondly, the enzyme is stimulated by either membrane or phosphatidylcholine vesicles and, thirdly, the sequence homologies with the interfacial binding domains of lipases as described above are suggestive of a membrane interaction. Other calcium dependent enzymes such as phospholipase C and protein kinase C undergo a translocation from the cytosol to the membrane following cellular activation (Melloni et al., 1985; Creutz et al., 1985). Recently it has been shown that following activation of human polymorphonuclear leukocytes with ionophore A23187, 5-lipoxygenase enzyme activity and 5-lipoxygenase protein is lost from the cytosol and inactive 5-lipoxygenase protein appears in the 100,000 xg membrane preparation (Rouzer and Kargman, 1988). These results are consistent with the hypothesis that enzyme activation involves translocation to the membrane where the enzyme becomes activated, synthesizes leukotrienes and undergoes suicide inactivation. Consistent with this was the fact that cytosolic 5-lipoxygenase had retained its activity suggesting that this enzyme pool was not used for leukotriene synthesis. Also consistent with the hypothesis was the fact that varying the ionophore concentration resulted in varying amounts of leukotriene synthesis which could be correlated with varying amounts of enzyme translocation (Rouzer and Kargman, 1989). Also terminating leukotriene biosynthesis during ongoing cell activation by addition of excess EDTA resulted in termination of enzyme translocation (Rouzer and Kargman, 1989).

Inhibition of 5-Lipoxygenase Translocation by MK-886

Further support for the hypothesis that 5-lipoxygenase translocation is an essential step in leukotriene biosynthesis has come from studies on the leukotriene biosynthesis inhibitor, MK-886, which has been shown to inhibit the 5-lipoxygenase translocation process (Rouzer and Kargman, 1989). MK-886 (L-663,536; (3-[1-(4-chlorobenzyl)-3-t-butyl-thio-5-isopropylindol-2-yl]-2,2-dimethyl propanoic acid) has been shown to be a potent inhibitor of leukotriene biosynthesis in a variety of intact leukocyte preparations (Gillard et al., 1989). The compound is highly selective for the 5-lipoxygenase pathway having no effect on cyclooxygenase or 12- or 15-lipoxygenase. The potent <u>in vitro</u> effects on leukotriene biosynthesis are also observed <u>in vivo</u> both biochemically, through inhibition of leukotriene levels in inflammatory exudates and rat bile, and functionally, through inhibition of antigen-induced bronchoconstriction in the rat and the squirrel monkey (Gillard et al., 1989). In contrast to the potent inhibitory activity on leukotriene synthesis in intact cells, MK-886 shows no inhibitory activity on either broken cell preparations or purified 5-lipoxygenase preparations.

When studied on 5-lipoxygenase translocation in intact human polymorphonuclear leukocytes, MK-886 caused a concentration-dependent inhibition of the translocation of 5-lipoxygenase to the membrane which was correlated with the inhibition of leukotriene biosynthesis (Rouzer and Kargman, 1989). Similarly using a range of structural analogues of MK-886 it was possible to correlate their potency as 5-lipoxygenase translocation inhibitors with their ability to inhibit leukotriene biosynthesis in intact cells. One alternative explanation is that translocation has nothing to do with cell activation but simply reflects deposition of inactive enzyme after leukotriene biosynthesis.

Evidence against this are experiments in which addition of MK-886 after activation of the cells with ionophore A23187, and thus after enzyme translocation had occured, resulted in the release back into the cytosol of inactivated, translocated enzyme (Rouzer and Kargman, 1989).

In conclusion the results reviewed in the present article are consistant with the hypothesis that activation of leukocytes results in 5-lipoxygenase translocation from the cytosol to the membrane fraction and induction of leukotriene biosythesis. This process is inhibited by MK-886 which appears to represent a new class of therapeutic agent which may function as a specific translocation inhibitors. Further experiments are needed to define the exact nature of the interaction of both MK-886 and 5-lipoxygenase with cell membranes.

References

Anderson, M.E., Allison, R.D., and Meister, A., 1982, Interconversion of leukotrienes catalyzed by purified γ-glutamyl transpeptidase: concomitant formation of leukotriene D_4 and γ-glutamyl amino acids, Proc. Natl. Acad. Sci. U.S.A., 79:1088.

Bach, M.K., Brashler, J.R., and Morton, Jr., D.R., 1984, Solubilization and characterization of the leukotriene C_4 synthetase of rat basophil leukemia cells: a novel, particulate glutathione S-transferase, Arch. Biochem. Biophys., 230:455.

Creutz, C.E., Dowling, L.G., Kyger, E.M., and Franson, R.C., 1985, Phosphotidylinositol-specific phospholipase C activity of Chromaffin granule-binding proteins, J. Biol. Chem., 260:7171.

Crooke, S.T., Mattem, M., Sarau, H.M., Winkler, J.P., Balcarek, J., Wong, A. and Bennett, R.F., 1989, The signal transduction system of the leukotriene D_4 receptor, Trends Pharmacol. Sci., 10:103.

Dixon, R.A.F., Jones, R.E., Diehl, R.E., Bennett, C.D., Kargman, S., and Rouzer, C.A., 1988, Cloning of the cDNA for human 5-lipoxygenase, Proc. Nat. Acad. Sci. U.S.A., 85, 416.

Ford-Hutchinson, a.W., 1985, Leukotrienes: Their formation and role as inflammatory mediators, Fed. Proc., 44:25.

Ford-Hutchinson, A.W., Bray, M.A., Doig, M.V., Shipley, M.E., and Smith, M.J.H., 1980, Leukotriene B: a potent chemokinetic and aggregating substance released from polymorphonuclear leukocytes, Nature, 286:264.

Gillard, J., Ford-Hutchinson, A.W., Chan, C., Charleson, S., Denis, D., Foster, A., Fortin, R., Leger, S., McFarlane, C.S., Morton, H., Piechuta, H., Riendeau, D., Rouzer, C.A., Rokach, J., Young, R., MacIntyre, D.E., Peterson, L., Bach, T., Eiermann, G., Hopple, S., Humes, J., Hupe, L., Luell, S., Metzger, J., Meurer, R., Miller, D.K., Opas, E., and Pacholok, S., 1989, L-663,536 (MK-886) (3-[1-(4-chlorobenzyl)-3-t-butyl-thio-5-isopropylindol-2-yl]-2,2-dimethylpropanoic acid) a novel, orally active leukotriene biosynthesis inhibitor, Can. J. Physiol. Pharmacol. In press.

Goldman, D.W., Chang, F.H., Gifford, L.A., Goetzl, E.J., and Bourne, H.R., 1985, Pertussis toxin inhibition of chemotactic factor-induced calcium mobilization and function in human polymorphonuclear leukocytes, J. Exp. Med. 162:145.

Goldman, D.W., and Goetzl, E.J., 1982, Specific binding of leukotriene B_4 to receptors on human polymorphonuclear leukocytes, J. Immunol., 129:1600.

Hogaboom, G.K., Cook, M., Newton, J.F., Varrichio, A., Shorr, R.G.L., Sarau, H.M. and Crooke, S.T., 1986, Purification, characterization and structural properties of a single protein from rat basophilic leukemia (RBL-1) cells possessing 5-lipoxygenase and leukotriene A_4 synthase activities, Mol. Pharmacol., 30:510.

Melloni, E., Pontremoli, S., Michetti, M., Sacco, O., Sparatore, B., Salamino, F., and Horecker, B.L., 1985, Binding of protein kinase C to neutrophil membranes in the presence of Ca^{2+} and its activation by a Ca^{2+}-requiring proteinase, Proc. Natl. Acad. Sci. U.S.A. 82:6435.

Orning, L., and Hammarstrom, S., 1980, Inhibition of leukotriene C and leukotriene D biosynthesis, J. Biol. Chem., 255:8023.

Pong, S.S., and DeHaven, R.N., 1983, Characterization of a leukotriene D_4 receptor in guinea pig lung, Proc. Natl. Acad. Sci. U.S.A. 80:7415

Radmark, O., Shimizu, T., Jornvall, H., and Samuelsson, B., 1984, Leukotriene A_4 hydrolase in human leukocytes. Purification and properties, J. Biol. Chem., 259:12339.

Rouzer, C.A., and Kargman, S., 1988, Translocation of 5-lipoxygenase to the membrane in human leukocytes challenged with ionophore A23187. J. Biol. Chem. 263:10980.

Rouzer, C.A. and Kargman, S., 1989, The role of membrane translocation in the activation of human leukocyte 5-lipoxygenase in "New Trends Lipid Mediators Res." (eds. Zor, U., Noar, Z. and Danon, A.) Karger, Basel, vol. 3, In press.

Rouzer, C.A., Matsumoto, T., and Samuelsson, B., 1986, Single protein from human leukocytes possesses 5-lipoxygenase and leukotriene A_4 synthase activities, Proc. Natl. Acad. Sci. U.S.A., 83:857.

Rouzer, C.A., Rands, E., Kargman, S., Jones, R.E., Register, R.B., and Dixon, R.A.F., 1988, Expression of cloned cDNA for human leukocyte 5-lipoxygenase in 143.98.2 osteosarcoma cells, J. Biol. Chem., 263:10135.

Rouzer, C.A., and Samuelsson, B., 1985, On the nature of the 5-lipoxygenase reaction in human leukocytes: Enzyme purification and requirement for multiple stimulatory factors, Proc. Natl. Acad. Sci. U.S.A. 82:6040.

Rouzer, C.A., Shimizu, T., and Samuelsson, B., 1985, On the nature of the 5-lipoxygenase reaction in human leukocytes: Characterization of a membrane-associated stimulatory factor. Proc. Natl. Acad. Sci. U.S.A. 82:7505.

Samuelsson, B., 1983, Leukotrienes: mediators of immediate hypersensitivity reactions and inflammation, Science 220:568.

Sun, F.F., Chau, L-Y., Spur, B., Corey, E.J., Lewis, R.A., and Austen, K.F., 1986, Identification of a high affinity leukotriene C_4-binding protein in rat liver cytosol as glutathione S-transferase, J. Biol. Chem., 261:8540.

Ueda, N., Kaneko, S., Yoshimoto, T., and Yamamoto, S., 1986, Purification of arachidonate 5-lipoxygenase from porcine leukocytes and its reactivity with hydroperoxyeicosatetraenoic acids. J. Biol. Chem., 261:7982.

Winkler, J.D., Sarau, H.M., Foley, J.J., Mong, S., and Crooke, S.T., 1988, Leukotriene B_4-induced homologous desensitization of calcium mobilization and phosphoinositide metabolism in U-937 cells, J. Pharmacol. Exp. Ther., 246:204.

EICOSANOID SYNTHESIS IN RESIDENT MACROPHAGES :

ROLE OF PROTEIN KINASE C

Volkhard Kaever, Hans-Jürgen Pfannkuche, Klaus Wessel, Henning Sommermeyer and Klaus Resch

Division of Molecular Pharmacology
Medical School Hannover
D-3000 Hannover 61, F.R.G.

ABSTRACT

In this study we investigated the effects of different agents known as activators or inhibitors of the calcium and phospholipid-dependent protein kinase (PKC) on the eicosanoid production in resident mouse peritoneal macrophages and in the murine macrophage-like cell line HA38. PKC activators such as the phorbol ester 12-O-tetradecanoyl 13-acetate (TPA) or the synthetic cell-permeable diacylglycerols 1,2-dioctanoylglycerol (DiC_8) and 1-oleoyl-2-acetylglycerol (OAG) stimulated arachidonic acid (AA) release from prelabelled cells and also decreased AA incorporation into cellular phospholipids within minutes. This rapid enlargement in the amount of free AA available for eicosanoid synthesis resulted in a significantly enhanced cellular prostaglandin E_2 (PGE_2) production, whereas leukotriene C_4 (LTC_4) could not be detected under these experimental conditions. Increasing the intracellular calcium level by simultaneous addition of calcium ionophores together with PKC activators in suboptimal concentrations led to a synergistic production of PGE_2 and LTC_4. PKC inhibitors such as 1-(5-isoquinolinesulfonyl)-2-methylpiperazine (H-7), sphingosine, tamoxifen, and staurosporine, which exert their inhibitory action on PKC by different mechanisms, totally abolished TPA effects, whereas changes induced by DiC_8 or OAG were only partially reversed. Aluminum fluoride (AlF_4^-) as an activator of regulatory guanine nucleotide binding proteins (G-proteins) elevated the intracellular concentration of inositolphosphates (IP) and diacylglycerols (DAG), thus leading to an activation of PKC and an increase in the intracellular calcium level, and subsequent enhanced eicosanoid production. AlF_4^--mediated PGE_2 and LTC_4 synthesis again was reversed by using PKC inhibitors. Preincubation of the cells with pertussis toxin (PT) did not inhibit but even enhanced PGE_2 production stimulated by zymosan or AlF_4^-. Depletion of PKC activity by prolonged incubation of macrophages with TPA resulted in an extensive abrogation of the AlF_4^- or zymosan-induced PGE_2 secretion. These data provide further evidence that PKC is centrally involved in the regulation of macrophage eicosanoid synthesis initiated by different stimuli.

INTRODUCTION

Many physiological stimuli as for example, immune complexes or complement factors, which are capable of rapidly inducing eicosanoid synthesis in macrophages (Scott et al., 1983; Hartung et al., 1983) enhance phospholipase C (PLC)-mediated breakdown of phosphatidylinositol-4,5-bisphosphate (PIP_2), which leads to a short-term elevation of the intracellular calcium level (via IP_3) and translocation and activation of PKC (via DAG) (Berridge, 1984). This prompted us to study the influence of various agents that modulate PKC activity on the prostaglandin and leukotriene synthesis in these cells. It was shown earlier that resident macrophages respond to direct activators of the PKC such as synthetic diacylglycerols (Pfannkuche et al., 1986; Emilsson et al., 1986) or the phorbol ester TPA (Brune et al., 1978) with enhanced prostaglandin synthesis.

Besides the cellular equipment with AA-metabolizing enzymes, eicosanoid production is limited by the availability of their common precursor, free AA, which has to liberated from membrane phospholipids (Irvine, 1982). The intracellular concentration of AA is regulated not only by deacylating enzymes like phospholipase A_2 (PLA_2), but also by an effective fatty acid reacylating system consisting of acyl-CoA synthetase and lysophosphatide acyltransferase (LAT) activities (Hill and Lands, 1968). PKC activators increased the intracellular levels of AA available for eicosanoid synthesis and this effect was reversed by different PKC inhibitors (Kaever et al., 1989; Pfannkuche et al., 1989). In this report additional data are given, which support the hypothesis that the induction of macrophage eicosanoid synthesis by many stimuli is regulated by PKC.

EXPERIMENTAL

Macrophage preparation and cultivation

Resident macrophages were collected by peritoneal lavage from untreated 6 - 8 weeks old DBA/2 mice and cultured in DMEM medium at a cell density of 5 x 10^5 cells/ml in appropriate culture dishes as described elsewhere in detail (Pfannkuche et al., 1986). After preincubation for 2 hours the non-adherent cells were removed by washing. The adherent cells, which were shown to consist predominately of resident macrophages, were maintained under the conditions specified in the different experiments.

The macrophage-like HA38 cells were grown in MEM medium/10 % FCS (Wessel et al., 1989). For the different experiments the cells were seeded at a cell density of 2 x 10^5 cells/ml and allowed to adhere for 12 hours.

Arachidonic acid release of labelled macrophages

Cells were incubated in flat-bottomed tubes with [^{14}C]AA (50 nCi, Amersham Buchler, Braunschweig, sp. act. 60 mCi/mmol) in DMEM/2 % FCS for 20 hours. After this labelling period the macrophages were washed and provided with DMEM/0.3 % BSA. In the presence of PKC activators and inhibitors the cells were further incubated for 60 min. AA release was determined by measuring the radioactivity in the supernatants. Macrophage lipids were extracted and separated by thin layer chromatography as described (Kaever et al., 1988a).

Arachidonic acid incorporation into phospholipids

Cells were incubated in flat-bottomed tubes with [^{14}C]AA (25 nCi) and unlabelled AA (10^{-5} M, NuChek Prep, Elysian) in medium supplemented with bovine serum albumin (0.1 %) for 30 min in the presence or absence of PKC activators. In the case of PKC inhibitors these agents were added 15 min prior to the incubation with AA. Lipids were extracted as described for AA release (see above).

ADP-ribosylation of membrane proteins by pertussis toxin

Adherent macrophages were scrapped off and disrupted by sonication (3 x 10 sec bursts, 50 W) in 20 mM Hepes, pH 7.4. The membrane fraction was sedimented at 100,000 x g for 60 min at 4 °C and resuspended in the same buffer. [^{32}P]ADP-ribosylation was carried out as follows. Pertussis toxin (PT) (List Biological Laboratories, Campbell) was activated for 30 min at 37 °C with 20 mM DTT and 1 mM ATP. After activation the toxin was diluted to a final concentration of 0.5 µg/ml with 25 mM Tris/HCl pH 7.5, 1 mM EDTA, 20 mM DTT and 0.1 % (w/v) BSA. The ADP-ribosylation mixture contained 25 mM Tris/HCl pH 7.5, 1 mM EDTA, 5 mM ATP, 1 mM GTP, 0.1 % (w/v) Lubrol PX, 1 µM NAD^+, 10 µCi [^{32}P]NAD^+/assay (New England Nuclear, 800 Ci/mmol), 100 µg of membrane protein, and ± 0.17 µg/ml activated PT. After 30 min at 30 °C the reaction was stopped by adding TCA to a final concentration of 10 % (w/v). After centrifugation (16,000 x g for 10 min) the particulate fraction was resuspended in 50 µl Laemmli sample buffer with 5 % 2-mercaptoethanol (Laemmli, 1970). Proteins were separated on 10 % SDS-polyacrylamide gel electrophoresis. After coomassie staining the gels were dried. Autoradiography was performed by exposing the gels to Kodak X-Omat AR films at -80 °C.

Determination of eicosanoids

PGE_2 and LTC_4 were determined in the supernatants by double antibody radioimmunoassay as described previously (Kaever et al., 1988b). The detection limits were about 20 pg/ml for PGE_2 or 100 pg/ml for LTC_4, respectively. The employed agents did not interfere with the specific ligand binding in the different assays.

RESULTS

Characterization of the cellular systems

Collection of mouse peritoneal exsudate cells as source of resident macrophages likewise included the preparation of additional cells such as lymphocytes (50 - 60 % of total cells), polymorphonuclear leukocytes (less than 2 % of total cells), and sometimes few erythrocytes. The portion of macrophages was always greater than 90 % by including a plastic adherence step (2 hours under serumfree conditions) and washing of the macrophage cultures prior to the start of the experiments. As additional system consisting of absolutely homogenous cells the macrophage-like cell line HA38, achieved by virus transformation (malignant histiocytosis sarcoma virus) of murine precursor cells was employed (Wessel et al., 1989). Both cell types showed typical macrophage characteristics, i.e. lysozyme production, eicosanoid synthesis, and phagocytosis of zymosan particles, along with the presence of typical cell surface antigens (Mac-1, F4/80).

Optimal conditions for the induction of PGE_2 synthesis in peritoneal macrophages by protein kinase C activators

We first investigated the influence of fetal calf serum (FCS) addition to resident macrophage cultures on basal and TPA-induced PGE_2 synthesis. Short-term experiments (addition of stimuli for up to 4 h) were performed under serum-free conditions, as FCS (≥ 5 %) led to a significant increase in the basal PGE_2 secretion and a concomitantly decrease in the stimulation factor induced by TPA. Under those experimental conditions maximal TPA (10^{-7} M)-stimulated PGE_2 synthesis was measured after 2 hours with halfmaximal TPA concentrations ranging from 5×10^{-9} M - 10^{-8} M. Similar results were obtained with the cell-permeable diacylglycerols OAG and DiC_8 showing halfmaximal PGE_2 stimulation at 25 - 50 µg/ml. When macrophages had to be preincubated for longer time periods (up to 24 h), 2 % FCS was included, which did not substantial elevate PGE_2 control levels but entirely maintained cellular responsiveness to PKC activators.

Effects of PKC activators and inhibitors on AA release

After the 20 h labelling period the highest amount of $[^{14}C]$AA was found in the phosphatidylcholine (PC) fraction (about 40 % of total incorporated AA), whereas lesser amounts were determined in phosphatidylethanolamine (PE) (25 %), phosphatidylinositol (PI) (17 %) and other lipids. $[^{14}]$AA release was determined in the presence of BSA in order to prevent unspecific binding, metabolization, or reacylation. Under this condition PKC activators induced more than 2-fold increases in $[^{14}C]$AA found in the culture supernatants. Contrary, the radioactivity in the different phospholipids was decreased to various degrees with only PE exhibiting nonsignificant changes.

The PKC inhibitor H-7 dose-dependently abolished the TPA- or DiC_8-stimulated AA release without significantly changing the basal AA release in a concentration range comparable to that needed for PKC inhibition (12.5 - 100 µM). Sphingosine, known as potent PKC inhibitor, too, exhibited similar effects on AA release of prelabelled macrophages. In contrast to H-7, not only the PKC activator-induced but also the basal AA release was clearly decreased by sphingosine (2.5 - 10 µM). Whereas TPA effects were totally reversed, the DiC_8-induced AA release was only partially antagonized by both inhibitors. Prolongation of the preincubation time with the PKC inhibitors did not improve their inhibitory action.

Effects of PKC activators and inhibitors on AA incorporation

The addition of TPA or DiC_8 largely decreased the incorporation of AA into the major labelled phospholipid PC. The PKC inhibitor H-7 was able to totally (in the case of TPA) or at least partially (in the case of DiC_8) restore the suppressed incorporation rates. At higher H-7 concentrations (≥ 50 μg/ml) even control cells revealed higher reacylation activities.

Effects of PKC activators and inhibitors on macrophage eicosanoid synthesis

As mentioned above, TPA, OAG, or DiC_8 caused an immediate and dose-dependent secretion of PGE_2 as main cyclooxygenase product, whereas an increase in lipoxygenase products could not be detected. This stimulated PGE_2 synthesis was decreased by both PKC inhibitors H-7 and sphingosine in the same concentration range as seen when investigating their inhibitory effects on PKC activity and AA incorporation, or their stimulatory actions on AA release, respectively.

Simultaneous addition of TPA (10^{-8} M) together with the calcium ionophore A23187 (0.005 - 0.05 μg/ml) resulted in a synergistic rise in PGE_2 produced compared to the effects of both substances added alone. At higher concentrations of A23187 (0.05 - 1 μg/ml) this agent alone induced increased production of PGE_2 and LTC_4. Simultaneous addition of TPA led to a dramatically elevation of LTC_4 synthesis, whereas PGE_2 secretion was even diminished.

Effects of PKC activators on enzymes involved in the fatty acid deacylation/reacylation cycle

Incubation of the macrophages with TPA under conditions in which eicosanoid synthesis was induced, did not lead to measurable changes in the PLA_2 or LAT activities determined in the homogenates of treated cells. When diacylglycerols were employed the activity of the membrane-bound PLA_2 was enhanced, whereas diminished LAT activities were observed. These effects, however, could be prevented by extensive washing of the cells prior to homogenization.

Effect of pertussis toxin on PGE_2 synthesis

In further experiments we studied the possible involvement of regulatory G-proteins in the PKC-mediated eicosanoid production. Fig. 1 shows the effect of PT on the control and zymosan-stimulated PGE_2 release. PT had no effect when added simultaneously with zymosan, but increased the zymosan-induced PGE_2 synthesis after preincubation of the cells with the toxin for at least 30 min. It has to be noted that this enhancing effect was not detectable in all experiments, but in no case an inhibitory action was obvious. As can be seen in the inset of Fig. 1 PT led to the ADP-ribosylation of a membrane protein in the range of 39 kDa.

Effect of aluminum fluoride on macrophage eicosanoid synthesis

Another way to influence G-protein activities depends on the usage of AlF_4^-. In macrophages this agent evidently increased the PI-specific PLC activity as an elevation of the intracellular amounts of inositol phosphates and diacyl-

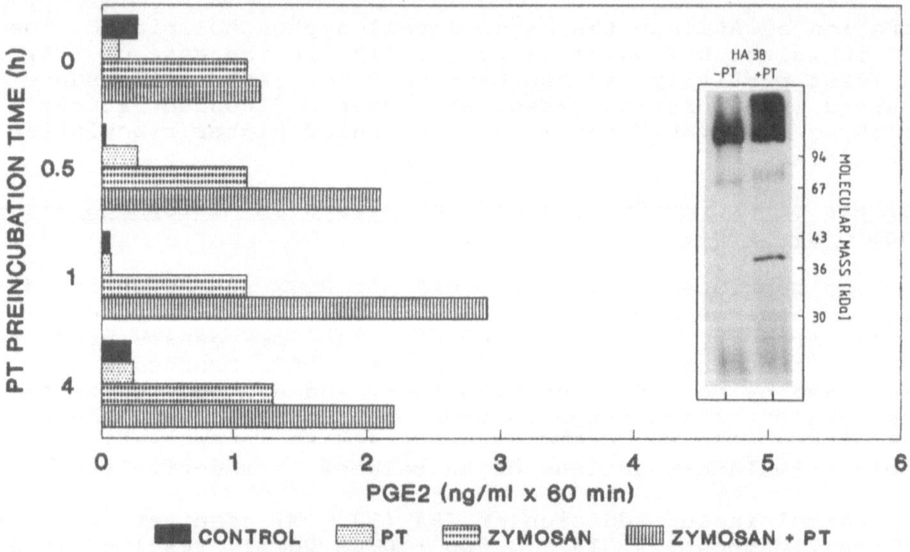

Fig. 1. Effect of pertussis toxin on macrophage eicosanoid synthesis. Cells were preincubated with or without PT (100 ng/ml) for the indicated times. PGE_2 was determined in the supernatants by RIA after an additional incubation time of 60 min in the presence or absence of zymosan (200 µg/ml) as stimulus. In the inset ADP-ribosylation of membrane proteins by PT is shown.

glycerols were detected. This led to enhanced calcium levels and activation and translocation of the PKC. Subsequently, PGE_2 and LTC_4 synthesis was significantly increased within minutes. PKC inhibitors (H-7, tamoxifen, or staurosporine) were able to reverse the AlF_4^--induced eicosanoid synthesis.

Effect of PKC depletion on macrophage PGE_2 synthesis

Prolonged incubation of macrophages with rather low concentrations of TPA leads to a total loss of specific PKC activities either cytosolic or membrane-bound. In Fig. 2 the effects of such PKC depletion on the PGE_2 production is shown. When the cells were preincubated with TPA (10^{-11} - 10^{-8} M) for 20 hours a repeated addition of TPA (10^{-8} M) in contrast to the control cells did not stimulate any enhanced PGE_2 production. The zymosan- or AlF_4^--induced stimulation of the PGE_2 synthesis was dose-dependently decreased by TPA preincubation (Fig.2).

DISCUSSION

Eicosanoids are important mediators of many physiological and pathophysiological processes. The amount of specific prostaglandin or leukotriene synthesized is depending not only on the cellular equipment with AA metabolizing enzymes but furthermore on the kind of the used stimulus. In comparison to other cells macrophages belong to the most active

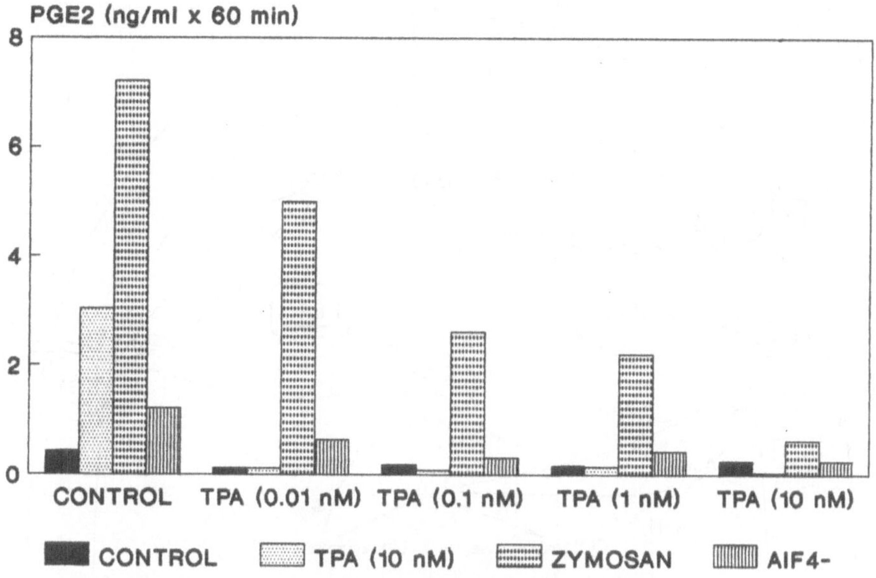

Fig. 2. Effect of protein kinase C depletion on macrophage eicosanoid synthesis. Adherent cells were incubated with the indicated concentrations of TPA for 20 h in DMEM/2 % FCS. After washing TPA (10^{-7} M), zymosan (200 µg/ml), or AlF_4^- (20 mM NaF + 10 µM $AlCl_3$) were added and the amount of PGE_2 produced after 60 min was determined by RIA.

producers of the whole eicosanoid spectrum. For example, they synthesize and secrete these AA metabolites in response to phagocytic stimuli, immune complexes, complement factors, lipopolysaccharide, or endotoxin (Leslie and Detty, 1986) and thus are deeply involved in inflammatory and allergic reactions.

Signal transduction mechanisms that mediate eicosanoid synthesis in macrophages are not well defined so far. More than 50 cell surface receptors have been described that are coupled to different second messenger systems in these cells (Hamilton and Adams, 1987), and which certainly influence each other. Modulation of PKC seems to play an important role in the regulation of eicosanoid synthesis in resident macrophages (Pfannkuche et al., 1989) as well as during macrophage activation induced by LPS or IFN_{gamma} (Uhing and Hamilton, 1989).

Our own experiments were performed with resident macrophages or HA38 macrophage-like cells, as it is known that in vivo preactivation leads to a greatly diminished capacity for eicosanoid synthesis in activated cells (Scott et al., 1982). Our findings using PKC activators and inhibitors demonstrate that changes in eicosanoid production can take place not only at the level of the cyclooxygenase/lipoxygenase system, but also by affecting the enzymes of the fatty acid deacylation/reacylation cycle. PKC activators were able to enhance the pool of free AA available for subsequent eicosa-

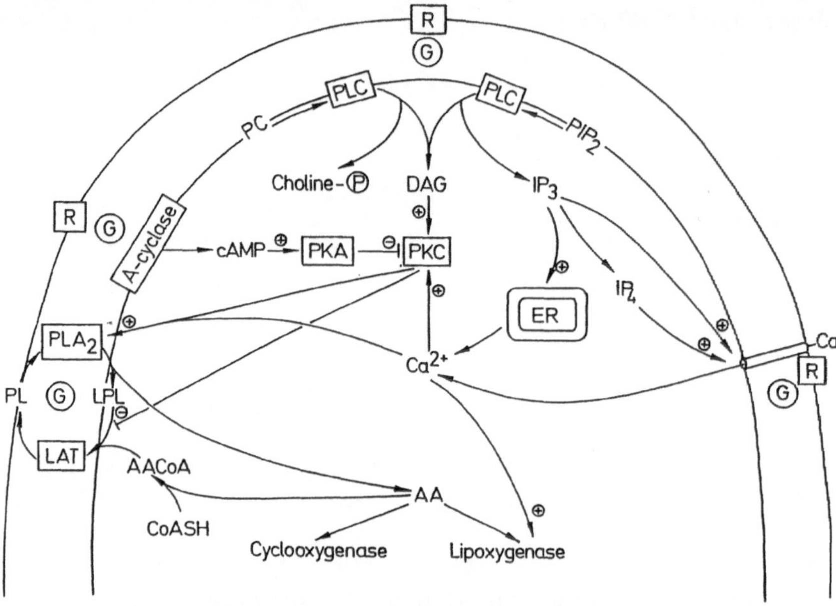

Fig. 3. A simplified model illustrating some mechanisms involved in the regulation of macrophage eicosanoid synthesis. Abbreviations are: R, receptor; G, guanine nucleotide binding regulatory protein; PLC, phospholipase(s) C; A-cyclase, adenylate cyclase; PLA_2, phospholipase A_2; LAT, lysophosphatide acyltransferase; PKA, cAMP-dependent protein kinase(s); PKC, calcium and phospholipid-dependent protein kinase(s); PL, phospholipid; LPL, lysophospholipid; DAG, diacylglycerol(s); PIP_2, phosphatidylinositol-4,5 bisphosphate; IP_3, inositol-1,4,5 trisphosphate; IP_4, inositol-1,3,4,5 tetrakisphosphate; AA, arachidonic acid; AACoA, coenzym A-activated AA; ER, endoplasmic reticulum.

noid synthesis, but the molecular basis of their action on PLA_2 or LAT are hithero unknown. Recently we reported that DiC_8 and OAG, in contrast to TPA, revealed a direct inhibitory effect on LAT independent of PKC (Goppelt-Strübe et al., 1987). It was therefore necessary to use different PKC inhibitors in order to further strengthen the concept of PKC involvement in the regulation of eicosanoid synthesis. Whereas all TPA effects studied were totally abolished by PKC inhibitors, used in concentrations that had no adverse effects on cell viability, diacylglycerol-induced changes were only partially reversed.

As unaltered PLA_2 and LAT activities were measured in homogenates of control or TPA-treated cells, even after cell homogenization in buffers containing phosphatase inhibitors, it is tempting to speculate that regulatory proteins may be involved in the PKC-mediated eicosanoid synthesis. Such proteins could be lost or uncoupled during cell disintegration. As determined by using monoclonal antibodies macrophages contain stimulatory as well as inhibitory G-proteins (data not shown). The G-protein activator AlF_4^- enhanced AA availability and intracellular calcium levels, both prerequisites for the observed PGE_2 and LTC_4 secretion, and these effects were totally reversed by PKC inhibitors. These data suggest that preferentially PLC and subsequent PKC are activated by AlF_4^-. PT led to ADP-ribosylation of a membrane protein (Fig. 1). In human monocytes a PT-sensitive G-protein was reported to play an important role in the activation by chemoattractants (Verghese et al., 1986). In our hands, PT preincubation did enhance zymosan-induced PGE_2 production, although a reduction would be expected by PT-mediated desactivation of the specific PI-PLC (Cockcroft, 1987). Similar, an enhanced AA release and an increased bradykinin-stimulated PGE_2 synthesis by PT treatment of RAW264.7 macrophages or fibroblasts, respectively, was reported (Burch and Axelrod, 1987; Moss et al., 1988).

PKC depletion by prolonged treatment of peritoneal macrophages with TPA resulted in a total inhibition of PKC activator-induced PGE_2 synthesis (Fig. 2) and also specific protein phosphorylation (Wijkander and Sundler, 1989). Under the same conditions the AlF_4^-- or zymosan-stimulated PGE_2 synthesis was decreased, too, but higher concentrations of TPA were needed to achieve extensive unresponsiveness. This may reflect a direct activation of PLA_2, LAT, or adenylate cyclase by AlF_4^- or zymosan and negative feedback control (down-regulation) by PKC.

Molecular cloning studies clearly revealed that PKC exists as a family of at least seven subspecies (Nishizuka, 1988; Ohno et al., 1988), and differential down-regulation of different PKC isozymes by TPA has been reported (Ase et al., 1988). Further work therefore has to be focussed on studies concerning the involvement of distinct PKC subspecies (i.e. calcium-independent but diacylglycerol-dependent ones) in the regulation of macrophage eicosanoid synthesis.

ACKNOWLEDGEMENTS

We wish to thank H. Hartmann and A. Garbe for their skilful technical assistance. This study was supported by grant Ka 730/1-2 from the Deutsche Forschungsgemeinschaft.

REFERENCES

Ase, K., Berry, N., Kikkawa, U., Kishimoto, A., and Nishizuka, Y., 1988, Differential down-regulation of protein kinase C subspecies in KM3 cells, FEBS Lett., 26:396.

Berridge, M.J., 1984, Inositol trisphosphate and diacylglycerol as second messengers, Biochem. J., 220:345.

Brune, K., Glatt, M., Kälin, H., and Peskar, B.A., 1978, Pharmacological control of prostaglandin and thromboxane release from macrophages, Nature, 274:261.

Burch, R.M., and Axelrod, J., 1987, A GTP-binding protein (G-protein) regulates phospholipase A_2 in RAW264.7 macrophages, Fed. Proc., 46:703.

Cockcroft, S., 1987, Polyphosphoinositide phosphodiesterase: regulation by a novel guanine nucleotide binding protein, G_p, TIPS, 12:75.
Emilsson, A., Wijkander, J., and Sundler, R., 1986, Diacylglycerol induces deacylation of phosphatidylinositol and mobilization of arachidonic acid in mouse macrophages, Biochem. J., 239:685.
Goppelt-Strübe, M., Pfannkuche, H.-J., Gemsa, D., and Resch, K., 1987, The diacylglycerols dioctanoylglycerol and oleoylacetylglycerol enhance prostaglandin synthesis by inhibition of the lysophosphatide acyltransferase, Biochem. J., 247:773.
Hamilton, T.A., and Adams, D.O., 1987, Molecular mechanisms of signal transduction in macrophages, Immunology Today, 8:151.
Hartung, H.P., Bitter-Suermann, D., and Hadding, U., 1983, Induction of thromboxane release from macrophages by anaphylatoxic peptide C_{3a} of complement and synthetic hexapeptide C_{3a} 72-77, J. Immunol., 130:1345.
Hill, E.E., and Lands, W.E.M., 1968, Incorporation of long-chain and polyunsaturated acids into phosphatidate and phosphatidylcholine, Biochim. Biophys. Acta, 152:645.
Irvine, R.F., 1982, How is the level of free arachidonic acid controlled in mammalian cells?, Biochem. J., 204:3.
Kaever, V., Firla, U., and Resch, K., 1988a, Sulfhydryl reagents as model substances for eicosanoid research, Eicosanoids, 1:49.
Kaever, V., Goppelt-Strübe, M., and Resch, K., 1988b, Enhancement of eicosanoid synthesis in mouse peritoneal macrophages by the organic mercury compound thimerosal, Prostaglandins, 35:885.
Kaever, V., Pfannkuche, H.-J., Wessel, K., and Resch, K., 1989, Regulation of eicosanoid synthesis in mouse peritoneal macrophages by protein kinase C, Agents Actions, 26:175.
Laemmli, U., 1970, Cleavage of structural proteins during the assembly of the head of bacteriophage T_4, Nature, 227:680.
Leslie, C.C., and Detty, D.M., 1986, Arachidonic acid turnover in response to lipopolysaccharide and opsonized zymosan in human monocyte-derived macrophages, Biochem. J., 236:251.
Moss, J., Hom, B.E., Hewlett, E.L., Tsai, S., Adamik, R., Halpern, J.L., Price, S.R., and Manganiello, V.C., 1988, Mechanism of enhanced sensitivity to bradykinin in pertussis toxin-treated fibroblasts: toxin increases bradykinin-stimulated prostaglandin formation, Mol. Pharmacol., 34:279.
Nishizuka, Y., 1988, The molecular heterogeneity of protein kinase C and its implication for cellular regulation, Nature, 334:661.
Ohno, S., Akita, Y., Konno, Y., Imajoh, S., and Suzuki, K., 1988, A novel phorbol ester receptor/protein kinase, nPKC, distantly related to the protein kinase C family, Cell, 53:731.
Pfannkuche, H.-J., Kaever, V., and Resch, K., 1986, A possible role of protein kinase C in regulating prostaglandin synthesis of mouse peritoneal macrophages, Biochem. Biophys. Res. Commun., 139:604.

Pfannkuche. H.-J., Kaever, V., Gemsa, D., and Resch, K., 1989, Regulation of prostaglandin synthesis by protein kinase C in mouse peritoneal macrophages, Biochem. J., 260:in press.
Scott; W.A., Pawlowski, N.A., Andreach, M., and Cohn, Z.A., 1982, Resting macrophages produce distinct metabolites from exogenous arachidonic acid, J. Exp. Med., 155:535.
Scott, W.A., Rouzer, C.A., and Cohn, Z.A., 1983, Leukotriene C_4 release by macrophages, Fed. Proc., 42:129.
Uhing, R.J., and Adams, D.O., 1989, Molecular events in the activation of murine macrophages, Agents Actions, 26:9.
Verghese, M.W., Smith, C.D., Charles, L.A., Jakoi, L., and Snyderman, R., 1986, A guanine nucleotide regulatory protein controls polyphosphoinositide metabolism, Ca^{2+} mobilization, and cellular responses to chemoattractants in human monocytes, J. Immunol., 137:271.
Wessel, K., Kaever, V., Ostertag, W., Bitter-Suermann, D., and Resch, K., 1989, A virus-transformed macrophage-like cell line as tool for eicosanoid research, J. Leukocyte Biol., in press.
Wijkander, J., and Sundler, R., 1989, A role for protein kinase C-mediated phosphorylation in the mobilization of arachidonic acid in mouse macrophages, Biochim. Biophys. Acta, 1010:78.

MODULATION OF THE S-K^+ CHANNEL OF APLYSIA SENSORY NEURONS BY LIPOXYGENASE METABOLITES OF ARACHIDONIC ACID AND cAMP

Steven A. Siegelbaum, Andrea Volterra and N. Buttner

Department of Pharmacology, Center for Neurobiology and Behavior, Howard Hughes Medical Institute, Columbia University. 722 W. 168 St., New York, NY 10032

INTRODUCTION

Ion channels are modulated by a variety of neurotransmitters and hormones acting through a number of second messenger pathways (see Kaczmarek and Levitan, 1987 for review). Modulation of ion channel activity leads to many changes in neuronal function, including alterations in resting potential, input resistance, action potential duration, threshold, spike accomodation and neurotransmitter release. We have been studying the role that channel modulation plays in controlling the activity of mechanoreceptor sensory neurons in the marine snail Aplysia californica. Our results show that the sensory neurons are subject to antagonistic modulatory actions by the transmitter 5-HT and the neuropeptide FMRFamide.

5-HT AND FMRFamide PRODUCE OPPOSITE ACTIONS

5-HT, acting through cAMP-dependent protein phosphorylation, produces a series of excitatory actions in the sensory neurons (Fig. 1A). It causes a slow depolarization associated with a decrease in membrane conductance, an increase in action potential duration and excitability, and an increase in transmitter release from the sensory neuron terminals (presynaptic facilitation). Presynaptic facilitation is thought to be the cellular mechanism for a simple form of learning in which the sensory neurons participate, behavioral sensitization of the gill and siphon-withdrawal reflex (Kandel and Schwartz, 1982).

In contrast to 5-HT, FMRFamide, acting through the lipoxygenase metabolites of arachidonic acid, produces a series of inhibitory actions in the sensory neurons (Fig. 1B). These actions include a slow hyperpolarization associated with an increase in membrane conductance, a decrease in action potential duration, and a decrease in transmitter release from the presynaptic terminals (presynaptic inhibition) (Abrams et al., 1984; Belardetti et al., 1987). Presynaptic inhibition produces an inhibition in the gill and siphon-withdrawal reflex (Mackey et al., 1987).

What are the ionic mechanisms responsible for these opposing modulatory actions? Voltage clamp experiments have shown that 5-HT leads to a decrease in a novel type of K^+ current termed the serotonin-sensitive (or S) current (Klein, Camardo and Kandel, 1982). Single channel recording experiments have identified the S-K^+ channel as a background K^+ channel whose gating is relatively voltage-independent (Siegelbaum et al., 1982). 5-HT decreases the net S current by causing prolonged all-or-none closures of single S-K^+ channels (Fig. 2A), leading to a decrease in the number of active channels (N_f) in the membrane. These single channel results provide independent evidence that 5-HT acts through an intracellular messenger since application of 5-HT to the bath closes S channels under the pipette that have no direct access to the transmitter.

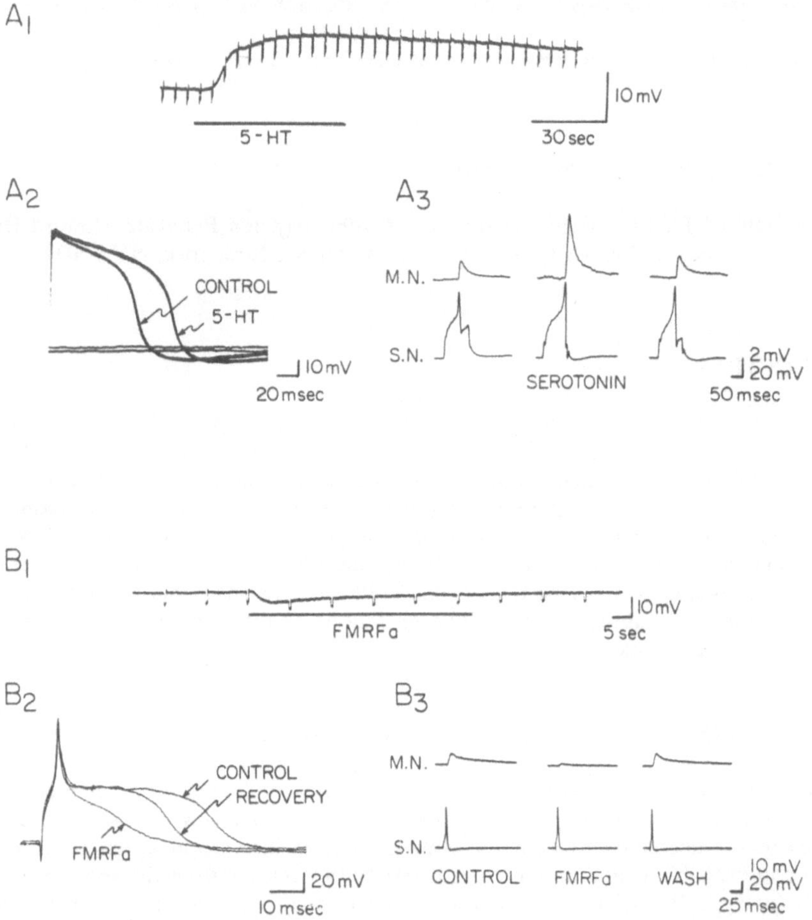

Figure 1. Dual modulatory actions of 5-HT and FMRFamide on membrane potential in Aplysia sensory neurons. A, Serotonin (10 uM) causes a slow depolarization of the resting potential with a decrease in conductance (seen as an increase in the voltage response to constant negative current pulses) (A1), an increase in the action potential duration (A2), and an enhancement in sensory-motor transmission (A3). Top trace in A3 shows fast excitatory post-synaptic potentials (EPSPs) in the motor neuron in response to action potential stimulation in the sensory neuron (bottom trace). B, FMRFamide (10 uM) produces a slow hyperpolarization of the sensory neuron resting potential accompanied by an increase in membrane conductance (B1), a reversible decrease in action potential duration (B2), and a reduction in the fast EPSP in follower motor neuron (upper traces) in response to firing an action potential in sensory neuron (lower traces) (B3).

FMRFamide, in contrast, causes both an increase in K^+ current (Ocorr and Byrne, 1985; Belardetti et al., 1987; Brezina et al., 1987a) and a decrease in calcium current (Brezina et al., 1987b; Edmonds et al., 1988). Does FMRFamide increase the net K^+ current by modulating the same S-K^+ channel targetted by 5-HT? Fig. 2B shows the results of a patch clamp experiment demonstrating that FMRFamide does indeed modulate single S-K^+ channels, leading to an increase in channel activity (see also Belardetti et al., 1987; Brezina et al., 1987a). Thus, a single type of ion channel is subject to opposing modulatory actions by separate transmitters. However, the action of FMRFamide is not the mirror image of that of 5-HT, as the peptide does not increase the number of active S channels but rather increases the probability, p_O, that an active channel is in the open state (Belardetti et al., 1987).

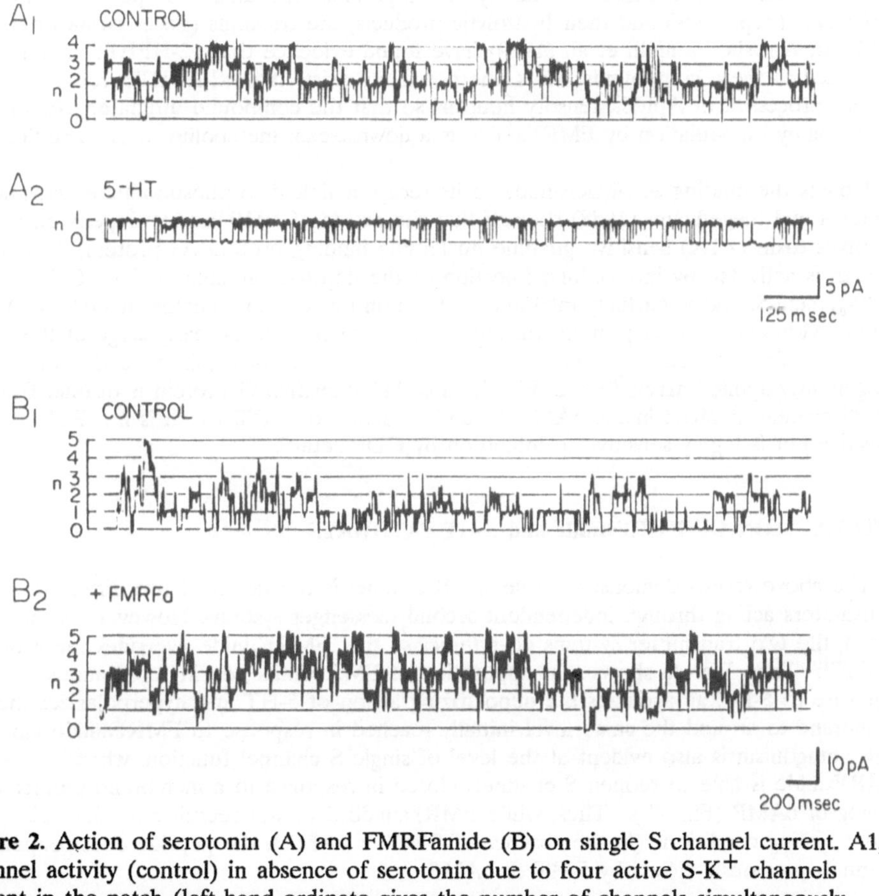

Figure 2. Action of serotonin (A) and FMRFamide (B) on single S channel current. A1, Channel activity (control) in absence of serotonin due to four active S-K^+ channels present in the patch (left-hand ordinate gives the number of channels simultaneously open). A2, after application of serotonin (100 uM) three of the four channels close. B1, Control trace from a patch containing five S channels. Application of 2 uM FMRFamide produces a significant increase in channel activity (B2). The maximal number of open channels does not change. Single channel currents recorded from two cell-attached patches. Transmitters applied to bath outside of patch pipette.

LIPOXYGENASE METABOLITES MEDIATE ACTION OF FMRFAMIDE

The action of FMRFamide appears to depend on a second messenger system since, like 5-HT, application of the transmitter to the bath modulates the activity of channels under the patch pipette. In a biochemical and biophysical study, we obtained evidence that the action of FMRFamide is mediated by the 12-lipoxygenase metabolites of arachidonic acid (Piomelli et al., 1987). Thus, FMRFamide (but not 5-HT) leads to the release of 5-HETE and 12-HETE from sensory neurons labelled with ^{3}H-arachidonic acid. Exogenous application of arachidonic acid (but not 11-eicosamonenoic acid) mimics the action of FMRFamide. Morevoer, the action of FMRFamide (but not 5-HT) is blocked by the phospholipase inhibitor bromophenacyl bromide and the lipoxygenase inhibitor NDGA. In contrast, the cyclooxygenase inhibitor indomethacin does not block the response to FMRFamide. A role for lipoxygenase metabolites is further supported by the finding that 12-hydroperoxyeicosatetraenoic acid (12-HPETE) mimics the action of FMRFamide on both resting potential and action potential duration (Fig. 3). In contrast, neither the 5-lipoxygenase metabolite 5-HPETE nor the hydroxy acids 5- and 12-hydroxyeicosatetraenoic acid have any significant effect.

12-HPETE is metabolized to a family of compounds, including two epoxy-hydroxy-derivatives (hepoxilins) and their hydrolytic products, the trioxilins (Pace-Asciak et al., 1983). In Aplysia, Piomelli et al. (1989) have found evidence that 12-HPETE is indeed converted to these compounds. Thus, the possibility exists that 12-HPETE could be further processed in Aplysia sensory neurons so that the compound ultimately responsible for S channel modulation by FMRFamide is a downstream metabolite of 12-HPETE.

How is the binding of FMRFamide to its receptor linked to phospholipase activation? Volterra and Siegelbaum (1988) showed that the action of FMRFamide depens on a pertussis toxin (PTX) sensitive guanine nucleotide binding protein (G protein). This G protein is activated by intracellular injection of the non-hydrolyzable analog of GTP, GTP-gamma-S, and is partially inhibited by the non-hydrolyzable analog of GDP, GDP-beta-S. Moreover, this G protein was shown to be involved at an early stage of the cascade, before release of arachidonic acid, since PTX did not block the action of exogenously applied arachidonate. Finally, this PTX-sensitive G protein is distinct from the G protein involved in the cAMP-dependent action of 5-HT, which is not PTX-sensitive but is highly sensitive to inhibition by GDP-beta-S.

INTERACTION OF FMRFamide and 5-HT PATHWAYS

The above results demonstrate that the S channel is the target of two different transmitters acting through independent second messenger systems. However, as shown in Fig. 4, the two transmitter systems do interact so that FMRFamide overrides the action of 5-HT. Thus, Fig. 4a shows that application of FMRFamide, during continuous exposure to 5-HT, antagonizes the depolarizing action of 5-HT and hyperpolarizes the membrane to around the same level initially reached in response to FMRFamide alone. This antagonism is also evident at the level of single S channel function, where FMRFamide is able to reopen S channels closed in response to a membrane permeant analog of cAMP (Fig. 4b). Thus, while FMRFamide does not recruit new channels under resting conditions (in the absence of 5-HT or cAMP), the peptide does reopen channels previously closed by 5-HT or cAMP.

These results suggest that FMRFamide has two actions on S channel function. First, it increases the open probability of normally active channels (in the absence of 5-HT or cAMP). Second, it antagonizes the effects of 5-HT and cAMP. Interestingly, in smooth muscle cells, an increase in the M K^+ current with cAMP and epinephrine is completely antagonized by ACh (Sims et al., 1988), suggesting a similar mode of action.

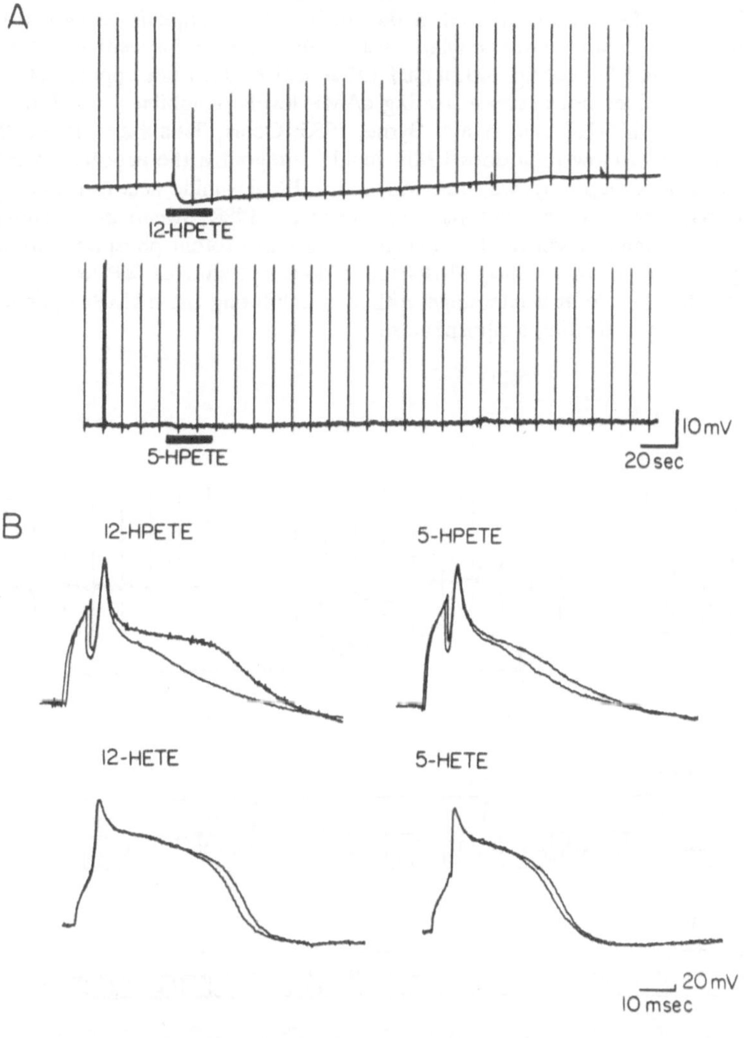

Figure 3. Effect of lipoxygenase metabolites on membrane and action potentials. A, effect of 1.5 uM 12-HPETE and 5-HPETE on resting potential and stimulated action potentials (brief upward spikes) in sensory neuron in culture. 12-HPETE inhibited spike firing for about 2 min (during this period the smaller spikes are subthreshold). B, effect of 60 uM 12-HPETE, 5-HPETE, 12-HETE and 5-HETE on action potential duration recorded from sensory neuron in intact abdominal ganglion in presence of 50 mM TEA. (From Piomelli et al., 1987).

FMRFamide DECREASES PROTEIN PHOSPHORYLATION

From the experiments summarized above, the dual modulation of the S channel by 5-HT and FMRFamide involves distinct G proteins that activate distinct second messenger cascades. While the G protein involved in the 5-HT cascade appears related to G_s (based on its ability to stimulate adenylate cyclase and its PTX insensitivity), the FMRFamide activated G protein, although PTX sensitive, does not appear related to G_i. Thus, FMRFamide does not decrease resting cAMP levels or inhibit activation of adenylate cyclase with 5-HT (Ocorr and Byrne, 1985; Ocorr, Tabata and Byrne, 1985; Volterra et al., 1988). How then does FMRFamide antagonize the action of 5-HT? Using 2-D gel autoradiography, we have found that FMRFamide acts to decrease protein phosphorylation in the sensory neurons (Volterra et al., 1988; Sweatt et al., submitted). In addition, FMRFamide reverses the normal increase in protein phosphorylation produced by 5-HT or cAMP. Since FMRFamide does not alter cAMP levels, it must inhibit the cAMP cascade at a late stage, either by inhibiting the cAMP-dependent protein kinase or by activating a phosphatase.

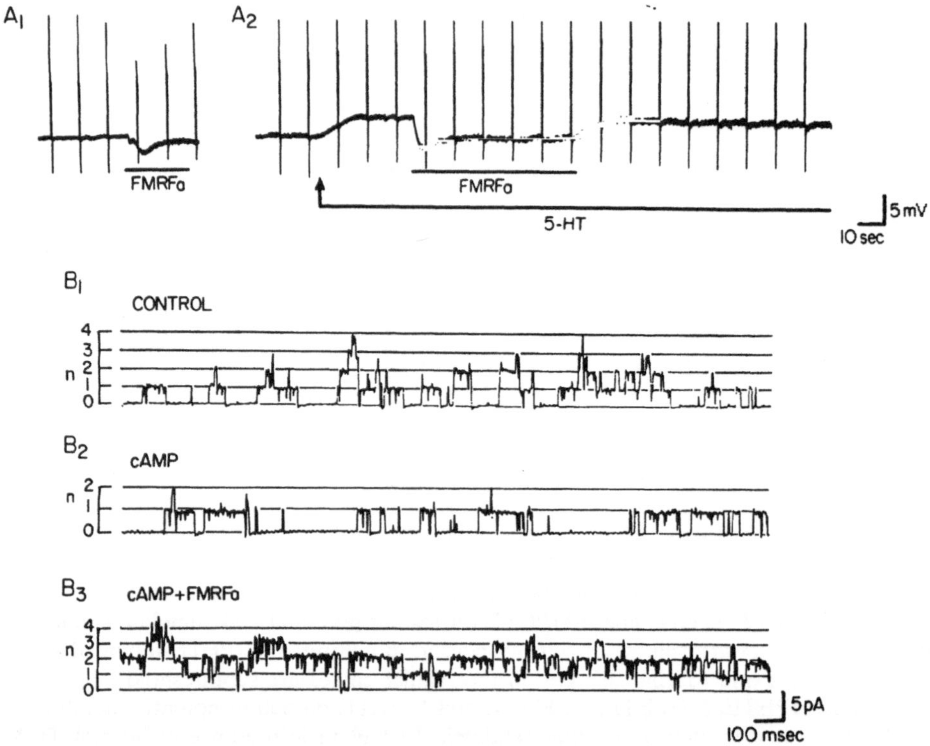

Figure 4. FMRFamide antagonizes S channel closure produced by serotonin and cAMP. A, Effects on membrane potential. Resting potential shows a slow hyperpolarizing response to 2 uM FMRFamide alone (A1). In presence of 5-HT (1 uM), response to FMRFamide is enhanced (A2). B, Single channel recording from cell-attached patch. Under control conditions four S channels active (B1); after application of 0.1 mM dibutyryl-cAMP to the bath, two of four channels close (B2); subsequent application of 20 uM FMRFamide, in presence of dibutyryl-cAMP, reopens closed channels (B3). (From Belardetti et al., 1987).

HOW DOES 12-HPETE LEAD TO THE INCREASE IN S CHANNEL ACTIVITY?

One final question is the mechanism whereby 12-HPETE leads to an increase in S channel opening. Four possible modes of action would appear likely. First, 12-HPETE (or a metabolite) could bind to a receptor and activate a G protein. The activated G protein could then lead to S channel modulation, perhaps through a direct G protein-gating of the channel (Codina et al., 1987; Logothetis et al., 1987). Such a model has recently been proposed to account for the activation of a K^+ channel in cardiac atrial cells by 5-lipoxygenase metabolites (Kurachi et al., 1989). Second, a 12-lipoxygenase metabolite could activate a protein kinase, leading to S channel phosphorylation at some site distinct from the cAMP phosphorylation site. Third, there could be a dephosphorylation of the channel, perhaps due to activation of a phosphoprotein phosphatase. Fourth, 12-HPETE or a metabolite could directly act at the level of the S channel or the membrane.

To distinguish among these possibilities, we have applied various arachidonate metabolites to cell-free inside-out and outside-out patches of membrane to test the ability of the metabolites to increase S channel activity in these in vitro conditions in the absence of ATP or GTP (Buttner et al., 1989). We find that 12-HPETE produces a consistent increase in S channel activity in inside-out and outside-out patches. On average, 30 uM 12-HPETE causes a 3.4 fold increase in S channel activity. Arachidonic acid (up to 70 uM) has no effect on single channel activity, while 12-HETE produces only a small effect (<2.0 fold change).

As there is no GTP in the solutions bathing these patches, it is unlikely that 12-HPETE modulates S channel function through activation of a G protein. Also since the solutions lack ATP, it is unlikely that a phosphorylation reaction is required. Finally, since the increases in S channel activity are readily reversible, a dephosphorylation reaction can be ruled out since it should produce an irreversible increase in S channel activity (due to the lack of ATP for rephosphorylation). Thus, we conclude that 12-HPETE or a metabolite may act directly on the S channel or a protein tightly coupled to the S channel in the cell-free patches.

Belardetti et al. (1988) report a similar modulation of S channel activity in cell-free inside-out patches with the epoxy-alcohol metabolites of 12-HPETE. As the epoxy-alcohol compounds were effective at around 500 nM, a concentration similar to that used in our outside-out patches, it is not clear whether 12-HPETE is indeed the final active metabolite responsible for S channel modulation or whether it must be further processed in the cell-free patches to a downstream compound.

ARACHIDONIC ACID MODULATION OF CHANNEL ACTIVITY IN OTHER SYSTEMS

Recently, several other studies have suggested that arachidonic acid and its metabolites may play important roles as modulators of ion channel activity in a number of other systems. Thus, Kurachi et al. (1989) and Kim et al. (1989) report that 5-lipoxygenase metabolites activate a potassium channel in cardiac atrial muscle cells that is indistinguishable from the acetylcholine-activated K^+ channel (KACh) in these cells. Moroever, Kim et al. have obtained evidence that the activation of this channel by purified beta-gamma subunits of G proteins is mediated by phospholipase A_2 activation, arachidonic acid release and subsequent metabolism by 5-lipoxygenase. Other studies suggest that arachidonic acid itself, as well as other unsaturated fatty acids, can serve as a modulator of other types of K^+ channels independently of further metabolism. For example, Kim and Clapham (1989) and Ordway, Walsh and Singer (1989) have found that fatty acids activate a K^+ channel (distinct from KACh) in cardiac and smooth muscle cells, respectively.

One of the more intriguing aspects of the eicosanoids as neuronal modulators is their ability to be released from the cells in which they are produced. This allows them to act as first messengers by binding to receptors in or on neighboring cells. Such a mechanism has been proposed to account for a puzzling finding regarding long-term potentiation (LTP) in the hippocampus. Here, post-synaptic activity due to activation of the NMDA-type glutamate receptors in the CA1 pyramidal neurons is required for induction of LTP (see Nicoll et al., 1988). However, LTP is expressed, at least partially, as an increase in transmitter release from the pre-synaptic terminals. Might the eicosanoid pathway provide a means for communication from the post-synaptic cell to the pre-synaptic cell? Recent experiments show that activation of NMDA receptors leads to the release of various lipoxygenase metabolites in mammalian brain (Dumuis et al., 1988; Pellerin and Wolfe, 1988). Moreover, Williams and Bliss (1988) have shown that NDGA can block the induction of LTP in the hippocampus. Whatever the final outcome in this controversial field, it is clear that the lipoxygenase metabolites of arachidonic acid are an important class of compounds for ion channel modulation in a variety of systems.

REFERENCES

Abrams TW, Castellucci VF, Camardo JS, Kandel ER, Lloyd PE (1984). Two endogenous neuropeptides modulate the gill and siphon withdrawal reflex in Aplysia by presynaptic facilitation involving cAMP-dependent closure of a serotonin-sensitive potassium channel. Proc. Natl. Acad. Sci. USA. 81: 7956-7960.

Belardetti F, Kandel ER, and Siegelbaum SA (1987). Neuronal inhibition by the peptide FMRFamide involves opening of S K^+ channels. Nature 325:153-156.

Belardetti F, Campbell W, Chabert P, Flack JR, Mioskowski C, Rosolowsky M (1988). A novel molecular mechanism for presynaptic inhibition: the arachidonic acid derivatrive hepoxilin-A directly acts on the serotonin-sensitive "S" K^+-channel to increase its opening in Aplysia sensory neurons. Soc. Neurosci. Abstr. 14, 1207.

Brezina V, Eckert R, Erxleben C (1987a). Modulation of potassium conductances by an endogenous neuropeptide in neurones of Aplysia californica. J. Physiol. 382: 267-290.

Brezina V, Eckert R, and Erxleben C (1987b). Suppression of calcium current by an endogenous neuropeptide in neurones of Aplysia californica. J. Physiol. (Lond) 388, 565-595.

Buttner N, Volterra A, Siegelbaum SA, (1989). Control of S-K^+ channel activity by arachidonic acid metabolites in cell-free inside-out patches from Aplysia sensory neurons. Biophys. J. 55:197a.

Dumuis A, Sebben M, Haynes L, Pin JP, and Bockaert J (1988). NMDA receptors activate the arachidonic acid cascade system in striatal neurons. Nature 336, 68-70.

Edmonds BW, Kandel ER, and Klein M (1988). Different functional roles for two components of calcium current in Aplysia sensory neurons. Soc. Neurosci. Abstr. 14, 1206.

Kaczmarek LK, Levitan IB (1987). **Neuromodulation.** Oxford University Press, New York.

Kandel, ER and Schwartz, JH (1982). Molecular biology of learning: modulation of transmitter release. Science 218:433-443.

Kim D, and Clapham DE (1989). Potassium channels in cardiac cells activated by arachidonic acid and phospholipids. Science (In press).

Kim, D, Lewis DL, Graziadei L, Neer EJ, Bar-Sagi D, and Clapham DE (1989). G-protein $\beta\gamma$ subunits activate the cardiac muscarinic K+ channel via phospholipase A_2. Nature 337, 557-560.

Klein M, Camardo JS, and Kandel ER (1982). Serotonin modulates a specific potassium current in the sensory neurons that show presynaptic facilitation in Aplysia. Proc. Natl. Acad. Sci. USA 79, 5713-5717.

Kurachi Y, Ito H, Sugimoto T, Shimuzu T, Miki I, and Ui M (1989). Arachidonic acid metabolites as intracellular modulators of the G protein-gated cardiac K^+ channel. Nature 337, 555-557.

Mackey, S.L., Glanzman, D.L., Small, S.A., Dyke, A.M., Kandel, E.R. and Hawkins, R.D. (1987). Tail shock produces inhibition as well as sensitization of the siphon-withdrawal reflex of Aplysia: Possible behavioral role for presynaptic inhibition mediated by the peptide Phe-Met-Arg-Phe-NH_2. Proc. Natl. Acad. Sci. USA 84, 8730-8734.

Nicoll RA, Kauer JA, Malenka RC (1988). The current excitement in long term potentiation. Neuron 1:97-103.

Ocorr KA, Byrne JH (1985). Membrane responses and changes in cAMP levels in Aplysia sensory neurons produced by serotonin, tryptamine, FMRFamide and small cardioactive peptide B (SCP_B). Neurosci.Lett. 55: 113-118.

Ocorr KA, Tabata M, Byrne JH (1985). Cyclic AMP, a common biochemical locus for the effects of 5-HT and SCP_B but not FMRFamide in tail sensory neurons of Aplysia. Soc. Neurosci. Abstr. 11: 481.

Pace-Asciak CR, Granstrom E, Samuelsson B (1983). Arachidonic acid epoxides: Isolation and structure of two hydroxy epoxide intermediates in the formation of 8,11,12- and 10,11,12-trihydroxyeicosatrienoic acid. J. Biol. Chem. 258: 6835-6840

Pellerin L, Wolfe LS (1988). Glutamate induces 12-HETE synthesis through NMDA receptor activation in rat cerebral cortex. Soc. Neurosci. Abstr. 14:131.

Piomelli D, Volterra A, Dale N, Siegelbaum SA, Kandel ER, Schwartz JH, Belardetti F (1987). Lipoxygenase metabolites of arachidonic acid as second messengers for presynaptic inhibition in Aplysia sensory neurons. Nature 328: 38-43.

Piomelli D, Shapiro E, Zipkin R, Schwartz JH, Feinmark SJ (1989). Formation and action of 8-hydroxy-11,12-epoxy-5,9,14-eicosatrienoic acid in Aplysia: A possible second messenger in neurons. Proc. Natl. Acad. Sci. USA (In press).

Siegelbaum SA, Camardo, JS, Kandel ER (1982). Serotonin and cyclic AMP close single K^+ channels in Aplysia sensory neurones. Nature 299: 413-417.

Sims SM, Singer JJ, and Walsh JV Jr. (1988). Antagonistic adrenergic-muscarinic regulation of M current in smooth muscle cells. Science 239, 190-193.

Volterra A, Siegelbaum SA (1988). Role of two different guanine nucleotide binding proteins in the antagonistic modulation of the S-type K^+ channel by cAMP and arachidonic acid metabolites in Aplysia sensory neurons. Proc. Natl. Acad. Sci. USA. 83:7810-7814.

Volterra A, Siegelbaum SA, Kandel ER, and Sweatt JD (1988). FMRFamide antagonizes S current modulation and protein phosphorylation produced by 5-HT and cAMP. Soc. Neurosci. Abstr. 14, 152.

Williams JH, Bliss TV (1988). Induction but not maintenance of calcium-induced long-term potentiation in dentate gyrus and area CA1 of the hippocampal slice is blocked by nordihydroguaiaretic acid. Neurosci. Lett. 88:81-85.

INHIBITION OF LEUKOTRIENE D_4 MEDIATED RELEASE OF PROSTACYCLIN USING ANTISENSE DNA

Mike A. Clarke*, Terresa M. Conway, Mike Cook, Lynne Webb*, Janice Dispoto, Seymore MOng, Robert Short**, Sheldon Steiner***, David Littlejohn, and S.T. Cooke****

Smith Kline and French, Philadelphia, PA
*Washington University, St. Louis, MO
***University of KY, Lexington, KY
Current Address:
*Schering-Plough Research, Bloomfield, NJ
**A.T. Biochem, Malvern, PA
****ISIS, San Diego, CA

We have chosen to study the response of endothelial cells during agonist stimulation as a model system for understanding the biochemical mechanisms of signal transduction because of the central role played by the endothelium in mediating inflammation and other important biological phenomenon. The response of the endothelium to inflammatory stimuli has recently received a great deal of attention due, at least in part, to the realization that it is not just a passive, vascular tissue, but an active cellular component of the inflammatory response. Inflammatory stimuli may directly effect the endothelium causing increased vascular permeability and enhanced adhesiveness of polymorphonuclear leukocytes and macrophages. A variety of inflammatory stimuli effect the endothelium by direct induction of the synthesis of several autocoids by endothelial cells (Clark, 1986a; Clark, 1986b; Clark, 1987a). Some of these stimuli cause the endothelium to develop large gaps between adjacent cells leading to plasma exudation and margination of lymphocytes, macrophages and polymorphonuclear leukocytes (PMN) into the extravascular space. Many of the aforementioned endothelial responses are regulated by the synthesis of oxygenated ecoisanoids which is demonstrated by the salulatory effect of anti-inflammatory drugs which attenuate ecoisanoids synthesis and prevent these endothelial cell responses.

Our early studies demonstrated that stimulation of the bovine endothelial cell line, CPAE', with LTD_4 resulted in a substantial augmentation and prostacyclin production after 5 min (Table 1) (Clark, 1986a; Clark, 1986b). Pretreatment of CPAE cells with either cycloheximide or actinomycin D resulted in complete ablation of the augmentation of prostacycline accumulation although no affects on the conversion of exogenously added arachidonic acid to prostacyclin were observed. Our data confirmed earlier reports demonstrating that the availability of free arachidonic acid was the rate limiting step in eicosanoid formation (Flower, 1976). To determine if all stimuli which increase eicosanoid synthesis require RNA and protein synthesis cells were also treated with bradykinin and A23187. It was found that all of these stimuli increase prostacyclin production; however, they appeared to do so by distinct pathways (Table 1).

To identify the biochemical mechanisms responsible for the release of arachidonic acid following LTD_4 treatment, we next measured phospholipase A_2 and C activities in these cells as a function of leukotriene D_4 treatment. We found that LTD_4 increased phospholipase A_2 activity but had no effect on phospholipase C activity (Table 2) (Clark, 1986a). To determine the specificity of this response, cells treated with bradykinin were also assayed for phospholipase activity and it was observed that these cells had increased PLC activity. These data support the concept that multiple pathways exist which can ultimately produce arachidonic acid for oxygenation pathways. Some stimuli such as LTD_4 appear to activate PLA_2 whereas other such as bradykins appear to activate PLC.

TABLE 1

EFFECTS OF PROTEIN AND RNA SYNTHESIS
INHIBITORS ON PROSTACYCLIN SYNTHESIS IN CPAE CELLS

Culture Conditions		INHIBITOR	
	None	Cycloheximide	Actinomycin D
Control	2.4 ± .6	1.4 ± .5	2.1 ± .3
+LTD_4 (1 uM)	4.6 ± .2	1.9 ± .3**	1.9 ± .2**
+Arachidonic Acid (0.1 mM)	3.7 ± .2	3.2 ± .3	3.5 ± .2

[a] Values expressed are ng/ml/10^5 cells/10 min of 6-keto-$PGF_{1\alpha}$. Values significantly differ (**$p < 0.05$) from the control samples.

Cells were pretreated with either saline or saline containing cycloheximide (100 ug/ml) or saline containing actinomycin D (10 ug/ml) for 30 min at 37°C. Next the cells were rinsed with fresh saline (containing the appropriate inhibitor) and treated with LTD_4 or arachidonic acid for 10 min at 37°C. The supernatants were then collected and assayed for prostacyclin by radioimmunoassay. The concentrations of cycloheximide and actinomycin D_3 used were sufficient to inhibit by 95% the incorporation of [^{3}H]-Leu and [^{3}H]-U respectively yet did not affect cell viability as determined by trypan blue dye exclusion during the course of these experiments.

TABLE 2

Activity	Substrate	Inhibitor	Control	+LTD_4 (1 uM)	+(5R,6S) -LTD_4 (1 uM)
PLA_2	PC	None	7.4 ± 3.5	32.0 ± 3.1	7.6 ± 3.4
		Cycloheximide	8.1 ± 5.1	5.2 ± 3.7	
		Actinomycin D	6.4 ± 1.4	6.4 ± 1.4	
	PI	None	BG	BG	BG
		Cycloheximide	BG	BG	
		Actinomycin D	BG	BG	
	PE	None	2.1 ± 1.5	1.7 ± 2.0	2.0 ± 1.0
		Cycloheximide	4.7 ± 3.0	4.0 ± 2.0	
		Actinomycin D	1.9 ± 0.5	3.6 ± 1.2	
PLC	PC	None	6.2 ± 8.0	9.5 ± 3.0	6.4 ± 4.0
		Cycloheximide	7.1 ± 3.5	6.1 ± 2.9	
		Actinomycin D	7.6 ± 2.5	7.2 ± 2.2	
	PI	None	38.7 ± 5.2	48.0 ± 2.3	31.0 ± 3.0
		Cycloheximide	37.0 ± 3.0	45.0 ± 13.0	
		Actinomycin D	37.0 ± 1.4	42.0 ± 5.0	
	PE	None	BG	BG	ND
		Cycloheximide	BG	BG	
		Actinomycin D	BG	BG	

Effects of leukotriene D_4 on phospholipase A_2 and phospholipase C activities in endothelial cells (CPAE). Cells were treated for 3 min with leukotriene D_4 (1 uM) or (5R,6S)-leukotriene D_4 (1 uM) and sonicated and the phospholipase activies were measured. The assays for phospholipase A_2 activity and for phospholipase C activity were done in cell-free preparations. In some of the experiments, cycloheximide (100 ug/ml) or actinomycin D (10 ug/ml) was added to the cells 30 min prior to the addition of leukotriene D_4. Values expressed are in picomoles of product produced per min/mg of protein. The data represent the results from three experiments assayed in triplicate (mean ± S.D.). BG indicates no activity detected and ND indicates the experiment was not performed. For experimental details see reference 7. The substrates used were 1-palmitoyl-2-[^{14}C]-arachidonyl-*sn*-glycero-3-[choline-methyl-^{3}H]-phospho choline, 1-palmitoyl-2-[14C]-arachidonyl-sn-glycero-2-phospho-[3H]-ethanolamine and 1-stearoyl-2-[^{14}C]-arachidonyl-*sn*-glycero-3-phospho-[^{3}H]-myoinositol.

Next, we directly measured phospholipase A_2 activity in CPAE cell sonicates as a function of time after leukotriene D_4 treatment. A 3-fold increase in phospholipase A_2 activity was observed within 2 min of LTD_4 stimulation. Phospholipase A_2 activation directly preceded the formation of 6-keto $PGF_1\alpha$ synthesis (Clark, 1986a) in these cells (Figure 1). Thus, the 3-fold increase in measurable phospholipase A_2 activity in vitro was observed just prior to the accumulation of oxygenated eicosanoid metabolites in the cell supernatants. To determine the concordance between the increase in measurable phospholipase A_2 activity and the accumulation of 6-keto PGF1α the concentration dependence of these phenomenon and their specificity was examined (Clark, 1986). The biologically active isomer of leukotriene D_4 (5S, 6R)

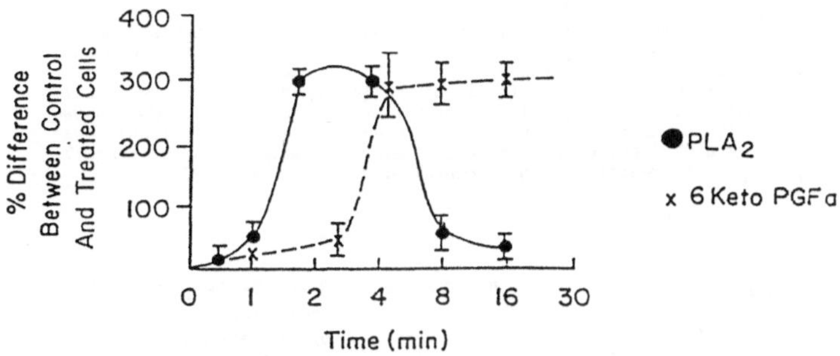

Figure 1. Effect of Leukotriene D_4 on the Induction of Phospholipase A_2 Activity and Prostanoid Synthesis in Endothelial Cells. Cells were treated with leukotriene D_4 (1 um) for the indicated times and assayed in triplicate in a cell-free preparation for phospholipase A_2 (PLA_2) activity using 1-palmitoyl-2-[^{14}C]-arachidonyl-sn-glycero-3-[choline-methyl-H]phosphocholine as substrate. Prostanoid synthesis was quantitated at the indicated times by radioimmunoassay. The values in these experiments were determined as percent increases above control values for phospholipase A_2 activity (°) or 6-keto PFG production (X). The data shown represent the (mean ± S.D.) from three experiments assayed in triplicate. The control values for PLA_2 and prostacyclin production are 7.4 pmol of lysophospholipid produced per min per mg of cell protein and 2.4 ng produced per ml per 10^5 cells per 10 min respectively.

resulted in a statistically significant increase in both measurable phospholipase A_2 activity as well as prostycyclin accumulation. This phenomenon was dose dependent and increases in both of these responses were observed at 100 nM and 100 mM leukotriene D_4 (Figure 2). This effect was saturable since increasing the LTD_4 concentration above 1 mM had no additional effect on either of these responses. Treatment of the CPAE cell line with (5R, 6S) LTD_4, the biologically inactive isomer, failed to increase either measurable phospholipase A_2 or eicosanoid accumulation even when present at 1 nM concentration. This result suggested that these responses were receptor mediated. This was confirmed using specific receptor antagonists. Addition of cycloheximide or actinomycin D completely attenuated the observable increase in phospholipase A_2 activity induced by LTD_4. Furthermore, accumulation of phospholipase C activity utilizing phosphatidylcholine, phosphatidylethanolamine or phosphatidylinositol was not enhanced by leukotriene D_4 treatment (Table 2). To determine whether the observed increases in measurable enzymic activity were related to alterations in the maximal velocity or substrate binding, Lineweaver-Burk analysis of the kinetic data was employed (Figure 3). Treatment of endothelial cells with leukotriene D_4 for 3 min resulted in an increase in the maximum velocity of fatty acid release from phosphatidylcholine substrate while no alteration in the binding affinity was observed. These results demonstrated that phospholipase A_2 activity is specifically augmented by treatment of the CPAE cell line with leukotriene D_4 in a fashion which is concentration dependent, saturable, stereospecific and is temporally related to prostacyclin synthesis (Clark, 1986a). Taken together, these data suggest that a protein is made soon after leukotriene D_4 treatment which enhances phospholpase A_2 maximum velocity. This protein was postulated as either a phospholipase or a protein which can accelerate measurable phospholipase A_2 activity (either by activation or ablation of inhibition).

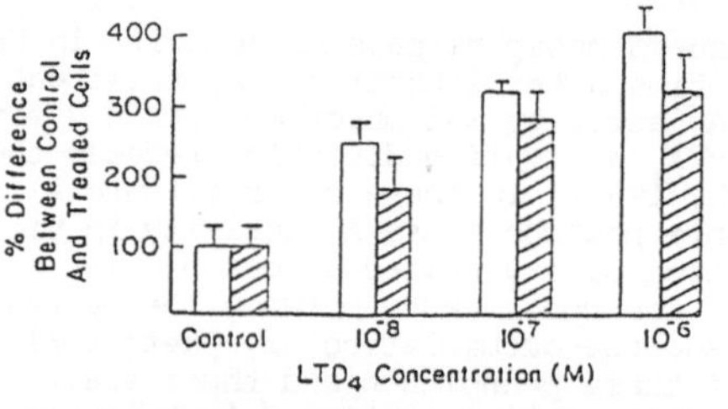

Figure 2. Effect of Increasing Concentrations of Leukotriene D_4 on Phospholipase A_2 Activity and on 6-Keto-$PGF_{1\alpha}$ Production. Endothelial cells were treated with leukotriene D_4 for 3 min at the indicated concentrations and assayed for phospholipase activity in a cell-free preparation with phosphatidylcholine (1-pal-mitoyl-2-[^{14}C]-arachidonyl-sn-glycero-3-phosphocholine) as substrate. The synthesis of 6-Keto-PGF_1 was determined by radioimmunoassays (New England Nuclear) following a 10-min treatment of cell lines with leukotriene D_4. Values in these experiments were expressed as percent increase above controls. Data are presented as a percent of control where the control levels of 6-Keto-PGF were 2.0 ng produced in 10 min by 10^5 endothelial cells and the control levels of phospholipase A_2 activity were 7.4 pmol of phosphatidylcholine hydrolyzed per min/mg of protein. The data shown represent results obtained from three experiments assayed in triplicate (mean ± S.D.). **, $p < 0.05$. For details on experimental protocol, see reference 7. The phospholipase A_2 reaction product contained from 2000 to 8000 cpm.

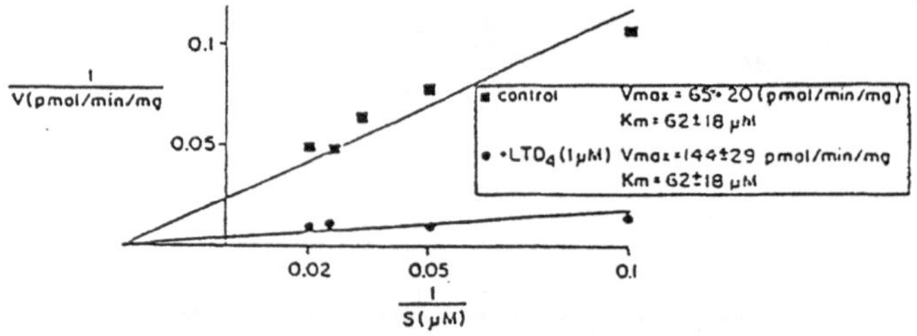

Figure 3. Lineweaver Burk Analysis of Phospholipase A_2 Activity of Endothelial Cells Treated with Leukotriene D_4. Cells were treated with leukotriene D_4 (1 um) for 3 min, and the phosphlipase A_2 activity was analyzed in a cell-free preparation using varying concentrations of phosphatidylcholine as substrate. The data shown represent the results from three experiments assayed in triplicate (mean ± S.D.). For experimental details see reference 7. The substrate used was 1-palmitoyl-2-[^{14}C]-arachidonyl-sn-glycero-3-[choline-methyl-^{3}H]-phosphocholine.

Since phospholipase activating proteins are abundant constituents of insect venomns (Haberman, 1972) we speculated that a similar protein might be synthesized after agonist stimulation in the CPAE cell line. To directly explore this possibility we generated antibodies to mellitin complexes which were subsequently injected intradermally into New Zealand white rabbits. The resulting antibodies were covalently bound to a HPLC immuno affinity column and subsequently CPAE cytosol was purcolated through the column. The column was extensively washed and specifically bound protein was eluted by application of pH3.1 Na acetate buffer. A single peak of stimulatory activity was found (Figure 4) (Clark, 1987b). To further purify this polypeptide, gel filtration chromatography was utilized. The fractions containing stimulatory activity were loaded onto a 1 x 24 cm HPLC gel filration column and eluates were again assayed for phospholipase A_2 stimulatory activity. The majority of stimulatory activity eluted as a single peak with an apparant molecular mass of 43 kD. The column eluant which contained the potent activating activity of phospholipase A_2 had absolutely no effect on phospholipase C activity utilizing either phosphatidylcholine or phosphatidylinositol as substrate (Table 3). To determine if the phospholipase stimulatory activity was a protein, aliquots of PLAP obtained from gel filtration chromatography were treated with trypsin (10 mg/ml), DNAse (30 mg/ml) or RNAse (30 mg/ml) for 30 min at 37° or boiled for 10 min. RNAse and DNAse treatment did not affect the stimulatory activity but trypsin digestion and boiling completely abolished stimulatory activity (Table 4) (Clark, 1987b). Taken together, these studies suggest that phospholipase activating activity was due to a protein constituent eluting from the gel filtration column. Accordingly, attempts were made at visualizing this protein on a silver stained sodium dodecyl sulfate-polyacrylamide gel.

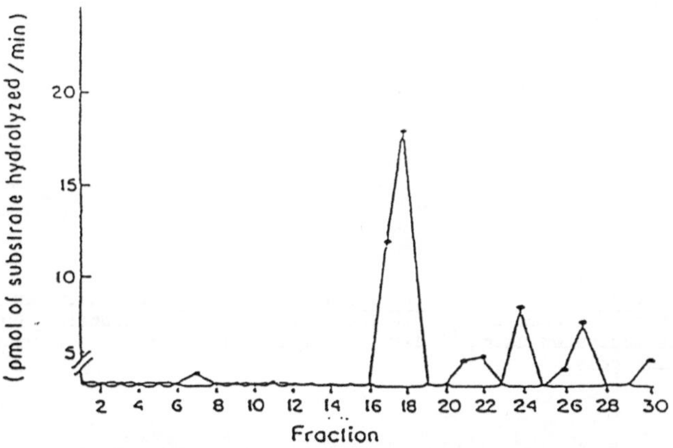

Figure 4. Affinity Chromatograph of PLAP. Affinity-purified anti-mellitin antibodies were immobilized on high performance affinity columns. CPAE cell-free sonicates were then loaded onto the column and the column was washed extensively prior to elution with pH 3.1 sodium acetate buffer. Fractions were collected and aliquots assayed in triplicate for stimulatory activity by mixing cellular sonicates containing endogenous phospholipase A_2, substrate and 100 ul of column eluent for 60 min at 37°C. Results are expressed as pmol hydrolyzed above control values (i.e., cell sonicate + substrate alone) which were 4 pmol/min.

TABLE 3

Activity	Phospholipase C		PLA_2	
	PC	PI	PC	PE
Sonicate Only	4.9 ± 0.6	2.1 ± 1.1	8.2 ± 0.9	2.6 ± 0.6
PLAP Only	.1	.1	.1	.1
Sonicate + PLAP	3.6 ± 0.7	1.6 ± 0.1	46.0 ± 0.3	2.1 ± 0.9

Enzyme and substrate specificity of PLAP activity. The effects of 10 units of PLAP on phospholipase A_2 and C activity were assayed using either phosphatidylcholine (PC) as substrate. Results shown are the mean ± S.D. for three separate experiments performed in triplicate and are expressed as pmol of reaction product produced/min/mg of protein.

TABLE 4

	PLAP Pretreatment	pmol substrate hydrolyzed/min/mg protein
Sonicate Only		10.0 ± 2.0
PLAP Only		0.1
Sonicate + PLAP	None	24.0 ± 3.1
	Trypsin	10.5 ± 1.6
	RNase	22.2 ± 4.0
	DNase	23.0 ± 2.1
	Boiling	11.5 ± 3.0

DNase, RNase and trypsin effects on PLAP activity. One unit of PLAP was treated with either RNase (30 ug/ul), DNase (10 ug/uL) or trypsin (10 ug/uL) for 30 min at 37°C or boiled for 10 min before being assayed for stimulatory activity. The results above are the mean S.D. of three separate experiments, each performed in triplicate and are expressed as pmol of reaction product produced/min/mg of protein.

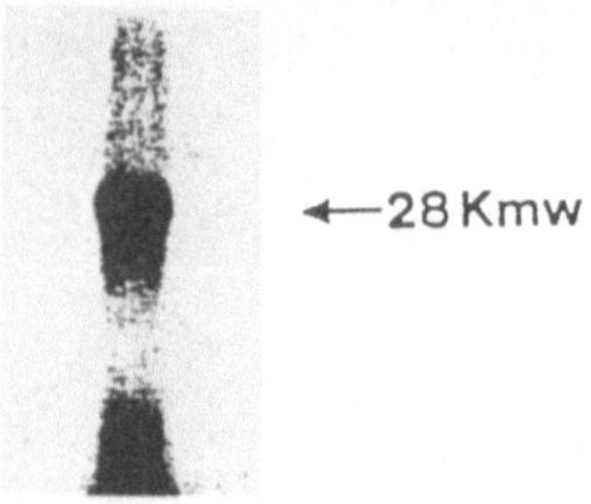

Figure 5. Polyacrylamide Gel Electrophoresis of PLAP. Polyacrylamide gel electrophoresis was used to analyze the results from the protein purification. Samples were iodinated using ^{125}I-Na and chloramine T, and electrophoresed on 12% acrylamide gels.

Although the sensitivity of the silver stain was such that approximately 10 ng of standard proteins could easily be visualized, no protein bands in the PLAP fractions were detected. Therefore to assess the molecular weight of the protein constituent, aliquots of the active fraction from gel filtration chromatography were iodinated using ^{125}I-Na and chloramine T prior to SDS-PAGE electorphoresis. As can be seen in figure 5, a single predominant protein band at 28 kD was present. The observed difference between the apparent molecular weight at which PLAP eluted from the gel filtration, and that visualized by the iodinated protein after autoradiography is likely due to the presence of aggregates formed in the gel filtration since detergent was necessary for successful elution of PLAP from this column (Clark, 1987b).

To directly assess the mechanism responsible for PLAP activation of phospholipase A_2, kinetic experiments were performed and analyzed by Lineweaver-Burk plots. As can be seen in figure 8, treatment of phospholipase A_2 with PLAP resulted in a 5-fold increase in V_{max} although no apparent alteration in substrate affinity for the phospholipase A_2 was present. Taken together, these results demonstrate that the CPAE cell line contains a protein constituent with an apparent molecular weight of 28 kD which cross-reacts with anti-mellitin antibodies and is a potent activator of phospholipase A_2. In order to directly determine whether LTD_4 treatment of CPAE cells resulted in an increase in PLAP formation, subsequent experiments were performed during which CPAE cells were briefly exposed to LTD_4 and PLAP formation was assessed by ELISA using anti-melliten antibodies (Figure 7). These experiments demonstrated a rapid induction of PLAP formation (within 2 min) followed by a rapid decline to basline level after 4 min. These effects were stereospecific

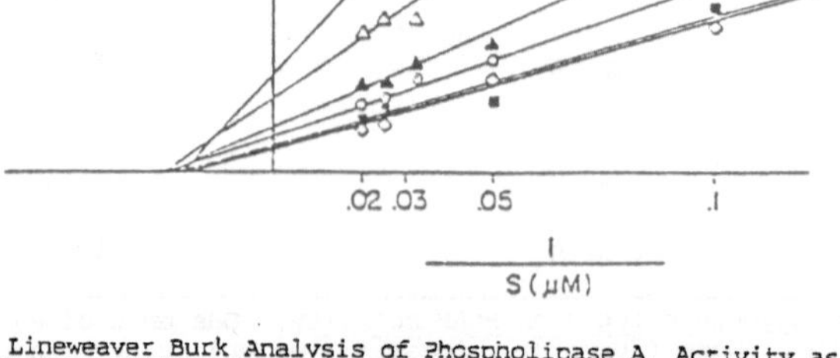

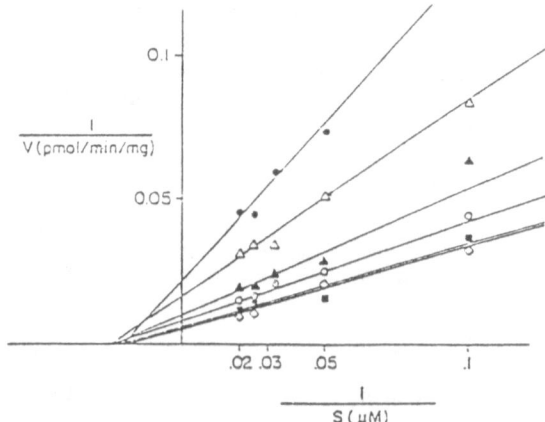

Figure 6. Lineweaver Burk Analysis of Phospholipase A_2 Activity as a Function of PLAP Concentration. Increasing concentrations of PLAP in units: 0 units, K_m = 49 V = 45; Δ 0.5 units, K_m = 33 V_{max} = 64; Δ 1.0 units, K_m = 45 V_{max} = 106; 1.5 units, K_m = 41 V_{max} 131; [] 4.0 units K_m = 50 V_{max} 222; 6.0 units K_m = 64 V_{max} = 192 were added to the cell-free sonicate (1 mg/ml), and phospholipase A_2 activity was assayed using increasing concentrations of phosphatidylcholine as a substrate.

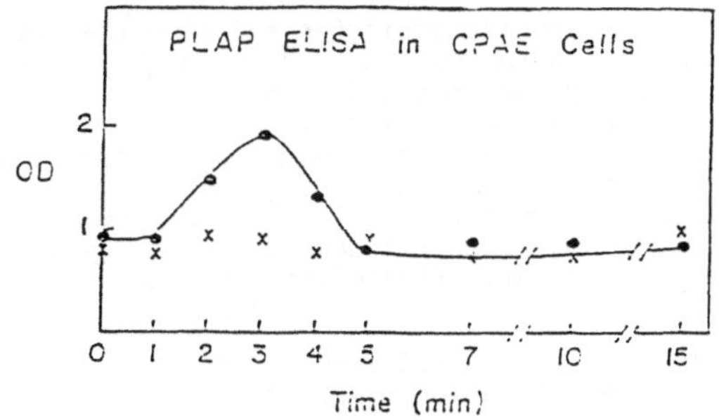

Figure 7. ELISA Assays for PLAP in Endothelial Cells Treated with Leukotriene D_4. Cells were treated (in quadruplicate) with LTD_4 (1 um) for the indicated periods of time prior to the addition of formalin (3% V/V). The cells were then permeabilized (0.05% Tween 20) and mellitin antibodies were then allowed to react with the cells. The amount of antibody bound was determined using peroxidase conjugated second antibody. In the experiments using antisense, synthetic DNA complimentary to the 5 prime end of the open reading frame or irrevelent DNA was added to the cells 25 μM for 4 hr prior to the start of the experiment.

since the inactive isomer (5R, 6S) LTD_4 failed to increase PLAP synthesis. To determine if the increase in PLAP synthesis induced by LTD_4 was specific or alternatively resulted from a general increase in total protein synthesis, the rate of actin synthesis was measure utilizing ELISA assays with anti-actin antibodies. Results from these control experiments demonstrated that actin synthesis did not increase during LTD_4 stimulation, supporting the hypothesis that agonist stimulation results in a selective increase in PLAP formation.

As a first step in determining the primary amino acid sequence of PLAP and to provide tools for future molecular biological analysis of PLAP induction, a cDNA library was screened utilizing anti-mellitin antibodies to isolate clones containing inserts encoding PLAP. After screening 5×10^6 recombinants a full-length cDNA clone was isolated. This clone hybridized to a 2.5 kb mRNA by Northern blot analysis. Furthermore, in preliminary experiments exposure of the CPAE cell line to LTD_4 resulted in an increase in the amount of hybridizable RNA (Figure 8). We have sequenced the insert in this clone

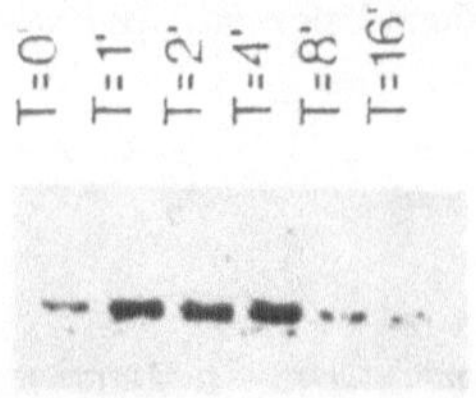

Figure 8. Northern Blot Analysis for the Induction of PLAP mRNA. Endothelial cells were treated with LTD_4 (1 um) for the indicated periods of time prior to the addition of hot (65°) phenol and total RNA was then extracted. The resulting RNA was then electrophoresed on an agarose gel (10 ug/lane). The gel was blotted and probed using the immunopositive PLAP cDNA clone which had been labeled with ^{32}P.

```
 1 gigsvlkvlttglpaliswikrk 23
        *****   * **    *
57 ESPLIAKVLTTE.PPIITPVRRT 78
```

Figure 9. The predicted amino-acid sequence of PLAP was deduced from the cDNA sequence (capital letters) and compared with the amino-acid sequence of mellitin (lower case letters). The region having greatest homology (as determined using the Wisconsin Genetics Group Data Base) is shown above.

utilizing dideoxy sequencing methodologies and have identified regions in the PLAP which are homologous with melliten. A region having a high degree of homology was found (Fig. 9) using the Wisconsin Data Base software. We have had this peptide synthesized and have shown that this peptide has phospholipase A_2 stimulatory activity, although it is much less potent than the intact native protein it is more potent than mellitin (Figure 10). In control experiments we have synthesized a peptide corresponding to the C terminal portion of PLAP. This portion of the molecule does not have any homology with melliten and this peptide was unable to stimulate phospholipase A_2 activity.

Antisense experiments were performed in order to assess the role of PLAP protein synthesis in the activation of phospholipase A_2 in endothelial cells in response to inflammatory stimuli. It has recently been shown that direct addition of synthetic antisense DNA to cultured cells specifically disrupts the synthesis of protein (Grahm, 1973; Heikkila, 1987). We synthesized a single strand of antisense DNA homologous to the first 20 bases of the 5 prime end of the open reading frame of the PLAP cDNA clone. This DNA was then added to the cells at a concentration of 25 mM for approximately 4 hrs prior to the start of the experiment. The addition of antisense to the cells successfully abolished the induction of PLAP protein synthesis following LTD_4 treatment as determined by ELISA assays (Figure 7). When these cultures were treated with LTD_4 (1 μM) phospholipase A_2 activation (Table 5) as well as the release of ^{3}H-arachidonic acid release (Table 6) and the production of prostacyclin (Table 7) were inhibited.

TABLE 5

AFFECTS OF ANTISENSE DNA ON LTD_4 ACTIVATION OF PLA_2

Treatment	Control	+LTD_4
None	7.0 ± 0.5	22.0 ± 2.0
Antisense DNA	6.0 ± 1.0	9.1 ± 1.1
Control DNA	6.0 ± 1.4	19,0 ± 2.0

Antisense or irrevelant DNA was added to the cells prior to the measurement of phospholipase A_2 activity as described in Table 2. All values are expressed as pmol of reaction product produced per min per mg of protein.

TABLE 6

EFFECTS OF ANTISENSE DNA ON LTD_4 INDUCED RELEASE OF ARACHIDONIC ACID

Treatment	Control	+LTD_4
None	4680	11680
Antisense DNA	4755	5516
Control DNA	4910	11180

Cells were incubated with ^{3}H-arachidonic acid 10 μCi/ml overnight next antisense or irrelevant DNA was added to the culture 25 μM for 4 hr prior to the addition of LTD_4. The amount of radiolabeled secreted into the supernatant following 10 min of treatment was quantitated. All values are expressed as cpm released by 10^5 cells following 10 min of treatment.

TABLE 7
EFFECTS OF ANTISENSE DNA ON LTD_4 INDUCED PGI_2 SYNTHESIS

Treatment	Control	$+LTD_4$
None	2.4 ± 0.6	4.3 ± 0.4
Antisence DNA	2.2 ± 0.5	2.1 ± 0.4
Control DNA	2.1 ± 0.6	4.4 ± 0.5

Cells were incubated with antisense DNA or irrelevent DNA 25 μM for 4 hr prior to the addition of LTD_4 (1 μM). The amount of prostacyclin released by the cells were quantitated by radioimmunoassays.

From these experiments we conclude that the mechanism by which leukotriene D_4 increases prostacyclin production in these bovine endothelial cells requires that the cells first synthesize PLAP mRNA, an event inhibitable by actinomycin D treatment. Next the PLAP mRNA must be translated in order for mature PLAP protein to be made. We can successfully abolish the translation of PLAP mRNA by using either the chemical inhibitor cycloheximid or the more specific PLAP antisense DNA technology. Irrespective of the type of inhibitor used it appears that PLAP is a prerequisite for the cells to successfully release arachidonic acid following leukotriene stimulation and that the cellular response to the increased eicosanoid biosynthesis following leukotriene treatment is largely determined by the ability of the cells to selectively express PLAP protein. Clearly other pathways are utilized by cells in order to regulate eicosanoid production. For example, bradykinin appears to increase prostacyclin production in the same cells by a pathway that is completely independent of PLAP and phospholipase A_2 and clearly cells such as platelets must produce eicosanoids by PLAP-independent mechanisms. Nonetheless, we conclude that PLAP plays a major role in LTD_4 signal transduction.

References

Clark M, Littlejohn D, Mong S, Crooke ST, (1986a) "The effects of leukotriene bradykinis and calcium ionophore on endothelial cell prostacyclin production" Prostaglandins 31:157-166.

Clark MA, Conway TM, Bennett CF, Crooke ST, Stadell JM (1986b) "Islet activating protein inhibits LTDy and LTCy but not Bradykinis or calcium ionophore induced prostaglandin synthesis in bovie endothelial cells" Proc Natl Acad Sci USA 83:7320-7324.

Clark MA, Littlejohn D, Conway TM, Mong S, Stiener S, Crooke ST. (1986c) "LTDy treatment of bovine aortic endothelial cells and murine smooth muscle cells results in an increase in phospholipase A_2 activity." J Biol Chem 261:10713-10718.

Clark MA, Simon PL, Chen MJ, Bomalaski JS. (1986d) "Response of the endothelium to inflammatory stimuli." Ann Reports Med Chem 22:235-243.

Clark MA, Chen MJ, Crooke ST, Bomalaski JS. (1987a) "Tumor necrosis factor induces phopholipase A2 activity and synthesis of a phospholipase A2 activity protein in endothelial cells." Biochem J 250:125-132.

Clark MA, Conway TM, Shorr RGL, Crooke ST. (1987b) "Identification and isolation of a mammalian protein which is antigenically and functionally related to the phospholipase A_2 stimulatory peptide mellitin" J Biol Chem 262:4402-4406.

Clark MA, Conway TM, Crooke ST. (1987c) "HPLC analysis of ^3H-arachidopic acid metabolites produced by smooth muscle and endothelial cells in response to LTDy" J Liq Chromatogr 10:2707-2719.

Flower RJ, Blackwell GJ. (1976) "The importance of phospholipase A_2 in prostaglandin synthesis" Biochem Pharmacol 25:2707–2719.

Grahm FL, Vandereb AJ. (1973) "Inhibition of SV40 antigen production by antisense DNA" Virology 52:456–463.

Haberman E. (1972) "The biochemistry and pharmacology of bee and wasp venom" Science 117:314–322.

Heikkila R, Schwab G, Wickson E, Pluzink DH, Watt R, Neekers L. (1987) "The role of C-MYC in lymphocyte proliferation: inhibition of S phase entry by C-myc antisense oligodeoprynucleotide" Nature 328:445–449 441–447.

Flower RJ, Blackwell GJ. (1976) "The importance of phospholipase A_2 in prostaglandin synthesis" Biochem Pharmacol 25:285–291 (1976).

SECOND MESSENGERS OF INSULIN ACTION

SECOND MESSENGERS OF INSULIN ACTION

Alan R. Saltiel
The Rockefeller University
New York, N.Y. 10021

Few hormones have received the attention accorded to insulin. It was one of the first peptide hormones to be purified, sequenced and undergo structural analysis. The physiology of insulin action was evaluated in the 1950's (Cahill 1971, Levine 1982). Studies on the binding of insulin to its cell surface receptor and the subsequent purification and cloning of the receptor (Rosen 1987, Goldfine 1987) served as a model of hormone-receptor interaction for other peptide hormones. Despite this intensive scrutiny, our understanding of the molecular events that link the insulin receptor to the regulation of cellular metabolism lags far behind that of other peptide hormones. One likely explanation for this slow progress may lie in the complicated nature of insulin action. The wide number and broad diversity of responses makes it improbable that a single mechanism could account for all of the actions of insulin. Rather, it is more likely that insulin action involves a network of interrelated and independent pathways with differing levels of divergence regarding mechanisms of regulation.

As it became clear that the well recognized mechanisms of signal transduction (i.e. cyclic nucleotides, ion channels) were not primarily responsible for explaining the actions of insulin, many investigators focused on the role of protein phosphorylation. Insulin simultaneously produces both the dephosphorylation of some proteins (i.e. glycogen synthetase, pyruvate dehydrogenase, hormone-sensitive lipase) along with the phosphorylation of other proteins (i.e. ribosomal S6, ATP citrate lyase). The dephosphorylation reactions induced by insulin result in the regulation of carbohydrate and lipid metabolism. Although it is not clear whether insulin-induced serine phosphorylation leads to changes in the catalytic activities of any enzymes, processes such as glucose transport and protein synthesis may require insulin-induced kinase reactions (Denton 1986). These and other observations have led to the emerging concept that two basic pathways, activation of protein phosphatases and kinases, may mediate many of insulin's actions.

Two major hypotheses have been proposed for coupling the insulin receptor to intracellular changes in protein phosphorylation: i) a phosphorylation cascade, the generation of a low molecular weight, diffusable second messenger. Evaluation of the phosphorylation cascade hypothesis has centered on site-directed mutagenesis and anti-receptor antibody experiments which suggest that the receptor tyrosine kinase is necessary for the full expression of insulin's activities (Chou et al. 1987, Ellis et al. 1986, Morgan and Roth 1987). The search for an insulin-dependent "second messenger" has been underway since the early 1970's. An insulin-sensitive substance was first detected in skeletal muscle that could acutely modulate glycogen synthetase activity *in vitro*. Similar kinds of extracts or substances of elusive chemical identity were subsequently identified in a variety of cell types, reported capable of regulating the activities of several insulin-sensitive enzymes (Jarett and Kiechle 1984). It is important to note that these two pathways need not be mutually exclusive, and in fact they may operate synergistically to coordinate a pattern of protein phosphorylation in response to insulin.

Biological characterization of a putative insulin second messenger

A low molecular weight substance was identified that was released from hepatic plasma membranes in response to insulin (Saltiel and Cuatrecasas 1986). The purification of this enzyme-modulating activity relied mainly on ion exchange, molecular sizing and phase partitioning procedures. The substance was negatively charged, insoluble in organic solvents and not adsorbed to reversed phase columns, indicating a relatively high degree of polarity. The chemical properties inferred from chromatographic behavior and susceptibility to specific chemical modification suggested an oligosaccharide-phosphate structure.

Initial studies on the biological activities of this substance focused on the modification of the activity of the low Km cAMP phosphodiesterase in fat cell membranes. The activity of this cAMP phosphodiesterase was activated acutely by the modulator, as reflected by an increase in the Vmax of the enzyme, with no appreciable effect on its Km. The purified substance could also modify *in vitro* other insulin-sensitive enzymes assayed in subcellular fractions, including adenylate cyclase, pyruvate dehydrogenase, phospholipid methyltransferase, cAMP dependent protein kinase and acetyl CoA carboxylase (Saltiel and Cuatrecasas 1988). Although the precise biochemical mechanism(s) by which this substance elicits its effects on these enzymes is unclear, the regulation of the activity of each of these enzymes might be explained by alterations in the state of phosphorylation of the enzyme or closely related regulatory factors. The specific regulation of protein phosphatase activitiy was observed in lysates from fat, brain and liver, although it is unknown whether the modulator produces the direct allosteric regulation of specific protein phosphatases.

Preliminary compositional analyses of the enzyme-modulating substance suggested the existence of inositol as a component. Several well known inositol phosphate-containing compounds were evaluated, but none exhibited significant enzyme-modulating activity, and they did not share the chemical properties, chromatographic or electrophoretic behavior, or insulin sensitivity of the enzyme modulator. These results suggested that the enzyme modulator might be an unusual derivative containing inositol phosphate. A potential clue was identified when a novel glycosylated derivative of inositol was found in certain cell surface proteins. This novel molecular species was shown to result from a covalent bond between certain proteins and phosphatidylinositol (PI) (Low and Saltiel 1988). This unusual linkage at the carboxy terminus of these proteins serves as an anchor for attachment to the plasma membrane. The protein is coupled via an amide bond to ethanolamine which is then attached through a phosphodiester linkage to an oligosaccharide that exhibits a terminal non-N-acetylated hexosamine glycosidically linked to the inositol ring of PI. The membrane-bound form of the protein can be converted to a water soluble form that contains a C-terminal glycosyl-inositol phosphate by digestion with a bacterial PI-specific phospholipase C (PLC), with the simultaneous liberation of diacylglycerol.

To evaluate the possibility that the insulin-dependent enzyme modulator might arise from the phosphodiesteratic hydrolysis of a structurally similar glycolipid, the PI-specific bacterial phospholipase C was added to liver plasma membranes, and the release of the modulator into the medium was assayed. In this series of experiments, PI-PLC was found to reproduce the effect of insulin in facilitating the generation of the enzyme modulator. The PI-PLC digestions generated a substance which was chromatographically, electrophoretically and chemically identical to that produced by insulin treatment, suggesting a basic inositol phosphate-glycan structure. Moreover, a potential precursor of the PI-PLC-generated substance could be extracted from liver membranes and chromatographically resolved from other known phosphoinositides. This precursor glycolipid was identified in a number of cell types and appeared to contain PI linked to a glycan though a non-N-substituted hexosamine. These experiments suggested that the enzyme modulator was produced as a result of a hormone-stimulated hydrolysis of this novel membrane-associated glycosyl-PI.

Considerable interest has been focused on the precise structure and biosynthesis of the glycosyl-PI precursor, and on its possible relationship to the glycosyl-PI protein anchor. Comparision of compositional analyses from several laboratories has indicated certain conserved and variant features. The precursor glycolipid appears to contain a basic core structure of PI-hexosamine. The T lymphocyte or BC3H1 cell-derived glycolipid appears to contain diacylglycerol as the major glycerolipid moiety (Saltiel et al. 1986, Saltiel et al. 1987, Gaulton et al. 1988, Farese et al. 1988), although in hepatoma cells a 1,2 alkylacylglycerol structure has been suggested (Mato et al. 1987). More recent studies have indicated another possible structural variation in the liver-derived glycosyl-PI, the presence of significant but variable amounts of *chiro*inositol, which perhaps accounts for the apparent lack of [^{3}H]*myo*inositol labeling in hepatoma cells (Mato et al. 1987, Larner et al. 1988). Another minor structural variation regards the presence of galactosamine in lieu of glucosamine (Larner et al. 1988). Distal to the hexosamine there appear to be considerable differences in glycan composition reported by different laboratories. Whether these apparently different structures represent molecules with distinct cellular functions or subcellular localizations remains to be determined.

The insulin-sensitive glycosyl-PI appears to exhibit considerable similarity to the glycosyl-PI protein anchor. The two types of glycolipid contain similar glycerolipid domains, sensitivity to PI-PLC and nitrous acid, the presence of inositol, nonacetylated glucosamine and a variable glycan region. However, the insulin-sensitive glycosyl-PI apparently lacks two of the features commonly observed in the protein anchor, ethanolamine and amino acids. Additionally, the molecular size of the insulin sensitive glycosyl-PI is smaller than similar molecules bound to protein. Despite the similarities between the insulin-sensitive glycosyl-PI and the cell surface protein anchor, the topological distribution of the insulin-sensitive glycolipid in the plasma membrane is uncertain. Some studies have suggested a cytoplasmic orientation, since treatment of cells with PI-PLC (presumably exhaustive) did not block the insulin-induced intracellular accumulation of the inositol glycan. Gaulton and Pawlowski (Gaulton and Pawlowski 1988) have reported that fluorescently labeled monoclonal antibodies to the insulin-sensitive glycosyl-PI show cytoplasmic staining, indicating an intracellular localization. In contrast, Alvarez et al. (Alvarez et al. 1988) suggested an extracellular location for the lipid, since 85% of the total cellular glycosyl-PI was apparently surface amidinated with isethionyl acetimidate. Although the free release from cells of the inositol glycan in response to insulin has not been reported, the exogenous addition of PI-PLC to intact cells produces an insulin-like activity, exogenous addition of purified inositol glycan to intact cells mimics many of the effects of insulin, and insulin may lead to the release from the cell surface of certain glycosyl-PI anchored proteins (see below). Thus, the issue of topology of the relevant glycolipid must remain open until definitive evidence is available.

The possibility that the protein-bound and free forms of glycosyl-PI are located on opposite sides of the plasma membrane leads to further uncertainty regarding their respective biosynthetic processing. One possibility to explain this apparent dilemma is that the early stages of glycosylation of PI occur on the cytoplasmic aspect of the endoplasmic reticulum. Upon attaining a certain level of glycosylation, a fraction of the glycolipid molecules ultimately destined for protein anchoring might be translocated from the cytoplasmic face of the endoplasmic reticulum membrane to the luminal side, in analogy to the translocation of the $(Man)_5(GlcNAc)_2$-lipid utilized for N-linked glycosylation of proteins. This translocation step may serve to segregate further biosynthetic modifications of the lipid molecules ultimately destined for protein attachment from those that will remain on the cytoplasmic face. Alternately, a final processing event, such as addition of a terminal sugar-phosphate, may serve to segregate those molecules not destined for translocation. In either case, the subsequent membrane trafficking to the cell surface might then result in a cytoplasmically-oriented free glycolipid and a cell surface-oriented protein-anchored glycolipid.

Regulation of Glycosyl-PI Hydrolysis

While insulin is known to cause increased labeling of several phospholipids, it has not been found to immediately stimulate the hydrolysis of PI or the polyphosphoinositides, and it does not induce calcium mobilization through the generation of inositol trisphosphate. In contrast, insulin does stimulate hydrolysis of glycosyl-PI, with the simultaneous production of the [^{3}H]inositol glycan and diacylglycerol that contains predominantly saturated fatty acids, but little if any arachidonic acid. The rapid production of this specific species of labeled diacylglycerol is not observed with agonists known to stimulate the hydrolysis of polyphosphoinositides (Saltiel et al. 1987). Thus, the unique species of diacylglycerol and the inositol glycan probably arise from the specific, insulin-sensitive hydrolysis of the free glycosyl-PI.

The production by insulin of a structurally distinct species of diacylglycerol in the absence of calcium mobilization may help explain the perplexing role of protein kinase C in insulin action. A number of conflicting observations regarding kinase C may be accomodated by invoking the involvement of distinct chemical forms or metabolic pools of diacylglycerol. Most agonists that cause kinase C activation do so by stimulating the hydrolysis of the polyphosphoinositides, leading to the generation of inositol phosphates and diacylglycerol that contains arachidonate in the C2 position. The absence of phosphoinositide turnover in response to insulin, as well as the scarcity of arachidonate in the insulin-generated diacylglycerol may cause a selective activation of kinase C depending upon the cell type, extent of activation, enzyme compartmentalization, substrate specificity or susceptibility to proteolysis. Perhaps the most interesting possibility includes the selective activation of isoforms of kinase C by structurally distinct diacylglycerols. Multiple forms of the enzyme were predicted by the cloning of multiple cDNA's, and several isozymes have now been chromatographically resolved. Some evidence suggests that these isoforms may exhibit distinct regulatory properties, especially with regard to calcium and diacylglycerol sensitivity as well as substrate specificity. Moreover, these isoforms may exhibit different tissue and/or subcellular distributions, or may be differentially susceptible to down regulation or proteolysis. Thus, the selective activation of protein kinase C or a fraction of isozymes may explain the apparent discrepancies between the biological actions of phorbol esters and insulin.

The observation that insulin specifically stimulated the hydrolysis of glycosyl-PI led to the search for a specific phospholipase C. Such an enzyme was isolated from a plasma membrane fraction of liver, using as an assay the liberation of diacylglycerol from the glycosyl-PI anchored variant surface glycoprotein from <u>T. brucei</u> (Fox et al. 1987). The catalytic activity appears to reside in a single polypeptide with an apparent molecular weight of about 52,000 daltons. The enzyme is calcium-independent, and is specific for glycosyl-PI; no hydrolysis of PI, PIP_2 or other phospholipids is observed under a variety of conditions. Although there is still no information available on the primary sequence or subcellular localization of the mammalian glycosyl-PI PLC, the cDNA for an enzyme with similar specificity was cloned and sequenced from <u>T. brucei</u> (Hereld et al. 1988). Interestingly, both immunohistochemical data and the predicted amino acid sequence indicate a cytoplasmic orientation for this parasitic enzyme.

How is the regulation of glycosyl-PI hydrolysis coupled to the activity of the insulin receptor? Although a phospholipase C capable of catalyzing this reaction has been purified, it has thus far been difficult to directly demonstrate the activation of this enzyme by insulin. Recent studies with anti-receptor antibodies or site directed mutagenesis indicate that the tyrosine kinase activity of the receptor may be necessary for the expression of all of the biological actions of insulin (Rosen 1987). In cells transfected with mutant insulin receptors that lack tyrosine kinase activity, glycosyl-PI hydrolysis is not observed in response to insulin, whereas cells transfected with wild type receptors respond normally (Rosen, personal communication). This suggests that the activation of the glycosyl-PI-specific phospholipase C by the receptor might occur as a consequence of a tyrosine kinase-induced cascade, possibly leading to changes in the state of phosphorylation of the enzyme or an associated regulatory factor. Alternatively, the autophosphorylation of the receptor on tyrosine residues could catalyze an intramolecular conformational change that initiates a membrane coupling event, involving noncovalent interactions with a regulatory factor. The latter possibility is supported by recent studies demonstrating that certain monoclonal antibodies to the insulin receptor which do not stimulate the receptor tyrosine kinase activity retain other insulin-mimetic properties regarding metabolic regulation (Hawley 1989). Perhaps these antibodies induce a conformational change in the receptor similar to that caused by autophosphorylation, resulting in some intramembrane coupling event. The coupling factor might be a specific GTP-binding protein, which in turn could activate the PLC. The involvement of a G protein in insulin action has been suggested in studies demonstrating that pertussis toxin (Elks 1987, Goren et al. 1985, Luttrell et al. 1988) or antibodies to the GTP-binding *ras* p21 protein (Korn et al. 1987) can block certain actions of insulin. Moreover, insulin inhibited the pertussis toxin-catalyzed ADP-ribosylation of a 41,000 Mr substrate in liver membranes (Rothenberg and Kahn 1988). Certain of the G proteins are relatively good substrates for the insulin receptor kinase *in vitro* (Zick et al. 1986). Although the direct phosphorylation of a G protein on tyrosine residues in response to insulin has not been observed *in vivo*, these *in vitro* data suggest at least the possibility of a high affinity interaction between certain G proteins and the receptor.

Hormone-stimulated release of glycosyl-PI anchored proteins

Another result of the insulin-induced hydrolysis of glycosyl-PI might be the release of glycosyl-PI-anchored proteins. Insulin has been reported to cause reduced levels of cellular alkaline phosphatase activity in BC3H1 cells (Romero et al. 1988). Like PI-PLC, insulin caused the release of the glycosyl-PI anchored heparan sulfate proteoglycan from rat hepatocytes (Ishihara et al. 1987). A form of lipoprotein lipase may be directly or indirectly anchored to the cell surface by a glycosyl-PI anchor, since enzyme activity or LPL protein labeled metabolically with [^{3}H]glucosamine or [^{32}P]orthophosphate can be specifically immunoprecipitated from the media of 3T3-L1 cells after treatment with PI-PLC. Additionally, membrane-associated lipoprotein lipase can be labeled at the cell surface with biotin and subsequently solubilized with PI-PLC. The kinetics of release of lipoprotein lipase activity from 3T3-L1 cells by insulin and PI-PLC are identical, indicating that the acute phase release by insulin may be due to activation of a glycosyl-PI-specific phospholipase (Chan et al. 1988). The observation that lipoprotein lipase is itself anchored or is tightly coupled to a glycosyl-PI anchored protein is of special significance, since this represents the first example of a protein anchored in this fashion that is known to be released from cells in response to hormones. Tissue or circulating levels of certain of the glycosyl-PI anchored proteins are altered in diabetic states, including alkaline phosphatase, 5'-nucleotidase, lipoprotein lipase and heparan sulfate proteoglycan. However, it is unclear whether insulin will lead to the release of all accessible glycosyl-PI anchored proteins, or only a specific subset. The exploration of this issue may help to resolve whether there are distinct hormonally-sensitive and insensitive "structural" pools of glycosyl-PI, similar to what has been proposed for metabolic pooling of the inositol lipids. Although it is not yet known whether the insulin-induced release of these glycosyl-PI anchored proteins is due to a phospholipase C, the hydrolysis of glycolipid molecules on opposite sides of the membrane in response to insulin also raises questions concerning the polarity or orientation of the relevant hydrolytic enzymes. If the free glycosyl-PI is located on the cytoplasmic face of the plasma membrane, it may be necessary to invoke at least two separate enzymes, located on different sides of the membrane. One possibility is that the hormone-sensitive, cell surface-oriented enzyme is a specific phospholipase D (Low and Prasad 1988).

Inositol glycan as a second messengers of insulin action

Despite the progress made in identifying the structure and biogenesis of the inositol glycan, it is still premature to regard this compound as a second messenger for any of the actions of insulin. The apparent insulin-dependency, rapidity and extent of the generation of the inositol-glycan are consistent with the properties expected for second messengers. However, many questions remain concerning precise chemical structure as well as the nature of the insulin-mimetic properties. Thus far, the actions of the inositol glycan have been explored mainly in subcellular assays, so that the extent to which this molecule reproduces the actions of insulin in intact cells remains unclear. In recent studies, these issues have been addressed by evaluating the effects of the inositol glycan in fat and liver cells. Purified preparations of these compounds mimic the lipogenic (Saltiel and Sorbara-Cazan 1987) and anti-lipolytic (Kelly et al. 1987) actions of insulin, as well as the regulation of phosphorylase a (Alvarez et al. 1987), pyruvate kinase (Alvarez et al. 1987), cAMP levels (Alvarez et al. 1987), and pyruvate dehydrogenase (Gottschalk and Jarett 1988). Moreover, they produce specific protein phosphorylation patterns similar to those produced by insulin in intact fat or liver cells (Alemany et al. 1987). However, the compound does not appear to modulate glucose transport activity or ribosomal S6 kinase. An oligosaccharide with chemical and chromatographic properties similar to the inositol glycan has recently been isolated from conditioned media of Reuber hepatoma cells (Witters and Watts 1988). This substance stimulates both [^{3}H]thymidine uptake and activation of acetyl CoA carboxylase in a manner kinetically indistinguishable from and not additive with insulin. In contrast, this glycan does not stimulate amino acid uptake or tyrosine aminotransferase induction in hepatoma cells nor glucose transport in 3T3-L1 cells. It is especially interesting to note that the insulin-mimetic actions of the inositol glycan appear to be limited to those anabolic activities that are 1) more or less insulin-specific (i.e. are generally not reproduced by other growth factors like EGF or PDGF) and 2) are probably mediated by enhanced dephosphorylation (i.e. stimulation of protein phosphatase). Thus, the selective ability of the inositol glycan to mimic only a subset of the actions of insulin provides further evidence for diverse pathways of signal transduction in the actions of the hormone.

Although the inositol glycan appears to be a promising candidate for a second messenger of insulin action, a number of issues remain to be resolved. The ultimate proof of a role for this compound as a second messenger will critically depend, among other things, on the determination of its precise structure. This may be complicated by the chromatographic resolution of multiple species of these molecules, perhaps reflecting the existence of distinct forms with different enzyme-modulating functions or subcellular distributions. It will also be important to produce these compounds in large quantity and homogeneous form, perhaps by organic synthesis, in order to reevaluate each of their biological activities in detail. Moreover, the biosynthetic route, mode of production and degradation, relationship to the insulin receptor, and the precise biochemical actions of these molecules need further exploration. It will be especially important to determine the topological distribution of the glycosyl-PI precursor in the plasma membrane. More detailed molecular characterization of the glycosyl-PI-specific phospholipase C will be necessary. Development of inhibitors or neutralizing antisera to this enzyme, and eventually site-directed mutagenesis studies, should help to define the functional role of this reaction in the pleiotropic actions of insulin.

A hypothetical model is presented illustrating the hydrolysis of glycosyl-phosphoinositides in plasma membranes. The interaction of insulin with its receptor causes the activation of the receptor tyrosine kinase, probably the initial signal for receptor function. The activated receptor is then coupled by an unknown mechanism that may involve an intermediate G protein to the stimulation of one or more phospholipases C specific for glycosyl-PI. This enzyme or group of enzymes then catalyzes the hydrolysis of a free glycosyl-PI that might be on the cytoplasmic side of the plasma membrane, resulting in the intracellular generation of the enzyme-modulating inositol phosphate glycan and a species of diacylglycerol that may cause a selective activation of the protein kinases C. Similarly, the activation of the insulin receptor might also cause the release of anchored proteins like heparan sulfate proteoglycan, alkaline phosphatase or lipoprotein lipase, through an unknown mechanism that may involve an anchor degrading enzyme.

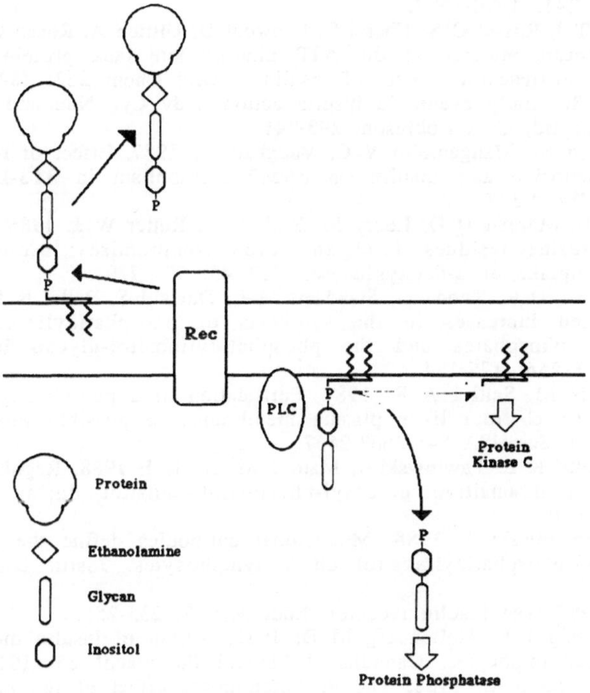

Figure 1. The metabolism of glycosyl-PI in insulin action

References

Alemany S, Mato J M, Stralfors P: 1987. Phosphodephospho control by insulin is mimicked by a phospho-oligosaccharide in adipocytes. Nature 330: 77-79.

Alvarez J F, Cabello M A, Feliu J E, Mato J M: 1987. A phospho-oligosaccharide mimics insulin action on glycogen phosphorylase and pyruvate kinae activities in isolated rat hepatocytes. Biochem Biophys Res Comm 147: 765-771 .

Alvarez J F, Varela I, Ruiz-Albusac J M, Mato J M: 1988. Localisation of the insulin-sensitive phosphatidylinositol glycan at the outer surface of the cell membrane. Biochem Biophys Res Comm 152: 1455-1462.

Cahill Jr G F: 1971. Physiology of insulin in man. Diabetes 20: 785-797.

Chan B L, Lisanti M R, Rodriguez-Boulan E, Saltiel A R: 1988. Lipoprotein lipase is anchored to the cell surface by glycosyl-phosphatidylinositol: regulation of release by insulin. Science 241: 1670-1672.

Chou, C D, Cull T J, Russel O S, Cherzi R, Lebwohl D, Ullrich A, Rosen O M: 1987. Human insulin receptors mutated at the ATP binding site lack protein kinase activity and fail to mediate postreceptor effects of insulin. J Biol Chem 262: 1842-1847.

Denton R M: 1986. Early events in insulin action. Adv Cyc Nuc and Protein Phos Res, vol 26. P Greengard, G A Robinson, 293-341.

Elks M L, Jackson M, Manganiello V C, Vaughan M: 1987. Effect of NG-(2-2-phenylisopropyl) adenosine and insulin on cAMP metabolism in 3T3-L1 adipocytes. Am J Physiol 252: C342-C348.

Ellis L, Clauser E, Morgan D O, Edery M, Roth R A, Rutter W J: 1986. Replacement of insulin receptor tyrosine residues 1162 and 1163 compromises insulin-stimulated kinase activity and uptake of 2-deoxyglucose. Cell 45: 721-729.

Farese R V, Cooper D R, Konda G, Standaert M L, Davis J S, Pollet R J: 1988. Insulin provokes co-ordinated increases in the synthesis of phosphatidylinositol, phosphatidylinositol phosphates and the phosphatidylinositol-glycan in BC3H-1 myocytes. Biochem J 256: 175-184.

Fox J A, Soliz N M, Saltiel A R: 1987. Purification of a phosphatidylinositol-glycan specific phospholipase C from liver plasma membranes; a possible target of insulin action. Proc Natl Acad Sci USA 84: 2663-2667.

Gaulton G N, Kelly K L, Pawlowski J, Mato J M, Jarett J: 1988. Regulation and function of an insulin-sensitive glycosyl-phosphatidylinositol during T lymphocyte activation. Cell 53: 970.

Gaulton G N, Pawlowski J: 1988. Monoclonal antibodies define the regulation and location of glycosyl-phosphatidylinositol on T lymphocytes. Joslin Diabetes Symposium.

Goldfine I D: 1987. The insulin receptor. Endo Rev 8: 235-255.

Goren J J, Northrup J K, Hollenberg M D: 1985. Action of insulin modulated by pertussis toxin in rat adipocytes. Canadian J Physiol Pharmacol 63: 1017-1022.

Gottschalk W K, Jarett L: 1988. The insulinomimetic effect of the polar head group of an insulin-sensitive glycophospholipid on pyruvate dehydrogenase in both subcellular and whole cell assays. Arch Biochem Biophys 261: 175-185.

* Hawley D M, Maddux B A, Patel R G, Wong K-Y, Mamula P W, Firestone G L, Burnetti A, Verspohl E, Goldfine I D: 1989. Insulin receptor monoclonal antibodies that mimic insulin action without activating tyrosine kinase. J Biol Chem 264: 2438-2444.

* Hereld D, Hart G W, Englund P T: 1988. cDNA encoding VSG lipase, the glycosyl-phosphatidylinositol-specific phospholipase C. Proc Natl Acad Sci USA 85: 8914-8918.

Ishihara M, Fedarko N S, Conrad H E: 1987. Involvement of phosphatidylinositol and insulin in the coordinate regulation of proteoheparan sulfate metabolism and hepatocyte growth. J Biol Chem 262: 4708-4716.

Jarett L, Kiechle P L: 1984. Intracellular mediators of insulin action. Vit and Horm 41: 51-78.

Kasuga M, Karlsson F A, Kahn C R: 1982. Insulin stimulates the phosphorylation of the 95,000 dalton subunit of its own receptor. Science 215: 185-187.

Kelly K L, Mato J M, Merida I, Jarett L: 1987. Glucose transport and antilipolysis are differentially regulated by the polar head group of an insulin sensitive glycophospholipid. Proc Natl Acad Sci USA 84: 6404-6407.

Korn L J, Siebel C W, McCormick F, Roth R A: 1987. *Ras* p21 as a potential mediator of insulin action in *Xenopus* oocytes. Science 236: 840-843.

Larner J, Huang L C, Schwartz C F W, Oswald A S, Shen T-Y, Kinter M, Tang G, Zeller K: 1988. Rat liver insulin mediator which stimulates pyruvate dehydrogenase phosphatase contains galactosamine and D-chiroinositol. Biochem Biophys Res Comm 151: 1416-1426.

Levine R: 1982. Insulin: the effects and mode of action of the hormone. Vitamin and Hormone 39: 145-173.

Low M G, Prasad A R S: 1988. A phospholipase D specific for the phosphatidylinositol anchor of cell surface proteins is abundant in plasma. Proc Natl Acad Sci USA 85: 980-984.

Low M G, Saltiel A R: 1988. Structural and functional roles of glycosyl-phosphatidylinositol in membranes. Science 239: 268-275.

Luttrell L M, Hewlett E L, Romero G, Rogol A K: 1988. Bordetella pertussis toxin treatment attenuates some of the effects of insulin in BC3H-1 murine myocytes. J Biol Chem 263: 6134-6141.

Mato J M, Kelly K L, Abler A, Jarett L, Corkey B E, Cashel J A, Zopf D: 1987. Partial structure of an insulin-sensitive glycophospholipid. Biochem Biophys Res Comm 146: 764-772.

Mato J M, Kelly K L, Abler A, Jarret L: 1987. Identification of a novel insulin-sensitive glycophospholipid from H35 hepatoma cells. J Biol Chem 262: 2131-2137.

Morgan D O, Roth R A: 1987. Acute insulin action requires insulin receptor kinase activity: Introduction of an inhibitory monoclonal antibody into mammalian cells blocks the rapid effects of insulin. Proc Natl Acad Sci USA 84: 41-45.

* Romero G, Luttrell L, Rogol A, Zeller K, Hewlett E, Larner J: 1988. Phosphatidylinositol-glycan anchors of membrane proteins: potential precursors of insulin mediators. Science 240: 509-511.

Rosen O M. Personal communication.

Rosen O M: 1987. After insulin binds. Science 237: 1452-1458.

Rothenberg P L, Kahn C R: 1988. Insulin inhibits pertussis toxin-catalyzed ADP-ribosylation of G-proteins. J Biol Chem 263: 15546-15552.

Saltiel A R, Cuatrecasas P: 1986. Insulin stimulates the generation from hepatic plasma membranes of modulators derived from an inositol glycolipid. Proc Natl Acad Sci USA 83: 5793-5797.

Saltiel A R, Cuatrecasas P: 1988. In search of a second messenger for insulin. Am J Physiol 255: C1-C11.

Saltiel A R, Fox J A, Sherline P, Cuatrecasas P: 1986. Insulin stimulated hydrolysis of a novel glycolipid generates modulators of cAMP phosphodiesterase. Science 233: 967-972.

Saltiel A R, Sherline P, Fox J A: 1987. Insulin-stimulated diacylglycerol production results from the hydrolysis of a novel phosphatidylinositol glycan. J Biol Chem 262: 1116-1121.

Saltiel A R, Sorbara-Cazan LR: 1987. Inositol glycan mimics the action of insulin on glucose utilization in rat adipocytes. Biochem Biophys Res Comm 149: 1084-1092.

Witters L A, Watts T D: 1988. An autocrine factor from Reuber hepatoma cells that stimulates DNA synthesis and acetyl CoA carboxylase: Characterization of biologic activity and evidence for a glycan structure. J Biol Chem 263: 8027-8036.

Zick Y, Sagi-Eisenberg R, Pines M, Gierschik P, Spiegel A M: 1986. Multisite phosphorylation of the α subunit of transducin by the insulin receptor kinase and protein kinase C. Proc Natl Acad Sci USA 83: 9294.

CONTROL BY TSH OF THE PRODUCTION BY THYROID CELLS OF AN INOSITOL PHOSPHATE-GLYCAN, PUTATIVE MEDIATOR OF THE cAMP-INDEPENDENT EFFECTS IN THE GLAND

L. Martiny, B. Thuilliez, B. Lambert, F. Antonicelli,
C. Jacquemin* and B. Haye

Biochimie Faculté des Sciences 51062 Reims France
*INSERM U.96 Hôpital de Bicêtre 94275 Le Kremlin-Bicêtre France

INTRODUCTION

In thyroid tissue, an effect of thyrotropin (TSH) was shown to affect the turnover rates of phospholipids (Morton and Schwartz, 1953) and that of phosphatidylinositol in particular (Freinkel, 1957) several years before the same hormone was found to stimulate cyclic AMP accumulation (Klainer et al., 1962). However, progress in the field of cyclic AMP biochemistry was so rapid, that there was a tendency to consider cyclic AMP to be the sole mediator of all the effects of TSH (Pastan and Macchia, 1967). But, the "phospholipid effect" of TSH was clearly demonstrated to be cyclic AMP independent in 1970 (Jacquemin and Haye, 1970 ; Scott et al., 1970). This was followed by the discovery of a TSH-dependent cytosolic phospholipase C acting on phosphatidylinositol (Haye and Jacquemin, 1974), and of the transient formation of diacylglycerol (Igarashi and Kondo, 1980). These observations correlate well with the first "phosphatidylinositol cycle" proposed by Michell (1975).

Interest in thyroid phosphoinositide metabolism was stimulated by the identification of two potential second messengers formed by hydrolytic products of phosphatidylinositol bis-phosphate, inositol tris-phosphate (Berridge, 1983) and diacylglycerol (Nishizuka, 1984). But, attempts to demonstrate that TSH stimulated this pathway in thyroid cells were not very convincing (Graff et al., 1987 ; Bone et al., 1986 ; Taguchi and Field, 1988) except in human tissue (Laurent et al., 1987). Nevertheless, inositol tris-phosphate formation was clearly promoted by cholinergic muscarinic and α_1-adrenergic agonists in all these cells.

In 1982, we observed that chronic treatment of pig thyroid cells in culture with 0.1 mU/ml TSH for 4 days produced a large specific increase in the turnover rate of phosphatidylinositol (Gerard et al., 1982). This "chronic phospholipidic effect" of TSH was cyclic AMP-dependent in contrast to the "acute effect" previously described and could also be produced by chronic treatment with dibutyryl cyclic AMP, prostaglandin E_2 or forskolin (Gerard et al., 1982 ; Haye et al., 1985). We therefore studied the incorporation of [^{3}H]-inositol into the various phosphoinositides and the basal rate of inositol phosphate release, in TSH-or forskolin-treated cells. There was a very great increase in phosphatidylinositol labelling which exceeded that of phosphatidylinositol mono- and bis-phosphate by a much greater margin than could be expected from a precursor-product relationship.

Moreover, although the release of inositol monophosphate was increased by this chronic treatment, it is unlikely to be the sole product of the steadily increased turnover of phosphatidylinositol. We have therefore explored the possibility that other metabolites of phosphatidylinositol, which were not identified by the analytical procedure previously used, are produced. This study focusses on two candidates, an inositol phosphate-glycan and its lipophilic precursor, phosphatidylinositol-glycan, which have been found to be mediators of certain insulin effects in several types of cells (Saltiel and Cuatrecasas, 1986 ; Saltiel et al., 1986 ; Mato et al., 1987 ; Alemany et al., 1987).

RESULTS

Labelling, isolation and characterization of phosphatidylinositol- and inositol phosphate-glycans

Pig thyroid cells were isolated and cultured as previously described (Gerard et al., 1982). The cells were incubated with 0.1 mU/ml TSH for 24h before adding the putative labelled precursor ([^{3}H]-glucosamine, [^{3}H]-inositol, 20 µCi/20 x 10^6 cells). Two days later, the cells were washed, resuspended in 3 ml Earle-hepes buffer (pH 7.2) and incubated for 20 min at 37°C. The reaction was stopped by adding $HClO_4$, final concentration 5% (v/v) and the cells were removed by centrifugation at 1200 g. Cell pellet and supernatant were analysed separately.

The cells were extracted with 2.5 ml chloroform-methanol-0.45 M HCl (20:40:13) (v/v). The lower phase, containing phosphatidylinositol-glycan, was evaporated under a stream of nitrogen. The residue was dissolved in 60 µl of chloroform-methanol-0.1 M HCl (75:25:2) (v/v) and chromatographed on oxalate-impregnated silica gel-TLC in chloroform- methanol-4 M NH_4OH containing 0.1 % 1,2-cyclohexanediaminetetraacetic acid (CDTA) (45:35:10) (v/v) (Igarashi and Chambaz, 1987). Five-millimeter-wide bands were scraped off and counted for radioactivity. Lipid markers were included and visualized by iodine staining (Fig 1).

The supernatants were neutralized with 9 M KOH, clarified by centrifugation at 3000 g, and run onto Dowex AG1 X8 (200-400 mesh) columns. The columns were washed with 9 ml H_2O and 10 ml 0.1 M ammonium formate and inositol phosphate-glycan was eluted with 0.1 M formic acid (Kelly et al., 1987 b). Formic acid was removed by lyophilization and the concentration of inositol phosphate-glycan was measured by reacting the non-acetylated amino-group of the glucosamine moiety with fluorescamine (Merida et al., 1983). A molecular weight of 1500 Da was assumed.

Effects of chronic and acute TSH treatment on inositol phosphate-glycan synthesis

The incorporation of [^{3}H]-glucosamine into inositol phosphate-glycan and its immediate precursor phosphatidylinositol-glycan in cells treated for 3 days with TSH and forskolin, and control cells was measured. The amount of labelled inositol phosphate-glycan in the medium was more than doubled by chronic treatment with 0.1 mU/ml TSH or 10^{-5} M forskolin : the radioactivity released by 20 x 10^6 cells in a 17-min incubation after removal of radioactive glucosamine, was 3889 cpm for control cells, 9573 cpm for TSH-treated cells and 10459 cpm for forskolin-treated cells. The amount of radioactive phosphatidylinositol-glycan decreased from 665 cpm to 362 cpm in

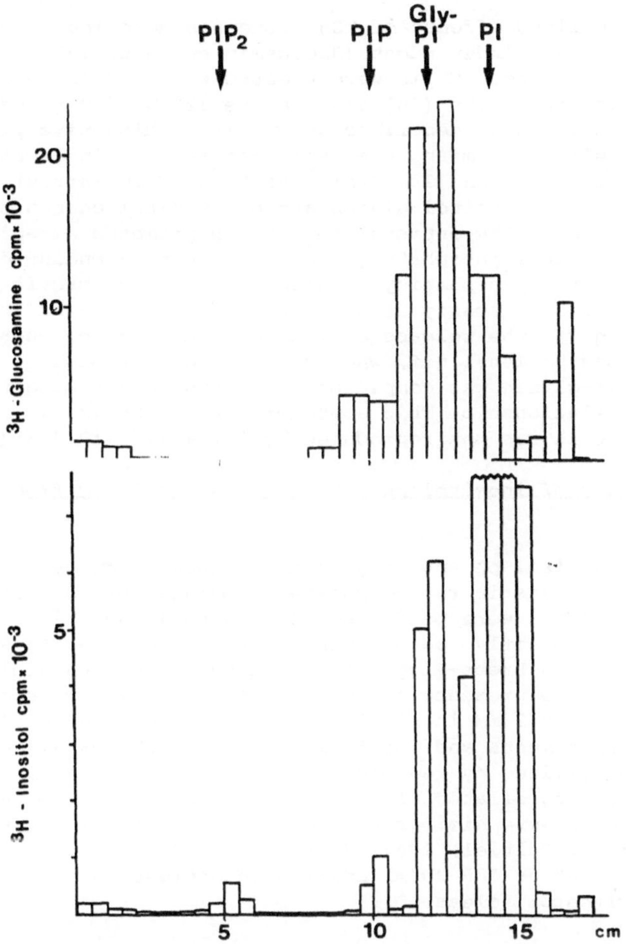

Fig. 1. Thin-layer chromatographic analysis of [^{3}H]-inositol and [^{3}H]-glucosamine labelled phosphatidylinositol-glycan in thyroid cells. The organic phase obtained from 3 day TSH-treated cells labelled with [^{3}H]-glucosamine (upper part) or [^{3}H]-inositol (lower part) as described, was analysed by TLC in chloroform : methanol : NH4OH (45:35:10) containing 0.1 % CDTA. PIP2 : phosphatidylinositol 4,5 bis-phosphate, PIP : phosphatidylinositol 4-phosphate, PI : phosphatidylinositol, gly-PI : phosphatidylinositol-phosphate. PI culminated at 97500 cpm at the 29th strip.

control cells, but increased from 231 cpm to 662 cpm in TSH-treated cells and from 155 cpm to 387 cpm in forskolin-treated cells over the same 17-min period. The radioactivity of inositol phosphate-glycan was always greater than that of phosphatidylinositol-glycan, suggesting that the pool of phosphatidylinositol-glycan was resplenished at the expense of labelled phosphatidylinositol. In these conditions, the increased labelling of inositol phosphate-glycan could be due to the increased turnover of phosphatidylinositol elicited by the chronic treatment of the cells with TSH or forskolin, in a cyclic AMP-dependent mechanism. The increased labelling of inositol phosphate-glycan was probably related to the "chronic phospholipid effect" of TSH described previously (Gerard et al., 1982).

The "acute phospholipid effect" of TSH, long ago described (Freinkel, 1957) which is cyclic AMP independent (Jacquemin et Haye, 1970 ; Scott et al., 1970) was also examined. Cells were treated with TSH for 3 days and labelled for the last 48 h with [^{3}H]-glucosamine (20 µCi/20 x 10^6 cells). These cells were processed as described above. The radioactive precursor was removed by washing with medium and the cells resuspended in fresh medium, were incubated for 15 min. TSH (10 mU/ml) was then added (arrow) and the labelling of phosphatidylinositol-glycan and of inositol phosphate-glycan were followed for 20 min. The introduction of TSH promoted a rapid increase in the release of inositol phosphate-glycan and a corresponding decrease in its immediate precursor, phosphatidylinositol-glycan. The previously observed quantitative discrepancy between the radioactivity of the two pools was again found (Fig 2). The release of radioactive inositol phosphate-glycan was also found to level off, when the radioactive pool of phosphatidylinositol-glycan was completely depleted, confirming their strict precursor-product relationship. This acute effect of TSH is cyclic AMP-independent, because it was not reproduced by forskolin (10^{-4} M) (not shown).

Biological effects of inositol phosphate-glycan isolated from thyroid cells

The effects of inositol phosphate-glycan and insulin on cyclic AMP accumulation and lipolysis in rat adipocyte stimulated by isoproterenol were measured. The experiments were performed as previously described (Lambert and Jacquemin, 1964). Briefly, adipocytes isolated according to Rodbell (1964) were incubated in Krebs-Ringer bicarbonate buffer containing fatty acid-free albumin in the presence or absence of 10^{-6} M isoproterenol, insulin (100 µU/ml) or 0.2 mM inositol phosphate-glycan. Incubation was for 1 h in lipolysis experiments and for 6 min in cyclic AMP experiments. Cyclic AMP accumulation determined by radioimmunoassay (Cailla et al., 1973), and glycerol release were correlated with the total triacylglycerol content of the cells. The results were similar to those reported for insulin mediators (Kelly et al., 1987 a; Saltiel, 1987) (not shown). 0.2 mM inositol phosphate glycan was as efficient as 100 µU/ml insulin in decreasing both cyclic AMP accumulation and glycerol release.

The acute effects of inositol phosphate-glycan on cyclic AMP accumulation in thyroid cells were more complex and depended on the presence or absence of TSH in the test, and on the physiological state of the cells, i.e. control or TSH treated cells. The lowest concentration of inositol phosphate-glycan produced a slight stimulation of both acute TSH-stimulated (10 mU/ml) and unstimulated control cells, whereas higher concentrations were inhibitory (not shown). Inositol phosphate-glycan alone induced little or no cyclic AMP accumulation by 3 day TSH-treated cells, but produced a very strong concentration-dependent inhibition in the presence of TSH (10mU/ml) or of forskolin (10^{-4} M) (Fig 3).

The effect of inositol phosphate-glycan on protein-bound iodine (PBI), a characteristic thyroid cell parameter (Gerard et al., 1981) was also tested (Fig 4). Three-day TSH-treated cells were incubated with [^{125}I]- iodide (1µCi/10^6 cells) and challenged for 45 min with increasing concentrations of TSH and/or inositol phosphate-glycan. The [^{125}I]-PBI was determined after trichloracetic acid precipitation. The chronic treatment with TSH caused a very large increase in PBI, which was further enhanced by the 45-min challenge with a low concentration of TSH (0.1 mU/ml). Higher hormone concentrations were inhibitory ; the PBI level after a 45 min challenge with 10 mU/ml TSH was only 15 % of the basal. A low concentration (10^{-7} M) of inositol phosphate-glycan did not affect PBI, in the presence or absence of 0.1 mU/ml TSH. Higher concentrations of inositol phosphate-glycan evoked a

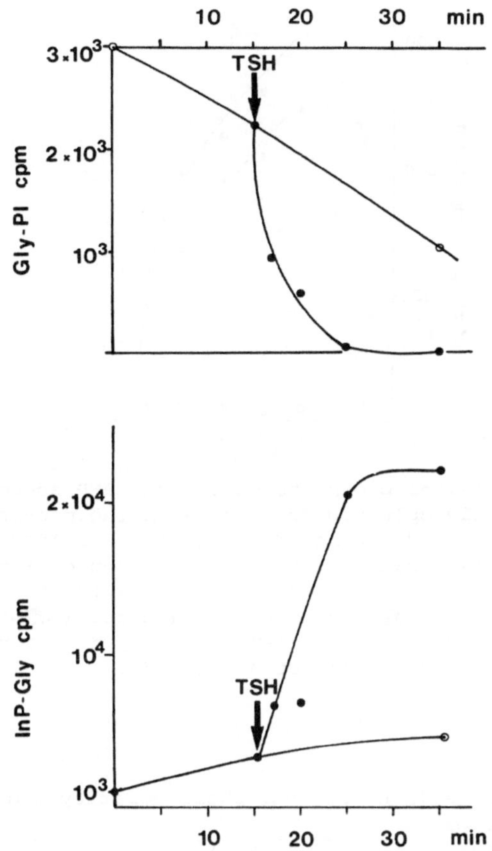

Fig. 2 Kinetic study of the acute action of TSH on the production of inositol phosphate-glycan from phosphatidylinositol-glycan in [^{3}H]-glucosamine labelled thyroid cells. Three day TSH (0.1mU/ml)-treated cells, labelled for the last 48 h with [^{3}H]-glucosamine (20 µCi/20 x 10^6 cells) were processed as described. After 15 min incubation, TSH (10 mU/ml) was added and a kinetic study was performed (—●—). The amounts of phosphatidyl-inositol-glycan (Gly-Pi) (upper part) and of inositol phosphate-glycan (InP-Gly) (lower part) measured in a representative experiment (among 3 with similar results) were presented. Open symbols : control experiment.

concentration-dependent decrease in PBI. These results suggest that the inhibitory effect of the highest concentrations of TSH could be mediated by endogenously produced inositol phosphate-glycan.

DISCUSSION

The present results provide a new insight into the complex biosignalling of TSH. Chronic treatment of thyroid cells with a moderate concentration of hormone (0.1 mU/ml ≈ 1.2 x 10^{-10} M) which does not desensitize the

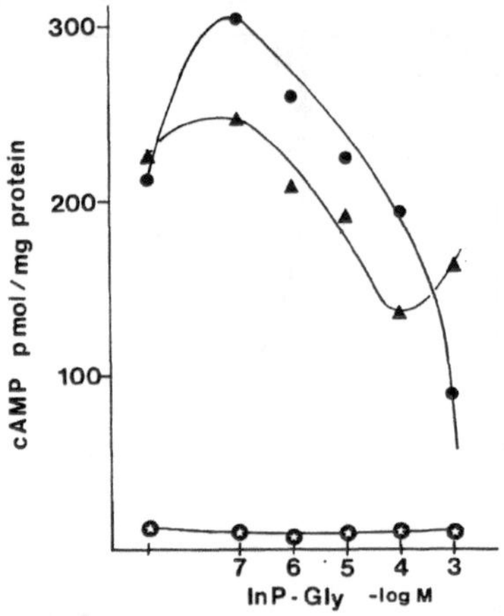

Fig. 3 Effects of inositol phosphate-glycan on cAMP accumulation by thyroid cells. Cells cultured for 3 days in the presence of 0.1 mU/ml TSH were washed and incubated for 5 min with 10^{-3} M isobutylmethylxanthine, with increasing concentrations of inositol phosphate-glycan (InP-gly) supplemented with 10 mU/ml TSH (-●-) or 10^{-4} M forskolin (-▲-) or not supplemented (-○-). Reactions were stopped with 1 M $HClO_4$ and cAMP was assayed by a radioimmunological method described by Cailla et al. (1973).

receptors, allows the cells to recover their polarity and to form follicle-like structures.

Such treatment also has long-term effects on several biochemical parameters, enhancing the sensitivity of adenylate cyclase to acute stimulation by TSH, PGE_2 (Gerard et al., 1982), and forskolin (Haye et al., 1985), causing a basal increase in protein-bound iodine (PBI) (Gerard et al., 1981), an increase in GTP concentration (Aublin et al., 1986), a decrease in protein kinase A activity (Breton et al., 1988), and an increase in the basal turnover of phosphatidylinositol (Gerard et al., 1982). Our data show that this last effect is partly related to the increased synthesis of phosphatidylinositol-glycan and to the increased release of inositol phosphate-glycan. The fact that all these modifications can also be induced by chronic treatment with forskolin or dibutyryl cyclic AMP, suggests that they are initiated by a sustained, moderate increase in cyclic AMP concentration. However, the steady increase in inositol phosphate-glycan release from phosphatidylinositol-glycan, is stimulated by acute treatment with TSH via a cyclic AMP-independent pathway, because it is not mimicked by forskolin. The steps of this pathway remain unknown. Two distinct sets of TSH receptors have not been directly demonstrated, although binding experiments (Pekonen and Weintraub, 1980) suggest this possibility. Perhaps the receptor responsible for the release of inositol phosphate-glycan possess a tyrosine protein-kinase activity, as does the insulin receptor. The biological role of this new mediator of TSH also remains unknown. It produces concentration-dependent decreases in the cyclic AMP accumulation promoted by TSH and in protein-bound iodine. The biological significance of this latter effect is difficult to interpret because protein-bound iodine is regulated by two mutually exclusive pathways : the iodination and exocytosis of thyroglobulin and its resorption and proteolysis leading to the secretion of iodothyronines. Our present working hypothesis is that inositol

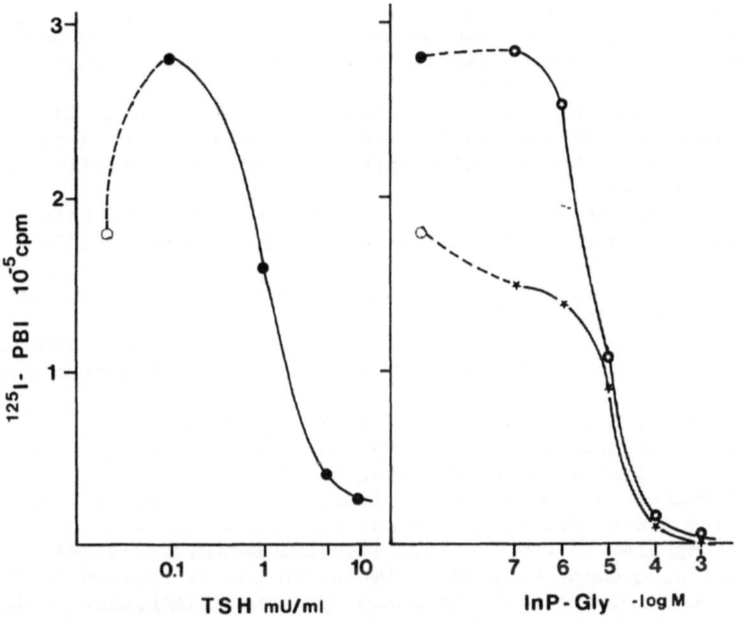

Fig. 4 Effects of inositol phosphate glycan on protein-bound iodine (PBI) in thyroid cells. Cells cultured for 3 days in the presence of 0.1 mU/ml TSH were washed and incubated for 45 min with [^{125}I] iodide (1 μCi/10^6 cells). Incubation media were supplemented with increasing concentrations of TSH (—●—) (part A) or with increasing concentrations of inositol phosphate-glycan supplemented (—✪—) or not (—▲—) with 0.1 mU/ml TSH (part B). At the end of the incubation, samples were precipitated with trichloracetic acid and [^{125}I]-PBI radioactivity was determined. Basal PBI : (—○—).

phosphate-glycan controls the expression of cyclic-AMP-dependent effects, allowing a switch from one pathway to the other.

REFERENCES

Alemany, S., Mato, J. M. and Stralfors, P. 1987, Phospho-dephospho-control by insulin is mimicked by a phospho-oligosaccharide in adipocytes Nature, 330, 77

Aublin, J. L., Lambert, B., Haye, B., Jacquemin, C. and Champion, S. 1986, Relationships between the GTP content and the TSH-stimulated adenylate cyclase activity of cultured thyroid cells. Mol. Cell. Endocrinol., 45:11

Berridge, M. J., 1983, Rapid accumulation of inositol trisphosphate reveals that agonists hydrolyse polyphosphoinositides instead of phosphatidylinositol. Biochem. J., 212:849

Bone, E. A., Alling, D. W. and Grollman, E. F. 1986, Norepinephrine and thyroid-stimulating hormone induce inositol phosphate accumulation in FRTL-5 cells. Endocrinology, 119:2193

Breton, M. F., Haye, B., Omri, B., Jacquemin, C. and Pavlovic-Hournac, M. 1988, Decrease in cAMP-dependent protein kinase activity in suspension cultures of porcine thyroid cells exposed to TSH or forskolin. Mol. Cell. Endocrinol., 55:243

Cailla, H., Racine-Weisbuch, H. and Delaage, M. 1973, Adenosine 3',5' cyclic monophosphate assay at 10^{-15} mole level. Anal. Biochem. 56:396

Freinkel, N. 1957, Pathways of thyroidal phosphorus metabolism : the effect of pituitary thyrotropin upon the phospholipids of the sheep thyroid gland. Endocrinology, 61:448

Gerard, C., Haye, B.and Jacquemin, C. 1981, Effects of arachidonate on cultured pig thyroid cells and their stimulation by thyrotropin. FEBS Letters, 132:23

Gerard, C., Haye, B., Jacquemin, C. and Mauchamp, J. 1982, Chronic and acute effects of thyrotropin on phosphatidylinositol turnover in cultured porcine thyroid cells. Biochem. Biophys. Acta, 710:359

Graff, I., Mockel, J., Laurent, E., Erneux, C. and Dumont, J.E. 1987, Carbachol and sodium fluoride, but not TSH, stimulate the generation of inositol phosphates in the dog thyroid. FEBS Letters, 210:204

Haye, B. and Jacquemin, C. 1974, Stimulation par la thyréostimuline de la production de l'inositol 1-2 phosphate cyclique. Biochimie, 56:1283

Haye, B. , Aublin, J. L., Champion, S., Lambert, B. and Jacquemin, C. 1985 Chronic and acute effects of forskolin on isolated thyroid cell metabolism. Mol. Cell. Endocrinol., 43:41

Igarashi, Y. and Kondo, Y. 1980, Acute effect of thyrotropin on phosphatidylinositol degradation and transient accumulation of diacylglycerol in isolated thyroid follicles. Biochem. Biophys. Res. Commun., 97:759

Igarashi, Y. and Chambaz, E. M. 1987, A novel inositol glycophospholipid (IGPL) and the serum dependence of its metabolism in bovine adrenocortical cells. Biochem. Biophys. Res. Commun., 145:249

Jacquemin, C. and Haye, B. 1970, Action de la TSH sur le métabolisme des phospholipides thyroidiens in vitro. Effet biphasique non reproduit par cAMP et DBC. Bull. Soc. Chim. Biol., 52:153

Kelly, K. L., Mato, J. M., Merida, I. and Jarett, L. 1987a, Glucose transport and antilipolysis are differentially regulated by the polar head group of an insulin-sensitive glycophospholipid. Proc. Natl. Acad. Sci. USA, 84:6404

Kelly, K. L., Merida, I., Wong, E. H. A., DiCenzo, D. and Mato, J. M. 1987b, A phospho-oligosaccharide mimics the effect of insulin to inhibit isoproterenol-dependent phosphorylation of phospholipid methyltransferase in isolated adipocytes. J. Biol. Chem., 262:15285

Klainer, L. M., Chi, Y. M., Friedberg, S. L., Rall, T. W., and Sutherland, E. W. (1962). Adenyl cyclase : IV The effect of neurohormones on the formation of adenosine 3',5'-phosphate by preparations from brain and other tissues. J. Biol. Chem. 237:1239.

Lambert, B. and Jacquemin, C. 1983, Restoration by insulin of the responsiveness of stimulated adipocytes to adenosine. FEBS Letters, 155:31

Laurent, E., Mockel, J., Van Sande, J., Graff, I. and Dumont, J. E. 1987, Dual activation by thyrotropin of the phospholipase C and cyclic AMP cascades in human thyroid. Mol. Cell. Endocrinol., 52:273

Mato, J. M., Kelly, K. L., Abler, A. and Jarett, L., 1987, Identification of a novel insulin-sensitive glycophospholipid from H35 hepatoma cells. J. Biol. Chem., 262:2131

Merida, I., Corrales, F. J., Clemente, R., Ruiz-Albusac, J. M., Villalba, M. and Mato, J.M. 1983, Different phosphorylated forms of an insulin-sensitive glycosylphosphatidylinositol from rat hepatocytes. FEBS Letters, 155:31

Michell, R. H., 1975, Inositol phospholipids and cell surface receptor function. Biochim. Biophys. Acta. 415:81

Morton, M. E. and Schwartz, J. R. 1953, The stimulation in vitro of phospholipid synthesis in thyroid tissue by thyrotropic hormone Science, 117:103

Nishizuka, Y. 1984, The role of protein kinase C in cell surface signal transduction and tumour promotion. Nature, 308:693

Pastan, I. and Macchia, V. 1967 Mechanism of thyroid stimulating hormone action : studies with dibutyryl 3',5'-adenosine monophosphate and lecithinase C. J. Biol. Chem., 242:5757

Pekonen, F. and Weintraub, B. D. 1980, Salt-induced exposure of high affinity thyrotropin receptors on human and porcine thyroid membranes. J. Biol. Chem., 255:8121

Rodbell, M. 1964, Metabolism in isolated fat cells. I. Effects of hormones on glucose metabolism and lipolysis. J. Biol. Chem., 239:375

Saltiel, A. R. and Cuatrecasas, P. 1986, Insulin stimulates the generation from hepatic plasma membranes of modulators derived from an inositol glycolipid. Proc. Natl. Acad. Sci. USA, 83:5793

Saltiel, A. R., Fox, J. A., Sherline, P. and Cuatrecasas, P, 1986, Insulin generates an enzyme modulator from hepatic plasma membranes : regulation of adenosine 3',5'-monophosphate phosphodiesterase, pyruvate dehydrogenase, and adenylate cyclase. Science, 233:967

Saltiel, A. R., 1987 Insulin generates an enzyme modulator from hepatic plasma membranes : regulation of adenosine 3',5'-monophosphate phosphodiesterase, pyruvate dehydrogenase, and adenylate cyclase. Endocrinology, 120:967

Scott, T. W., Freinkel, N., Klein, J. M. and Nitzan, M. 1970, Metabolism of phospholipids, neutral lipids and carbohydrates in dispersed porcine thyroid cells : comparative effects of pituitary thyrotropin and dibutyryl 3',5'-adenosine monophosphate on the turnover of individual phospholipids in isolated cells from pig thyroid. Endocrinology, 87:854

Taguchi, M. and Field, J. B. 1988. Effects of thyroid-stimulating hormone, carbachol, norepinephrine, and adenosine 3',4'-monophosphate on polyphosphatidylinositol phosphate hydrolysis in dog thyroid slices. Endocrinology, 123:2019

ACKNOWLEDGEMENTS: We thank Mrs A. Lefevre and M. Sorgue and Mr M. Bahloul for the preparation of the manuscript. This work was supported by grants from the Fondation de la Recherche Médicale and INSERM (CRE N° 874009).

INDEX

Acetylcholine, 88, 221, 308, 339
Acetyl coenzyme A
  carboxylase, 392, 396
  synthetase, 360
Acrosome reaction, 218
Actin, 126, 128, 129, 292, 249
Actinomycin D, 126-130, 381, 382, 389
Adenine nucleotide, 75, 194, 345
Adenosine
  A_2-receptor, 318
  diphosphate(ADP), 126-129, 344
    ribose, 194, 201
    ribosyltransferase, 194, 195 202, 208
    ribosylation, 33, 34, 155, 166, 174, 177, 189, 193-211, 293, 296, 361, 363, 364, 367, 395
      factor, 193-198
  monophosphate, cyclic (cAMP), 33, 55, 73-82, 141-145, 149, 176, 177, 217, 218, 278, 289, 323, 327, 330, 331, 371-379, 401-409
    8-bromo -, 143, 149, 177
    dibutyryl -, 143, 146, 149, 177, 179, 327, 376, 406
    phosphodiesterase, 141-152, 392
  triphosphatase (ATPase), 125, 126, 129, 303, 306, 323, 345, 349, 354
  triphosphate (ATP), 30, 94, 126-130, 175, 186, 187, 195, 255, 270, 271, 278, 377
Adenylate cyclase, 107, 116, 120, 122, 153, 141-153, 163-172, 392, 406
  see Adenylyl cyclase
Adenylyl cyclase, 15, 16, 50, 51, 65, 73-85, 95, 96, 99, 101, 174, 176, 185, 193, 195, 196, 202, 277, 279, 289, 292
Adipocyte, 404
Adrenal medulla, 333
  and stress response, 333
Aequorin, 323
Agmatine, 194, 195
A-kinase, 230, 323, 331
Alkaline phosphatase, 395
Aluminum fluoride, 253, 271, 363, 367
Amiloride, 94, 156, 158, 317
Amine ligand, 13
γ-Aminobutyric acid (GABA), 108-112
Anaphylaxis slow-reacting substances (SRS-A), 21
Angiotensin, 144, 146
  I, 334
  II, 66, 304, 323, 333-341
  III, 336
  disorders, 313
Antibody, 5, 8, 9, 134
Antiphosphotyrosine affinity matrix, 265
Aplysia californica (snail), 113, 215, 217, 219, 371-379
Arachidonic acid, 21, 39, 40, 43, 44, 98, 235, 275-277, 281-285, 291-293, 343-344, 347-348, 353-355, 360-363, 365, 367, 371-379, 381, 389, 394
ARF, see Adenosine diphosphate ribosylation factor
Assembly, macromolecular, 1-10
  and cell activation, 1-10
  and real-time analysis, 1-10
Autocoids, 381
Autophosphorylation, 31-33, 394, 395

Baclofen, 109
Basophil, 313
B-cell, human
  virus-transformed, 205
Benzamide, 206
4,5-Biphosphate, 42
Blood platelet, *see* Platelet
Bombesin, 62, 64

Botulinum toxin
  C2, 6, 202
  D, 155, 156
Bradykinin, 270, 271, 333-341, 381, 389
p-Bromophenacyl bromide, 282-283
Burst, oxidative, 154

Calciosome, 216
Calcium, 1, 8, 40-44, 49, 52-56, 61, 66, 68-70, 83, 94-97, 107-116, 144, 155, 173-178, 202, 213-225, 227, 231, 235, 245-248, 253-258, 264, 265, 269-270, 275, 289, 305-307, 314-319, 323-335, 344, 354, 355, 360, 364, 394
  ionophore A*23187*, 144, 147, 277, 282-284, 305, 316, 355, 356, 363, 381
  mobilization, 333-352
Calf serum, fetal, 362
Calmedin, 54
Calmodulin, 68, 163
Capitance pool of calcium, 215-216
Carbachol, 65, 67, 99
Carcinoma, epidermoid, human cell line, A-*431*, 39, 264, 265-270
Catechol hydroxyl group, 11
Catecholamine, 333, 334
  agonist, 11, 12
Cell
  activation, 1-10
  coupling enzyme, 126-130
  coupling process, 125
  endothelial, 381-390
  G-protein, *see* G-protein
  and HIV (human immunodeficiency virus), 222
  lacrimal, 219
  ligand receptor, 1-10
  line
    A-*431*, 39, 264-270
    Astrocytoma cell line *1321* N*1*, 67
    HA-*38*, 360, 365
    HeLa, 203
    HL-*60*, 22, 24,203-208, 305, 306
    Jurkat, 175-180, 254-257
      LBRM-*33*, 176
    NR*6*, 40, 44
    promyelocytic, 22
      C*6*, 136, 137
    RBL-*243*, 313-322
    S-*49* lymphome, 117, 120
    U-*937*, 22, 24, 65, 235-243
    *3*T*3*, 39, 40, 44, 135, 136, 264, 265-271, 395, 396
  motility, *see* ATPase, dynein ATPase, kinesin ATPase, tubulin

Cell (continued)
  permeabilization, 3, 4, 318-319, 346, 347
  transformation, 201-211
    of rat bone, fetal, 205
    by virus, 205
  of umbilical vein, 221
  see specific cells
Channel ionic, 93-106
  see separate elements
CHAPS (detergent), zwitterionic, 23, 55
Charybodotoxin, 94
Chemotactic factor, *see* Factor, chemotactic
Chiroinositol, 393
Cholate, 194
Cholecystokinin-8-sulfate, 62
Choleragen, *see* Cholera toxin
Cholera toxin, 24, 83-85, 149, 155, 156, 163, 166, 175, 193-195, 201-202, 290-293, 304, 308, 315-318, 335-339
Chromaffin cell, 319, 333-341
Cobalt, 111
Cobra venom phospholipase A_2, 284
Colchicine, 166
Collagen, 229
Colony-stimulating factor, 153-162
Competence, cellular, 264
Complex, ternary, *see* G-protein-ligand-receptor
Compound *48/80*, 313, 314
Computer simulation, 75
  and KINSIM, 75-78
Concanavalin A, 53
Coupling enzyme, 126-130
CTX, 176-181
Cyclohexamine, 236, 237
Cycloheximide, 154, 381, 382, 389
Cyclooxygenase, 42, 43, 275, 276, 277, 281, 283, 354, 355
Cytochalasin B, 154, 282, 304, 306
Cytochrome P-*450*, 276, 277
Cytoplast, 154
Cytoskeleton, 165, 248, 249
Cytotoxin I, 66

DDT, *see* Dithiothreitol
Decapeptide, 117-122
Degranulation, 313, 314
Deoxycholate, 293, 294
Dephosphorylation, 30, 32, 391, 396
Desensitization, 61, 141-152
1,2-Diacylglycerol, 39, 40, 43, 44, 49, 61, 144, 173-174, 213, 218, 221, 228, 230, 235, 245, 246, 253-257, 264, 304-307, 314, 316, 323, 354, 360-367, 392-394, 397, 401
  lipase, 281

Dictyostelium (slime mold), 218
Diglyceride kinase, 302, 303
Dihydropyridine, 94, 107
Diltiazen, 347-349
Dimethylsulfoxide, 207, 306
Dimyristoyl phosphatidylcholine, 193, 194
1,2-Diolein, 255
Dipalmitoyl phosphatidylcholine, 283
Dipeptidase, 353
2,3-Diphosphoglycerate, 228
Diphtheria toxin, 202
Dithiothreitol (DDT), 13-16
DNA, 29, 41, 110, 122, 127, 128, 189, 203, 245, 246, 254, 307, 388, 394
  antisense-, 381-390
  gyrase, 125
  polymerase, 125
Dual second messenger hypothesis, 213
Dynein ATPase, 125, 128, 129

EGF, *see* Growth factor, epidermal
EGTA, *see* Ethyleneglycol-bis-(b-aminoethyl ether)-N,N,N;N-tetraacetic acid
Eicosanoid, 275-281, 284, 353-357, 359-369, 378, 381, 382, 389
Elongation factor, 127-129, 202
Endoperoxide, 276, 277
Endothelin, 336-337
Endothelium
  cell line CPAE, 381-390
  inflammation, 381
Enkephalin, 112, 113
Eosinophil, human, 302
Epidermal growth factor, *see* Growth factor, epidermal
Epinephrine, 304, 339, 344, 347, 349
Epitope, 134
Epstein-Barr virus, 205
Erythrocyte, human, 96, 99, 166, 202
Erythroleukemia, murine, 208
Escherichia coli, 119, 122
Ethanol, 305, 306
Ethanolamine, 392, 393, 397
5'-(N-Ethylcarboxamido)adenosine, 318
Ethyleneglycol-bis-(b-amino-ethyl ether)-N,N,N',N'-tetraacetic acid, 42
N-Ethylmaleimide, 79
Exocytosis, 175, 314, 317, 339

F-actin, 6
Fibroblast, 67, 248, 264, 305
  transformed, 222, 246-250
Ficoll, 154
Flow cytometry, 9
Fluorescamine, 402
Fluorescein, 4, 5, 8, 9
Fluorescence, 2, 5, 6, 8, 247-249, 324
Fluoroalbuminate, 176
Formic acid, 291, 402
Formyl-Met-Leu-Phe peptide, 1, 8, 9, 153-157, 303-307
Forskolin, 50, 143, 149, 177, 330, 401-406
Fura-*2*(fluorescence probe), 40, 69, 230, 323-332, 343-348

GABA, *see* γ-Aminobutyric acid
Genistein, *see* Tyrosine kinase
Glioma membrane, 167-170
  C*6* cells, 164-168
Glucagon, 141-149, 323-332
Glucocorticoid, 202, 275
Glucosamine, 395, 402-404
L-Glutamic acid, 52, 353
γ-Glutamyl transferase, 353
Glutathione-S-transferase, 277, 354
Glycan, 397
Glycerol, 404
Glycerol phosphoinositide, 236, 237
Glycerophosphoinositol bisphosphate, 228, 229
Glycogen synthetase, 391
Glycolipid, 393
Glycosyl-inositol phosphate, 392-397
Glycosyl-PI, *see* Glycosyl-inositol phosphate
G-protein(guanine nucleotide-binding regulatory protein), 1-13, 17, 18, 33, 61, 66, 67, 75, 78, 81, 83-106, 115-123, 133, 153-168, 175, 185-200, 221, 229, 235, 253-262, 279, 289-300, 307, 314-318, 339, 343, 348, 353, 363, 367, 374-377, 395, 397
  activation, 11, 100
  antibody, 115-123
    as probe, 117-122
  complex, ternary, 2-7, *see* receptor-effector
  coupling the receptor to the effector, 93-106, 115
  family of, 83-85
  functions, 96-97, 115-116, 288
  and guanine nucleotide, 83-91
  ion channel, 88-89, 93-114
  and ligand receptor, 1-3

G-protein (continued)
and neurosecretion, 111-112
and pertussis toxin, 33, 95
phosphorylation, 185-192
as receptor-effector complex, 93-106, 115
in retina, 86
and signal transduction, 115
structure, primary, 95-96, 115-116
subunits, 83-89, 93-106, 116, 119, 163, 164, 169, 289, 295
see Transducin, Tubulin
Gi-protein, 79, 81, 133-140, 202
in cytoplasm, 133-140
distribution, intracellular, 133-140
immunofluorescence, 133-140
α-subunit, 133-140
Granulocyte, 52, 65, 207, 302-307
-macrophage colony-stimulating factor, 153-162
Growth factor, 263-274
epidermal, 34, 39-47, 263-264
kinase-deficient, 263
and mitogenesis, 39
platelet-derived, 41-44
receptor, 263
regulation 267-268
stimulation, 268-271
Guanidine compounds, 194
Guanine diphosphate, 110
Guanine nucleotide, 2-6, 23, 83-91, 109,115, 155-160, 178, 193-196, 202, 253, 289, 292, 345
Guanosine diphosphate, 343-352
Guanosine monophosphate, cyclic, phosphodiesterase, 83, 86, 95, 99, 116, 118, 126, 289
Guanosine 5'-O
thiodiphosphate, 315
thiotriphosphate, 315
Guanosine triphosphatase, 115, 126, 128, 156-157, 343
Guanosine triphosphate (GTP), 23-24, 83, 109, 110, 115, 128, 130, 165, 174, 176, 178, 195, 196, 345
Guanylate cyclase, 153, 202

Hamster kidney cell, 202
HeLa cell, 203
Hematoma cell, 396
Hemocyanin of keyhole limpet, 134, 186
Heparan sulfate proteoglycan, 395, 397
Hepatocyte, 141-152, 185, 221, 304, 306, 323-332, 395
Hexapeptide, 3, 8
Histamine, 221, 314, 315
Histocompatibility complex molecules, 173
Histone, 203
Hormone-binding affinity, 30
12-HPETE, *see* 12-Hydroxyperoxy eicosatetraenoic acid
5-HT, 371-374
12-Hydroxyperoxyeicosatetraenoic acid, 277, 374-377
12-Hydroxyheptadecatrienoic acid, 374
Hydroxylamine, 201, 206
Hyperphosphorylation, 174

IBMX(inhibitor), 143, 145
Iloprost, 74-79, 279
Immunoblot analysis, 118-122, 247, 248
Immunocytochemistry, 249
Immunocytofluorescence, 248, 250
Immunofluorescence, 133-140
microscopy, indirect, 135-137
Immunoglobulin E, 313-319
receptor, 313-319
aggregation, 314-317
see Mast cell
Immunoprecipitation, 119, 121, 148, 187-190, 255, 265-267
Indomethacin, 42, 43, 284
Inhibition, presynaptic, 371
Inositide, 213
Inositol, 314-316, 392
glycan, 395, 396
monophosphate, 402
phosphate, 41-43, 49-57, 63-69, 175, 176, 214, 217, 218, 221, 227-233, 235, 240, 257-259, 268, 293, 318, 323, 330, 331, 339, 392, 394
glycan, 401-409
kinase, 55
phosphatase, 55
receptor, 50-57
phospholipid, 43, 144, 146, 148, 257, 314-318
hydrolysis, 315
polyphosphate, 39, 40, 42, 227, 230, 315
1,3,4,5-tetrakisphosphate, 50, 213-225, 235-243, 315
kinase, 213-217
3-phosphatase, 221
1,4,5-triphosphate, 1, 43, 49, 61, 213-229, 253, 264, 307, 314-317, 323, 346, 401
3-kinase, 220, 222, 227, 229, 231

Insulin, 146-149, 185, 196, 394-396, 404
action, 391-399
cell response, 29
glucose transport, 33
-like growth factor, 264
pathophysiology, 29
receptor, 29-38
antibody, 31, 33
autophosphorylation, 31-33
covalent, 29-38
kinase, 31-33, 185, 395
noncovalent, 29-38
subunits, 29
resistance, 31, 147, 148
second messenger, 391-399
and signal transduction, 29-33
Integrin, 249
Interleukin-2, 173, 253, 255, 257
Internalization analysis of ligand, fluorescent, 5
Iodine, protein-bound, 404, 406
Ion channel, 193, 314, 317, 371-379
see separate elements
Isethionyl acetimide, 393
Isobutylmethylxanthine, 405
Isoproterenol, 16, 18, 120, 404

Keyhole limpet hemocyanin, 134, 186
Kidney cell, 305
Kinesin ATPase, 125, 128, 129
KINSIM computer program, 75-78

Lactate dehydrogenase, 255
Lanthanum, 345-350
Lectins, 253-262, *see* separate compounds
Leukemia
basophilic, 313-322
promyelocytic, acute, 22
Leukocyte, human, polymorphonuclear, 21-27, 355
Leukotriene, 21-27, 275, 277, 353, 364
A_4, 353
B_4, 353
C_4, 353, 354
D_4, 235-243, 353, 354, 381-390
E_4, 353
antibody, monoclonal, 23
dihydroxy activity, 21
distribution, cellular, 21
enzymes, 353
epoxytriene, 277
and leukocyte, human, 21-27
receptor, 21-27, 235, 242, 353-356
sulfidopeptide activity, 21
Ligand receptor, 1-10
Limulus photoreceptor, 217, 218
Leukotriene (continued)
Lipid, cellular, 236-237
Lipocortin, 275
Lipoprotein lipase, 395, 397
Liposome, 55, 56
Lipoxygenase, 21, 275-277, 281, 353-357, 371-379
Lithium chloride, 63
therapy, 51
Lymphocyte, *see* B-cell, T-cell
Lymphoma cell line, 176
Lysophosphatide acyltransferase, 360
Lysophospholipase, 281, 303
Lysozyme, 362

Macrophage, 21, 281-287, 359-369
-granulocyte colony-stimulating factor, 153-162
Magnesium ion, 129, 130
Manganese ion, 345-346, 349
Manolide, 283-285
Manoalogue, 283-285
Mast cell, 313-319
Melittin, 282, 293, 384-388
Membrane, synaptic, 167-170
channel, *see* Channel, ionic
Messenger, intracellular, 61
second, 391-399
hypothesis, 213-225
Methacholine, 62
α-Methylmannoside, 55
Microscopy by indirect immunofluorescence, 135-137
Microtubule, *see* Cytoskeleton
Mitogenesis, 39
MK-*886*, 355, 356
Monocyte, 207
Mouse, 264-271
cell, lacrimal, 219, 221
tumor, myeloid, 204
Mutagenesis, site-directed, 11-19
Myeloid tumor, *see* Tumor
Myoinositol, 43, 63, 64, 227, 228, 237, 257, 393
phosphate, 291
Myosin, 126-129
light-chain kinase, 230, 257
phosphorylation, 316
NAD, *see* Nicotinamide adenine dinucleotide
Neomycin, 290, 292, 295
Neuron, sensory, 108-113
in *Aplysia californica*, 113, 371-379
Neuroblastoma cell, 214, 305, 307
Neuropeptide
FMRFamide, 371-375
Y, 113

Neurosecretion, 107-114
and calcium ion channel, 107-114
Neurotransmitter, 49, 107, 164
Neutrophil, 1-10, 116, 153-162, 174, 217, 235, 303, 304
Nicotinamide, 201
adenine dinucleotide, 194, 201, 203, 206, 208
Nicotine, 338
Nifedipine, 349
Norepinephrine, 107-113
Northern blot analysis, 145, 387
5'-Nucleotidase, 395
Nucleotide
binding site, 129-130
cyclic, 154
phosphodiesterase, 148-149
triphosphate, 83

Obesity, insulin-resistant, 147
1-Oleolyl-2-acetylglycerol, 255
Oncogene, 208
Oöcyte, 196, 214, 219
Opsins, 86
Ornithine decarboxylase, 175
Orthophosphate, 395
Ovalbumin, 168

Pansorbin, 121
Parotid cell, 67, 68
PDGF, *see* Growth factor, platelet-derived
Peptide
antibody, 117-122
antigen, synthetic, 186
vasoactive, *see* separate compounds
Permeabilization, 3, 4, 318-319, 346, 347
Peroxidase, 277
Pertussis toxin, 6, 23, 24, 33, 79, 83-85, 95, 107, 109-112, 116, 120, 145, 146, 155-158, 163, 166, 174, 188, 202, 203, 235, 243, 279, 290-293, 305, 314, 315,318, 335-339, 343, 353, 354, 361, 363, 364, 374, 395
Phagocytosis, 362
Phenylephrine, 324-331
Phenylphosphate, 265
β-Phorbol-12,13-dibutyrate, 228, 246-249, 255-257
β-Phorbol diester, 228, 305
Phorbol ester, 31, 50, 51, 66, 67, 79, 144, 146, 174, 185, 207, 229, 230, 245-249, 255, 255, 306-307, 360, 363-367, 394
Phorbol 12-myristate 13-acetate (PMA), 158, 175-176, 185-189, 229, 305, 316, 318, 335-339
Phosphatase, 31, 376, 377, 392
Phosphatidic acid, 229, 301-305
Phosphatidylalcohol, 303
Phosphatidylcholine, 144, 221, 222, 235, 249, 254, 302, 354, 355, 362, 383, 384
Phosphatidylethanolamine, 302, 362
Phosphatidylinositide, 264
Phosphatidylinositol, 42, 63, 64, 291, 292, 302, 303, 362, 384, 392, 401-406
4,5-biphosphate, 264, 307, 315, 360
"cycle", 401
kinase, 33
phosphate, 253
triphosphate, 235-243
Phosphatidylserine, 246, 291
Phosphoamino acid, 265
Phosphocholine, 302, 305
Phosphodiesterase, 74-76, 79, 206, 221, 302
Phosphohydrolase, 308
Phosphoinositidase C, 42, 43, 222, 227
Phosphoinositide, 39, 49-59, 61-72, 154, 173-177, 291, 305, 307, 315, 394
Phospholipase, 96, 374
A_2, 42, 43, 88, 116, 163, 235, 254, 275-287, 289-300, 307, 355, 360, 381-389
C, 1, 33, 34, 50, 61-70, 83, 116, 146, 173, 185, 202, 235, 245-248, 253, 254, 257, 263-274, 277, 281, 302, 305, 314-318, 323,330, 348, 349, 354, 360, 381, 383, 330-397, 401
D, 254, 301-311, 316, 395
Phospholipid, 116, 118, 235-246, 254, 275-276, 281, 302, 303, 360, 361, 401-404
methyltransferase, 392
Phosphomonoesterase, 227, 228, 315
Phosphonic acid, 307
Phosphopeptide, 267
Phosphorylase, 396
Phosphorylation, 32, 33, 146-149, 153, 174, 178, 179, 185-192, 253-262, 316, 319, 391, 396
cascade hypothesis, 391
Phosphoserine, 31, 265
Phosphothreonine, 31
Phosphotyrosine, 31, 32, 249, 267
phosphate, 31
Photoaffinity labelling, 13
Photoreceptor, boving, 290
Photochemagglutinin, 173, 177-180, 255-260

Pituitary cell, 246-248
PLAP protein, 387-389
Platelet, 69, 75-81, 185-192, 203, 227-233, 277, 279, 304, 343-352
  activating factor (PAF), 229, 282, 343, 344, 347-349
  -derived growth factor, 263-264
PMA, *see* Phorbol *12*-myristate *13*-acetate
Polarization, 9, *see* Fluorescence
Poly(ADP-ribose), 206, 208
Poly(ADP-ribosyl)polymerase, 208
Poly(ADP-ribosyl)synthetase, 203
Polymyxin B, 67
Polyoma virus, 205
Polypeptide CD-3, 173
  antibody, 173
Potassium
  channel, 88-89, 94-99, 116, 289, 371-379
  chloride, 336
  ion, 318, 371-379
  phosphate, 203
Preadipocyte, 208
Progesterone, 196
Prolactin, 246, 248
Promyelocyte, neutrophilic, 204
Prostacyclin, 276-279, 381-390
Prostaglandin, 43, 73-82, 275-277, 282, 285, 360, 364, 401, 405
Prostanoid production, 282
Protein, 13, 125, 396, 397
  kinase, 54, 65, 68, 94, 376, 377, 392
    A, 143, 146, 149, 406
    C, 31, 39, 44, 49, 50, 61, 65,68, 79, 81, 107, 222, 227-233, 235, 245-262, 264 279, 306-308, 314-319, 339, 354, 355, 359-369, 394, 397
  see G-protein
Proto-oncogene, 156, 174
Pseudomonas exotoxin A, 202
P-substance, 111, 112
PTX, *see* Pertussis toxin islet-activating factor
Purkinje cell, 51, 52
Pyruvate dehydrogenase, 392, 396
Pyruvate kinase, 396

Quinacrine, 282, 283

Ram spermatozoön, 218
ras proteins, 33, 115, 278, 279, 395
Rat, 65, 147, 204, 219, 246-248, 302, 304, 306, 313-322
  leukemia, 313-322
Real time, *see* Time, real
rec A protein, 125-128
Receptor
  β-adrenergic, 11-19, 65-68
    conformation, 11-19
    and disulfide bond, intramolecular, 13, 16-17
    modification, post-translational, 17-18
    and mutagenesis, site-directed, 11-19
    mutant, 14-16
    and reducing agents, 14-16
    and residues, essential, 11-19
    structure, model of, 12
      cys residues, 13-16
  for agonist affinity site, 14-15
  catalysis by, 100-101
  desensitization, 6, 64-70
  dopaminic, 13
  and effector coupling, 93-106, *see* G-protein
  for glutamate, 52
  glycosylated, 13, 17
  for inositol phosphate, 52-57
  for insulin, 29-38
  for leukotriene, 21-27
  for messenger, second, 49-59
  muscarinic, 13, 63, 66-68
  non-glycosylated, 13
  phosphorylation, 55, 66, 68
  for serotonin, 13
  for T-cell antigen, 173-183
Retina, bovine
  rod, 289-300
Retinoic acid, 207
Reuber hematoma cell, 396
Rhodopsin, 11, 13, 83, 86, 116, 126, 293
Rho transcription
  termination factor, 125
Ribonucleic acid, *see* RNA
Ribosome, 127, 203
Ribosylation, *see* ADP
RNA, messenger, 51, 86, 87, 127, 148, 156, 381, 387, 389
Ro-*2074*(inhibitor), 145
Rod outer segment (ROS), 290, 291, 293-297
ROS, *see* Rod

Saccharomyces cerevisiae, 88-89
  mating pheromone, 89
Saponin, 166, 168, 188, 288, 230, 346, 347
Sarcoma virus histiocytosis, malignant, 362
Scatchard plot, 30
Sea urchin, 214
Serine phosphorylation, 32, 391
Serotonin, 314, 371-379

Signal
output, 173-183
transduction, 9, 149-150, 163-172
pathway, biochemical, 23-24, *see* G-protein
and tubulin G-protein, 163-172
Silver stain, 386
Simian virus-*40*, *see* SV*40*
S-K^+ channel, 371-379
Serotonin-sensitive potassium ion channel, *see* S-k^+
Slow reacting substance of anaphylaxis, *see* Anaphylaxis
Snail, *see Aplysis californica*
Snake venom phosphodiesterase, 206
Sodium chloride, 193, 194
Sodium fluoride, 120, 143, 270, 271
Sodium ion, 126, 156, 157, 289, 337, 339
Sodium nitroprusside, 203
Spectrofluorometry, 9
Sphingosine, 66, 362
Squid retina, 218
Staphylococcus aureus
protein A, *see* Pansorbin
α-toxin, 175
Staurosporine, 229-231, 364
Streptolysin-O, 255, 256
and permeabilization, 255
Substance P, 62, 111, 112
Superoxide, 153, 154
SV*40*, 248, 249

Talin, 249, 250
Tamoxifen, 364
TCA, *see* Trichloracetic acid
T-cell, human
activation, 255
antigen receptor, 173-183
antigens, 253-262
phosphorylation, 255-259
signal output, 173-183
growth factor receptor, 173
hybridoma, 253, 256
inositol phosphate production, 258, 259
leukemia cell line, 255-257
TH-glucagon, *see* ([1-N-α-trinitrophenylhistidine 12-homoarginine] glucagon)
5'-(3-0)-Thiotriphosphate, 290
Thrombin, 67, 68, 188, 221, 227-231, 304, 344-350
Thromboxane A_2, 229, 231, 276, 277
Thymocyte, nutrine, 253
Thyroid gland, 401-409
cell, 401-409
Thyrotropin, 202, 401-409
-releasing hormone, 246-248
Time, real, analysis
of assembly, macromolecular, 1-10
technology, 8
T-lymphocyte, *see* T-cell
Transducin, 11, 33, 83, 86, 118, 125-132, 175, 289-295
Transformation, malignant, 208
Transphosphatidylation, 303, 305
Triacylglycerol, 404
Trichloracetic acid, 230
(1-N-α-Trinitrophenylhistidine 12-homoarginine)glucagon, 144
Trypanosoma brucei, 394
TSH, *see* Thyrotropin
Tubulin, 115, 125-129, 164-168
function, 166
as G-protein, 165-166, 168-170
purification, 167
structure, 165
Tumor
myeloid, murine, 204
cell line *32*D, 40
necrosis factor, 202
Tyrosine, 32, 33
aminotransferase, 396
kinase, 30-32, 39, 51, 157-158, 174, 176-181, 230, 263, 264, 391, 394-397
phosphorylation, 32, 33, 157-160, 263-274
protein kinase, 174, 406

U-*46619*(calcium agonist), 346

Vanadate, 129, 376
Vasopressin, 65, 144, 146, 148, 248, 249, 304, 323-331
Verapamil, 347-349
Vinblastin, 166
Vinculin, 249, 250
Voltage clamp, 371

Western blot analysis, 32, 86, 148, 168, 186, 189
Wheat germ agglutinin, 31

Xenopus oöcyte, 196, 214, 219
Yeast pheromone, 88-89

Zymosan, 154, 362-367